Fundamentals of Acoustic Field Theory and Space-Time Signal Processing

Fundamentals of Acoustic Field Theory and Space-Time Signal Processing

Lawrence J. Ziomek
Department of Electrical and Computer Engineering
Naval Postgraduate School
Monterey, California

CRC Press
Boca Raton   Ann Arbor   London   Tokyo

Library of Congress Catalog-in-Publication Data

Ziomek, Lawrence J.
    Fundamentals of acoustic field theory and space-time signal processing/
Lawrence J. Ziomek
        p.   cm.
    Includes bibliographical references and index.
    ISBN 0-8493-9455-4
    1. Sound-waves—Transmission.   2. Underwater acoustics.   3. Signal
processing.   4. System analysis.   5. Linear systems.   I. Title.
QC233.Z66 1994
534'.2—dc20   94-27368
                                                                              CIP

No claim to original U.S. Government works
International Standard Book Number 0-8493-9455-4
Library of Congress Card Number 94-27368
Printed in the United States of America 1 2 3 4 5 6 7 8 9 0
Printed on acid-free paper

To my mother
LOTTIE,

my wife
VIRGINIA,

and my daughters
NICOLE and KIRSTEN,

and to the loving memory of my father
CASIMIR

Table of Contents

Preface

The science of acoustics encompasses many specialty areas. This book concentrates on fundamental topics in the areas of *linear* air and underwater acoustics (as opposed to nonlinear acoustics) and space-time signal processing. Air and underwater acoustics are devoted to the mathematical modeling and measurement of the propagation, radiation, and scattering of acoustic (sound) waves in fluid and elastic media. In a more general sense, air and underwater acoustics can be considered as interdisciplinary branches of science involving biomedical engineering (medical ultrasonics—acoustic imaging), electrical engineering (electroacoustic transducer design, array theory, space-time signal processing, etc.), mechanical engineering (noise and vibration), ocean engineering (underwater acoustics), acoustical oceanography, and physics.

As a consequence of the interdisciplinary nature of air and underwater acoustics, this book is divided into two parts. Part I, entitled *Acoustic Field Theory*, is based, in part, on lecture notes that I wrote for a course by the same name. Part II, entitled *Space-Time Signal Processing*, is based on the major revision of several chapters from my first book, *Underwater Acoustics—A Linear Systems Theory Approach* (Academic Press, Orlando, Florida, 1985). These revised chapters were used as lecture notes for a course in Sonar Systems Engineering. Both courses, Acoustic Field Theory (using Part I) and Sonar Systems Engineering (using Part II), are taught in the Department of Electrical and Computer Engineering at the Naval Postgraduate School to students from the Departments of Electrical and Computer Engineering, Oceanography, and Physics. The students enrolled in these courses have diverse undergraduate degrees and backgrounds. The topics covered in Parts I and II are based on the philosophy that it is just as important for the person who is interested in space-time acoustic signal processing to understand the fundamentals of acoustic wave propagation as it is for the person who is interested in air or underwater acoustics to understand the fundamentals of aperture theory, array theory, and signal processing. Starting from first principles whenever possible and practical, and using a consistent and mainly standard notation from the first chapter to the last, this book develops, in detail, basic results that are useful in a variety of air and underwater acoustic applications. All derivations are presented in a clear, logical, step-by-step fashion so that the student is aware of all assumptions and approximations that are made. Some derivations are provided in appendices so that the main flow of discussion is not interrupted. Figures and examples are included in all chapters, and homework problems are included in Chapters 1 through 8.

This book is suitable for advanced college seniors and first-year graduate students. It is also intended to be used as a reference or self-study book by practicing engineers, scientists, and researchers in the fields of linear air and underwater acoustics and space-time acoustic signal processing.

Part I of this book is intended to be used as a textbook for a course in acoustics by students in the fields of biomedical engineering, electrical engineering, mechanical engineering, ocean engineering, acoustical oceanography, and physics. The propagation, radiation, and scattering of sound waves in fluids is emphasized. Part I includes Chapters 1 through 5. Using the fundamentals of fluid mechanics, Chapter 1 presents a detailed derivation of the *linear wave equation* with an arbitrary source distribution for an inhomogeneous fluid medium whose speed of sound is a function of position. The wave equation derivation is based on the Navier-Stokes equation of motion, and as a result, includes the effects of viscosity. Chapters 2 through 4 are devoted to wave propagation, radiation, and scattering in a homogeneous fluid medium with a constant speed of sound in the rectangular, cylindrical, and spherical coordinate systems, respectively. Chapter 5 returns to wave propagation in inhomogeneous fluid media and discusses the approximate solution of the linear wave equation using the WKB approximation, three-dimensional ray acoustics, and the parabolic equation approximation.

As can be seen from the Table of Contents, Chapters 2 through 4 in Part I are provided with a mix of topics for greater flexibility in teaching. The first two or three sections in Chapters 2 through 4 are, in general, important and of interest to both air and underwater acousticians. However, some of the remaining sections are of interest mainly to air acousticians, whereas others are of interest mainly to underwater acousticians. Therefore, depending on which sections are used per chapter in Part I, it is possible to teach either an air or an underwater acoustics course. For example, at the Naval Postgraduate School, Part I (Acoustic Field Theory) is used to teach a course in acoustics emphasizing topics in underwater acoustics. Material from the following chapters is taught in one academic quarter (11 weeks of instruction with 4 lecture hours per week): Chapter 1, Sections 1.5 and 1.6; Chapter 2, Sections 2.1, 2.2, 2.3, 2.6, 2.7, and 2.8; Chapter 3, Sections 3.1, 3.6, 3.8, 3.9; Chapter 4, Sections 4.1 and 4.2; and all of Chapter 5. When using Part I for a course in acoustics, the recommended prerequisites are the standard undergraduate sequence of calculus and physics courses, and proficiency in working with complex numbers, complex exponentials, and Fourier transforms.

Part II is intended to be used as a textbook in space-time acoustic signal processing. Taking an acoustics course from Part I of this book would be desirable, but is not absolutely necessary. Whenever a result from wave propagation theory is used in order to derive an equation for space-time signal processing, the appropriate section in Part I is always referenced. Part II includes Chapters 6 through 9. Chapters 6 and 7 cover the fundamentals of complex aperture theory and array theory, respectively. Complex aperture theory is discussed before array theory, analogous to the general practice of discussing the theory of continuous-time signals before discussing discrete-time signals. Once the general principles of aperture theory are developed, array-theory results can be derived quickly and easily. Chapter 8 covers basic topics in signal processing. And finally, Chapter 9 deals with the fundamentals of linear, time-variant, space-variant filters and how *linear systems theory* can be used to model the propagation of small-amplitude acoustic signals. Since the wave equation for small-amplitude acoustic signals is *linear*, a fluid medium can be thought of as acting like a *linear filter*. The purpose of Chapter 9 is to demonstrate that the seemingly unrelated disciplines of wave propagation, array theory, and signal processing can be brought together using linear systems theory.

At the Naval Postgraduate School, Part II (Space-Time Signal Processing) is used to teach a course in Sonar Systems Engineering. All the material from Chapters 6 through 8 is taught in one academic quarter (11 weeks of instruction with 4 lecture hours per week). When using Part II for a course in space-time acoustic signal processing, the recommended prerequisites are introductory-level courses in communication theory, digital signal processing, and probability and random processes.

I would like to express my appreciation to my students over the years for their many thought-provoking questions and comments concerning early drafts of my course notes which led to the publication of this book. Their questions and comments motivated me to make many revisions that improved the quality of the book. In addition, I thank those reviewers who took the time and energy to provide constructive comments and suggestions, most of which have been implemented. Finally, I would also like to acknowledge the support that I have received from the Naval Postgraduate School Foundation Research Program, the Defense Advanced Research Projects Agency (DARPA), the Office of Naval Research (ONR), the Naval Sea Systems Command (NAVSEA), and the Program Executive Office for Undersea Warfare, Advanced Systems and Technology Office (PEO-USW, ASTO). Their support of my research efforts in underwater acoustics is responsible in many different ways for the material contained in this book.

Lawrence J. Ziomek
Monterey, California

Part I

ACOUSTIC FIELD THEORY

Chapter 1

The Linear Wave Equation and Fundamental Acoustical Quantities

Sound waves in fluids, such as air and water, are *longitudinal* waves, that is, the fluid molecules (mass) move back and forth in the direction of wave propagation, producing adjacent regions of compression and rarefaction (expansion). Fluids have the two major characteristics known as *elasticity* and *mass density* that are responsible for acoustic (sound) wave propagation. The elasticity, or *bulk modulus*, of a fluid, which is equal to the reciprocal of its *compressibility*, causes the fluid to resist being compressed or expanded. It enables the fluid to restore itself to the original (equilibrium) state that it was in before the application of any forces, and determines the potential energy of an infinitesimal volume element of fluid. The mass density of a fluid provides inertia and determines the kinetic energy of an infinitesimal volume element of fluid.

In the absence of any applied forces, an infinitesimal volume element of fluid which is at rest at time t and position $\mathbf{r} = (x, y, z)$ has an *equilibrium* or *ambient density* of ρ_0 kilograms per cubic meter, is under an *ambient pressure* of p_0 pascals (newtons per square meter), and is at an *ambient temperature* of T_0 degrees Celsius (°C). The presence of an acoustic (sound) wave produces changes in pressure, velocity, density, and temperature in the fluid, each change being proportional to the amplitude of the wave. The *change in pressure* from the equilibrium or ambient value p_0 that is due to the presence of an acoustic wave is called the *acoustic pressure p*. If the amplitude of the acoustic wave is large, so that p is not small in comparison with the ambient pressure p_0, then nonlinear effects become important. As a result, a nonlinear wave equation must be used to describe the propagation of large-amplitude acoustic waves. This type of wave

3

propagation falls into the general area of acoustics known as *nonlinear acoustics*. However, if the amplitude of the acoustic wave is small, so that $p/p_0 \ll 1$, then nonlinear effects become negligible. As a result, a linear wave equation adequately describes the propagation of small-amplitude acoustic waves. This type of wave propagation falls into the general area of acoustics known as *linear acoustics*. In this chapter, we shall be concerned with the derivation of the *linear wave equation*. Indeed, this entire book is devoted to the study of fundamental problems in linear acoustics. In order to derive the wave equation, we need equations of motion, continuity, and state.

1.1 Equations of Motion

In this section we shall derive the equations of motion for an infinitesimal volume element of fluid using Newton's second law of motion. Although all fluids are composed of millions of molecules in constant motion, we shall treat a fluid as a *continuum*, that is, as an infinitely divisible substance, and not concern ourselves with the behavior of individual molecules. As a result of the continuum assumption, each fluid property can be treated as a continuous function of time and space.

Consider the infinitesimal volume element of fluid dV illustrated in Fig. 1.1-1, where $dV = dx\, dy\, dz$. There are two kinds of forces acting on dV, known as *surface forces* and *body forces*. Surface forces include all forces acting on the surfaces of dV through direct physical contact with the surfaces of other fluid elements (particles). Surface forces are the components of *stress*, and include both *normal* forces and *tangential* (*shear*) forces. Body forces are developed without direct physical contact between the surfaces of fluid elements (particles) and are distributed over the entire volume of a fluid. Examples of body forces in a fluid medium are gravitational and electromagnetic forces. Before we can apply Newton's second law of motion, we need to define the *stress tensor*, that is, the nine components of stress at a point. We also need to define a gravitational body force; it is the only kind of body force that we shall consider.

Figure 1.1-2 illustrates an infinitesimal surface area vector

$$\mathbf{dS} = dS\,\hat{n}, \tag{1.1-1}$$

where dS is equal to the surface area ABC and $\hat{n}$ is a unit vector, normal (perpendicular) to the surface, pointing in the conventional, outward direction away from the surface. Equation (1.1-1) can also be expressed as

$$\mathbf{dS} = dS_X\,\hat{x} + dS_Y\,\hat{y} + dS_Z\,\hat{z}, \tag{1.1-2}$$

where the magnitudes of dS_X, dS_Y, and dS_Z are equal to the surface areas OBC, OAC, and OAB, respectively, which are equal to the projections of dS onto the YZ, XZ, and XY planes, respectively. Note that dS_X, dS_Y, and dS_Z lie in planes that are *perpendicular* to the X, Y, and Z axes, respectively (see Fig. 1.1-3). Also

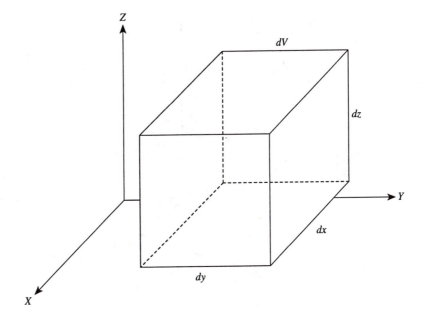

Figure 1.1-1. Infinitesimal volume element of fluid, $dV = dx\,dy\,dz$.

shown in Fig. 1.1-2 is an infinitesimal force vector

$$\mathbf{dF} = dF_X\,\hat{x} + dF_Y\,\hat{y} + dF_Z\,\hat{z} \tag{1.1-3}$$

applied to the surface dS at the point indicated in Fig. 1.1-2. The *stress tensor* σ_{ij}, $i, j = 1, 2, 3$ or, in other words, the ijth component of stress at a point, is defined as follows:

$$\sigma_{ij} \triangleq \frac{dF_i}{dS_j}, \qquad i, j = 1, 2, 3, \tag{1.1-4}$$

where the index values 1, 2, and 3 correspond to X, Y, and Z, respectively. As can be seen from Eq. (1.1-4), σ_{ij} has the units of force per unit area (newtons per square meter). The first subscript, i, indicates that the stress is acting in a direction that is *parallel* to the ith axis. The second subscript, j, indicates that the stress is acting on a plane that is *perpendicular* to the jth axis. Therefore,

$$\sigma_{12} = \frac{dF_1}{dS_2} \tag{1.1-5}$$

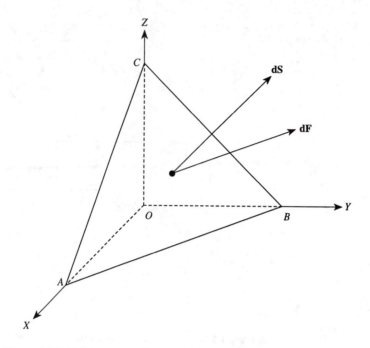

Figure 1.1-2. Infinitesimal surface-area and force vectors **dS** and **dF**, respectively.

is equivalent to

$$\sigma_{XY} = \frac{dF_X}{dS_Y},$$
(1.1-6)

which is the force per unit area in the X direction acting on a plane perpendicular to the Y axis. Note that the stress tensor is *symmetric*, that is,

$$\sigma_{ij} = \sigma_{ji}, \qquad i, j = 1, 2, 3.$$
(1.1-7)

As a result, there are six *independent* components of stress.

Since the components of both the infinitesimal surface-area vector and force vector given by Eqs. (1.1-2) and (1.1-3), respectively, can be either positive or negative in value, the same is true of the various components of stress. For example, if σ_{XY} is positive, then dF_X and dS_Y are either both positive or both negative. Similarly, if σ_{XY} is negative, then either dF_X is positive and dS_Y is negative, or dF_X is negative and dS_Y is positive. A negative value of dF_X and dS_Y simply indicates vector components pointing in the negative X and Y directions, respectively.

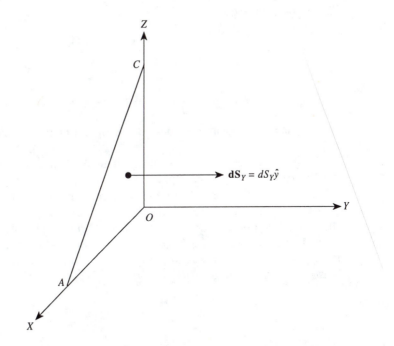

Figure 1.1-3. Illustration of the infinitesimal surface area vector $\mathbf{dS}_Y$.

The three components of stress σ_{XX}, σ_{YY}, and σ_{ZZ} are known as *normal stresses*. If the value of any one of the normal stresses is *positive*, then that component is known as a *tensional stress*, or simply as a *tension*. For example, if $\mathbf{dF}_X = dF_X\,\hat{x}$ and $\mathbf{dS}_X = dS_X\,\hat{x}$, then $\sigma_{XX} = dF_X/dS_X$ is a tension in the X direction. If the value of any one of the normal stresses is *negative*, then that component is known as a *compressive stress*, or simply as a *pressure*. For example, if $\mathbf{dF}_X = -dF_X\,\hat{x}$ and $\mathbf{dS}_X = dS_X\,\hat{x}$, then $\sigma_{XX} = -dF_X/dS_X$ is a pressure in the X direction. The six components of stress σ_{XY}, σ_{XZ}, σ_{YX}, σ_{YZ}, σ_{ZX}, and σ_{ZY} are known as *tangential stresses* or *shearing stresses*. When all the shearing stresses are equal to *zero*, and all the normal stresses are equal to each other and are *negative*, then the stress distribution is known as a *hydrostatic pressure*.

The *gravitational body force per unit volume of fluid* is given by

$$\mathbf{B}_g = \rho\mathbf{g} \tag{1.1-8}$$

where ρ is the *density* of the fluid in kilograms per cubic meter and $\mathbf{g}$ is the *local gravitational acceleration vector*. The components of $\mathbf{g}$ have units of meters per square second. Therefore, the infinitesimal gravitational body force $\mathbf{dF}_g$ acting on

dV can be expressed as

$$\mathbf{dF}_g = \mathbf{B}_g \, dV \tag{1.1-9}$$

or

$$\mathbf{dF}_g = \rho \mathbf{g} \, dV. \tag{1.1-10}$$

We are now in a position to derive the equations of motion for the fluid particle dV.

Consider the infinitesimal volume element of fluid dV with infinitesimal mass dm illustrated in Fig. 1.1-1. First we shall derive an expression for the *total force* (in newtons) acting on dV in the X direction. If the stresses acting in the X direction at the *center* of dV are σ_{XX}, σ_{XY}, and σ_{XZ}, then the stresses acting in the X direction on each side (surface) of dV can be obtained by using the first two terms in a Taylor series expansion of the stresses at the center of dV. For example, if we denote the stresses acting in the X direction on the right side, left side, top, bottom, front, and back of dV as σ_{XR}, σ_{XL}, σ_{XT}, σ_{XB}, σ_{XF}, and σ_{XBK}, respectively, then

$$\sigma_{XR} = \sigma_{XY} + \frac{dy}{2}\frac{\partial}{\partial y}\sigma_{XY}, \tag{1.1-11}$$

$$\sigma_{XL} = -\left(\sigma_{XY} - \frac{dy}{2}\frac{\partial}{\partial y}\sigma_{XY}\right), \tag{1.1-12}$$

$$\sigma_{XT} = \sigma_{XZ} + \frac{dz}{2}\frac{\partial}{\partial z}\sigma_{XZ}, \tag{1.1-13}$$

$$\sigma_{XB} = -\left(\sigma_{XZ} - \frac{dz}{2}\frac{\partial}{\partial z}\sigma_{XZ}\right), \tag{1.1-14}$$

$$\sigma_{XF} = \sigma_{XX} + \frac{dx}{2}\frac{\partial}{\partial x}\sigma_{XX}, \tag{1.1-15}$$

and

$$\sigma_{XBK} = -\left(\sigma_{XX} - \frac{dx}{2}\frac{\partial}{\partial x}\sigma_{XX}\right), \tag{1.1-16}$$

where σ_{XX}, σ_{XY}, and σ_{XZ} are assumed to be positive. Note that the stresses given by Eqs. (1.1-11), (1.1-13), and (1.1-15) are assumed to be positive, whereas the stresses given by Eqs. (1.1-12), (1.1-14), and (1.1-16) are assumed to be negative. The positive stresses correspond to forces acting in the positive X direction on surfaces (planes) whose unit normal vectors are pointing in the positive Y, Z, and X directions, respectively. Similarly, the negative stresses correspond to forces acting in the positive X direction on surfaces (planes) whose unit normal vectors are pointing in the negative Y, Z, and X directions, respectively. In order to obtain the *total surface force* (in newtons) acting on dV in the X direction, denoted by dF_{S_X}, we must *sum* all the individual surface forces (in newtons) acting in the X direction. Therefore, with the use of Eqs. (1.1-11) through (1.1-16), and by multiplying the stresses (in newtons per square meter) by the appropriate surface areas $dx\,dz$, $dx\,dy$, and $dy\,dz$ in square meters, we can write that

$$dF_{S_X} = \left(\sigma_{XY} + \frac{dy}{2}\frac{\partial}{\partial y}\sigma_{XY}\right)dx\,dz - \left(\sigma_{XY} - \frac{dy}{2}\frac{\partial}{\partial y}\sigma_{XY}\right)dx\,dz$$

$$+ \left(\sigma_{XZ} + \frac{dz}{2}\frac{\partial}{\partial z}\sigma_{XZ}\right)dx\,dy - \left(\sigma_{XZ} - \frac{dz}{2}\frac{\partial}{\partial z}\sigma_{XZ}\right)dx\,dy$$

$$+ \left(\sigma_{XX} + \frac{dx}{2}\frac{\partial}{\partial x}\sigma_{XX}\right)dy\,dz - \left(\sigma_{XX} - \frac{dx}{2}\frac{\partial}{\partial x}\sigma_{XX}\right)dy\,dz, \quad (1.1\text{-}17)$$

or

$$dF_{S_X} = \left(\frac{\partial}{\partial x}\sigma_{XX} + \frac{\partial}{\partial y}\sigma_{XY} + \frac{\partial}{\partial z}\sigma_{XZ}\right)dV. \qquad (1.1\text{-}18)$$

Next, let us compute the *total body force* (in newtons) acting on dV in the X direction. We shall assume that only a gravitational body force is present. Therefore, from Eq. (1.1-10), we can write that the *gravitational body force* acting on dV in the X direction is given by

$$dF_{g_X} = \rho g_X \, dV, \qquad (1.1\text{-}19)$$

where g_X is the component of the gravitational acceleration vector in the X direction.

The *total force* (in newtons) acting on dV in the X direction, denoted by dF_X, can be obtained by adding Eqs. (1.1-18) and (1.1-19). Doing so yields

$$dF_X = dF_{S_X} + dF_{g_X} = \left(\frac{\partial}{\partial x}\sigma_{XX} + \frac{\partial}{\partial y}\sigma_{XY} + \frac{\partial}{\partial z}\sigma_{XZ} + \rho g_X\right)dV. \quad (1.1\text{-}20)$$

Similarly, the total forces acting on dV in the Y and Z directions can be expressed as follows:

$$dF_Y = dF_{S_Y} + dF_{g_Y} = \left(\frac{\partial}{\partial x}\sigma_{YX} + \frac{\partial}{\partial y}\sigma_{YY} + \frac{\partial}{\partial z}\sigma_{YZ} + \rho g_Y\right)dV \quad (1.1\text{-}21)$$

and

$$dF_Z = dF_{S_Z} + dF_{g_Z} = \left(\frac{\partial}{\partial x}\sigma_{ZX} + \frac{\partial}{\partial y}\sigma_{ZY} + \frac{\partial}{\partial z}\sigma_{ZZ} + \rho g_Z\right)dV. \quad (1.1\text{-}22)$$

Example 1.1-1 (Calculation of the ambient pressure in a fluid)

In this example we shall derive an equation that can be used to calculate the ambient pressure in air as a function of height, and the ambient pressure in water as a function of depth. Referring to Fig. 1.1-1, let the XY plane ($z = 0$) represent the ocean surface. Therefore, a *positive* value of z corresponds to a *height* above the ocean surface and a *negative* value of z corresponds to a *depth* below the ocean surface.

The gravitational body force acting on dV in the *negative* Z direction, pushing down on dV, is given by

$$dF_{g_z} = -\rho_0 g_z \, dV \tag{1.1-23}$$

where ρ has been replaced by the ambient density ρ_0, and g_z is the component of the gravitational acceleration vector in the Z direction. Since $dV = dx \, dy \, dz$, Eq. (1.1-23) can be rewritten as

$$dP = \frac{dF_{g_z}}{dx \, dy} = -\rho_0 g_z \, dz \tag{1.1-24}$$

where dP is an infinitesimal pressure. Integrating both sides of Eq. (1.1-24) yields

$$\int_{p_0(0)}^{p_0(z)} dP = -\int_0^z \rho_0 g_z \, dz, \tag{1.1-25}$$

or

$$\boxed{p_0(z) = p_0(0) - \int_0^z \rho_0 g_z \, dz,} \tag{1.1-26}$$

where $p_0(z)$ is the ambient pressure at z, and $p_0(0)$ is the ambient pressure at the ocean surface $z = 0$. If both ρ_0 and g_z are constants, then Eq. (1.1-26) reduces to

$$\boxed{p_0(z) = p_0(0) - \rho_0 g_z z.} \tag{1.1-27}$$

Typical values for the parameters in Eq. (1.1-27) are as follows: $p_0(0) = 1$ atm $= 1.013 \times 10^5$ Pa, $g_z = 9.8$ m/sec^2, $\rho_0 = 1.21$ kg/m^3 for air, $\rho_0 = 998$ kg/m^3 for fresh water, and $\rho_0 = 1026$ kg/m^3 for sea water.

Equations (1.1-26) and (1.1-27) correctly predict that the ambient pressure decreases as the height above the ocean surface increases ($z > 0$), and that the ambient pressure increases as the depth below the ocean surface increases ($z < 0$) (see Problem 1-1). Finally, the *gauge pressure* p_g, which is simply the difference between $p_0(z)$ and $p_0(0)$, is given by

$$p_g(z) = -\int_0^z \rho_0 g_z \, dz,$$ (1.1-28)

and if both ρ_0 and g_z are constants, then

$$p_g(z) = -\rho_0 g_z z.$$ (1.1-29)

Returning to the derivation of the equations of motion for the fluid particle dV, let us assume that dV, with mass dm, is in motion with velocity vector

$$\mathbf{U} = U_X \hat{x} + U_Y \hat{y} + U_Z \hat{z}.$$ (1.1-30)

Newton's *second law of motion* states that

$$d\mathbf{F} = dm \, \frac{d}{dt}\mathbf{U},$$ (1.1-31)

where

$$d\mathbf{F} = dF_X \hat{x} + dF_Y \hat{y} + dF_Z \hat{z}.$$ (1.1-32)

Since

$$\mathbf{U} = \mathbf{U}(t, \mathbf{r}) = \mathbf{U}(t, x, y, z),$$ (1.1-33)

the *total derivative* of $\mathbf{U}$ is given by

$$\frac{d}{dt}\mathbf{U} = \frac{\partial}{\partial t}\mathbf{U}\frac{dt}{dt} + \frac{\partial}{\partial x}\mathbf{U}\frac{dx}{dt} + \frac{\partial}{\partial y}\mathbf{U}\frac{dy}{dt} + \frac{\partial}{\partial z}\mathbf{U}\frac{dz}{dt},$$ (1.1-34)

$$= \frac{\partial}{\partial t}\mathbf{U} + U_X\frac{\partial}{\partial x}\mathbf{U} + U_Y\frac{\partial}{\partial y}\mathbf{U} + U_Z\frac{\partial}{\partial z}\mathbf{U},$$ (1.1-35)

or

$$\frac{d}{dt}\mathbf{U} = \frac{\partial}{\partial t}\mathbf{U} + (\mathbf{U} \cdot \nabla)\mathbf{U},$$ (1.1-36)

where

$$\nabla = \frac{\partial}{\partial x}\hat{x} + \frac{\partial}{\partial y}\hat{y} + \frac{\partial}{\partial z}\hat{z}$$ (1.1-37)

is the *gradient* in the rectangular coordinate system, and the *speeds* (in meters per second) of the fluid particle dV in the X, Y, and Z directions are given by

$$U_X = \frac{dx}{dt},$$ (1.1-38)

$$U_Y = \frac{dy}{dt},$$ (1.1-39)

and

$$U_Z = \frac{dz}{dt},$$ (1.1-40)

respectively. Equation (1.1-36) represents the *total acceleration* of an infinitesimal volume element of fluid dV. The first term on the right-hand side of Eq. (1.1-36) represents a *local acceleration* when the velocity vector $\mathbf{U}$ is a function of time. The second term on the right-hand side of Eq. (1.1-36) represents a *convective acceleration*, that is, the fluid particle is accelerated (or decelerated) when it is convected into a region of higher (or lower) velocity even if the velocity vector $\mathbf{U}$ is not a function of time. For example, in the case of *steady fluid flow* through a nozzle, in which, by definition, the fluid velocity vector is *not* a function of time, a fluid particle will accelerate as it moves through the nozzle, since the particle is convected into a region of higher velocity. Equation (1.1-36) can also be written in scalar form as follows:

$$\frac{d}{dt}U_X = \frac{\partial}{\partial t}U_X + U_X\frac{\partial}{\partial x}U_X + U_Y\frac{\partial}{\partial y}U_X + U_Z\frac{\partial}{\partial z}U_X,$$ (1.1-41)

$$\frac{d}{dt}U_Y = \frac{\partial}{\partial t}U_Y + U_X\frac{\partial}{\partial x}U_Y + U_Y\frac{\partial}{\partial y}U_Y + U_Z\frac{\partial}{\partial z}U_Y,$$ (1.1-42)

$$\frac{d}{dt}U_Z = \frac{\partial}{\partial t}U_Z + U_X\frac{\partial}{\partial x}U_Z + U_Y\frac{\partial}{\partial y}U_Z + U_Z\frac{\partial}{\partial z}U_Z,$$ (1.1-43)

or

$$\frac{d}{dt}U_X = \frac{\partial}{\partial t}U_X + \mathbf{U} \cdot \nabla U_X,$$ (1.1-44)

$$\frac{d}{dt}U_Y = \frac{\partial}{\partial t}U_Y + \mathbf{U} \cdot \nabla U_Y,$$ (1.1-45)

$$\frac{d}{dt}U_Z = \frac{\partial}{\partial t}U_Z + \mathbf{U} \cdot \nabla U_Z.$$ (1.1-46)

Newton's second law of motion given by Eq. (1.1-31) can also be expressed in scalar form as follows:

$$dF_X = dm \frac{d}{dt} U_X, \tag{1.1-47}$$

$$dF_Y = dm \frac{d}{dt} U_Y, \tag{1.1-48}$$

$$dF_Z = dm \frac{d}{dt} U_Z. \tag{1.1-49}$$

Therefore, with the use of Eqs. (1.1-20) through (1.1-22), and since

$$dm = \rho \, dV, \tag{1.1-50}$$

Eqs. (1.1-47) through (1.1-49) can be rewritten as:

$$\rho \frac{d}{dt} U_X = \frac{\partial}{\partial x} \sigma_{XX} + \frac{\partial}{\partial y} \sigma_{XY} + \frac{\partial}{\partial z} \sigma_{XZ} + \rho g_X, \tag{1.1-51}$$

$$\rho \frac{d}{dt} U_Y = \frac{\partial}{\partial x} \sigma_{YX} + \frac{\partial}{\partial y} \sigma_{YY} + \frac{\partial}{\partial z} \sigma_{YZ} + \rho g_Y, \tag{1.1-52}$$

$$\rho \frac{d}{dt} U_Z = \frac{\partial}{\partial x} \sigma_{ZX} + \frac{\partial}{\partial y} \sigma_{ZY} + \frac{\partial}{\partial z} \sigma_{ZZ} + \rho g_Z. \tag{1.1-53}$$

Equations (1.1-51) through (1.1-53) are the differential equations of motion for a fluid particle dV in the X, Y, and Z directions, respectively, for any fluid that satisfies the continuum assumption and is subjected to gravitational body forces.

In order to proceed further, we need to rewrite Eqs. (1.1-51) through (1.1-53) by expressing the nine components of stress in terms of the nine components of the *rate-of-strain tensor*. For *real* (*viscous*) *fluids*, that is, for fluids that possess *viscosity* or resistance to fluid flow,

$$\boxed{\sigma_{ij} = -P\delta_{ij} + d_{ij}, \quad i, j = 1, 2, 3,} \tag{1.1-54}$$

where

$$\boxed{P \triangleq -\tfrac{1}{3}(\sigma_{11} + \sigma_{22} + \sigma_{33})} \tag{1.1-55}$$

is defined as the normal stress (pressure) in an *ideal* (*nonviscous*) *fluid* at rest or in motion, or as the normal stress (pressure) in a *real fluid* at rest;

$$\delta_{ij} = \begin{cases} 1, & i = j, \\ 0, & i \neq j, \end{cases} \qquad (1.1\text{-}56)$$

is the *Kronecker delta*; and d_{ij} is the *deviatoric stress tensor* and is chosen so that

$$d_{11} + d_{22} + d_{33} = 0. \qquad (1.1\text{-}57)$$

Note that an ideal fluid is one whose viscosity is zero. Recall that the index values 1, 2, and 3 correspond to X, Y, and Z, respectively. For *Newtonian fluids* (i.e., fluids in which the shear stress is directly proportional to the rate of deformation of the fluid), the deviatoric stress tensor is given by

$$\boxed{d_{ij} = \left(\mu_v - \tfrac{2}{3}\mu\right)\Delta\delta_{ij} + 2\mu\varepsilon_{ij}, \qquad i, j = 1, 2, 3,} \qquad (1.1\text{-}58)$$

where μ_v and μ are the *positive coefficients of volume (bulk) viscosity* and *viscosity* (or *shear viscosity*) of the fluid, respectively, with units of pascal seconds (Pa-sec),

$$\boxed{\Delta = \nabla \cdot \mathbf{U} = \frac{\partial}{\partial x}U_X + \frac{\partial}{\partial y}U_Y + \frac{\partial}{\partial z}U_Z} \qquad (1.1\text{-}59)$$

is called the *local rate of expansion* (in inverse seconds) and is a measure of the rate of change in volume per unit volume of a fluid particle dV as it moves, and

$$\boxed{\varepsilon_{ij} \triangleq \frac{1}{2}\left(\frac{\partial}{\partial x_j}U_i + \frac{\partial}{\partial x_i}U_j\right), \qquad i, j = 1, 2, 3,} \qquad (1.1\text{-}60)$$

is the definition of the *rate-of-strain tensor* (in inverse seconds) where $U_1 = U_X$, $U_2 = U_Y$, $U_3 = U_Z$, $x_1 = x$, $x_2 = y$, and $x_3 = z$. The term *strain* refers to the relative change in dimensions or shape of a body (fluid element dV in our case) that is subjected to stress. Note that

$$\Delta = \varepsilon_{11} + \varepsilon_{22} + \varepsilon_{33}. \qquad (1.1\text{-}61)$$

Also note that the rate-of-strain tensor is *symmetric*, that is,

$$\varepsilon_{ij} = \varepsilon_{ji}, \qquad i, j = 1, 2, 3. \qquad (1.1\text{-}62)$$

As a result, there are six *independent* rate-of-strain components. Therefore, upon substituting Eq. (1.1-58) into Eq. (1.1-54), we obtain the following stress tensor for Newtonian fluids:

$$\sigma_{ij} = \left[-P + \left(\mu_v - \tfrac{2}{3}\mu \right) \Delta \right] \delta_{ij} + 2\mu \varepsilon_{ij}, \qquad i, j = 1, 2, 3. \qquad (1.1\text{-}63)$$

Note that Eq. (1.1-63) reduces to

$$\sigma_{ij} = -P\delta_{ij}, \qquad i, j = 1, 2, 3 \qquad (1.1\text{-}64)$$

for ideal fluids ($\mu_v = 0$ and $\mu = 0$), and for real fluids at rest ($\mu_v \neq 0$, $\mu \neq 0$, $\Delta = 0$, and $\varepsilon_{ij} = 0$, $i, j = 1, 2, 3$). Also note that Eq. (1.1-63) is only valid for *isotropic* Newtonian fluids, since the coefficients μ_v and μ do not depend on direction, that is, they do not depend on the indices i and j.

Equation (1.1-63) is the expression that we were looking for in order to continue the derivation of the equations of motion for dV. Substituting Eqs. (1.1-59) and (1.1-60) into Eq. (1.1-63) yields the following set of equations:

$$\sigma_{XX} = -P + \left(\mu_v - \tfrac{2}{3}\mu \right) \nabla \cdot \mathbf{U} + 2\mu \frac{\partial}{\partial x} U_X, \qquad (1.1\text{-}65)$$

$$\sigma_{XY} = \sigma_{YX} = \mu \left(\frac{\partial}{\partial y} U_X + \frac{\partial}{\partial x} U_Y \right), \qquad (1.1\text{-}66)$$

$$\sigma_{XZ} = \sigma_{ZX} = \mu \left(\frac{\partial}{\partial z} U_X + \frac{\partial}{\partial x} U_Z \right), \qquad (1.1\text{-}67)$$

$$\sigma_{YY} = -P + \left(\mu_v - \tfrac{2}{3}\mu \right) \nabla \cdot \mathbf{U} + 2\mu \frac{\partial}{\partial y} U_Y, \qquad (1.1\text{-}68)$$

$$\sigma_{YZ} = \sigma_{ZY} = \mu \left(\frac{\partial}{\partial z} U_Y + \frac{\partial}{\partial y} U_Z \right), \qquad (1.1\text{-}69)$$

$$\sigma_{ZZ} = -P + \left(\mu_v - \tfrac{2}{3}\mu \right) \nabla \cdot \mathbf{U} + 2\mu \frac{\partial}{\partial z} U_Z. \qquad (1.1\text{-}70)$$

Next, upon substituting Eqs. (1.1-65) through (1.1-70) into Eqs. (1.1-51) through (1.1-53), we obtain

$$
\rho \frac{d}{dt} U_X = -\frac{\partial}{\partial x} P + \frac{\partial}{\partial x} \left[(\mu_v - \tfrac{2}{3}\mu) \nabla \cdot \mathbf{U} + 2\mu \frac{\partial}{\partial x} U_X \right]
$$

$$
+ \frac{\partial}{\partial y} \left[\mu \left(\frac{\partial}{\partial y} U_X + \frac{\partial}{\partial x} U_Y \right) \right]
$$

$$
+ \frac{\partial}{\partial z} \left[\mu \left(\frac{\partial}{\partial z} U_X + \frac{\partial}{\partial x} U_Z \right) \right] + \rho g_X,
$$

(1.1-71)

$$
\rho \frac{d}{dt} U_Y = -\frac{\partial}{\partial y} P + \frac{\partial}{\partial x} \left[\mu \left(\frac{\partial}{\partial y} U_X + \frac{\partial}{\partial x} U_Y \right) \right]
$$

$$
+ \frac{\partial}{\partial y} \left[(\mu_v - \tfrac{2}{3}\mu) \nabla \cdot \mathbf{U} + 2\mu \frac{\partial}{\partial y} U_Y \right]
$$

$$
+ \frac{\partial}{\partial z} \left[\mu \left(\frac{\partial}{\partial z} U_Y + \frac{\partial}{\partial y} U_Z \right) \right] + \rho g_Y,
$$

(1.1-72)

and

$$
\rho \frac{d}{dt} U_Z = -\frac{\partial}{\partial z} P + \frac{\partial}{\partial x} \left[\mu \left(\frac{\partial}{\partial z} U_X + \frac{\partial}{\partial x} U_Z \right) \right]
$$

$$
+ \frac{\partial}{\partial y} \left[\mu \left(\frac{\partial}{\partial z} U_Y + \frac{\partial}{\partial y} U_Z \right) \right]
$$

$$
+ \frac{\partial}{\partial z} \left[(\mu_v - \tfrac{2}{3}\mu) \nabla \cdot \mathbf{U} + 2\mu \frac{\partial}{\partial z} U_Z \right] + \rho g_Z
$$

(1.1-73)

which are the *Navier-Stokes equations of motion* for an infinitesimal volume element of fluid dV in the X, Y, and Z directions, respectively. If the coefficients μ_v and μ are *constants*, then Eqs. (1.1-71) through (1.1-73) can be represented by the following single vector Navier-Stokes equation:

$$
\rho \frac{d}{dt} \mathbf{U} = -\nabla P + \rho \mathbf{g} + \left(\mu_v + \tfrac{1}{3}\mu \right) \nabla (\nabla \cdot \mathbf{U}) + \mu \nabla^2 \mathbf{U}
$$

(1.1-74)

or, since

$$\nabla^2 \mathbf{U} = \nabla(\nabla \cdot \mathbf{U}) - \nabla \times (\nabla \times \mathbf{U}), \qquad (1.1\text{-}75)$$

$$\rho \frac{d}{dt} \mathbf{U} = -\nabla P + \rho \mathbf{g} + \left(\mu_v + \tfrac{4}{3}\mu\right)\nabla(\nabla \cdot \mathbf{U}) - \mu\nabla \times (\nabla \times \mathbf{U}). \qquad (1.1\text{-}76)$$

The Navier-Stokes equation of motion given by Eq. (1.1-76) can be decomposed into two separate equations by taking advantage of the fact that *any* vector function of position, such as the velocity vector **U** of the fluid particle *dV*, can always be uniquely separated into a *longitudinal* or *irrotational* part $\mathbf{U}_L$, for which

$$\nabla \times \mathbf{U}_L = \mathbf{0}, \qquad (1.1\text{-}77)$$

and a *transverse* or *rotational* part $\mathbf{U}_T$, for which

$$\nabla \cdot \mathbf{U}_T = 0. \qquad (1.1\text{-}78)$$

Equation (1.1-78) indicates that transverse or rotational fluid flow is *incompressible*, since the local rate of expansion given by Eq. (1.1-78) is equal to zero [see Eq. (1.1-59)]. *Incompressible fluid flow* is defined as the density of the fluid medium ρ being equal to a *constant* (i.e., a function of neither time nor space). Note that the *rotation* $\boldsymbol{\omega}$ of an infinitesimal volume element of fluid *dV* is given by

$$\boldsymbol{\omega} = \tfrac{1}{2}\nabla \times \mathbf{U}_T, \qquad (1.1\text{-}79)$$

which is the *angular velocity vector* of *dV*. In general, a fluid particle may rotate about all three coordinate axes. Also note that the *vorticity* **V** is given by

$$\mathbf{V} = 2\boldsymbol{\omega} = \nabla \times \mathbf{U}_T. \qquad (1.1\text{-}80)$$

Therefore, if we express **U** as

$$\mathbf{U} = \mathbf{U}_L + \mathbf{U}_T, \qquad (1.1\text{-}81)$$

then substituting Eq. (1.1-81) into Eq. (1.1-76), and using Eqs. (1.1-75), (1.1-77), and (1.1-78), yields the following two equations:

$$\rho \frac{d}{dt}\mathbf{U}_L = -\nabla P + \left(\mu_v + \tfrac{4}{3}\mu\right)\nabla^2 \mathbf{U}_L + \rho \mathbf{g}_L \qquad (1.1\text{-}82)$$

and

$$\rho \frac{d}{dt}\mathbf{U}_T = \rho\mathbf{g}_T - \mu \nabla \times (\nabla \times \mathbf{U}_T).$$ (1.1-83)

Note that ∇P appearing in Eq. (1.1-82) is a longitudinal or irrotational vector function of position, since the curl of the gradient of *any* scalar function of position (pressure P in this case) is equal to the zero vector, that is,

$$\nabla \times \nabla P = \mathbf{0}.$$ (1.1-84)

Therefore, the gradient of any scalar function of position is longitudinal (irrotational). Also note that the gravitational body force per unit volume of fluid, $\rho\mathbf{g}$, was treated as a general vector function of position, and as a result, was separated into a longitudinal or irrotational part $\rho\mathbf{g}_L$ and a transverse or rotational part $\rho\mathbf{g}_T$. By referring to Eqs. (1.1-82) and (1.1-83), it can be seen that $\mathbf{U}_L$ and $\mathbf{U}_T$ can be solved for separately. Their solutions can then be combined according to Eq. (1.1-81) to give the total fluid-particle velocity vector.

The fluid-particle velocity vectors $\mathbf{U}_L$ and $\mathbf{U}_T$ can be expressed as

$$\mathbf{U}_L = \nabla\varphi$$ (1.1-85)

and

$$\mathbf{U}_T = \nabla \times \mathbf{\Phi},$$ (1.1-86)

where φ and $\mathbf{\Phi}$ are *scalar* and *vector velocity potentials*, respectively, both with units of square meters per second. Therefore, with the use of Eqs. (1.1-85) and (1.1-86), the fluid-particle velocity vector given by Eq. (1.1-81) can be rewritten as

$$\mathbf{U} = \nabla\varphi + \nabla \times \mathbf{\Phi}.$$ (1.1-87)

Note that since the curl of the gradient of *any* scalar function of position is equal to the zero vector, we can write that [see Eq. (1.1-77)]

$$\nabla \times \nabla\varphi = \nabla \times \mathbf{U}_L = \mathbf{0},$$ (1.1-88)

and as a result, taking the curl of both sides of Eq. (1.1-87) yields

$$\nabla \times \mathbf{U} = \nabla \times \nabla \times \mathbf{\Phi} = \nabla \times \mathbf{U}_T.$$ (1.1-89)

Also, since the divergence of the curl of *any* vector function of position is equal to zero, we can write that [see Eq. (1.1-78)]

$$\nabla \cdot \nabla \times \mathbf{\Phi} = \nabla \cdot \mathbf{U}_T = 0, \tag{1.1-90}$$

and as a result, taking the divergence of both sides of Eq. (1.1-87) yields

$$\nabla \cdot \mathbf{U} = \nabla \cdot \nabla \varphi = \nabla \cdot \mathbf{U}_L. \tag{1.1-91}$$

From our discussion it is clear that, in general, we must deal with the superposition of two different types of fluid motion. *Fluid motion due to acoustic (sound) waves is longitudinal (irrotational) and compressible.* Fluid motion due to *internal waves* in the ocean, for example, which are volume gravity waves generated within the ocean mass, is transverse (rotational) and incompressible. Therefore, the equation of motion given by Eq. (1.1-82) is to be used in the derivation of the linear wave equation.

Finally, for an ideal (nonviscous) fluid, that is, one is which $\mu_v = 0$ and $\mu = 0$, the Navier-Stokes equation of motion given by Eq. (1.1-76) reduces to

$$\boxed{\rho \frac{d}{dt} \mathbf{U} = -\nabla P + \rho \mathbf{g},} \tag{1.1-92}$$

which is *Euler's equation of motion*.

1.2 Equation of Continuity

In this section we shall derive the equation of continuity or, in other words, the partial differential equation for the *conservation of mass*. Consider the infinitesimal volume element of fluid dV illustrated in Fig. 1.1-1. Let us assume that dV is fixed in space, that is, it is not allowed to move. Also assume that the volume of dV is kept constant. If we denote the mass flow per unit cross-sectional area in the positive X direction at the *center* of dV as

$$J_X = \rho U_X, \tag{1.2-1}$$

with units of kilograms per second per square meter (kg/sec-m^2), then the mass flow per unit cross-sectional area in the positive X direction at the front and back sides (surfaces) of dV, denoted as J_{XF} and J_{XBK}, respectively, can be obtained by using the first two terms in a Taylor series expansion of J_X at the center of dV. For example,

$$J_{XF} = J_X + \frac{dx}{2} \frac{\partial}{\partial x} J_X \tag{1.2-2}$$

and

$$J_{XBK} = J_X - \frac{dx}{2} \frac{\partial}{\partial x} J_X, \tag{1.2-3}$$

where J_X is assumed to be positive. Mass flow per unit cross-sectional area is also referred to as *mass flux*. Next, let us assume that there is a net *influx* of mass into the constant volume dV in the positive X direction, that is, more mass flows into the back side of dV than flows out of the front side. With the use of Eqs. (1.2-1) through (1.2-3), the net influx of mass into the constant volume dV in the positive X direction can be expressed as

$$(J_{XBK} - J_{XF})\, dy\, dz = -\frac{\partial}{\partial x}(\rho U_X)\, dV, \tag{1.2-4}$$

where $dV = dx\, dy\, dz$. Similarly, the net influx of mass into the constant volume dV in the positive Y and Z directions can be expressed as

$$-\frac{\partial}{\partial y}(\rho U_Y)\, dV$$

and

$$-\frac{\partial}{\partial z}(\rho U_Z)\, dV,$$

respectively. Since it was assumed that there is a net influx of mass into the constant volume dV in the positive X, Y, and Z directions, there will be a corresponding net increase in the value of the density of the fluid measured at the center of dV per unit time. Therefore, since the mass within dV increases with time, we can write that

$$-\left[\frac{\partial}{\partial x}(\rho U_X) + \frac{\partial}{\partial y}(\rho U_Y) + \frac{\partial}{\partial z}(\rho U_Z)\right] dV = +\left(\frac{\partial}{\partial t}\rho\right) dV. \tag{1.2-5}$$

Now consider the presence of an acoustic (sound) source. When an electro-acoustic transducer vibrates, fluid mass will flow due to the compression and expansion of the fluid medium. If fluid mass is being added inside dV at a rate of $\rho Q\, dV$, where Q is the volume flow rate per unit volume of fluid due to the sound source (i.e., Q has units of inverse seconds), then Eq. (1.2-5) can be rewritten as

$$-\left[\frac{\partial}{\partial x}(\rho U_X) + \frac{\partial}{\partial y}(\rho U_Y) + \frac{\partial}{\partial z}(\rho U_Z)\right] dV + \rho Q\, dV = +\left(\frac{\partial}{\partial t}\rho\right) dV \tag{1.2-6}$$

or

$$\boxed{\frac{\partial}{\partial t}\rho + \nabla \cdot (\rho \mathbf{U}) = \rho Q.} \tag{1.2-7}$$

Equation (1.2-7) is the *equation of continuity* (*conservation of mass*). An alternative expression for the equation of continuity can be obtained as follows: since

$$\nabla \cdot (\rho \mathbf{U}) = \mathbf{U} \cdot \nabla \rho + \rho \nabla \cdot \mathbf{U}, \tag{1.2-8}$$

substituting Eq. (1.2-8) into Eq. (1.2-7) yields

$$\boxed{\frac{d}{dt}\rho + \rho \nabla \cdot \mathbf{U} = \rho Q,} \tag{1.2-9}$$

where [see Eqs. (1.1-34) through (1.1-36)]

$$\frac{d}{dt}\rho = \frac{\partial}{\partial t}\rho + (\mathbf{U} \cdot \nabla)\rho = \frac{\partial}{\partial t}\rho + \mathbf{U} \cdot \nabla \rho. \tag{1.2-10}$$

In the case of *steady fluid flow*, all fluid properties are, by definition, independent of time. Since the density ρ is only a function of spatial coordinates in the case of steady fluid flow, Eq. (1.2-7) reduces to

$$\nabla \cdot (\rho \mathbf{U}) = \rho Q, \tag{1.2-11}$$

and in the absence of a sound source,

$$\nabla \cdot (\rho \mathbf{U}) = 0. \tag{1.2-12}$$

1.3 Equation of State and the Speed of Sound

The equation of state for a fluid describes the thermodynamic behavior of the fluid. In this section, we shall present various equations of state that relate the pressure, density, and temperature of a fluid. Before proceeding further, we need to introduce several new expressions. The pressure P, which is the *instantaneous* or *total pressure* in pascals (Pa), can be decomposed as follows:

$$\boxed{P = p_0 + p,} \tag{1.3-1}$$

where p_0 is the *equilibrium* or *ambient pressure* and p is called the *acoustic pressure*. The acoustic pressure represents the *change* in pressure in the fluid medium from the ambient value due to the presence of an acoustic (sound) wave (see Fig. 1.3-1). The density ρ is the *instantaneous* or *total density* of the fluid medium in kilograms per cubic meter. The *relative change* in the density of the fluid due to the presence of an acoustic wave is measured by the *condensation s*,

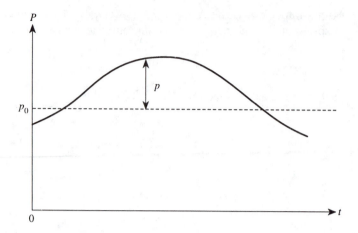

Figure 1.3-1. Illustration of the total pressure P, the ambient pressure p_0, and the acoustic pressure $p = P - p_0$ as a function of time at a fixed point in space.

which is given by

$$s = \frac{\rho - \rho_0}{\rho_0},$$
(1.3-2)

where ρ_0 is the *equilibrium* or *ambient density* of the fluid. Note that the condensation s is dimensionless. Also note that $\rho_0 s$, which is equal to the numerator $\rho - \rho_0$ of the condensation, which is called the *acoustic density*, represents the *change* in density of the fluid from the ambient value due to the presence of an acoustic wave.

For fluids with very high *thermal conductivity*, the temperature of the fluid is practically unchanged by the passage of a *small-amplitude* sound wave. In this case, the fluid remains approximately at a constant temperature and the compression of the fluid is called *isothermal*. As a result, it can be shown that

$$p \approx s/\kappa_T,$$
(1.3-3)

where κ_T is the *isothermal compressibility* of the fluid with units of inverse pascals. The compressibility of a fluid is defined as minus the ratio of the volume strain to the change in pressure. Volume strain is defined as the ratio of a change in volume to the original volume. A minus sign is included in the definition of compressibility because an increase in pressure always causes a decrease in volume, so that the compressibility will be positive. Equation (1.3-3) is the *isothermal equation of state*. However, if there is insignificant exchange of thermal energy from one particle of fluid to another during the passage of a small-amplitude sound wave, the *entropy* of the fluid remains nearly constant, but the temperature does not. In this case, the compression of the fluid is called *adiabatic*. In this case, it can be shown that

$$\boxed{p \approx s/\kappa_E,}$$

(1.3-4)

where κ_E is the *adiabatic compressibility* of the fluid with units of inverse pascals. Equation (1.3-4) is the *adiabatic equation of state*. As can be seen from Eqs. (1.3-3) and (1.3-4), the relationship between the acoustic pressure p and the condensation s is approximately *linear* for *small-amplitude* acoustic signals, that is, when $p/p_0 \ll 1$ and $|s| \ll 1$.

For *perfect gases*,

$$\kappa_T = 1/p_0$$

(1.3-5)

and

$$\kappa_E = \frac{\kappa_T}{\gamma} = \frac{1}{\gamma p_0}$$

(1.3-6)

where γ is the ratio of the specific heat at constant pressure to the specific heat at constant volume of the gas.

For *liquids*,

$$\kappa_E = \frac{\kappa_T}{\gamma},$$

(1.3-7)

where γ is the ratio of the specific heat at constant pressure to the specific heat at constant volume of the liquid. The specific heat at constant pressure and the specific heat at constant volume of a fluid (gas or liquid) are defined as the heat necessary to raise the temperature of a unit mass of fluid one degree when the fluid is kept at a constant pressure and at a constant volume, respectively.

Since the *elasticity*, or *bulk modulus*, of a fluid is equal to the reciprocal of the compressibility of the fluid, Eqs. (1.3-3) and (1.3-4) can be rewritten as

$$\boxed{p \approx B_T s}$$

(1.3-8)

and

$$\boxed{p \approx B_E s,}$$

(1.3-9)

where

$$B_T = 1/\kappa_T$$

(1.3-10)

is the *isothermal bulk modulus* in pascals,

$$B_E = 1/\kappa_E \tag{1.3-11}$$

is the *adiabatic bulk modulus* in pascals, and

$$B_E = \gamma B_T. \tag{1.3-12}$$

The *compression of a fluid due to the passage of an acoustic (sound) wave is best described as being adiabatic.* Therefore, the adiabatic equation of state, as given by either Eq. (1.3-4) or Eq. (1.3-9), will be used in the derivation of the linear wave equation.

We shall conclude our discussion in this section by presenting the following formulas for the speed of sound in meters per second. For *perfect gases*, the *isothermal speed of sound* is given by

$$c = \sqrt{\frac{1}{\kappa_T \rho_0}} = \sqrt{\frac{B_T}{\rho_0}} = \sqrt{\frac{p_0}{\rho_0}} \tag{1.3-13}$$

and the *adiabatic speed of sound* is given by

$$c = \sqrt{\frac{1}{\kappa_E \rho_0}} = \sqrt{\frac{B_E}{\rho_0}} = \sqrt{\frac{\gamma p_0}{\rho_0}}. \tag{1.3-14}$$

A typical value for the adiabatic speed of sound in air is $c = 343$ m/sec (see Problem 1-2).

For *liquids*, the *isothermal speed of sound* is given by

$$c = \sqrt{\frac{1}{\kappa_T \rho_0}} = \sqrt{\frac{B_T}{\rho_0}} \tag{1.3-15}$$

and the *adiabatic speed of sound* is given by

$$c = \sqrt{\frac{1}{\kappa_E \rho_0}} = \sqrt{\frac{B_E}{\rho_0}} = \sqrt{\frac{\gamma B_T}{\rho_0}}. \tag{1.3-16}$$

A typical value for the adiabatic speed of sound in sea water is $c = 1500$ m/sec (see Problem 1-2).

And finally, the relationships between the acoustic pressure and the condensation given by Eqs. (1.3-8) and (1.3-9) can be summarized by the following single equation:

$$p \approx \rho_0 s c^2$$ (1.3-17)

where $\rho_0 s$ is the acoustic density and the speed of sound c is given by Eqs. (1.3-13) through (1.3-16), whichever one is appropriate.

1.4 Derivation of the Linear Wave Equation

In order to derive the linear wave equation, we must combine the equations of motion, continuity, and state. Let us begin with the Navier-Stokes equation of motion for a fluid particle given by Eq. (1.1-82), which is rewritten below for convenience:

$$\rho \frac{d}{dt} \mathbf{U}_L = -\nabla P + \left(\mu_v + \tfrac{4}{3}\mu \right) \nabla^2 \mathbf{U}_L + \rho \mathbf{g}_L.$$ (1.1-82)

The fluid-particle velocity vector $\mathbf{U}_L$, which is the *instantaneous* or *total* longitudinal (irrotational) fluid-particle velocity vector associated with acoustic (sound) wave motion in fluids, can be decomposed as follows:

$$\mathbf{U}_L = \mathbf{u}_0 + \mathbf{u},$$ (1.4-1)

where $\mathbf{u}_0$ is the *equilibrium* or *ambient fluid velocity vector* and $\mathbf{u}$, which is called the *acoustic fluid velocity vector*, represents the change in velocity of the fluid medium from the ambient value due to the presence of an acoustic wave. Next, let us assume that $\mathbf{u}_0 = \mathbf{0}$, that is, assume that the fluid medium is at rest in the absence of an acoustic wave. Then Eq. (1.4-1) reduces to

$$\mathbf{U}_L = \mathbf{u}.$$ (1.4-2)

Substituting Eq. (1.4-2) into Eq. (1.1-82) and making use of Eq. (1.1-36) yields

$$\rho \left[\frac{\partial}{\partial t} \mathbf{u} + (\mathbf{u} \cdot \nabla)\mathbf{u} \right] = -\nabla P + \left(\mu_v + \tfrac{4}{3}\mu \right) \nabla^2 \mathbf{u} + \rho \mathbf{g}_L.$$ (1.4-3)

Equation (1.4-3) is a *nonlinear partial differential equation* in $\mathbf{u}$ because of the convective acceleration term $(\mathbf{u} \cdot \nabla)\mathbf{u}$ on the left-hand side. Also, since $\mathbf{u}$ is longitudinal (irrotational),

$$\nabla^2 \mathbf{u} = \nabla(\nabla \cdot \mathbf{u}),$$ (1.4-4)

where use has been made of Eqs. (1.1-75) and (1.1-77). Substituting Eq. (1.4-4) into Eq. (1.4-3) yields

$$\rho \left[\frac{\partial}{\partial t} \mathbf{u} + (\mathbf{u} \cdot \nabla)\mathbf{u} \right] = -\nabla P + \mu' \nabla(\nabla \cdot \mathbf{u}) + \rho \mathbf{g}_L,$$ (1.4-5)

where

$$\mu' = \mu_v + \tfrac{4}{3}\mu.$$ (1.4-6)

By referring to Eq. (1.3-2), the instantaneous or total density ρ can be expressed as

$$\rho = \rho_0 s + \rho_0 = \rho_0(s + 1),$$ (1.4-7)

where $\rho_0 s$ is the acoustic density. Also, by referring to Eq. (1.3-1), the gradient of the instantaneous or total pressure P is given by

$$\nabla P = \nabla p_0 + \nabla p.$$ (1.4-8)

Note that in the absence of an acoustic wave, substituting $\mathbf{u} = \mathbf{0}$, $\rho = \rho_0$, and $P = p_0$ into Eq. (1.4-3) yields the *hydrostatic pressure equation*

$$\nabla p_0 = \rho_0 \mathbf{g}_L.$$ (1.4-9)

Equation (1.4-9) indicates that the equilibrium pressure p_0 is a function of position; otherwise its gradient would be equal to the zero vector. The equilibrium density ρ_0 is, in general, a function of position also. Substituting Eqs. (1.4-7) through (1.4-9) into Eq. (1.4-5) yields

$$\rho_0(s + 1) \left[\frac{\partial}{\partial t} \mathbf{u} + (\mathbf{u} \cdot \nabla)\mathbf{u} \right] = -\nabla p + \mu' \nabla(\nabla \cdot \mathbf{u}) + \rho_0 s \mathbf{g}_L.$$ (1.4-10)

Since we are dealing with small-amplitude acoustic signals,

$$|s| \ll 1,$$ (1.4-11)

and by neglecting second- and higher-order terms in the acoustic variables—that is, by neglecting those terms that involve the multiplication of the acoustic variables p, $\mathbf{u}$, and $\rho_0 s$ with each other or themselves—Eq. (1.4-10) reduces to

$$\rho_0 \frac{\partial}{\partial t}\mathbf{u} \approx -\nabla p + \mu' \nabla(\nabla \cdot \mathbf{u}) + \rho_0 s \mathbf{g}_L, \qquad (1.4\text{-}12)$$

where μ' is given by Eq. (1.4-6). Equation (1.4-12) represents a *linearized* version of the Navier-Stokes equation of motion which is valid for *small-amplitude* acoustic signals and will be used in the derivation of the linear wave equation.

   The next equation that we need to simplify is the equation of continuity given by Eq. (1.2-7), which is rewritten below for convenience:

$$\frac{\partial}{\partial t}\rho + \nabla \cdot (\rho \mathbf{U}) = \rho Q. \qquad (1.2\text{-}7)$$

Since we are dealing with acoustic wave motion which is longitudinal (irrotational), and since we assumed that the fluid medium is at rest in the absence of an acoustic wave [see Eq. (1.4-2)],

$$\mathbf{U} = \mathbf{U}_L = \mathbf{u}. \qquad (1.4\text{-}13)$$

Substituting Eq. (1.4-13) into Eqs. (1.2-7) and (1.2-8) yields

$$\frac{\partial}{\partial t}\rho + \nabla \cdot (\rho \mathbf{u}) = \rho Q \qquad (1.4\text{-}14)$$

and

$$\nabla \cdot (\rho \mathbf{u}) = \mathbf{u} \cdot \nabla \rho + \rho \nabla \cdot \mathbf{u}, \qquad (1.4\text{-}15)$$

respectively. If we refer back to Eq. (1.4-7) and assume that the equilibrium density ρ_0 is *not* a function of time, then

$$\frac{\partial}{\partial t}\rho = \rho_0 \frac{\partial}{\partial t}s. \qquad (1.4\text{-}16)$$

Next, if we substitute Eq. (1.4-7) into Eq. (1.4-15), then

$$\nabla \cdot (\rho \mathbf{u}) = \mathbf{u} \cdot \nabla[\rho_0(s + 1)] + \rho_0(s + 1)\nabla \cdot \mathbf{u}. \qquad (1.4\text{-}17)$$

With the use of Eq. (1.4-11), Eq. (1.4-17) reduces to

$$\nabla \cdot (\rho \mathbf{u}) \approx \mathbf{u} \cdot \nabla \rho_0 + \rho_0 \nabla \cdot \mathbf{u}. \qquad (1.4\text{-}18)$$

Similarly,

$$\rho Q = \rho_0(s + 1)Q \approx \rho_0 Q. \qquad (1.4\text{-}19)$$

Therefore, upon substituting Eqs. (1.4-16), (1.4-18), and (1.4-19) into Eq. (1.4-14), we obtain

$$\frac{\partial}{\partial t}s + \frac{\nabla \rho_0}{\rho_0} \cdot \mathbf{u} + \nabla \cdot \mathbf{u} \approx Q. \qquad (1.4\text{-}20)$$

Equation (1.4-20) is a simplified version of the equation of continuity which is valid for *small-amplitude* acoustic signals and will be used in the derivation of the linear wave equation.

The final equation that is required in order to begin the derivation of the linear wave equation is the equation of state given by Eq. (1.3-17), which is valid for *small-amplitude* acoustic signals and is rewritten below for convenience:

$$p \approx \rho_0 s c^2 \qquad (1.3\text{-}17)$$

where c is the *adiabatic* speed of sound.

We shall derive the linear wave equation in terms of the scalar velocity potential φ, where (see Section 1.1)

$$\mathbf{u} = \nabla \varphi. \qquad (1.4\text{-}21)$$

We begin by eliminating the acoustic pressure p from the linearized Navier-Stokes equation of motion given by Eq. (1.4-12). Taking the gradient of both sides of the equation of state given by Eq. (1.3-17) yields

$$\nabla p \approx \rho_0 s \nabla c^2 + c^2 \nabla(\rho_0 s), \qquad (1.4\text{-}22)$$

or

$$\nabla p \approx 2\rho_0 s c \nabla c + c^2 \nabla(\rho_0 s), \qquad (1.4\text{-}23)$$

since

$$\nabla c^2 = 2c \nabla c. \qquad (1.4\text{-}24)$$

Substituting Eq. (1.4-23) into Eq. (1.4-12) and taking the partial derivative with respect to time t of both sides of the resulting equation yields

$$\rho_0 \frac{\partial^2}{\partial t^2}\mathbf{u} \approx -2\rho_0 c \nabla c \frac{\partial s}{\partial t} - c^2 \nabla\!\left(\rho_0 \frac{\partial s}{\partial t}\right) + \mu' \frac{\partial}{\partial t}\nabla(\nabla \cdot \mathbf{u}) + \rho_0 \mathbf{g}_L \frac{\partial s}{\partial t}, \quad (1.4\text{-}25)$$

since μ' is constant, and where it is assumed that ρ_0, c, and $\mathbf{g}_L$ are *not* functions of time. The parameter μ' is constant, that is, it is a function of neither time nor space, because the coefficients μ_v and μ were assumed to be constants [see Eq. (1.4-6)]. Furthermore, since

$$\nabla\left(\rho_0 \frac{\partial s}{\partial t}\right) = \rho_0 \nabla \frac{\partial s}{\partial t} + \frac{\partial s}{\partial t}\nabla\rho_0, \tag{1.4-26}$$

upon substituting Eq. (1.4-26) into Eq. (1.4-25) and dividing both sides of the resulting equation by $\rho_0 c^2$, we obtain

$$\frac{1}{c^2}\frac{\partial^2}{\partial t^2}\mathbf{u} \approx \left[-2\frac{\nabla c}{c} - \frac{\nabla\rho_0}{\rho_0} + \frac{\mathbf{g}_L}{c^2}\right]\frac{\partial s}{\partial t} - \nabla\frac{\partial s}{\partial t} + \frac{\mu'}{\rho_0 c^2}\frac{\partial}{\partial t}\nabla(\nabla\cdot\mathbf{u}). \tag{1.4-27}$$

Equation (1.4-27) can be simplified further.

We shall show that the entire first term on the right-hand side of Eq. (1.4-27) involving the square brackets is negligible compared to the second term. It is important to note that in the case of a *homogeneous fluid medium* where the speed of sound c and the equilibrium density ρ_0 are constants, the gradients of c and ρ_0 are *zero vectors*. In order to estimate the magnitudes of the individual terms involving the condensation s, we shall represent s as a unit-amplitude, time-harmonic plane wave propagating in three-dimensional space as follows:

$$s = \exp(-j\mathbf{k}\cdot\mathbf{r})\exp(+j2\pi ft), \tag{1.4-28}$$

where $\mathbf{k}$ is the propagation vector with units of radians per meter and $\mathbf{r}$ is a position vector with units of meters (see Section 2.2.1). With the use of Eq. (1.4-28), it can be shown that

$$\left|\frac{\partial s}{\partial t}\right| = kc \tag{1.4-29}$$

and

$$\left|\nabla\frac{\partial s}{\partial t}\right| = k^2 c, \tag{1.4-30}$$

where $|\mathbf{k}| = k = 2\pi f/c$ is the wave number with units of radians per meter. In the case of *sea water*,

$$\frac{|\nabla c|}{c} \approx 1 \times 10^{-5}, \tag{1.4-31}$$

$$\frac{|\nabla\rho_0|}{\rho_0} \approx 5 \times 10^{-6}, \tag{1.4-32}$$

and

$$\frac{|\mathbf{g}_L|}{c^2} \approx 4 \times 10^{-6}. \tag{1.4-33}$$

Table 1.4-1 Approximate values of the terms involving the condensation s in Eq. (1.4-27) for several representative frequencies in the case of sea water.

f (Hz)	$2\dfrac{\|\nabla c\|}{c}\left\|\dfrac{\partial s}{\partial t}\right\|$	$\dfrac{\|\nabla \rho_0\|}{\rho_0}\left\|\dfrac{\partial s}{\partial t}\right\|$	$\dfrac{\|\mathbf{g}_L\|}{c^2}\left\|\dfrac{\partial s}{\partial t}\right\|$	$\left\|\nabla\dfrac{\partial s}{\partial t}\right\|$
1	1×10^{-4}	3×10^{-5}	3×10^{-5}	3×10^{-2}
50	6×10^{-3}	2×10^{-3}	1×10^{-3}	7×10^{1}
250	3×10^{-2}	8×10^{-3}	6×10^{-3}	2×10^{3}
1000	1×10^{-1}	3×10^{-2}	3×10^{-2}	3×10^{4}

Therefore, if we use the nominal value $c = 1500$ m/sec for sea water, then Eqs. (1.4-29) through (1.4-33) can be combined to form Table 1.4-1. The frequencies 50 and 250 Hz were chosen because they represent typical carrier frequency values used in long-range underwater acoustic pulse-propagation studies. In addition, when $f = 250$ Hz and $c = 1500$ m/sec, $k \approx 1$ rad/m.

As can be seen from Table 1.4-1, $\nabla(\partial s/\partial t)$ is the dominant term compared to the other terms in Eq. (1.4-27) involving the condensation s. As a result, Eq. (1.4-27) reduces to

$$\frac{1}{c^2}\frac{\partial^2}{\partial t^2}\mathbf{u} \approx -\nabla\frac{\partial s}{\partial t} + \frac{\mu'}{\rho_0 c^2}\frac{\partial}{\partial t}\nabla(\nabla \cdot \mathbf{u}), \qquad (1.4\text{-}34)$$

and upon substituting Eqs. (1.4-20) and (1.4-21) into Eq. (1.4-34), we obtain

$$\nabla(\nabla^2\varphi) + \nabla\left(\frac{\nabla\rho_0}{\rho_0}\cdot\nabla\varphi\right) + \frac{\mu'}{\rho_0 c^2}\nabla\frac{\partial}{\partial t}\nabla^2\varphi - \frac{1}{c^2}\nabla\frac{\partial^2}{\partial t^2}\varphi \approx \nabla Q. \quad (1.4\text{-}35)$$

In order to obtain the linear wave equation in terms of the velocity potential φ, we must simplify Eq. (1.4-35) further.

Since

$$\nabla\left(\frac{\mu'}{\rho_0 c^2}\frac{\partial}{\partial t}\nabla^2\varphi\right) = \frac{\mu'}{\rho_0 c^2}\nabla\left(\frac{\partial}{\partial t}\nabla^2\varphi\right) + \left(\frac{\partial}{\partial t}\nabla^2\varphi\right)\nabla\left(\frac{\mu'}{\rho_0 c^2}\right), \qquad (1.4\text{-}36)$$

and since μ' is constant,

$$\nabla\left(\frac{\mu'}{\rho_0 c^2}\right) = \mu'\nabla\left(\frac{1}{\rho_0 c^2}\right) = -\frac{\mu'}{\rho_0 c^2}\left(2\frac{\nabla c}{c} + \frac{\nabla\rho_0}{\rho_0}\right). \qquad (1.4\text{-}37)$$

Therefore, substituting Eq. (1.4-37) into Eq. (1.4-36) yields

$$\nabla\left(\frac{\mu'}{\rho_0 c^2}\frac{\partial}{\partial t}\nabla^2\varphi\right) = \frac{\mu'}{\rho_0 c^2}\left[\nabla - \left(2\frac{\nabla c}{c} + \frac{\nabla\rho_0}{\rho_0}\right)\right]\frac{\partial}{\partial t}\nabla^2\varphi. \qquad (1.4\text{-}38)$$

Also, since

$$\nabla\left(\frac{1}{c^2}\frac{\partial^2}{\partial t^2}\varphi\right) = \frac{1}{c^2}\nabla\left(\frac{\partial^2}{\partial t^2}\varphi\right) + \left(\frac{\partial^2}{\partial t^2}\varphi\right)\nabla\left(\frac{1}{c^2}\right) \qquad (1.4\text{-}39)$$

and

$$\nabla\left(\frac{1}{c^2}\right) = -2\frac{\nabla c}{c^3}, \qquad (1.4\text{-}40)$$

substituting Eq. (1.4-40) into Eq. (1.4-39) yields

$$\nabla\left(\frac{1}{c^2}\frac{\partial^2}{\partial t^2}\varphi\right) = \frac{1}{c^2}\left(\nabla - 2\frac{\nabla c}{c}\right)\frac{\partial^2}{\partial t^2}\varphi. \qquad (1.4\text{-}41)$$

The right-hand sides of Eqs. (1.4-38) and (1.4-41) can be simplified if we can estimate the magnitudes of the individual terms involving the velocity potential φ. Recall that in the case of a homogeneous fluid medium, the gradients of c and ρ_0 are *zero vectors*. If we represent φ as a unit-amplitude, time-harmonic plane wave propagating in three-dimensional space as was done for the condensation s [see Eq. (1.4-28)], then it can be shown that

$$\left|\frac{\partial^2}{\partial t^2}\varphi\right| = k^2 c^2, \qquad (1.4\text{-}42)$$

$$\left|\nabla\frac{\partial^2}{\partial t^2}\varphi\right| = k^3 c^2, \qquad (1.4\text{-}43)$$

$$\left|\frac{\partial}{\partial t}\nabla^2\varphi\right| = k^3 c, \qquad (1.4\text{-}44)$$

and

$$\left|\nabla\frac{\partial}{\partial t}\nabla^2\varphi\right| = k^4 c. \qquad (1.4\text{-}45)$$

Therefore, if we use the nominal value $c = 1500$ m/sec for sea water, then Eqs. (1.4-31) and (1.4-32), together with Eqs. (1.4-42) through (1.4-45), can be combined to form Tables 1.4-2 and 1.4-3.

Table 1.4-2 Approximate values of the terms on the right-hand side of Eq. (1.4-38) for several representative frequencies in the case of sea water.

f (Hz)	$\left\| \nabla \dfrac{\partial}{\partial t} \nabla^2 \varphi \right\|$	$2\dfrac{\|\nabla c\|}{c} \left\| \dfrac{\partial}{\partial t} \nabla^2 \varphi \right\|$	$\dfrac{\|\nabla \rho_0\|}{\rho_0} \left\| \dfrac{\partial}{\partial t} \nabla^2 \varphi \right\|$
1	5×10^{-7}	2×10^{-9}	6×10^{-10}
50	3	3×10^{-4}	7×10^{-5}
250	2×10^3	3×10^{-2}	9×10^{-3}
1000	5×10^5	2	6×10^{-1}

As can be seen from Table 1.4-2, $\nabla(\partial/\partial t)\nabla^2\varphi$ is the dominant term on the right-hand side of Eq. (1.4-38), and as a result, Eq. (1.4-38) reduces to

$$\nabla\left(\frac{\mu'}{\rho_0 c^2}\frac{\partial}{\partial t}\nabla^2\varphi\right) \approx \frac{\mu'}{\rho_0 c^2}\nabla\frac{\partial}{\partial t}\nabla^2\varphi. \tag{1.4-46}$$

Similarly, from Table 1.4-3 it can be seen that $\nabla(\partial^2\varphi/\partial t^2)$ is the dominant term on the right-hand side of Eq. (1.4-41), and as a result, Eq. (1.4-41) reduces to

$$\nabla\left(\frac{1}{c^2}\frac{\partial^2}{\partial t^2}\varphi\right) \approx \frac{1}{c^2}\nabla\frac{\partial^2}{\partial t^2}\varphi. \tag{1.4-47}$$

Note that in the case of a homogeneous fluid medium, Eqs. (1.4-46) and (1.4-47) would be exact and not approximate expressions. Substituting Eqs. (1.4-46) and (1.4-47) into Eq. (1.4-35) yields

$$\nabla\left[\nabla^2\varphi + \frac{\nabla\rho_0}{\rho_0}\cdot\nabla\varphi + \frac{\mu'}{\rho_0 c^2}\frac{\partial}{\partial t}\nabla^2\varphi - \frac{1}{c^2}\frac{\partial^2}{\partial t^2}\varphi\right] \approx \nabla Q. \tag{1.4-48}$$

Therefore, from Eq. (1.4-48) we obtain the following two versions of the *linear wave equation* for *small-amplitude* acoustic signals in terms of the scalar velocity

Table 1.4-3 Approximate values of the terms on the right-hand side of Eq. (1.4-41) for several representative frequencies in the case of sea water.

f (Hz)	$\left\| \nabla \dfrac{\partial^2}{\partial t^2} \varphi \right\|$	$2\dfrac{\|\nabla c\|}{c} \left\| \dfrac{\partial^2}{\partial t^2} \varphi \right\|$
1	2×10^{-1}	8×10^{-4}
50	2×10^4	2
250	3×10^6	5×10^1
1000	2×10^8	8×10^2

potential $\varphi(t, \mathbf{r})$ with units of square meters per second:

$$\left[1 + \frac{\mu'}{\rho_0(\mathbf{r})c^2(\mathbf{r})}\frac{\partial}{\partial t}\right]\nabla^2\varphi(t,\mathbf{r}) + \frac{\nabla\rho_0(\mathbf{r})}{\rho_0(\mathbf{r})} \cdot \nabla\varphi(t,\mathbf{r})$$

$$- \frac{1}{c^2(\mathbf{r})}\frac{\partial^2}{\partial t^2}\varphi(t,\mathbf{r}) \approx x_M(t,\mathbf{r}) \tag{1.4-49}$$

or, neglecting the term involving the gradient of the equilibrium density,

$$\left[1 + \frac{\mu'}{\rho_0(\mathbf{r})c^2(\mathbf{r})}\frac{\partial}{\partial t}\right]\nabla^2\varphi(t,\mathbf{r}) - \frac{1}{c^2(\mathbf{r})}\frac{\partial^2}{\partial t^2}\varphi(t,\mathbf{r}) \approx x_M(t,\mathbf{r}), \tag{1.4-50}$$

where μ' is given by Eq. (1.4-6), and where

$$x_M(t,\mathbf{r}) = Q(t,\mathbf{r}) \tag{1.4-51}$$

is the *input acoustic signal to the fluid medium* (or *source distribution*) in inverse seconds, and represents the volume flow rate per unit volume of fluid at time t and position $\mathbf{r} = (x, y, z)$. In the case of an ideal fluid where there is no viscosity, that is, $\mu' = 0$, Eqs. (1.4-49) and (1.4-50) reduce to

$$\nabla^2\varphi(t,\mathbf{r}) + \frac{\nabla\rho_0(\mathbf{r})}{\rho_0(\mathbf{r})} \cdot \nabla\varphi(t,\mathbf{r}) - \frac{1}{c^2(\mathbf{r})}\frac{\partial^2}{\partial t^2}\varphi(t,\mathbf{r}) \approx x_M(t,\mathbf{r}) \tag{1.4-52}$$

and

$$\nabla^2\varphi(t,\mathbf{r}) - \frac{1}{c^2(\mathbf{r})}\frac{\partial^2}{\partial t^2}\varphi(t,\mathbf{r}) \approx x_M(t,\mathbf{r}), \tag{1.4-53}$$

respectively. *For the remainder of this book, we shall use the linear wave equation given by* Eq. (1.4-50) *for viscous fluids and* Eq. (1.4-53) *for nonviscous fluids.*

Recall that the acoustic fluid velocity vector is related to the scalar velocity potential by the following equation [see Eq. (1.4-21)]:

$$\mathbf{u}(t,\mathbf{r}) = \nabla\varphi(t,\mathbf{r}). \tag{1.4-54}$$

What we also need is a relationship between the acoustic pressure and the scalar velocity potential. In Appendix 1A it is shown that

$$\left[1 + \frac{\mu'}{\rho_0(\mathbf{r})c^2(\mathbf{r})}\frac{\partial}{\partial t}\right]p(t,\mathbf{r})$$
$$\approx -\rho_0(\mathbf{r})\frac{\partial}{\partial t}\varphi(t,\mathbf{r}) - \mu'\frac{\nabla\rho_0(\mathbf{r})}{\rho_0(\mathbf{r})}\cdot\nabla\varphi(t,\mathbf{r}) + \mu'x_M(t,\mathbf{r})$$

(1.4-55)

or, neglecting the term involving the gradient of the equilibrium density,

$$\left[1 + \frac{\mu'}{\rho_0(\mathbf{r})c^2(\mathbf{r})}\frac{\partial}{\partial t}\right]p(t,\mathbf{r}) \approx -\rho_0(\mathbf{r})\frac{\partial}{\partial t}\varphi(t,\mathbf{r}) + \mu'x_M(t,\mathbf{r}),$$ (1.4-56)

where μ' is given by Eq. (1.4-6). In the case of an ideal fluid, both Eqs. (1.4-55) and (1.4-56) reduce to

$$p(t,\mathbf{r}) \approx -\rho_0(\mathbf{r})\frac{\partial}{\partial t}\varphi(t,\mathbf{r}).$$ (1.4-57)

The Helmholtz Equation

We shall finish our discussion in this section by deriving the time-independent Helmholtz equation, which will bring up the additional topics of complex wave numbers and dispersive fluid media. We begin by rewriting the linear wave equation given by Eq. (1.4-50) as follows:

$$\left[1 + \tau(\mathbf{r})\frac{\partial}{\partial t}\right]\nabla^2\varphi(t,\mathbf{r}) - \frac{1}{c^2(\mathbf{r})}\frac{\partial^2}{\partial t^2}\varphi(t,\mathbf{r}) \approx x_M(t,\mathbf{r}),$$ (1.4-58)

where

$$\tau(\mathbf{r}) = \frac{\mu'}{\rho_0(\mathbf{r})c^2(\mathbf{r})}$$ (1.4-59)

is the *viscous relaxation time* in seconds and can be considered as the time scale for viscous effects. Next, assume that the source distribution $x_M(t,\mathbf{r})$ has a *time-harmonic* dependence, that is,

$$x_M(t,\mathbf{r}) = x_{f,M}(\mathbf{r})\exp(+j2\pi ft)$$ (1.4-60)

where $x_{f,M}(\mathbf{r})$ is the spatial-dependent part of the source distribution and f is the frequency in hertz. A time-harmonic quantity is one whose value at any time and

Table 1.4-4 Approximate values of $\omega\tau(\mathbf{r})$ for several representative frequencies in the case of sea water.

f (Hz)	$\omega = 2\pi f$	$\omega\tau(\mathbf{r})$
1	6.283	1×10^{-11}
50	314.159	6×10^{-10}
250	1570.796	3×10^{-9}
1000	6283.185	1×10^{-8}
10,000	62,831.853	1×10^{-7}
50,000	314,159.265	6×10^{-7}

point in space depends on a single frequency component. If the source distribution has a time-harmonic dependence, then the resulting velocity potential also has a time-harmonic dependence, that is,

$$\varphi(t,\mathbf{r}) = \varphi_f(\mathbf{r}) \exp(+j2\pi ft), \qquad (1.4\text{-}61)$$

where $\varphi_f(\mathbf{r})$ is the spatial-dependent part of the velocity potential. Therefore, upon substituting Eqs. (1.4-60) and (1.4-61) into the time-dependent wave equation given by Eq. (1.4-58), we obtain the following time-independent *lossy Helmholtz equation*:

$$\nabla^2\varphi_f(\mathbf{r}) + K^2(\mathbf{r})\varphi_f(\mathbf{r}) \approx \frac{x_{f,M}(\mathbf{r})}{1 + j\omega\tau(\mathbf{r})} \qquad (1.4\text{-}62)$$

or

$$\boxed{\nabla^2\varphi_f(\mathbf{r}) + K^2(\mathbf{r})\varphi_f(\mathbf{r}) \approx x_{f,M}(\mathbf{r}),} \qquad (1.4\text{-}63)$$

where

$$K(\mathbf{r}) = \frac{\omega}{c(\mathbf{r})} \frac{1}{\sqrt{1 + j\omega\tau(\mathbf{r})}} \qquad (1.4\text{-}64)$$

is the *complex wave number* in radians per meter and

$$\omega = 2\pi f \qquad (1.4\text{-}65)$$

is the *angular frequency* in radians per second. By comparing the right-hand sides of Eqs. (1.4-62) and (1.4-63), it can be seen that the right-hand side of Eq. (1.4-63) was obtained by assuming that $\omega\tau(\mathbf{r}) \ll 1$. In the case of sea water, $\tau(\mathbf{r}) \approx 2 \times 10^{-12}$ sec. Therefore, by referring to Table 1.4-4, it can be seen that $\omega\tau(\mathbf{r}) \ll 1$ for the range of frequency values shown. Only at extremely high (ultrasonic) frequencies will $\omega\tau(\mathbf{r})$ take on significant values.

If we can solve the Helmholtz equation given by Eq. (1.4-63), then substituting the solution $\varphi_f(\mathbf{r})$ into Eq. (1.4-61) will yield a time-harmonic solution of the wave equation given by Eq. (1.4-58). Time-harmonic solutions are very important, because solutions of the wave equation for source distributions with arbitrary time dependence can be obtained from time-harmonic solutions by using Fourier transform techniques.

Next, let us rewrite the complex wave number given by Eq. (1.4-64) as follows:

$$K(\mathbf{r}) = k(\mathbf{r}) - j\alpha(\mathbf{r}), \qquad (1.4\text{-}66)$$

where $k(\mathbf{r})$ and $\alpha(\mathbf{r})$ are the *real wave number* and *real attenuation coefficient*, respectively. In addition to being real quantities, both $k(\mathbf{r})$ and $\alpha(\mathbf{r})$ are *positive*. In Appendix 1B it is shown that the real wave number in radians per meter is given by

$$k(\mathbf{r}) = \frac{\omega}{c(\mathbf{r})} \frac{1}{\sqrt{2}} \left[\frac{\sqrt{1 + [\omega\tau(\mathbf{r})]^2} + 1}{1 + [\omega\tau(\mathbf{r})]^2} \right]^{1/2} \qquad (1.4\text{-}67)$$

and that the real attenuation coefficient in *nepers per meter* is given by

$$\alpha(\mathbf{r}) = \frac{\omega}{c(\mathbf{r})} \frac{1}{\sqrt{2}} \left[\frac{\sqrt{1 + [\omega\tau(\mathbf{r})]^2} - 1}{1 + [\omega\tau(\mathbf{r})]^2} \right]^{1/2}. \qquad (1.4\text{-}68)$$

Note that the *neper* (Np) is a *dimensionless* unit like the radian. The real attenuation coefficient $\alpha(\mathbf{r})$ is responsible for the attenuation of sound waves in viscous fluids (see Example 2.2-3). In the case of an ideal fluid where there is no viscosity, that is, $\mu' = 0$, Eq. (1.4-59) reduces to

$$\tau(\mathbf{r}) = 0, \qquad (1.4\text{-}69)$$

and as a result,

$$K(\mathbf{r}) = k(\mathbf{r}), \qquad (1.4\text{-}70)$$

$$k(\mathbf{r}) = \frac{2\pi f}{c(\mathbf{r})}, \qquad (1.4\text{-}71)$$

and

$$\alpha(\mathbf{r}) = 0. \qquad (1.4\text{-}72)$$

Substituting Eq. (1.4-70) into Eq. (1.4-63) yields the *lossless Helmholtz equation*

$$\nabla^2 \varphi_f(\mathbf{r}) + k^2(\mathbf{r}) \varphi_f(\mathbf{r}) \approx x_{f,M}(\mathbf{r}), \qquad (1.4\text{-}73)$$

where $k(\mathbf{r})$ is given by Eq. (1.4-71).

Finally, we shall conclude this section with a brief discussion of the *phase speed* (in meters per second) of a traveling wave. The phase speed is defined as follows:

$$c_p(\mathbf{r}) \triangleq \frac{\omega}{k(\mathbf{r})}, \qquad (1.4\text{-}74)$$

and upon substituting Eq. (1.4-67) into Eq. (1.4-74), we obtain

$$c_p(\mathbf{r}) = c(\mathbf{r})\sqrt{2} \left[\frac{1 + [\omega\tau(\mathbf{r})]^2}{\sqrt{1 + [\omega\tau(\mathbf{r})]^2} + 1} \right]^{1/2}. \qquad (1.4\text{-}75)$$

Therefore, the real wave number given by Eq. (1.4-67) can be expressed as

$$k(\mathbf{r}) = \frac{\omega}{c_p(\mathbf{r})}. \qquad (1.4\text{-}76)$$

Note that in the case of real, viscous fluids, the phase speed $c_p(\mathbf{r})$ is *not* equal to the speed of sound $c(\mathbf{r})$ and that it is a function of frequency f [see Eq. (1.4-65)]. Since the phase speed is a function of frequency as a result of the viscous properties of the medium, this is known as *material* or *intrinsic dispersion* and the medium is said to be *dispersive*. However, in the case of *sea water*,

$$c_p(\mathbf{r}) \approx c(\mathbf{r}) \qquad (1.4\text{-}77)$$

since $\omega\tau(\mathbf{r}) \ll 1$ for the typical range of frequencies used in underwater acoustics (see Table 1.4-4). Finally, in the case of an ideal fluid where there is no viscosity, substituting Eq. (1.4-69) into Eq. (1.4-75) yields

$$c_p(\mathbf{r}) = c(\mathbf{r}). \qquad (1.4\text{-}78)$$

1.5   Time-Average Intensity Vector and Time-Average Power

Two of the most important quantities of interest in acoustics are the time-average radiated and scattered acoustic power. However, in order to compute the time-average power, we must first compute the time-average intensity vector. Intensity has units of watts per square meter.

We begin our discussion with the *instantaneous intensity vector* $\mathbf{I}(t,\mathbf{r})$, which is defined as

$$\mathbf{I}(t,\mathbf{r}) \triangleq \text{Re}\{p(t,\mathbf{r})\}\,\text{Re}\{\mathbf{u}(t,\mathbf{r})\}. \tag{1.5-1}$$

When $\mathbf{I}(t,\mathbf{r})$ is an *arbitrary* function of time, the *time-average intensity vector* $\mathbf{I}_{avg}(\mathbf{r})$ is defined as

$$\mathbf{I}_{avg}(\mathbf{r}) \triangleq \lim_{T \to \infty} \frac{1}{T} \int_{-T/2}^{T/2} \mathbf{I}(t,\mathbf{r})\, dt \tag{1.5-2}$$

where T is the length of the data record in seconds. However, if $\mathbf{I}(t,\mathbf{r})$ is *periodic* with *fundamental period* T_0 seconds, then

$$\mathbf{I}_{avg}(\mathbf{r}) \triangleq \frac{1}{T_0} \int_{-T_0/2}^{T_0/2} \mathbf{I}(t,\mathbf{r})\, dt. \tag{1.5-3}$$

Whereas Eq. (1.5-2) is a general result, Eq. (1.5-3) is applicable to the special case of *time-harmonic acoustic fields*. A time-harmonic field is one whose value at any time and point in space depends on a single frequency component. Equation (1.5-3) can be simplified further, as shall be demonstrated next.

Time-harmonic acoustic pressure and acoustic fluid velocity vector fields can be expressed as follows:

$$p(t,\mathbf{r}) = p_f(\mathbf{r})\exp(+j2\pi ft) \tag{1.5-4}$$

and

$$\mathbf{u}(t,\mathbf{r}) = \mathbf{u}_f(\mathbf{r})\exp(+j2\pi ft), \tag{1.5-5}$$

respectively, where

$$p_f(\mathbf{r}) = |p_f(\mathbf{r})|\exp[+j\alpha(\mathbf{r})], \tag{1.5-6}$$

$$\mathbf{u}_f(\mathbf{r}) = u_f(\mathbf{r})\hat{u} = |u_f(\mathbf{r})|\exp[+j\beta(\mathbf{r})]\,\hat{u}, \tag{1.5-7}$$

$\alpha(\mathbf{r})$ is the phase of the *complex* acoustic pressure $p_f(\mathbf{r})$, $u_f(\mathbf{r})$ is the *complex* acoustic fluid *speed*, $\hat{u}$ is the unit vector in the direction of the velocity vector $\mathbf{u}_f(\mathbf{r})$, $\beta(\mathbf{r})$ is the phase of the complex speed $u_f(\mathbf{r})$, $\mathbf{r}$ is a position vector, and f is the frequency in hertz. Substituting Eqs. (1.5-4) through (1.5-7) into Eq. (1.5-1) yields the instantaneous intensity vector

$$\mathbf{I}(t,\mathbf{r}) = |p_f(\mathbf{r})||u_f(\mathbf{r})|\cos[2\pi ft + \alpha(\mathbf{r})]\cos[2\pi ft + \beta(\mathbf{r})]\,\hat{u}, \quad (1.5\text{-}8)$$

or, since

$$\cos x \cos y = \tfrac{1}{2}[\cos(x+y) + \cos(x-y)], \quad (1.5\text{-}9)$$

$$\mathbf{I}(t,\mathbf{r}) = \tfrac{1}{2}|p_f(\mathbf{r})||u_f(\mathbf{r})|$$

$$\times\{\cos[4\pi ft + \alpha(\mathbf{r}) + \beta(\mathbf{r})] + \cos[\alpha(\mathbf{r}) - \beta(\mathbf{r})]\}\,\hat{u}. \quad (1.5\text{-}10)$$

Substituting Eq. (1.5-10) into Eq. (1.5-3) yields the time-average intensity vector

$$\mathbf{I}_{avg}(\mathbf{r}) = \tfrac{1}{2}|p_f(\mathbf{r})||u_f(\mathbf{r})|\cos[\alpha(\mathbf{r}) - \beta(\mathbf{r})]\,\hat{u}, \quad (1.5\text{-}11)$$

since

$$\frac{1}{T_0}\int_{-T_0/2}^{T_0/2}\cos[4\pi ft + \alpha(\mathbf{r}) + \beta(\mathbf{r})]\,dt = 0, \quad (1.5\text{-}12)$$

where

$$T_0 = \frac{1}{2f} \quad (1.5\text{-}13)$$

is the fundamental period of the instantaneous intensity vector given by Eq. (1.5-10), since $\mathbf{I}(t,\mathbf{r}) = \mathbf{I}(t + T_0,\mathbf{r})$. Finally, note that Eq. (1.5-11) can be rewritten as

$$\boxed{\mathbf{I}_{avg}(\mathbf{r}) = \tfrac{1}{2}\mathrm{Re}\{p_f(\mathbf{r})\mathbf{u}_f^*(\mathbf{r})\},} \quad (1.5\text{-}14)$$

where the asterisk * denotes complex conjugate. Equation (1.5-14) is the desired expression for the time-average intensity vector for time-harmonic acoustic fields propagating in either homogeneous or inhomogeneous fluid media, and in the absence or presence of boundaries. It can easily be verified that Eqs. (1.5-11) and (1.5-14) are equal by substituting Eqs. (1.5-6) and (1.5-7) into Eq. (1.5-14).

Now that we have expressions for the time-average intensity vector for both the arbitrary time-dependent and time-harmonic cases, we can compute the time-average acoustic power. The *time-average acoustic power* (in watts) is given by

$$\boxed{P_{avg} = \oint_S \mathbf{I}_{avg}(\mathbf{r})\cdot d\mathbf{S}} \quad (1.5\text{-}15)$$

where S is *any surface* enclosing the sound source,

$$\mathbf{dS} = dS\,\hat{n}, \tag{1.5-16}$$

dS is an infinitesimal element of surface area, and $\hat{n}$ is a unit vector normal to the surface S pointing in the conventional outward direction away from the enclosed volume, and hence the source.

1.6 Sound-Pressure Level, Source Level, Transmission Loss, and Sound-Intensity Level

Once the wave equation has been solved for a particular problem and the resulting acoustic pressure has been calculated, it is often convenient to express the values of acoustic pressure in terms of *sound-pressure levels*. The sound-pressure level (SPL) is defined as follows:

$$\mathrm{SPL} \triangleq 20\log_{10}\left(\frac{p_{\mathrm{rms}}}{P_{\mathrm{ref}}}\right) \mathrm{dB\ re\ } P_{\mathrm{ref}}, \tag{1.6-1}$$

where p_{rms} is the root-mean-square (rms) value of the acoustic pressure, P_{ref} is the rms reference pressure, and "re" means "relative to." For problems in air acoustics, $P_{\mathrm{ref}} = 20\ \mu\mathrm{Pa}$ (rms) is commonly used. However, for problems in underwater acoustics, $P_{\mathrm{ref}} = 1\ \mu\mathrm{Pa}$ (rms) is common. For *time-harmonic* acoustic fields,

$$p_{\mathrm{rms}} = \frac{\sqrt{2}}{2}\left|p_f(\mathbf{r})\right|, \tag{1.6-2}$$

where $|p_f(\mathbf{r})|$ is the magnitude of the spatial-dependent part of the acoustic pressure. For acoustic fields with *arbitrary* time dependence,

$$p_{\mathrm{rms}} = \left[\lim_{T\to\infty}\frac{1}{T}\int_{-T/2}^{T/2}\left|p(t,\mathbf{r})\right|^2 dt\right]^{1/2}, \tag{1.6-3}$$

where T is the length of the data record in seconds.

One way to compare different sound sources is to compare their *source levels*. The source level (SL) is defined as follows:

$$SL \triangleq 20 \log_{10} \left(\frac{\sqrt{2} \, P_0 / 2}{P_{\text{ref}}} \right) \text{dB re } P_{\text{ref}} \qquad (1.6\text{-}4)$$

where P_0 is the peak acoustic pressure amplitude measured at a distance of one meter from the source along its acoustic axis, that is, in the direction of the maximum response of the source. The difference between the SL and the SPL is referred to as the *transmission loss* (TL) in dB, that is,

$$TL = SL - SPL. \qquad (1.6\text{-}5)$$

Finally, the *sound-intensity level* (SIL) is defined as follows:

$$SIL \triangleq 10 \log_{10} \left(\frac{I_{\text{avg}}}{I_{\text{ref}}} \right) \text{dB re } I_{\text{ref}}, \qquad (1.6\text{-}6)$$

where I_{avg} is the magnitude of the time-average intensity vector and I_{ref} is the reference time-average intensity. For problems in air acoustics, $I_{\text{ref}} = 10^{-12} \text{ W/m}^2$ is commonly used, since it is approximately the time-average intensity of a 1-kHz pure tone that is just barely audible to a person with normal hearing. In the case of time-harmonic plane wave propagation in air, $I_{\text{ref}} = 10^{-12} \text{ W/m}^2$ corresponds approximately to $P_{\text{ref}} = 20 \ \mu\text{Pa}$ (rms).

Problems

1-1 With $p_0(0) = 1$ atm $= 1.013 \times 10^5$ Pa, $g_z = 9.8$ m/sec^2, $\rho_0 = 1.21$ kg/m^3 for air, and $\rho_0 = 1026$ kg/m^3 for sea water, find the ambient (equilibrium) pressure in:
(a) air at the following heights: $z = 50$ m, 100 m, 300 m, 1 km, and 2 km.
(b) sea water at the following depths: $z = -50$ m, -100 m, -300 m, -1 km, and -2 km.

1-2 Find the adiabatic speed of sound in:
(a) air with $p_0 = 1.013 \times 10^5$ Pa, $\rho_0 = 1.21$ kg/m^3, and $\gamma = 1.402$. Note that the ambient (equilibrium) pressure at sea level is 1 atm $= 1.013 \times 10^5$ Pa.
(b) fresh water with $\rho_0 = 998$ kg/m^3 and $B_E = 2.18 \times 10^9$ Pa.
(c) sea water with $\rho_0 = 1026$ kg/m^3 and $B_E = 2.28 \times 10^9$ Pa.

1-3 Compute the SPL in both air and water for the following values of the magnitude of the acoustic pressure for a time-harmonic acoustic field:
(a) 1 Pa.
(b) 1 atm = 1.013×10^5 Pa, which is equal to the ambient (equilibrium) pressure at sea level.

1-4 Compute the peak acoustic pressure amplitudes at 1 m from a sound source in both air and water for the following values of the SL:
(a) 100 dB re P_{ref}.
(b) 200 dB re P_{ref}.

1-5 A time-harmonic sound source in water has a SL of 180 dB re P_{ref}. Find the magnitude of the acoustic pressure at a receiver if the TL is 30 dB.

Appendix 1A

In this appendix we shall derive a relationship between the acoustic pressure p and the scalar velocity potential φ. We begin by rewriting the simplified version of the equation of continuity given by Eq. (1.4-20) as follows:

$$\nabla \cdot \mathbf{u} \approx Q - \frac{\partial}{\partial t}s - \frac{\nabla \rho_0}{\rho_0} \cdot \mathbf{u}. \tag{1A-1}$$

From the equation of state given by Eq. (1.3-17), we can write that

$$s \approx \frac{1}{\rho_0 c^2}p. \tag{1A-2}$$

If we assume, as before, that both the equilibrium density ρ_0 and the adiabatic speed of sound c are *not* functions of time, then substituting Eq. (1A-2) into Eq. (1A-1) yields

$$\nabla \cdot \mathbf{u} \approx Q - \frac{1}{\rho_0 c^2}\frac{\partial}{\partial t}p - \frac{\nabla \rho_0}{\rho_0} \cdot \mathbf{u}. \tag{1A-3}$$

Therefore, upon substituting Eqs. (1.4-21), (1A-2), and (1A-3) into the linearized version of the Navier-Stokes equation of motion given by Eq. (1.4-12), and since it was assumed that μ' is *constant*, that is, it is neither a function of time nor space, we obtain

$$\rho_0 \nabla \frac{\partial}{\partial t}\varphi \approx -\nabla p - \nabla\left(\frac{\mu'}{\rho_0 c^2}\frac{\partial}{\partial t}p\right) - \nabla\left(\mu'\frac{\nabla \rho_0}{\rho_0} \cdot \nabla\varphi\right) + \frac{\mathbf{g}_L}{c^2}p + \nabla(\mu'Q). \tag{1A-4}$$

Table 1A-1 Approximate values of the terms on the right-hand side of Eq. (1A-6) for several representative frequencies in the case of sea water.

f (Hz)	$\left\| \nabla \dfrac{\partial}{\partial t} \varphi \right\|$	$\dfrac{\|\nabla \rho_0\|}{\rho_0} \left\| \dfrac{\partial}{\partial t} \varphi \right\|$
1	3×10^{-2}	3×10^{-5}
50	7×10^{1}	2×10^{-3}
250	2×10^{3}	8×10^{-3}
1000	3×10^{4}	3×10^{-2}

Next, note that

$$\nabla\left(\rho_0 \frac{\partial}{\partial t}\varphi\right) = \rho_0 \nabla \frac{\partial}{\partial t}\varphi + \left(\frac{\partial}{\partial t}\varphi\right)\nabla\rho_0, \tag{1A-5}$$

or

$$\nabla\left(\rho_0 \frac{\partial}{\partial t}\varphi\right) = \rho_0\left(\nabla + \frac{\nabla\rho_0}{\rho_0}\right)\frac{\partial}{\partial t}\varphi. \tag{1A-6}$$

In order to simplify the right-hand side of Eq. (1A-6), we shall represent the velocity potential φ as a unit-amplitude, time-harmonic plane wave propagating in three-dimensional space, as was done for the condensation s [see Eq. (1.4-28)]. Therefore, it can be shown that [see Eqs. (1.4-29) and (1.4-30)]

$$\left|\frac{\partial}{\partial t}\varphi\right| = kc \tag{1A-7}$$

and

$$\left|\nabla \frac{\partial}{\partial t}\varphi\right| = k^2 c, \tag{1A-8}$$

where $k = 2\pi f/c$ is the wave number. Therefore, if we use the nominal value $c = 1500$ m/sec for sea water, then Eqs. (1A-7), (1A-8), and (1.4-32) can be combined to form Table 1A-1.

As can be seen from Table 1A-1, $\nabla(\partial\varphi/\partial t)$ is the dominant term on the right-hand side of Eq. (1A-6), and as a result, Eq. (1A-6) reduces to

$$\nabla\left(\rho_0 \frac{\partial}{\partial t}\varphi\right) \approx \rho_0 \nabla \frac{\partial}{\partial t}\varphi. \tag{1A-9}$$

Note that in the case of a homogeneous fluid medium, Eq. (1A-9) would be an exact expression—not approximate. Substituting Eq. (1A-9) into Eq. (1A-4) yields

$$\nabla\left(\rho_0\frac{\partial}{\partial t}\varphi\right) \approx -\nabla p - \nabla\left(\frac{\mu'}{\rho_0 c^2}\frac{\partial}{\partial t}p\right) - \nabla\left(\mu'\frac{\nabla\rho_0}{\rho_0}\cdot\nabla\varphi\right) + \frac{\mathbf{g}_L}{c^2}p + \nabla(\mu'Q).$$

(1A-10)

Next, let us simplify the right-hand side of Eq. (1A-10). Retaining all terms involving μ', we shall compare the magnitudes of ∇p and $\mathbf{g}_L p/c^2$. If we treat the acoustic pressure p as a unit-amplitude, time-harmonic plane wave propagating in three-dimensional space as was done for the condensation s [see Eq. (1.4-28)], then it can be shown that $|\nabla p| = k$ and $|\mathbf{g}_L p/c^2| = |\mathbf{g}_L|/c^2 \approx 4 \times 10^{-6}$ [see Eq. (1.4-33)]. If we use the nominal value $c = 1500$ m/sec for sea water, then $k \geq 4.189 \times 10^{-3}$ rad/m for $f \geq 1$ Hz, and as a result, $\mathbf{g}_L p/c^2$ is negligible compared to ∇p. Therefore, Eq. (1A-10) reduces to

$$\nabla\left(p + \frac{\mu'}{\rho_0 c^2}\frac{\partial}{\partial t}p\right) \approx -\nabla\left(\rho_0\frac{\partial}{\partial t}\varphi + \mu'\frac{\nabla\rho_0}{\rho_0}\cdot\nabla\varphi - \mu'Q\right),$$ (1A-11)

from which we obtain

$$\left[1 + \frac{\mu'}{\rho_0(\mathbf{r})c^2(\mathbf{r})}\frac{\partial}{\partial t}\right]p(t,\mathbf{r})$$

$$\approx -\rho_0(\mathbf{r})\frac{\partial}{\partial t}\varphi(t,\mathbf{r}) - \mu'\frac{\nabla\rho_0(\mathbf{r})}{\rho_0(\mathbf{r})}\cdot\nabla\varphi(t,\mathbf{r}) + \mu'x_M(t,\mathbf{r}),$$

(1A-12)

where μ' is given by Eq. (1.4-6) and use has been made of Eq. (1.4-51). In the case of an ideal fluid where there is no viscosity, that is, $\mu' = 0$, Eq. (1A-12) reduces to

$$p(t,\mathbf{r}) \approx -\rho_0(\mathbf{r})\frac{\partial}{\partial t}\varphi(t,\mathbf{r}).$$ (1A-13)

Appendix 1B

In this appendix we shall derive expressions for the real wave number $k(\mathbf{r})$ and the real attenuation coefficient $\alpha(\mathbf{r})$ by using Eqs. (1.4-64) and (1.4-66) for the complex wave number $K(\mathbf{r})$. Squaring both sides of Eqs. (1.4-64) and (1.4-66) yields

$$K^2(\mathbf{r}) = a - jab$$ (1B-1)

and

$$K^2(\mathbf{r}) = \left[k^2(\mathbf{r}) - \alpha^2(\mathbf{r}) \right] - j2k(\mathbf{r})\alpha(\mathbf{r}), \qquad \text{(1B-2)}$$

respectively, where

$$a = \left[\frac{\omega}{c(\mathbf{r})} \right]^2 \frac{1}{1 + \left[\omega\tau(\mathbf{r}) \right]^2} \qquad \text{(1B-3)}$$

and

$$b = \omega\tau(\mathbf{r}). \qquad \text{(1B-4)}$$

Therefore, since Eqs. (1B-1) and (1B-2) must be equal, equating real and imaginary parts yields

$$k^2(\mathbf{r}) - \alpha^2(\mathbf{r}) = a \qquad \text{(1B-5)}$$

and

$$2k(\mathbf{r})\alpha(\mathbf{r}) = ab. \qquad \text{(1B-6)}$$

Solving for $\alpha(\mathbf{r})$ from Eq. (1B-6) and squaring the result yields

$$\alpha^2(\mathbf{r}) = \frac{a^2 b^2}{4k^2(\mathbf{r})}. \qquad \text{(1B-7)}$$

Substituting Eq. (1B-7) into Eq. (1B-5) yields

$$k^4(\mathbf{r}) - ak^2(\mathbf{r}) = \frac{a^2 b^2}{4}, \qquad \text{(1B-8)}$$

and upon completing the square, we obtain

$$k^2(\mathbf{r}) = \frac{a}{2}\left(\sqrt{1 + b^2} + 1 \right). \qquad \text{(1B-9)}$$

Substituting Eqs. (1B-3) and (1B-4) into Eq. (1B-9) yields the following desired result for the real wave number in radians per meter:

$$k(\mathbf{r}) = \frac{\omega}{c(\mathbf{r})} \frac{1}{\sqrt{2}} \left[\frac{\sqrt{1 + \left[\omega\tau(\mathbf{r}) \right]^2} + 1}{1 + \left[\omega\tau(\mathbf{r}) \right]^2} \right]^{1/2}. \qquad \text{(1B-10)}$$

Similarly, substituting Eqs. (1B-9), (1B-3), and (1B-4) into Eq. (1B-5) yields the following desired result for the real attenuation coefficient in nepers per meter:

$$\alpha(\mathbf{r}) = \frac{\omega}{c(\mathbf{r})} \frac{1}{\sqrt{2}} \left[\frac{\sqrt{1 + [\omega\tau(\mathbf{r})]^2} - 1}{1 + [\omega\tau(\mathbf{r})]^2} \right]^{1/2}. \tag{1B-11}$$

Bibliography

J. R. Apel, *Principles of Ocean Physics*, Academic Press, Orlando, Florida, 1987.

C. A. Boyles, *Acoustic Waveguides*, Wiley, New York, 1984.

C. S. Clay and H. Medwin, *Acoustical Oceanography*, Wiley, New York, 1977.

R. W. Fox and A. T. McDonald, *Introduction to Fluid Mechanics*, 4th ed., Wiley, New York, 1992.

D. E. Hall, *Basic Acoustics*, Krieger, Melbourne, Florida, 1993.

L. E. Kinsler, A. R. Frey, A. B. Coppens, and J. V. Sanders, *Fundamentals of Acoustics*, 3rd ed., Wiley, New York, 1982.

J. A. Knauss, *Introduction to Physical Oceanography*, Prentice-Hall, Englewood Cliffs, New Jersey, 1978.

P. M. Morse and K. U. Ingard, *Theoretical Acoustics*, Princeton University Press, Princeton, New Jersey, 1987.

C. B. Officer, *Introduction to the Theory of Sound Transmission*, McGraw-Hill, New York, 1958.

A. D. Pierce, *Acoustics: An Introduction to Its Physical Principles and Applications*, Acoustical Society of America, Woodbury, New York, 1989.

S. Temkin, *Elements of Acoustics*, Wiley, New York, 1981.

Chapter 2

Wave Propagation in the Rectangular Coordinate System

2.1 Solution of the Linear Three-Dimensional Homogeneous Wave Equation

Most wave propagation problems are solved in either the rectangular, cylindrical, or spherical coordinate systems. The rectangular coordinate system is considered to be the easiest coordinate system to work in, since solutions of the wave equation for a homogeneous medium, for example, can then be expressed in terms of complex exponentials rather than special functions such as Bessel and Hankel functions and Legendre polynomials. However, the symmetry of the problem dictates which coordinate system is most appropriate. There are nevertheless wave propagation problems that are best treated in the rectangular coordinate system, which is the main focus of this chapter.

In this section we shall solve the following linear, three-dimensional, lossless, homogeneous wave equation in the rectangular coordinate system:

$$\nabla^2 \varphi(t, \mathbf{r}) - \frac{1}{c^2} \frac{\partial^2}{\partial t^2} \varphi(t, \mathbf{r}) = 0, \qquad (2.1\text{-}1)$$

where

$$\nabla^2 = \frac{\partial^2}{\partial x^2} + \frac{\partial^2}{\partial y^2} + \frac{\partial^2}{\partial z^2} \qquad (2.1\text{-}2)$$

is the *Laplacian* expressed in the rectangular coordinates (x, y, z) (see Fig. 2.1-1),

47

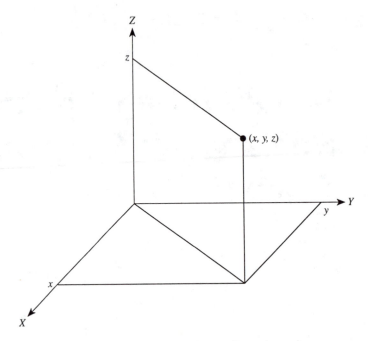

Figure 2.1-1. The rectangular coordinates (x, y, z).

$\varphi(t, \mathbf{r})$ is the *velocity potential* at time t and position $\mathbf{r} = (x, y, z)$ with units of square meters per second, and c is the *constant* speed of sound in meters per second. When the speed of sound is constant, the fluid medium is referred to as a *homogeneous medium*.

The wave equation given by Eq. (2.1-1) is called a *homogeneous wave equation*, meaning that there is no source distribution (i.e., no input or forcing function). Solutions of the inhomogeneous wave equation with arbitrary source distribution will be discussed in Section 2.7. Equation (2.1-1) describes the propagation of small-amplitude acoustic signals in ideal (nonviscous), homogeneous fluids (see Section 1.4). Later, in Example 2.2-3, we shall consider three-dimensional free-space propagation in a viscous fluid medium.

We begin the solution of Eq. (2.1-1) by assuming that the velocity potential $\varphi(t, \mathbf{r})$ has a time-harmonic dependence, that is,

$$\varphi(t, \mathbf{r}) = \varphi_f(\mathbf{r}) \exp(+j2\pi ft), \qquad (2.1\text{-}3)$$

where $\varphi_f(\mathbf{r})$ is the spatial-dependent part of the velocity potential and f is the frequency in hertz. A time-harmonic field is one whose value at any time and point in space depends on a single frequency component. Substituting Eq. (2.1-3) into the time-dependent wave equation given by Eq. (2.1-1) yields

$$\nabla^2 \varphi_f(\mathbf{r}) + k^2 \varphi_f(\mathbf{r}) = 0, \qquad (2.1\text{-}4)$$

which is the time-independent *lossless Helmholtz equation*, where

$$k = 2\pi f/c = 2\pi/\lambda \qquad (2.1\text{-}5)$$

is the *wave number* in radians per meter and

$$c = f\lambda, \qquad (2.1\text{-}6)$$

where λ is the *wavelength* in meters. If we can solve the Helmholtz equation given by Eq. (2.1-4), then substituting the solution $\varphi_f(\mathbf{r})$ into Eq. (2.1-3) will yield a time-harmonic solution of the wave equation given by Eq. (2.1-1), which is the main objective in this section. Time-harmonic solutions are very important, because solutions of the wave equation with arbitrary time dependence can be obtained from time-harmonic solutions by using Fourier transform techniques, as will be discussed in Section 2.2.3.

The solution of the Helmholtz equation given by Eq. (2.1-4) will be obtained by using *the method of separation of variables*, that is, we assume a solution of the form

$$\boxed{\varphi_f(\mathbf{r}) = \varphi_f(x, y, z) = X(x)Y(y)Z(z).} \qquad (2.1\text{-}7)$$

Substituting Eq. (2.1-7) into Eq. (2.1-4) and making use of Eq. (2.1-2) yields

$$\frac{1}{X(x)} \frac{d^2}{dx^2} X(x) + k^2 = -\frac{1}{Y(y)} \frac{d^2}{dy^2} Y(y) - \frac{1}{Z(z)} \frac{d^2}{dz^2} Z(z), \quad (2.1\text{-}8)$$

where the second-order partial derivatives with respect to x, y, and z have been replaced by ordinary second-order derivatives, since the functions $X(x)$, $Y(y)$, and $Z(z)$ depend on only one of the independent variables x, y, and z, respectively. Since the left-hand side of Eq. (2.1-8) is a function of x, and the right-hand side is a function of both y and z, equality is possible only if both sides of Eq. (2.1-8) are equal to a constant; call it κ^2. The constant κ^2 is referred to as a *separation constant*. Therefore, since

$$\frac{1}{X(x)} \frac{d^2}{dx^2} X(x) + k^2 = \kappa^2, \qquad (2.1\text{-}9)$$

Eq. (2.1-9) can be rewritten as

$$\frac{d^2}{dx^2}X(x) + k_X^2 X(x) = 0,$$

(2.1-10)

where

$$k_X^2 = k^2 - \kappa^2.$$

(2.1-11)

Equation (2.1-10) is a second-order, homogeneous (no input or forcing function) ordinary differential equation (ODE) with *exact* solution

$$\boxed{X(x) = A_X \exp(-jk_X x) + B_X \exp(+jk_X x),}$$

(2.1-12)

where A_X and B_X are complex constants in general, whose values are determined by satisfying boundary conditions. If k_X is *positive*, then the first term on the right-hand side of Eq. (2.1-12) represents a plane wave traveling in the *positive X* direction, whereas the second term represents a plane wave traveling in the *negative X* direction. The constant k_X is the propagation-vector component in the X direction with units of radians per meter. Plane waves and propagation-vector components will be discussed in detail in Section 2.2.1.

Before continuing further, let us show that the first term on the right-hand side of Eq. (2.1-12) is indeed a plane wave traveling in the positive X direction when k_X is positive. First, form the function

$$X(t, x) = A_X \exp(-jk_X x) \exp(+j2\pi ft)$$

(2.1-13)

or, rewriting,

$$X(t, x) = A_X \exp\left[-jk_X\left(x - \frac{2\pi f}{k_X}t\right)\right].$$

(2.1-14)

Next, let [see Eq. (2.1-5)]

$$k_X = 2\pi f/c_{p_X},$$

(2.1-15)

where c_{p_X} is the *phase speed* in the X direction in meters per second. Phase speed will be discussed in detail in Section 2.2.2. Substituting Eq. (2.1-15) into Eq. (2.1-14) yields

$$X(t, x) = A_X \exp\left[-jk_X\left(x - c_{p_X}t\right)\right].$$

(2.1-16)

Note that the phase speed appears as one of the terms in the *phase* of Eq. (2.1-16). Figure 2.1-2 is a plot of Eq. (2.1-16) which shows that as time t increases, the constant amplitude value A_X, and hence the plane wave itself, travels in the positive X direction. The value of A_X occurs at locations $x = c_{p_X}t$ along the

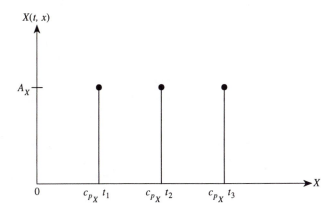

Figure 2.1-2. Plot of Eq. (2.1-16) where $t_3 > t_2 > t_1$.

positive X axis. At these locations, the *phase* of the complex exponential in Eq. (2.1-16) is equal to zero radians. However, if we had chosen $\exp(-j2\pi ft)$ as the time-harmonic dependence instead of $\exp(+j2\pi ft)$, then the first term on the right-hand side of Eq. (2.1-12) would represent a plane wave traveling in the *negative* X direction when k_X is positive.

Returning to the solution of Eq. (2.1-8), recall that the right-hand side of Eq. (2.1-8) must also be equal to the separation constant κ^2. Therefore,

$$-\frac{1}{Y(y)}\frac{d^2}{dy^2}Y(y) - \frac{1}{Z(z)}\frac{d^2}{dz^2}Z(z) = \kappa^2, \qquad (2.1\text{-}17)$$

or

$$-\frac{1}{Y(y)}\frac{d^2}{dy^2}Y(y) = \frac{1}{Z(z)}\frac{d^2}{dz^2}Z(z) + \kappa^2. \qquad (2.1\text{-}18)$$

Since the left-hand side of Eq. (2.1-18) is a function of y and the right-hand side is a function of z, equality is possible only if both sides of Eq. (2.1-18) are equal to a constant; call it k_Y^2. The constant k_Y^2 is also referred to as a separation constant. Then since

$$-\frac{1}{Y(y)}\frac{d^2}{dy^2}Y(y) = k_Y^2, \qquad (2.1\text{-}19)$$

Eq. (2.1-19) can be rewritten as

$$\frac{d^2}{dy^2}Y(y) + k_Y^2 Y(y) = 0. \qquad (2.1\text{-}20)$$

Equation (2.1-20) is a second-order, homogeneous ODE with *exact* solution

$$Y(y) = A_Y \exp(-jk_Y y) + B_Y \exp(+jk_Y y), \qquad (2.1\text{-}21)$$

where A_Y and B_Y are complex constants in general, whose values are determined by satisfying boundary conditions. If k_Y is *positive*, then the first term on the right-hand side of Eq. (2.1-21) represents a plane wave traveling in the *positive Y* direction, whereas the second term represents a plane wave traveling in the *negative Y* direction. The constant k_Y is the propagation-vector component in the Y direction with units of radians per meter.

Finally, recall that the right-hand side of Eq. (2.1-18) must also be equal to k_Y^2. Therefore,

$$\frac{1}{Z(z)} \frac{d^2}{dz^2} Z(z) + \kappa^2 = k_Y^2, \qquad (2.1\text{-}22)$$

or

$$\frac{d^2}{dz^2} Z(z) + k_Z^2 Z(z) = 0, \qquad (2.1\text{-}23)$$

where

$$k_Z^2 = \kappa^2 - k_Y^2. \qquad (2.1\text{-}24)$$

Equation (2.1-23) is a second-order, homogeneous ODE with *exact* solution

$$Z(z) = A_Z \exp(-jk_Z z) + B_Z \exp(+jk_Z z) \qquad (2.1\text{-}25)$$

where A_Z and B_Z are complex constants in general, whose values are determined by satisfying boundary conditions. If k_Z is *positive*, then the first term on the right-hand side of Eq. (2.1-25) represents a plane wave traveling in the *positive Z* direction, whereas the second term represents a plane wave traveling in the *negative Z* direction. The constant k_Z is the propagation-vector component in the Z direction with units of radians per meter.

Therefore, upon substituting Eqs. (2.1-12), (2.1-21), and (2.1-25) into Eq. (2.1-7), we obtain

$$\begin{aligned}
\varphi_f(x, y, z) = &[A_X \exp(-jk_X x) + B_X \exp(+jk_X x)] \\
\times &[A_Y \exp(-jk_Y y) + B_Y \exp(+jk_Y y)] \\
\times &[A_Z \exp(-jk_Z z) + B_Z \exp(+jk_Z z)],
\end{aligned} \qquad (2.1\text{-}26)$$

which is a solution of the Helmholtz equation given by Eq. (2.1-4). Also, substituting Eq. (2.1-24) into Eq. (2.1-11) yields

$$k_X^2 + k_Y^2 + k_Z^2 = k^2. \tag{2.1-27}$$

Equation (2.1-27) relates the wave number k, given by Eq. (2.1-5), to the propagation-vector components k_X, k_Y, and k_Z. And finally, upon substituting Eq. (2.1-26) into Eq. (2.1-3), we obtain a time-harmonic solution (in rectangular coordinates) of the linear, three-dimensional, lossless, homogeneous wave equation given by Eq. (2.1-1).

2.2 Free-Space Propagation

2.2.1 Time-Harmonic Solution of the Homogeneous Wave Equation

In this section we shall consider wave propagation in free space, that is, wave propagation in the absence of boundaries. As a result, there will be *no reflected waves*. This kind of wave propagation is also known as wave propagation in an *unbounded medium*. Therefore, if a plane wave is initially traveling in the positive X, Y, and Z directions in free space, then since there are no reflected waves traveling in the negative X, Y, and Z directions, we have [see Eq. (2.1-26)]

$$B_X = B_Y = B_Z = 0. \tag{2.2-1}$$

Substituting Eq. (2.2-1) into Eq. (2.1-26) yields

$$\varphi_f(x, y, z) = A \exp\left[-j(k_X x + k_Y y + k_Z z)\right] \tag{2.2-2}$$

or

$$\varphi_f(\mathbf{r}) = A \exp(-j\mathbf{k} \cdot \mathbf{r}), \tag{2.2-3}$$

where

$$A = A_X A_Y A_Z \tag{2.2-4}$$

is a complex constant in general,

$$\mathbf{k} = k_X \hat{x} + k_Y \hat{y} + k_Z \hat{z} \tag{2.2-5}$$

is the *propagation vector*, and

$$\mathbf{r} = x\hat{x} + y\hat{y} + z\hat{z} \tag{2.2-6}$$

is the position vector to a field point with rectangular coordinates (x, y, z). And upon substituting Eq. (2.2-3) into Eq. (2.1-3), we obtain

$$\boxed{\varphi(t, \mathbf{r}) = A \exp(-j\mathbf{k} \cdot \mathbf{r}) \exp(+j2\pi ft),} \tag{2.2-7}$$

which represents a time-harmonic plane wave traveling in the direction of the propagation vector $\mathbf{k}$. Note that if the dot product (also known as a scalar or inner product) appearing in Eq. (2.2-7) is set equal to a constant, that is, if

$$\mathbf{k} \cdot \mathbf{r} = \text{constant}, \tag{2.2-8}$$

then the phase of Eq. (2.2-7) as a function of spatial location $\mathbf{r} = (x, y, z)$ is constant at any given instant of time t. As Fig. 2.2-1 illustrates, Eq. (2.2-8) defines

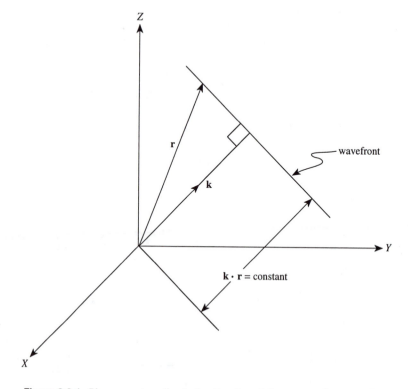

Figure 2.2-1. Plane wave traveling in the direction of the propagation vector **k**.

a *surface of constant phase* in three-dimensional space. Equation (2.2-8) is, in fact, the equation for a plane in three-dimensional space. When the acoustic field given by Eq. (2.2-7) is evaluated at any point on this surface, it will have the same value of phase at any given instant of time. Surfaces of constant phase are called *wavefronts*. Thus, in the case of a homogeneous medium where the speed of sound is constant, the wavefronts defined by Eq. (2.2-8) are *planar*. Hence the term "plane waves." Also note that the propagation vector **k** is *normal*, or perpendicular, to the wavefront.

The propagation vector **k** given by Eq. (2.2-5) can also be expressed as

$$\mathbf{k} = k\hat{n}, \tag{2.2-9}$$

where the wave number k is the magnitude of **k** and is given by either Eq. (2.1-5) or Eq. (2.1-27), and

$$\hat{n} = u\hat{x} + v\hat{y} + w\hat{z} \tag{2.2-10}$$

is the unit vector in the direction of **k**, where

$$u = \cos \alpha = \sin \theta \cos \psi, \tag{2.2-11}$$

$$v = \cos \beta = \sin \theta \sin \psi, \tag{2.2-12}$$

$$w = \cos \gamma = \cos \theta \tag{2.2-13}$$

are the dimensionless *direction cosines* with respect to the X, Y, and Z axes, respectively (see Fig. 2.2-2). Note that u, v, and w take on values between -1 and 1, since $0 \le \theta \le \pi$ and $0 \le \psi \le 2\pi$, and that

$$u^2 + v^2 + w^2 = 1. \tag{2.2-14}$$

Substituting Eq. (2.2-10) into Eq. (2.2-9) yields

$$\mathbf{k} = ku\hat{x} + kv\hat{y} + kw\hat{z}, \tag{2.2-15}$$

and upon equating the right-hand sides of Eqs. (2.2-5) and (2.2-15), we obtain

$$k_X = ku, \tag{2.2-16}$$

$$k_Y = kv, \tag{2.2-17}$$

$$k_Z = kw. \tag{2.2-18}$$

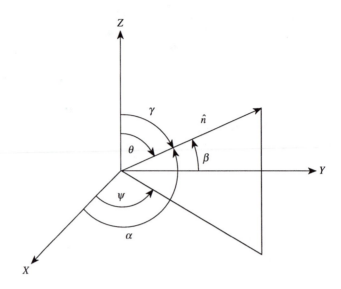

Figure 2.2-2. The unit vector $\hat{n}$ and the two sets of angles (θ, ψ) and (α, β, γ).

Equations (2.2-16) through (2.2-18) are the *propagation-vector components* in the X, Y, and Z directions, respectively, with units of radians per meter. With the use of Eq. (2.1-5), Eqs. (2.2-16) through (2.2-18) can also be expressed as

$$k_X = 2\pi f_X, \tag{2.2-19}$$

$$k_Y = 2\pi f_Y, \tag{2.2-20}$$

$$k_Z = 2\pi f_Z, \tag{2.2-21}$$

where

$$f_X = u/\lambda, \tag{2.2-22}$$

$$f_Y = v/\lambda, \tag{2.2-23}$$

$$f_Z = w/\lambda \tag{2.2-24}$$

are the *spatial frequencies* in the X, Y, and Z directions, respectively, with units of cycles per meter. The propagation-vector components are the spatial analog of the radian frequency $\omega = 2\pi f$ in radians per second, whereas the spatial frequencies are the spatial analog of frequency f in cycles per second.

Example 2.2-1 (**Time-average intensity vector**)
In this example we shall derive two expressions for the time-average intensity vector for a time-harmonic plane wave. Recall from Section 1.4 that the acoustic pressure $p(t, \mathbf{r})$ in pascals (Pa) and the acoustic fluid velocity vector $\mathbf{u}(t, \mathbf{r})$ in meters per second can be obtained from the scalar velocity potential $\varphi(t, \mathbf{r})$ in square meters per second as follows:

$$p(t, \mathbf{r}) = -\rho_0 \frac{\partial}{\partial t} \varphi(t, \mathbf{r}), \qquad (2.2\text{-}25)$$

where ρ_0 is the equilibrium (ambient) density of the fluid medium in kilograms per cubic meter, and

$$\mathbf{u}(t, \mathbf{r}) = \nabla \varphi(t, \mathbf{r}), \qquad (2.2\text{-}26)$$

where

$$\nabla = \frac{\partial}{\partial x} \hat{x} + \frac{\partial}{\partial y} \hat{y} + \frac{\partial}{\partial z} \hat{z} \qquad (2.2\text{-}27)$$

is the *gradient* expressed in the rectangular coordinates (x, y, z). Note that ρ_0 is constant, since we are dealing with a homogeneous medium. Substituting Eq. (2.2-7) into Eq. (2.2-25) yields

$$p(t, \mathbf{r}) = -j2\pi f \rho_0 \varphi(t, \mathbf{r}) = -jk\rho_0 c\varphi(t, \mathbf{r}), \qquad (2.2\text{-}28)$$

where $\varphi(t, \mathbf{r})$ is given by Eq. (2.2-7). Equation (2.2-28) can also be expressed as

$$p(t, \mathbf{r}) = p_f(\mathbf{r}) \exp(+j2\pi ft), \qquad (2.2\text{-}29)$$

where

$$p_f(\mathbf{r}) = -j2\pi f \rho_0 \varphi_f(\mathbf{r}) = -jk\rho_0 c\varphi_f(\mathbf{r}) \qquad (2.2\text{-}30)$$

and $\varphi_f(\mathbf{r})$ is given by Eq. (2.2-3). Equation (2.2-30) represents the spatial-dependent part of the acoustic pressure. And upon substituting Eq. (2.2-7) into Eq. (2.2-26), we obtain

$$\mathbf{u}(t, \mathbf{r}) = \nabla \left[\exp(-j\mathbf{k} \cdot \mathbf{r}) \right] A \exp(+j2\pi ft), \qquad (2.2\text{-}31)$$

or

$$\mathbf{u}(t, \mathbf{r}) = -jk\varphi(t, \mathbf{r}) \hat{n}, \qquad (2.2\text{-}32)$$

where $\varphi(t, \mathbf{r})$ is given by Eq. (2.2-7) and the unit vector $\hat{n}$ is given by

Eq. (2.2-10). Equation (2.2-32) can also be expressed as

$$\mathbf{u}(t,\mathbf{r}) = \mathbf{u}_f(\mathbf{r}) \exp(+j2\pi ft), \qquad (2.2\text{-}33)$$

where

$$\mathbf{u}_f(\mathbf{r}) = -jk\varphi_f(\mathbf{r})\hat{n} \qquad (2.2\text{-}34)$$

and $\varphi_f(\mathbf{r})$ is given by Eq. (2.2-3). Equation (2.2-34) represents the spatial-dependent part of the acoustic fluid velocity vector, which is in the same direction as the propagation vector $\mathbf{k}$ [see Eqs. (2.2-9) and (2.2-10)]. Note that if we rewrite Eq. (2.2-34) as

$$\mathbf{u}_f(\mathbf{r}) = u_f(\mathbf{r})\hat{n}, \qquad (2.2\text{-}35)$$

where

$$u_f(\mathbf{r}) = -jk\varphi_f(\mathbf{r}) \qquad (2.2\text{-}36)$$

is the complex acoustic fluid speed, then the acoustic pressure given by Eq. (2.2-30) can be expressed as

$$\boxed{p_f(\mathbf{r}) = \rho_0 c u_f(\mathbf{r}),} \qquad (2.2\text{-}37)$$

where $\rho_0 c$ is known as the *characteristic impedance* of the medium with units of rayls (1 rayl = 1 Pa-sec/m). The term "characteristic impedance" is used for two reasons. First, the equilibrium or ambient density ρ_0 and the speed of sound c characterize the properties of the fluid medium. Second, in analogy with electrical circuits, if we let the complex acoustic pressure $p_f(\mathbf{r})$ be the analog of voltage (which is an "across" variable), and the complex acoustic fluid speed $u_f(\mathbf{r})$ be the analog of current (which is a "through" variable), then the ratio

$$\boxed{Z = \frac{p_f(\mathbf{r})}{u_f(\mathbf{r})}} \qquad (2.2\text{-}38)$$

is known as the *specific acoustic impedance* of the medium in rayls for time-harmonic acoustic fields [see Problem 2-5 for a generalization of Eq. (2.2-38)]. In particular, for the special case of time-harmonic plane waves in an unbounded, ideal, homogeneous, fluid medium, substituting Eq. (2.2-37) into Eq. (2.2-38) yields

$$\boxed{Z = \rho_0 c,} \qquad (2.2\text{-}39)$$

which indicates that the specific acoustic impedance is equal to the characteristic impedance. This is *not* true for spherical waves, for example, as shall be shown in Section 4.2.2.

Next, recall from Section 1.5 that the time-average intensity vector for time-harmonic acoustic fields is given by

$$\mathbf{I}_{avg}(\mathbf{r}) = \tfrac{1}{2}\mathrm{Re}\{p_f(\mathbf{r})\mathbf{u}_f^*(\mathbf{r})\}. \tag{2.2-40}$$

Substituting Eqs. (2.2-30), (2.2-34), and (2.2-3) into Eq. (2.2-40) yields

$$\mathbf{I}_{avg}(\mathbf{r}) = \tfrac{1}{2}k^2\rho_0 c|A|^2\hat{n}, \tag{2.2-41}$$

where A is the amplitude of the velocity potential, and, with the use of Eqs. (2.2-30) and (2.2-3),

$$\boxed{\mathbf{I}_{avg}(\mathbf{r}) = \frac{|p_f(\mathbf{r})|^2}{2\rho_0 c}\hat{n}.} \tag{2.2-42}$$

Note that the time-average intensity vector is in the same direction as the propagation vector, and as a result, the acoustic energy contained in the time-harmonic plane wave travels in the direction $\hat{n}$ with speed of sound c. The magnitude of the time-average intensity vector is given by

$$I_{avg}(\mathbf{r}) = \tfrac{1}{2}k^2\rho_0 c|A|^2 \tag{2.2-43}$$

or

$$\boxed{I_{avg}(\mathbf{r}) = \frac{|p_f(\mathbf{r})|^2}{2\rho_0 c}} \tag{2.2-44}$$

with units of watts per square meter.

2.2.2 Group Speeds, Phase Speeds, and Wavelengths

The next topics of discussion are *group speeds*, *phase speeds*, and *wavelengths* in the X, Y, and Z directions. Besides being applicable to free-space propagation, group speeds, for example, are very useful for describing the propagation of normal modes in waveguides, including ocean waveguides. Recall from Example 2.2-1 that the energy in a plane wave travels with the speed of sound c in the direction $\hat{n}$ of the propagation vector $\mathbf{k}$. Also recall that $\mathbf{k}$ is normal (perpendicular) to the wavefront. Therefore, if we let the *velocity vector of the wavefront* be

given by

$$\boxed{\mathbf{c} = c\hat{n},}$$

(2.2-45)

then substituting Eq. (2.2-10) into Eq. (2.2-45) yields

$$\boxed{\mathbf{c} = c_{g_X}\hat{x} + c_{g_Y}\hat{y} + c_{g_Z}\hat{z},}$$

(2.2-46)

where

$$\boxed{\begin{aligned} c_{g_X} &= cu = ck_X/k, \\[6pt] c_{g_Y} &= cv = ck_Y/k, \\[6pt] c_{g_Z} &= cw = ck_Z/k \end{aligned}}$$

(2.2-47)

(2.2-48)

(2.2-49)

are the *group speeds* (in meters per second) in the X, Y, and Z directions, respectively, where u, v, and w are the dimensionless direction cosines given by Eqs. (2.2-11) through (2.2-13), respectively. The group speeds given by Eqs. (2.2-47) through (2.2-49) are the speeds with which *energy* travels in the X, Y, and Z directions, respectively. Since the direction cosines can take on values between -1 and 1, if any of the group speeds are negative, simply take the absolute value. A negative group speed indicates wave propagation in the negative X, Y, or Z direction. Also, since the magnitudes of the direction cosines are between 0 and 1, the magnitudes of the group speeds are between 0 and c.

As we know, a wavefront is a surface of constant phase traveling with the speed of sound c in the direction $\hat{n}$ of the propagation vector $\mathbf{k}$. In an ideal (nonviscous) fluid medium, the speed of sound c is equal to the phase speed c_p in the direction $\hat{n}$ (see Section 1.4). Since

$$k = 2\pi f/c,$$

(2.2-50)

let us express k_X using Eq. (2.1-15) as follows:

$$k_X = 2\pi f/c_{p_X},$$

(2.1-15)

where c_{p_X} is the phase speed in the X direction in meters per second [see Eq. (2.1-16)]. Substituting Eq. (2.2-50) into Eq. (2.2-16) yields

$$k_X = 2\pi fu/c,$$

(2.2-51)

and upon comparing Eqs. (2.1-15) and (2.2-51), and following the same procedure for k_Y and k_Z, we obtain

$$c_{p_X} = c/u = ck/k_X = 2\pi f/k_X, \tag{2.2-52}$$

$$c_{p_Y} = c/v = ck/k_Y = 2\pi f/k_Y, \tag{2.2-53}$$

$$c_{p_Z} = c/w = ck/k_Z = 2\pi f/k_Z \tag{2.2-54}$$

as the phase speeds (in meters per second) in the X, Y, and Z directions, respectively. The phase speeds given by Eqs. (2.2-52) through (2.2-54) are the speeds with which surfaces of constant phase travel in the X, Y, and Z directions, respectively. If any of the phase speeds are negative, simply take the absolute value. A negative phase speed indicates wave propagation in the negative X, Y, or Z direction. Note that the magnitudes of the phase speeds are between c and ∞. Also note that

$$c_{p_X} c_{g_X} = c^2, \tag{2.2-55}$$

$$c_{p_Y} c_{g_Y} = c^2, \tag{2.2-56}$$

$$c_{p_Z} c_{g_Z} = c^2. \tag{2.2-57}$$

Since the wavelength measured in the direction of the propagation vector **k** is λ meters and since

$$k = 2\pi/\lambda, \tag{2.2-58}$$

let us express k_X as

$$k_X = 2\pi/\lambda_X, \tag{2.2-59}$$

where λ_X is the wavelength (in meters) in the X direction. Substituting Eq. (2.2-58) into Eq. (2.2-16) yields

$$k_X = 2\pi u/\lambda, \tag{2.2-60}$$

and upon comparing Eqs. (2.2-59) and (2.2-60), and following the same procedure for k_Y and k_Z, we obtain

$$\lambda_X = \lambda/u = \lambda k/k_X = 2\pi/k_X, \tag{2.2-61}$$

$$\lambda_Y = \lambda/v = \lambda k/k_Y = 2\pi/k_Y, \tag{2.2-62}$$

$$\lambda_Z = \lambda/w = \lambda k/k_Z = 2\pi/k_Z \tag{2.2-63}$$

as the wavelengths (in meters) in the X, Y, and Z directions, respectively. If any of the wavelengths are negative, simply take the absolute value. A negative wavelength indicates wave propagation in the negative X, Y, or Z direction. Note that the magnitudes of the wavelengths are between λ and ∞. Also note that

$$c_{p_X} = f\lambda_X, \tag{2.2-64}$$

$$c_{p_Y} = f\lambda_Y, \tag{2.2-65}$$

$$c_{p_Z} = f\lambda_Z, \tag{2.2-66}$$

where $c = f\lambda$.

Example 2.2-2 (Two-dimensional free-space propagation)

In this example we shall consider a time-harmonic plane wave propagating in an unbounded fluid medium in the YZ plane as shown in Fig. 2.2-3. Since $\psi = \pi/2$ as shown in Fig. 2.2-3, the direction cosines given by Eqs. (2.2-11) through (2.2-13) reduce to

$$u = 0, \tag{2.2-67}$$

$$v = \sin\theta, \tag{2.2-68}$$

$$w = \cos\theta. \tag{2.2-69}$$

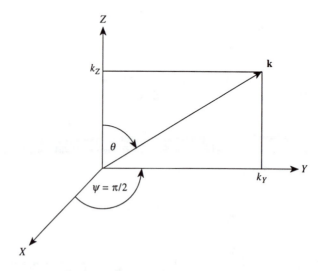

Figure 2.2-3. Wave propagation in the YZ plane.

As a result, the propagation-vector components given by Eqs. (2.2-16) through (2.2-18) become

$$k_X = 0, \tag{2.2-70}$$

$$k_Y = k \sin \theta, \tag{2.2-71}$$

$$k_Z = k \cos \theta. \tag{2.2-72}$$

Equation (2.2-70) indicates that there is no wave propagation in the X direction, as expected. Substituting Eqs. (2.2-70) through (2.2-72) into Eq. (2.2-5) yields

$$\mathbf{k} = k \sin \theta \, \hat{y} + k \cos \theta \, \hat{z}, \tag{2.2-73}$$

and upon substituting Eqs. (2.2-73) and (2.2-6) into Eq. (2.2-7), we obtain

$$\varphi(t, y, z) = A \exp\left[-jk(y \sin \theta + z \cos \theta)\right] \exp(+j2\pi ft). \tag{2.2-74}$$

Equation (2.2-74) is the velocity potential of a time-harmonic plane wave propagating in the YZ plane at an angle θ which is measured from the positive Z axis.

Let us compute the group speeds, phase speeds, and wavelengths in the X, Y, and Z directions next. Substituting Eqs. (2.2-67) through (2.2-69) into Eqs. (2.2-47) through (2.2-49), respectively, yields

$$c_{g_X} = 0, \tag{2.2-75}$$

$$c_{g_Y} = c \sin \theta, \tag{2.2-76}$$

$$c_{g_Z} = c \cos \theta. \tag{2.2-77}$$

Equation (2.2-75) indicates that there is no energy propagating in the X direction, as expected. Substituting Eqs. (2.2-67) through (2.2-69) into Eqs. (2.2-52) through (2.2-54), respectively, yields

$$c_{p_X} = \infty, \tag{2.2-78}$$

$$c_{p_Y} = c/\sin \theta, \tag{2.2-79}$$

$$c_{p_Z} = c/\cos \theta. \tag{2.2-80}$$

And substituting Eqs. (2.2-67) through (2.2-69) into Eqs. (2.2-61) through

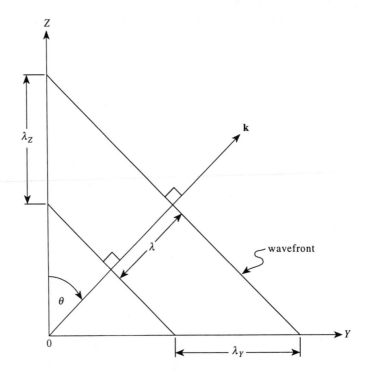

Figure 2.2-4. Wavelengths λ_Y and λ_Z in the Y and Z directions, respectively. Wavelength λ is measured in the direction of the propagation vector **k**.

(2.2-63), respectively, yields (see Fig. 2.2-4)

$$\lambda_X = \infty, \tag{2.2-81}$$

$$\lambda_Y = \lambda/\sin\theta, \tag{2.2-82}$$

$$\lambda_Z = \lambda/\cos\theta. \tag{2.2-83}$$

Equations (2.2-78) and (2.2-81) both indicate that the propagation-vector component in the X direction is zero, since

$$k_X = 2\pi f/c_{p_X} \to 0 \quad \text{as} \quad c_{p_X} \to \infty \tag{2.2-84}$$

and

$$k_X = 2\pi/\lambda_X \to 0 \quad \text{as} \quad \lambda_X \to \infty. \tag{2.2-85}$$

Finally, note that if $\theta = \pi/2$, then

$$u = 0, \qquad v = 1, \qquad w = 0 \tag{2.2-86}$$

$$k_X = 0, \qquad k_Y = k, \qquad k_Z = 0 \tag{2.2-87}$$

$$\varphi(t, y) = A \exp(-jky) \exp(+j2\pi ft), \tag{2.2-88}$$

$$c_{g_X} = 0, \qquad c_{g_Y} = c, \qquad c_{g_Z} = 0 \tag{2.2-89}$$

$$c_{p_X} = \infty, \qquad c_{p_Y} = c, \qquad c_{p_Z} = \infty \tag{2.2-90}$$

and

$$\lambda_X = \infty, \qquad \lambda_Y = \lambda, \qquad \lambda_Z = \infty. \tag{2.2-91}$$

Equations (2.2-86) through (2.2-91) correspond to wave propagation in the positive Y direction.

2.2.3 Solution of the Homogeneous Wave Equation for Arbitrary Time Dependence

The next topic to be discussed in this section is the solution of the linear, three-dimensional, lossless, homogeneous wave equation given by Eq. (2.1-1) for arbitrary time dependence. We begin by substituting Eqs. (2.2-9) and (2.1-5) into Eq. (2.2-7), which yields

$$\varphi(t, \mathbf{r}) = A \exp[+j2\pi f(t - \tau)], \tag{2.2-92}$$

where

$$\tau = \hat{n} \cdot \mathbf{r}/c \tag{2.2-93}$$

is the *travel time* or *time delay* in seconds. Equation (2.2-93) is the amount of time it takes for a plane wave transmitted from the origin at $t = 0$ in the $\hat{n}$ direction at speed c to reach the field point at position $\mathbf{r} = (x, y, z)$. Recall that the constant A is complex in general. If we associate the value of A with the single frequency f, that is, if we treat A as a function of f, then by using the *principle of linear superposition*, we can generalize the solution given by Eq. (2.2-92) as follows:

$$\varphi(t, \mathbf{r}) = \int_{-\infty}^{\infty} A(f) \exp[+j2\pi f(t - \tau)] \, df \tag{2.2-94}$$

where

$$A(f) = F_t\{a(t)\} = \int_{-\infty}^{\infty} a(t) \exp(-j2\pi ft) \, dt \tag{2.2-95}$$

is the Fourier transform or complex frequency spectrum of the *arbitrary* function of time $a(t)$. If we rewrite Eq. (2.2-94) as

$$\varphi(t,\mathbf{r}) = \int_{-\infty}^{\infty} A(f) \exp(-j2\pi f\tau) \exp(+j2\pi ft) \, df, \qquad (2.2\text{-}96)$$

then it is easy to see that

$$\varphi(t,\mathbf{r}) = F_f^{-1}\{A(f) \exp(-j2\pi f\tau)\}. \qquad (2.2\text{-}97)$$

As a result,

$$\boxed{\varphi(t,\mathbf{r}) = a(t - \tau) = a\left(t - \frac{\hat{n} \cdot \mathbf{r}}{c}\right),} \qquad (2.2\text{-}98)$$

where $a(t) = F_f^{-1}\{A(f)\}$. Equation (2.2-98), which is a solution of the wave equation given by Eq. (2.1-1), represents a general plane-wave field with arbitrary time dependence propagating in the $\hat{n}$ direction. Note that if we let $a(t) = A \exp(+j2\pi ft)$, then Eq. (2.2-98) reduces to the time-harmonic plane wave given by Eq. (2.2-92). See Problem 2-4 for a generalization of Eq. (2.2-98).

Example 2.2-3 (Three-dimensional free-space propagation in a viscous fluid medium)

In this example we shall consider wave propagation in an unbounded, viscous fluid medium; in other words, we shall consider the solution of the lossy Helmholtz equation given by

$$\nabla^2 \varphi_f(\mathbf{r}) + K^2 \varphi_f(\mathbf{r}) = 0, \qquad (2.2\text{-}99)$$

where the *complex wave number* K is given by

$$K = k - j\alpha \qquad (2.2\text{-}100)$$

where k and α are the *real wave number* (in radians per meter) and *real attenuation coefficient* (in nepers per meter), respectively (see Section 1.4). In addition to being real quantities, both k and α are *positive*, and α is, in general, a function of frequency, that is, $\alpha = \alpha(f)$.

The solution of Eq. (2.2-99) is actually trivial, since all we have to do is modify the results already obtained in Section 2.2.1 concerning three-dimensional free-space propagation in an ideal (nonviscous) fluid medium. Therefore, referring to Section 2.2.1, the solution of

Eq. (2.2-99) is given by

$$\varphi_f(\mathbf{r}) = A \exp(-j\mathbf{K} \cdot \mathbf{r}),$$
(2.2-101)

where A is a complex constant in general,

$$\mathbf{K} = K_X \hat{x} + K_Y \hat{y} + K_Z \hat{z}$$
(2.2-102)

is the *complex propagation vector*, and

$$\mathbf{r} = x\hat{x} + y\hat{y} + z\hat{z}$$
(2.2-103)

is the position vector to a field point with rectangular coordinates (x, y, z). And upon substituting Eq. (2.2-101) into Eq. (2.1-3), we obtain

$$\varphi(t, \mathbf{r}) = A \exp(-j\mathbf{K} \cdot \mathbf{r}) \exp(+j2\pi ft),$$
(2.2-104)

which represents a time-harmonic plane wave traveling in the direction of $\mathbf{K}$.

The complex propagation vector $\mathbf{K}$ can also be expressed as

$$\mathbf{K} = K\hat{n},$$
(2.2-105)

where

$$\hat{n} = u\hat{x} + v\hat{y} + w\hat{z}$$
(2.2-106)

is the unit vector in the direction of $\mathbf{K}$ and u, v, and w are the dimensionless direction cosines with respect to the X, Y, and Z axes, respectively [see Eqs. (2.2-11) through (2.2-13) and Fig. 2.2-2]. Upon substituting Eqs. (2.2-100) and (2.2-106) into Eq. (2.2-105) and comparing the result with Eq. (2.2-102), we obtain the following expressions for the *complex propagation-vector components* in the X, Y, and Z directions, respectively, with units of radians per meter:

$$K_X = Ku = k_X - j\alpha_X,$$
(2.2-107)

$$K_Y = Kv = k_Y - j\alpha_Y,$$
(2.2-108)

$$K_Z = Kw = k_Z - j\alpha_Z$$
(2.2-109)

where $k_X = ku$, $k_Y = kv$, and $k_Z = kw$ are the *real propagation-vector components* in the X, Y, and Z directions, respectively [see Eqs. (2.2-16) through (2.2-18)], with units of radians per meter, and

$$\alpha_X = \alpha u, \tag{2.2-110}$$

$$\alpha_Y = \alpha v, \tag{2.2-111}$$

$$\alpha_Z = \alpha w \tag{2.2-112}$$

are the *real attenuation coefficients* in the X, Y, and Z directions, respectively, with units of nepers per meter. Note that

$$K_X^2 + K_Y^2 + K_Z^2 = K^2, \tag{2.2-113}$$

$$k_X^2 + k_Y^2 + k_Z^2 = k^2, \tag{2.2-114}$$

$$\alpha_X^2 + \alpha_Y^2 + \alpha_Z^2 = \alpha^2, \tag{2.2-115}$$

since $u^2 + v^2 + w^2 = 1$ [see Eq. (2.2-14)]. Also note from Eq. (2.2-100) that

$$K^2 = k^2 - \alpha^2 - j2k\alpha. \tag{2.2-116}$$

An alternative expression for the solution of Eq. (2.2-99) will be obtained next. Substituting Eqs. (2.2-102), (2.2-103), and (2.2-107) through (2.2-109) into Eq. (2.2-101) yields

$$\varphi_f(x, y, z) = A \exp\left[-j(k_X x + k_Y y + k_Z z)\right]$$
$$\times \exp\left[-(\alpha_X x + \alpha_Y y + \alpha_Z z)\right]. \tag{2.2-117}$$

Equation (2.2-117) represents a plane wave propagating in three-dimensional space with magnitude decaying exponentially in the X, Y, and Z directions.

Finally, if we evaluate Eq. (2.2-101) in the direction of the complex propagation vector **K**, that is, if

$$\mathbf{r} = r\hat{n}, \tag{2.2-118}$$

then substituting Eqs. (2.2-105), (2.2-118), and (2.2-100) into Eq. (2.2-101) yields

$$\varphi_f(x, y, z) = A \exp(-jkr) \exp(-\alpha r) \qquad (2.2\text{-}119)$$

where

$$r = \sqrt{x^2 + y^2 + z^2}. \qquad (2.2\text{-}120)$$

If Eq. (2.2-119) is evaluated at the range $r = 1/\alpha$ meters, then the *magnitude* of the plane wave is $|A|/e$. Also, if we form the ratio of the magnitude of Eq. (2.2-119) at $r = 0$ to the magnitude at r, then

$$\frac{|A|}{|A|\exp(-\alpha r)} = \exp(+\alpha r), \qquad (2.2\text{-}121)$$

or, in decibels (dB),

$$20 \log_{10} \exp(+\alpha r) = (20 \log_{10} e)\alpha r \text{ dB}$$

$$= 8.686\alpha r \text{ dB}. \qquad (2.2\text{-}122)$$

Equation (2.2-122) is used to calculate the amount of attenuation in decibels that a plane wave will experience after propagating a distance of r meters from the source. From Eq. (2.2-122) it can also be seen that

$$\boxed{\alpha' = 8.686\alpha \text{ dB/m,}} \qquad (2.2\text{-}123)$$

and as a result,

$$\boxed{\alpha = \frac{\alpha'}{8.686} \text{ Np/m,}} \qquad (2.2\text{-}124)$$

where α is the real attenuation coefficient with units of nepers (Np) per meter, and α' is the real attenuation coefficient with units of decibels (dB) per meter. Equation (2.2-123) is used whenever it is necessary to convert α in Np/m to α' in dB/m. Similarly, Eq. (2.2-124) is used whenever it is necessary to convert α' in dB/m to α in Np/m. Note that only α in Np/m can be used in exponentials such as $\exp(-\alpha r)$.

2.3 Reflection and Transmission Coefficients

In Section 2.2 we considered free-space propagation, that is, wave propagation in the absence of boundaries. Now, in this section, we shall be concerned with acoustic wave propagation in the presence of a *massless* boundary between two

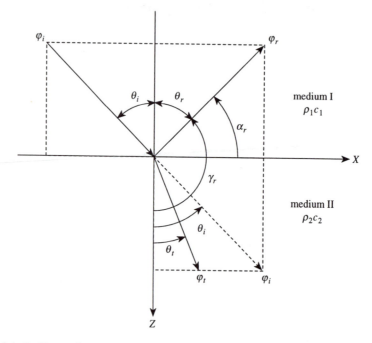

Figure 2.3-1. Incident, reflected, and transmitted plane waves at the boundary between two different fluid media.

different fluid media. The results obtained in this section have direct application to many practical problems in acoustics such as room acoustics, scattering from objects, and pulse propagation in ocean waveguides.

Consider a time-harmonic plane wave propagating in the XZ plane in fluid medium I with ambient density ρ_1, speed of sound c_1, and characteristic impedance $\rho_1 c_1$. The plane wave is incident upon a boundary that separates fluid medium I from fluid medium II with ambient density ρ_2, speed of sound c_2, and characteristic impedance $\rho_2 c_2$ (see Fig. 2.3-1). The angles of incidence, reflection, and transmission—θ_i, θ_r, and θ_t, respectively—take on values between 0° and 90°. The boundary is the XY plane where $z = 0$. Since the two fluid media have different characteristic impedances, that is, since there is an *impedance mismatch*, the incident wave separates into reflected and transmitted waves at the boundary. In this section, we shall derive plane-wave velocity potential, pressure, and time-average intensity reflection and transmission coefficients at the boundary between two fluid media.

At the boundary between two fluid media, the following two boundary conditions must be satisfied at all times and at all spatial locations on the boundary: (1) *continuity of acoustic pressure* and (2) *continuity of the normal component of the acoustic fluid velocity vector*. The first boundary condition implies that on the boundary, the total acoustic pressure in medium I must be equal to the total acoustic pressure in medium II so that no net force acts on the massless boundary. The second boundary condition implies that on the boundary, the total acoustic fluid *speed* in medium I in the direction perpendicular to the boundary must be equal to the total acoustic fluid *speed* in medium II in the direction perpendicular to the boundary so that the two fluid media maintain contact at all times and at all spatial locations on the boundary.

In order to apply the boundary conditions to the problem illustrated in Fig. 2.3-1, we need mathematical models for the incident, reflected, and transmitted time-harmonic plane waves. Without loss of generality, we assumed two-dimensional wave propagation in the XZ plane only. Therefore,

$$\varphi_i(t, x, z) = A_i \exp\left[-j\left(k_{X_i}x + k_{Z_i}z\right)\right] \exp(+j2\pi f_i t), \qquad (2.3\text{-}1)$$

$$\varphi_r(t, x, z) = A_r \exp\left[-j\left(k_{X_r}x + k_{Z_r}z\right)\right] \exp(+j2\pi f_r t), \qquad (2.3\text{-}2)$$

and

$$\varphi_t(t, x, z) = A_t \exp\left[-j\left(k_{X_t}x + k_{Z_t}z\right)\right] \exp(+j2\pi f_t t) \qquad (2.3\text{-}3)$$

are the scalar velocity potentials of the incident, reflected, and transmitted time-harmonic plane waves, respectively. The incident and reflected waves are in medium I, and the transmitted wave is in medium II. Note that all three waves are initially being modeled as traveling in the positive X and Z directions, which is not correct, since the reflected wave is traveling in the negative Z direction. However, once the propagation-vector components are evaluated corresponding to the problem illustrated in Fig. 2.3-1, the correct directions of wave propagation will be obtained. By referring to Eqs. (2.2-11), (2.2-13), (2.2-16), and (2.2-18), the incident, reflected, and transmitted propagation-vector components in the X and Z directions can be expressed as follows:

$$k_{X_i} = k_i u_i = \frac{2\pi f_i}{c_1} \sin \theta_i \cos \psi_i, \qquad (2.3\text{-}4)$$

$$k_{Z_i} = k_i w_i = \frac{2\pi f_i}{c_1} \cos \theta_i; \qquad (2.3\text{-}5)$$

$$k_{X_r} = k_r u_r = \frac{2\pi f_r}{c_1} \cos \alpha_r, \qquad (2.3\text{-}6)$$

$$k_{Z_r} = k_r w_r = \frac{2\pi f_r}{c_1} \cos \gamma_r; \qquad (2.3\text{-}7)$$

and, finally,

$$k_{X_t} = k_t u_t = \frac{2\pi f_t}{c_2} \sin \theta_t \cos \psi_t,$$

(2.3-8)

$$k_{Z_t} = k_t w_t = \frac{2\pi f_t}{c_2} \cos \theta_t;$$

(2.3-9)

where f_i, f_r, and f_t are the frequencies (in hertz) of the incident, reflected, and transmitted waves, respectively. By referring to Fig. 2.2-2 for the definition of ψ and inspecting Fig. 2.3-1, it can be seen that

$$\psi_i = \psi_r = \psi_t = 0.$$

(2.3-10)

It can also be seen from Fig. 2.3-1 that

$$\alpha_r = \frac{\pi}{2} - \theta_r$$

(2.3-11)

and

$$\gamma_r = \pi - \theta_r.$$

(2.3-12)

With the use of Eqs. (2.3-10) through (2.3-12), Eqs. (2.3-4) through (2.3-9) become

$$k_{X_i} = \frac{2\pi f_i}{c_1} \sin \theta_i,$$

(2.3-13)

$$k_{Z_i} = \frac{2\pi f_i}{c_1} \cos \theta_i;$$

(2.3-14)

$$k_{X_r} = \frac{2\pi f_r}{c_1} \cos\left(\frac{\pi}{2} - \theta_r\right) = \frac{2\pi f_r}{c_1} \sin \theta_r,$$

(2.3-15)

$$k_{Z_r} = \frac{2\pi f_r}{c_1} \cos(\pi - \theta_r) = -\frac{2\pi f_r}{c_1} \cos \theta_r;$$

(2.3-16)

and, finally,

$$k_{X_t} = \frac{2\pi f_t}{c_2} \sin \theta_t,$$

(2.3-17)

$$k_{Z_t} = \frac{2\pi f_t}{c_2} \cos \theta_t.$$

(2.3-18)

Substituting the propagation-vector components given by Eqs. (2.3-13) through (2.3-18) into Eqs. (2.3-1) through (2.3-3) yields

$$\varphi_i(t, x, z) = A_i \exp\left[-j\frac{2\pi f_i}{c_1}(x \sin \theta_i + z \cos \theta_i)\right] \exp(+j2\pi f_i t), \quad (2.3\text{-}19)$$

$$\varphi_r(t, x, z) = A_r \exp\left[-j\frac{2\pi f_r}{c_1}(x \sin \theta_r - z \cos \theta_r)\right] \exp(+j2\pi f_r t), \quad (2.3\text{-}20)$$

and

$$\varphi_t(t, x, z) = A_t \exp\left[-j\frac{2\pi f_t}{c_2}(x \sin \theta_t + z \cos \theta_t)\right] \exp(+j2\pi f_t t). \quad (2.3\text{-}21)$$

Note that the reflected wave is now correctly modeled as traveling in the negative Z direction. We are now in a position to apply the boundary conditions at $z = 0$. The first boundary condition is continuity of acoustic pressure, that is,

$$p_1(t, x, z)|_{z=0} = p_2(t, x, z)|_{z=0}, \quad (2.3\text{-}22)$$

where

$$p_1(t, x, z) = p_i(t, x, z) + p_r(t, x, z) \quad (2.3\text{-}23)$$

and

$$p_2(t, x, z) = p_t(t, x, z) \quad (2.3\text{-}24)$$

are the total acoustic pressures in media I and II, respectively. Since for time-harmonic acoustic fields [see Eq. (2.2-28)]

$$p_i(t, x, z) = -j2\pi f_i \rho_1 \varphi_i(t, x, z), \quad (2.3\text{-}25)$$

$$p_r(t, x, z) = -j2\pi f_r \rho_1 \varphi_r(t, x, z), \quad (2.3\text{-}26)$$

and

$$p_t(t, x, z) = -j2\pi f_t \rho_2 \varphi_t(t, x, z), \quad (2.3\text{-}27)$$

substituting Eqs. (2.3-23) through (2.3-27) and Eqs. (2.3-19) through (2.3-21) into Eq. (2.3-22) yields

$$f_i \rho_1 A_i \exp\left(-j2\pi f_i x \frac{\sin \theta_i}{c_1}\right) \exp(+j2\pi f_i t)$$

$$+ f_r \rho_1 A_r \exp\left(-j2\pi f_r x \frac{\sin \theta_r}{c_1}\right) \exp(+j2\pi f_r t)$$

$$= f_t \rho_2 A_t \exp\left(-j2\pi f_t x \frac{\sin \theta_t}{c_2}\right) \exp(+j2\pi f_t t). \quad (2.3\text{-}28)$$

Since the boundary condition represented by Eq. (2.3-28) must hold for all allowed values of t and x, that is, since Eq. (2.3-28) must be *independent* of t and x, this can only be possible if

$$\boxed{f_i = f_r = f_t = f} \tag{2.3-29}$$

and

$$\boxed{\frac{\sin \theta_i}{c_1} = \frac{\sin \theta_r}{c_1} = \frac{\sin \theta_t}{c_2}.} \tag{2.3-30}$$

Therefore, upon substituting Eqs. (2.3-29) and (2.3-30) into Eq. (2.3-28), we obtain

$$A_i + A_r = \frac{\rho_2}{\rho_1} A_t, \tag{2.3-31}$$

which is independent of t and x. Equation (2.3-29) indicates that the frequencies of the incident, reflected, and transmitted waves are equal. From Eq. (2.3-30), it can be seen that the angle of reflection θ_r is equal to the angle of incidence θ_i, that is,

$$\boxed{\theta_r = \theta_i.} \tag{2.3-32}$$

It can also be seen from Eq. (2.3-30) that the angle of transmission θ_t is related to the angle of incidence θ_i by

$$\boxed{\frac{\sin \theta_t}{c_2} = \frac{\sin \theta_i}{c_1},} \tag{2.3-33}$$

which is *Snell's law*.

The second boundary condition is continuity of the normal component of the acoustic fluid velocity vector, that is,

$$u_{n_1}(t, x, z)\big|_{z=0} = u_{n_2}(t, x, z)\big|_{z=0}, \tag{2.3-34}$$

where

$$u_{n_1}(t, x, z) = u_{n_i}(t, x, z) + u_{n_r}(t, x, z) \tag{2.3-35}$$

and

$$u_{n_2}(t, x, z) = u_{n_t}(t, x, z) \tag{2.3-36}$$

are the total acoustic fluid *speeds* in media I and II, respectively, normal (perpendicular) to the boundary. Since the unit vector $\hat{z}$ is perpendicular to the boundary, the normal component of the acoustic fluid velocity vector is given by

$$u_n(t, x, z) = \hat{z} \cdot \mathbf{u}(t, x, z), \tag{2.3-37}$$

and upon substituting Eqs. (2.2-32) and (2.2-10) into Eq. (2.3-37), we obtain

$$u_n(t, x, z) = -jk_z\varphi(t, x, z), \tag{2.3-38}$$

where $k_z = kw$ is the propagation-vector component in the Z direction. Therefore, with the use of Eqs. (2.3-14), (2.3-16), (2.3-18), (2.3-29), (2.3-32), and (2.3-38),

$$u_{n_i}(t, x, z) = -j\frac{2\pi f}{c_1}\varphi_i(t, x, z)\cos\theta_i, \tag{2.3-39}$$

$$u_{n_r}(t, x, z) = +j\frac{2\pi f}{c_1}\varphi_r(t, x, z)\cos\theta_i, \tag{2.3-40}$$

and

$$u_{n_t}(t, x, z) = -j\frac{2\pi f}{c_2}\varphi_t(t, x, z)\cos\theta_t, \tag{2.3-41}$$

where [see Eqs. (2.3-19) through (2.3-21)]

$$\varphi_i(t, x, z) = A_i \exp\left[-j\frac{2\pi f}{c_1}(x\sin\theta_i + z\cos\theta_i)\right]\exp(+j2\pi ft), \tag{2.3-42}$$

$$\varphi_r(t, x, z) = A_r \exp\left[-j\frac{2\pi f}{c_1}(x\sin\theta_i - z\cos\theta_i)\right]\exp(+j2\pi ft), \tag{2.3-43}$$

and

$$\varphi_t(t, x, z) = A_t \exp\left[-j\frac{2\pi f}{c_2}(x\sin\theta_t + z\cos\theta_t)\right]\exp(+j2\pi ft). \tag{2.3-44}$$

Substituting Eqs. (2.3-35), (2.3-36), and Eqs. (2.3-39) through (2.3-44) into Eq. (2.3-34) yields

$$-A_i \exp\left(-j2\pi fx\frac{\sin\theta_i}{c_1}\right)\frac{\cos\theta_i}{c_1} + A_r \exp\left(-j2\pi fx\frac{\sin\theta_i}{c_1}\right)\frac{\cos\theta_i}{c_1}$$

$$= -A_t \exp\left(-j2\pi fx\frac{\sin\theta_t}{c_2}\right)\frac{\cos\theta_t}{c_2}, \tag{2.3-45}$$

and upon substituting Snell's law given by Eq. (2.3-33) into Eq. (2.3-45), we obtain

$$A_i - A_r = \frac{c_1 \cos \theta_t}{c_2 \cos \theta_i} A_t.$$

(2.3-46)

Equations (2.3-31) and (2.3-46) are the result of satisfying both boundary conditions at $z = 0$. With the use of these two equations, we shall derive expressions for the plane-wave velocity-potential reflection and transmission coefficients next.

Velocity-Potential Coefficients

The *plane-wave velocity-potential reflection* and *transmission coefficients* are defined as follows:

$$R_{12} \triangleq \frac{A_r}{A_i}$$

(2.3-47)

and

$$T_{12} \triangleq \frac{A_t}{A_i},$$

(2.3-48)

where A_i, A_r, and A_t are the complex amplitudes of the incident, reflected, and transmitted velocity potentials, respectively, and the subscript 12 is meant to indicate that the boundary is between media I and II, and that the incident wave is in medium I, since the number 1 appears first. By adding Eqs. (2.3-31) and (2.3-46), and by making use of Eq. (2.3-48), we obtain the *plane-wave velocity-potential transmission coefficient*

$$T_{12} = \frac{2\rho_1 c_2 \cos \theta_i}{Z_2 \cos \theta_i + Z_1 \cos \theta_t},$$

(2.3-49)

where

$$Z_1 = \rho_1 c_1$$

(2.3-50)

and

$$Z_2 = \rho_2 c_2$$

(2.3-51)

are the characteristic impedances of media I and II, respectively. If we next divide both sides of Eq. (2.3-31) by A_i, and if we also make use of Eqs. (2.3-47) and

(2.3-48), then we obtain the following relationship between the plane-wave veloc-ity-potential reflection and transmission coefficients:

$$R_{12} = \frac{\rho_2}{\rho_1} T_{12} - 1. \tag{2.3-52}$$

Substituting Eq. (2.3-49) into Eq. (2.3-52) yields the *plane-wave velocity-potential reflection coefficient*

$$R_{12} = \frac{Z_2 \cos \theta_i - Z_1 \cos \theta_t}{Z_2 \cos \theta_i + Z_1 \cos \theta_t}, \tag{2.3-53}$$

where the characteristic impedances Z_1 and Z_2 are given by Eqs. (2.3-50) and (2.3-51), respectively. Equation (2.3-53) is also known as the *Rayleigh reflection coefficient*. Note that both the transmission and reflection coefficients given by Eqs. (2.3-49) and (2.3-53), respectively, are expressed in terms of the angles of incidence and transmission. Since

$$\cos \theta_t = \sqrt{1 - \sin^2 \theta_t}, \tag{2.3-54}$$

substituting Snell's law given by Eq. (2.3-33) into Eq. (2.3-54) yields

$$\cos \theta_t = \sqrt{1 - \left(\frac{c_2}{c_1} \sin \theta_i\right)^2}. \tag{2.3-55}$$

Therefore, with the use of Eq. (2.3-55), both the transmission and reflection coefficients can be expressed in terms of the angle of incidence θ_i only, which is preferred. If $c_2 < c_1$, then Eq. (2.3-55) will be *real* for all angles of incidence, and as a result, both the transmission and reflection coefficients will be *real* for all angles of incidence. However, if $c_2 > c_1$, then Eq. (2.3-55) will be *imaginary* for all angles of incidence greater than the *critical angle of incidence* θ_c, and as a result, both the transmission and reflection coefficients will be *complex* for $\theta_i > \theta_c$. The critical angle of incidence will be discussed in detail in Example 2.3-3.

Alternative expressions for the plane-wave velocity-potential transmission and reflection coefficients can be obtained as follows. Substituting Eqs. (2.3-50) and (2.3-51) into Eqs. (2.3-49) and (2.3-53), and multiplying both the numerator and denominator in Eqs. (2.3-49) and (2.3-53) by $2\pi f/(c_1 c_2)$, yields

$$T_{12} = \frac{2\rho_1 k_{z_1}}{\rho_2 k_{z_1} + \rho_1 k_{z_2}} \tag{2.3-56}$$

and

$$R_{12} = \frac{\rho_2 k_{z_1} - \rho_1 k_{z_2}}{\rho_2 k_{z_1} + \rho_1 k_{z_2}},$$

(2.3-57)

where

$$k_{z_1} = \frac{2\pi f}{c_1} \cos \theta_i$$

(2.3-58)

and

$$k_{z_2} = \frac{2\pi f}{c_2} \cos \theta_t$$

(2.3-59)

are the propagation-vector components in the Z direction in media I and II, respectively.

Before we discuss the derivations of the pressure and intensity coefficients, let us make the following important observation. If media I and II are *identical*, that is, if $\rho_2 = \rho_1$ and $c_2 = c_1$, then $Z_2 = Z_1$. If $c_2 = c_1$, then from Snell's law, $\theta_t = \theta_i$. Therefore, $T_{12} = 1$ (total transmission) and $R_{12} = 0$ (no reflection) as expected, since there is no impedance mismatch. It is important to note that $Z_2 = Z_1$ does *not* necessarily imply that $\rho_2 = \rho_1$ and $c_2 = c_1$. If $c_2 \neq c_1$, then $\theta_t \neq \theta_i$.

Pressure Coefficients
The *plane-wave pressure reflection* and *transmission coefficients* are defined as follows:

$$R'_{12} \triangleq \frac{A'_r}{A'_i}$$

(2.3-60)

and

$$T'_{12} \triangleq \frac{A'_t}{A'_i},$$

(2.3-61)

where A'_i, A'_r, and A'_t are the complex amplitudes of the incident, reflected, and transmitted acoustic pressures, respectively, and the subscript 12 is meant to indicate that the boundary is between media I and II and that the incident wave is in medium I, since the number 1 appears first. Upon substituting Eqs. (2.3-42) through (2.3-44) into Eqs. (2.3-25) through (2.3-27), and by making use of Eq. (2.3-29), it can be shown that

$$A'_i = -j2\pi f \rho_1 A_i,$$ (2.3-62)

$$A'_r = -j2\pi f \rho_1 A_r,$$ (2.3-63)

and

$$A'_t = -j2\pi f \rho_2 A_t.$$ (2.3-64)

Therefore, substituting Eqs. (2.3-62), (2.3-63), and (2.3-47) into Eq. (2.3-60) yields

$$\boxed{R'_{12} = R_{12}.}$$ (2.3-65)

Equation (2.3-65) indicates that *the plane-wave pressure and velocity-potential reflection coefficients are equal.*

Next, if we substitute Eqs. (2.3-62), (2.3-64), and (2.3-48) into Eq. (2.3-61), then

$$\boxed{T'_{12} = \frac{\rho_2}{\rho_1} T_{12}.}$$ (2.3-66)

Equation (2.3-66) indicates that *the plane-wave pressure and velocity-potential transmission coefficients are not equal.* With the use of Eqs. (2.3-49) and (2.3-56), Eq. (2.3-66) can be expressed as

$$\boxed{T'_{12} = \frac{2Z_2 \cos\theta_i}{Z_2 \cos\theta_i + Z_1 \cos\theta_t} = \frac{2\rho_2 k_{Z_1}}{\rho_2 k_{Z_1} + \rho_1 k_{Z_2}},}$$ (2.3-67)

where Z_1 and Z_2 are given by Eqs. (2.3-50) and (2.3-51), respectively.

Finally, by substituting Eqs. (2.3-65) and (2.3-66) into Eq. (2.3-52), we obtain the following relationship between the plane-wave pressure transmission and reflection coefficients:

$$T'_{12} = R'_{12} + 1.$$ (2.3-68)

Time-Average Intensity Coefficients

The *plane-wave time-average intensity reflection* and *transmission coefficients* are defined as follows:

$$\boxed{R''_{12} \triangleq \frac{I_{avg_r}}{I_{avg_i}}}$$ (2.3-69)

and

$$T_{12}'' \triangleq \frac{I_{\text{avg}_t}}{I_{\text{avg}_i}},$$

(2.3-70)

where I_{avg_i}, I_{avg_r}, and I_{avg_t} are the time-average intensities of the incident, reflected, and transmitted acoustic fields, respectively, and the subscript 12 is meant to indicate that the boundary is between media I and II, and that the incident wave is in medium I, since the number 1 appears first.

By referring to Example 2.2-1, the time-average intensities of the incident, reflected, and transmitted time-harmonic plane waves can be expressed as follows:

$$I_{\text{avg}_i} = \tfrac{1}{2}k_1^2 \rho_1 c_1 |A_i|^2,$$

(2.3-71)

$$I_{\text{avg}_r} = \tfrac{1}{2}k_1^2 \rho_1 c_1 |A_r|^2,$$

(2.3-72)

and

$$I_{\text{avg}_t} = \tfrac{1}{2}k_2^2 \rho_2 c_2 |A_t|^2$$

(2.3-73)

where

$$k_1 = 2\pi f / c_1$$

(2.3-74)

and

$$k_2 = 2\pi f / c_2$$

(2.3-75)

are the wave numbers in media I and II, respectively. Therefore, substituting Eqs. (2.3-71) and (2.3-72) into Eq. (2.3-69), and by making use of Eqs. (2.3-47) and (2.3-65),

$$R_{12}'' = |R_{12}'|^2 = |R_{12}|^2.$$

(2.3-76)

Equation (2.3-76) indicates that the plane-wave time-average intensity reflection coefficient is equal to the magnitude squared of either the plane-wave pressure or velocity-potential reflection coefficient.

Next, if we substitute Eqs. (2.3-71) and (2.3-73) into Eq. (2.3-70), and if we make use of Eqs. (2.3-48) and (2.3-66), then

$$T_{12}'' = \frac{Z_1}{Z_2}|T_{12}'|^2 = \frac{\rho_2 c_1}{\rho_1 c_2}|T_{12}|^2,$$

(2.3-77)

where Z_1 and Z_2 are given by Eqs. (2.3-50) and (2.3-51), respectively.

Finally, conservation of energy requires that the sum of the time-average power of the reflected and transmitted fields be equal to the time-average power of the incident field, that is,

$$P_{avg_r} + P_{avg_t} = P_{avg_i}.$$ (2.3-78)

In Appendix 2A it is shown that Eq. (2.3-78) can be rewritten as

$$R''_{12} + \frac{\cos \theta_t}{\cos \theta_i} T''_{12} = 1.$$ (2.3-79)

With the use of Eq. (2.3-76), and the *plane-wave time-average power transmission coefficient* $T'''_{12} = (\cos \theta_t / \cos \theta_i) T''_{12}$ derived in Appendix 2A, the following additional expression for the conservation of energy is obtained:

$$T'''_{12} = 1 - |R'_{12}|^2 = 1 - |R_{12}|^2.$$ (2.3-80)

From Eq. (2.3-80) it can be seen that *if the magnitude of either the plane-wave pressure or velocity-potential reflection coefficient is equal to one, then the plane-wave time-average power transmission coefficient is equal to zero.* This makes physical sense, because if an incident wave is totally reflected, no time-average acoustic power can propagate into medium II. The plane-wave time-average power reflection and transmission coefficients are defined and discussed in Appendix 2A.

Example 2.3-1 (Ideal rigid and pressure-release boundaries)
Two of the most common boundaries encountered in acoustic wave propagation problems are the ideal rigid and pressure-release boundaries. For example, in room acoustics, walls are typically modeled as ideal rigid boundaries (see Section 2.4). And in ocean acoustics, if a sound source is located in the ocean medium, then the ocean surface is typically modeled as an ideal pressure-release boundary and the ocean bottom is sometimes modeled as an ideal rigid boundary, although this model for the ocean bottom is not very realistic (see Section 3.8.1).
    Ideal rigid boundary. If an incident acoustic wave is in medium I (see Fig. 2.3-1), then an *ideal rigid boundary* is, by definition, one on which $Z_2 \gg Z_1$, so that

$$Z_2/Z_1 \gg 1.$$ (2.3-81)

Another definition that is sometimes used to describe an ideal rigid boundary is that $\rho_2 \gg \rho_1$ so that $\rho_2/\rho_1 \gg 1$. In the limit as $Z_2/Z_1 \to \infty$,

Eq. (2.3-53) reduces to

$$\boxed{R_{12} = 1,} \tag{2.3-82}$$

and as a result,

$$A_r = A_i \tag{2.3-83}$$

regardless of the angle of incidence θ_i. Therefore, the amplitude of the reflected wave is equal to the amplitude of the incident wave. Solving for the transmission coefficient T_{12} in Eq. (2.3-52) yields

$$T_{12} = \frac{\rho_1}{\rho_2}(1 + R_{12}), \tag{2.3-84}$$

and upon substituting Eq. (2.3-82) into Eq. (2.3-84), and in the limit as $\rho_2/\rho_1 \to \infty$ (which implies that $\rho_1/\rho_2 \to 0$), Eq. (2.3-84) reduces to

$$T_{12} = 0, \tag{2.3-85}$$

and as a result,

$$A_t = 0. \tag{2.3-86}$$

Since Eq. (2.3-82) indicates that the incident wave is *totally reflected*, Eq. (2.3-85) correctly indicates that there is *no transmitted wave*, and as a result, there is no acoustic field in medium II. In addition, since the magnitude of the plane-wave velocity-potential reflection coefficient is equal to one, by referring to Eq. (2.3-80) it can be seen that the plane-wave time-average power transmission coefficient $T_{12}''' = 0$. Therefore, for an ideal rigid boundary, no time-average acoustic power propagates into medium II, as expected.

Let us next compute the acoustic pressure and the normal component of the acoustic fluid velocity vector on the boundary. The acoustic pressure in medium I is given by

$$p_1(t, x, z) = p_i(t, x, z) + p_r(t, x, z), \tag{2.3-87}$$

where [see Eqs. (2.3-25), (2.3-26), and (2.3-29)]

$$p_i(t, x, z) = -j2\pi f \rho_1 \varphi_i(t, x, z), \tag{2.3-88}$$

$$p_r(t, x, z) = -j2\pi f \rho_1 \varphi_r(t, x, z), \tag{2.3-89}$$

and the incident and reflected velocity potentials are given by Eqs. (2.3-42) and (2.3-43), respectively. Substituting Eq. (2.3-83) into Eq. (2.3-43) and evaluating Eqs. (2.3-42) and (2.3-43) on the boundary, that is, at $z = 0$, yields

$$\varphi_r(t, x, 0) = \varphi_i(t, x, 0). \qquad (2.3\text{-}90)$$

Therefore,

$$p_r(t, x, 0) = p_i(t, x, 0), \qquad (2.3\text{-}91)$$

and as a result [see Eq. (2.3-87)],

$$p_1(t, x, 0) = p_2(t, x, 0) = 2p_i(t, x, 0), \qquad (2.3\text{-}92)$$

since the total acoustic pressure in medium I must be equal to the total acoustic pressure in medium II on the boundary. Equation (2.3-92) indicates that there is a *doubling* of the incident acoustic pressure on an ideal rigid boundary.

If we next evaluate Eqs. (2.3-39) and (2.3-40) at $z = 0$, then with the use of Eq. (2.3-90), it can be shown that

$$u_{n_r}(t, x, 0) = -u_{n_i}(t, x, 0). \qquad (2.3\text{-}93)$$

And upon evaluating Eq. (2.3-35) at $z = 0$ and using Eq. (2.3-93), we obtain

$$u_{n_1}(t, x, 0) = u_{n_2}(t, x, 0) = 0, \qquad (2.3\text{-}94)$$

since the normal components of the total acoustic fluid velocity vectors in media I and II must be equal on the boundary. Equation (2.3-94) indicates that the normal component of the acoustic fluid velocity vector is equal to *zero* on an ideal rigid boundary. Hence the term "rigid," implying no motion. Equation (2.3-94) is an example of a *Neumann boundary condition*. A Neumann boundary condition is one that specifies the allowed value for the normal component of the acoustic fluid velocity vector on a boundary. In the case of an ideal rigid boundary, the allowed value is zero.

Ideal pressure-release boundary.   If an incident acoustic wave is in medium I (see Fig. 2.3-1), then an *ideal pressure-release boundary* is, by definition, one on which $Z_2 \ll Z_1$, so that

$$Z_2/Z_1 \ll 1. \qquad (2.3\text{-}95)$$

Another definition that is sometimes used to describe an ideal pressure-release boundary is that $\rho_2 \ll \rho_1$, so that $\rho_2/\rho_1 \ll 1$. In the limit as $Z_2/Z_1 \to 0$, Eq. (2.3-53) reduces to

$$\boxed{R_{12} = -1 = \exp(+j\pi),} \qquad (2.3\text{-}96)$$

and as a result,

$$A_r = -A_i = A_i \exp(+j\pi) \qquad (2.3\text{-}97)$$

regardless of the angle of incidence θ_i. Therefore, the magnitude of the reflected wave is equal to the magnitude of the incident wave, but the reflected wave is 180° out of phase with the incident wave.

Let us next try to evaluate the plane-wave velocity-potential transmission coefficient for an ideal pressure-release boundary. Upon substituting Eq. (2.3-96) into Eq. (2.3-84), and in the limit as $\rho_2/\rho_1 \to 0$ (which implies that $\rho_1/\rho_2 \to \infty$), Eq. (2.3-84) reduces to $T_{12} = \infty \times 0$, which is an indeterminate form, and as a result, we can say nothing about the value of T_{12}. However, since the magnitude of the plane-wave velocity-potential reflection coefficient is equal to one, by referring to Eq. (2.3-80) it can be seen that the plane-wave time-average power transmission coefficient $T_{12}''' = 0$. Therefore, for an ideal pressure-release boundary, no time-average acoustic power propagates into medium II, as expected. Also, since $R_{12}' = R_{12}$ [see Eq. (2.3-65)], substituting $R_{12}' = -1$ into Eq. (2.3-68) yields $T_{12}' = 0$, that is, the plane-wave pressure transmission coefficient is equal to zero. As a result, there is no transmitted acoustic pressure, and hence, no time-average acoustic power in medium II. Let us next compute the acoustic pressure and the normal component of the acoustic fluid velocity vector on the boundary.

Substituting Eq. (2.3-97) into Eq. (2.3-43) and evaluating Eqs. (2.3-42) and (2.3-43) on the boundary at $z = 0$ yields

$$\varphi_r(t, x, 0) = -\varphi_i(t, x, 0). \qquad (2.3\text{-}98)$$

Therefore, by referring to Eqs. (2.3-88) and (2.3-89), and with the use of Eq. (2.3-98),

$$p_r(t, x, 0) = -p_i(t, x, 0), \qquad (2.3\text{-}99)$$

and as a result [see Eq. (2.3-87)],

$$p_1(t, x, 0) = p_2(t, x, 0) = 0. \qquad (2.3\text{-}100)$$

Equation (2.3-100) indicates that the acoustic pressure is equal to *zero* on an ideal pressure-release boundary. Hence the term "pressure release," implying zero pressure. Equation (2.3-100) is an example of a *Dirichlet boundary condition*. A Dirichlet boundary condition is one that specifies the allowed value for the acoustic pressure on a boundary. In the case of an ideal pressure-release boundary, the allowed value is zero.

If we next evaluate Eqs. (2.3-39) and (2.3-40) at $z = 0$, then with the use of Eq. (2.3-98), it can be shown that

$$u_{n_r}(t, x, 0) = u_{n_i}(t, x, 0). \qquad (2.3\text{-}101)$$

And upon evaluating Eq. (2.3-35) at $z = 0$ and using Eq. (2.3-101), we obtain

$$u_{n_1}(t, x, 0) = u_{n_2}(t, x, 0) = 2u_{n_i}(t, x, 0), \qquad (2.3\text{-}102)$$

which indicates that there is a *doubling* of the normal component of the incident acoustic fluid velocity vector on an ideal pressure-release boundary.

Example 2.3-2 (Normal incidence)
When the angle of incidence $\theta_i = 0°$ (see Fig. 2.3-1), the incident acoustic wave is traveling in a direction that is normal (perpendicular) to the boundary. This special case is known as *normal incidence*. If $\theta_i = 0°$, then the angle of reflection $\theta_r = \theta_i = 0°$ and the angle of transmission $\theta_t = 0°$ [see Eqs. (2.3-32) and (2.3-33)]. Therefore, with $\theta_i = 0°$ and $\theta_t = 0°$, Eqs. (2.3-49) and (2.3-53) reduce to

$$\boxed{T_{12} = \frac{2\rho_1 c_2}{Z_2 + Z_1}, \qquad \theta_i = 0°} \qquad (2.3\text{-}103)$$

and

$$\boxed{R_{12} = \frac{Z_2 - Z_1}{Z_2 + Z_1}, \qquad \theta_i = 0°,} \qquad (2.3\text{-}104)$$

where the characteristic impedances Z_1 and Z_2 are given by Eqs. (2.3-50) and (2.3-51), respectively. Note that in the special case of normal incidence, both the transmission and reflection coefficients are always real. Also note that the reflection coefficient given by Eq. (2.3-104) reduces to 1 and -1 in the case of ideal rigid and pressure-release boundaries, respectively; and as a result, the plane-wave time-average power transmission coefficient given by Eq. (2.3-80) reduces to zero, as expected.

Example 2.3-3 (**Grazing and critical angles of incidence**)

Grazing incidence. When the angle of incidence $\theta_i = 90°$ (see Fig. 2.3-1), the incident acoustic wave is traveling in a direction that is parallel to the boundary. This special case is known as *grazing incidence*. Therefore, with $\theta_i = 90°$, Eqs. (2.3-49) and (2.3-53) reduce to

$$\boxed{T_{12} = 0, \qquad \theta_i = 90°} \qquad (2.3\text{-}105)$$

and

$$\boxed{R_{12} = -1 = \exp(+j\pi), \qquad \theta_i = 90°.} \qquad (2.3\text{-}106)$$

Also, if $\theta_i = 90°$, then the angle of reflection $\theta_r = \theta_i = 90°$. In addition, since Eq. (2.3-105) indicates that there is no acoustic wave transmitted into medium II, the angle of transmission is irrelevant. And since the magnitude of the plane-wave velocity-potential reflection coefficient is equal to one, the plane-wave time-average power transmission coefficient given by Eq. (2.3-80) reduces to zero. Therefore, no time-average acoustic power propagates into medium II.

Critical angle of incidence. Let us next refer to Eq. (2.3-55), which is rewritten below for convenience:

$$\boxed{\cos \theta_t = \sqrt{1 - \left(\frac{c_2}{c_1}\sin \theta_i\right)^2}.} \qquad (2.3\text{-}55)$$

If $c_2 < c_1$, then

$$\frac{c_2}{c_1}\sin \theta_i < 1 \qquad (2.3\text{-}107)$$

for $0° \leq \theta_i \leq 90°$. The implication of Eq. (2.3-107) is that Eq. (2.3-55) will be real, and as we mentioned earlier, both the transmission and reflection coefficients will be real for all angles of incidence (see Fig. 2.3-2). However, if $c_2 > c_1$, then

$$\frac{c_2}{c_1}\sin \theta_i \geq 1 \qquad (2.3\text{-}108)$$

only if

$$\sin \theta_i \geq c_1/c_2 \qquad (2.3\text{-}109)$$

or, in other words,

$$\frac{c_2}{c_1}\sin\theta_i \geq 1, \qquad c_2 > c_1, \qquad \theta_c \leq \theta_i \leq 90°, \qquad (2.3\text{-}110)$$

where

$$\theta_c = \sin^{-1}(c_1/c_2), \qquad c_2 > c_1, \qquad (2.3\text{-}111)$$

is known as the *critical angle of incidence*. The implication of Eq. (2.3-110) is that Eq. (2.3-55) will be equal to zero when $\theta_i = \theta_c$, and it will be imaginary when $\theta_c < \theta_i \leq 90°$. Equation (2.3-55) being equal to zero implies that $\theta_t = 90°$, which means that the transmitted wave propagates along the boundary—it does not enter medium II. Therefore, in summary,

$$\cos\theta_t = \sqrt{1 - \left(\frac{c_2}{c_1}\sin\theta_i\right)^2}, \qquad \begin{array}{l} c_2 < c_1, \quad 0° \leq \theta_i \leq 90°, \\[2mm] c_2 > c_1, \quad 0° \leq \theta_i \leq \theta_c, \end{array} \qquad (2.3\text{-}112)$$

and

$$\cos\theta_t = \sqrt{-\left[\left(\frac{c_2}{c_1}\sin\theta_i\right)^2 - 1\right]}, \qquad c_2 > c_1, \quad \theta_c \leq \theta_i \leq 90°,$$

$$(2.3\text{-}113)$$

or, since $(\pm j)^2 = -1$,

$$\cos\theta_t = \pm j\sqrt{\left(\frac{c_2}{c_1}\sin\theta_i\right)^2 - 1}, \qquad c_2 > c_1, \quad \theta_c \leq \theta_i \leq 90°. \quad (2.3\text{-}114)$$

Choosing $-j$ in Eq. (2.3-114) yields

$$\cos\theta_t = -j\sqrt{\left(\frac{c_2}{c_1}\sin\theta_i\right)^2 - 1}, \qquad c_2 > c_1, \quad \theta_c \leq \theta_i \leq 90°.$$

$$(2.3\text{-}115)$$

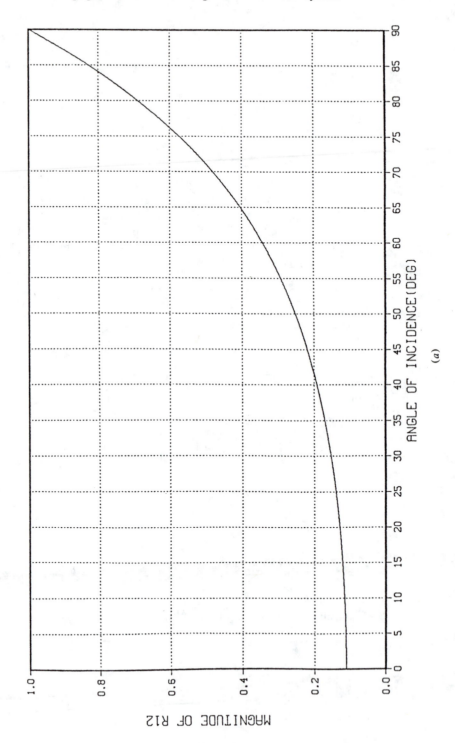

Figure 2.3-2. (a) Magnitude and (b) phase of the reflection coefficient R_{12} for $\rho_2/\rho_1 = \frac{6}{5}$ and $c_2/c_1 = \frac{2}{3}$.

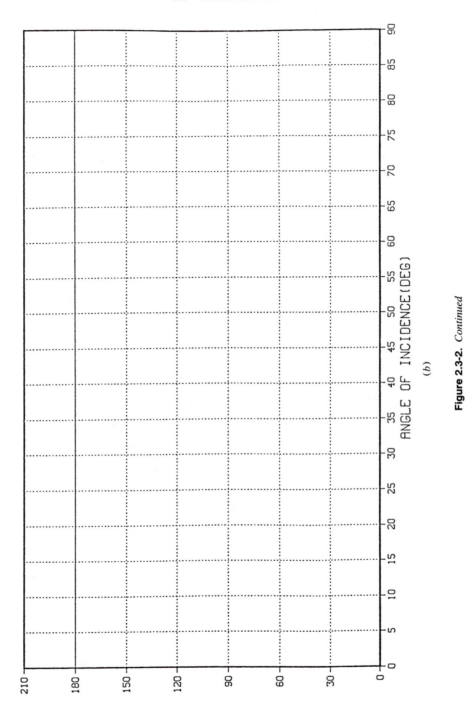

ANGLE OF INCIDENCE (DEG)

(b)

Figure 2.3-2. *Continued*

Before continuing further, we must justify our choice of $-j$ in Eq. (2.3-115). If we refer back to Eq. (2.3-44), then we can see that the transmitted plane wave is propagating in the positive X and Z directions. The "z term" is given by

$$\exp(-jk_{z_2}z) = \exp\left(-j\frac{2\pi f}{c_2}z\cos\theta_t\right), \qquad (2.3\text{-}116)$$

where use has been made of Eq. (2.3-59). Substituting Eq. (2.3-115) into Eq. (2.3-116) yields

$$\exp(-jk_{z_2}z) = \exp(-k_2 bz), \qquad c_2 > c_1, \quad \theta_c \leq \theta_i \leq 90°, \quad (2.3\text{-}117)$$

where k_2 is given by Eq. (2.3-75) and

$$\boxed{\; b = \sqrt{\left(\frac{c_2}{c_1}\sin\theta_i\right)^2 - 1}\;, \qquad c_2 > c_1, \quad \theta_c \leq \theta_i \leq 90°. \;} \qquad (2.3\text{-}118)$$

The right-hand side of Eq. (2.3-117) is known as an *evanescent wave* (i.e., a decaying exponential), since $\exp(-k_2 bz) \to 0$ as $z \to \infty$. If we had chosen $+j$ instead of $-j$, then we would have obtained a growing exponential that approaches infinity as z approaches infinity, which does not make physical sense.

Let us next examine the plane-wave velocity-potential reflection coefficient when the angle of incidence is between critical and grazing. Substituting Eqs. (2.3-115) and (2.3-118) into Eq. (2.3-53) yields

$$R_{12} = \frac{Z_2\cos\theta_i + jZ_1 b}{Z_2\cos\theta_i - jZ_1 b}, \qquad c_2 > c_1, \quad \theta_c \leq \theta_i \leq 90°. \quad (2.3\text{-}119)$$

If we let

$$\zeta = Z_2\cos\theta_i + jZ_1 b, \qquad (2.3\text{-}120)$$

then

$$R_{12} = \zeta/\zeta^* = \exp(+j2\angle\zeta), \qquad c_2 > c_1, \quad \theta_c \leq \theta_i \leq 90°, \quad (2.3\text{-}121)$$

where the asterisk * denotes complex conjugate. If we further let $\Phi = \angle\zeta$, then Eq. (2.3-121) can be rewritten as

$$\boxed{\; R_{12} = \exp(+j2\Phi), \qquad c_2 > c_1, \quad \theta_c \leq \theta_i \leq 90° \;} \qquad (2.3\text{-}122)$$

where

$$\Phi = \tan^{-1}\left[\frac{Z_1 b}{Z_2 \cos \theta_i}\right] \qquad (2.3\text{-}123)$$

and b is given by Eq. (2.3-118). Note that the magnitude of the reflection coefficient given by Eq. (2.3-122) is equal to one, implying *total reflection*. Therefore, for angles of incidence equal to or greater than the critical angle, the magnitudes of the reflected and incident waves are *equal* (see Fig. 2.3-3*a*). However, the reflected wave acquires a *phase shift* of 2Φ relative to the incident wave (see Fig. 2.3-3*b*). Also note that since [see Eqs. (2.3-111) and (2.3-118)]

$$b = 0, \qquad \theta_i = \theta_c \qquad (2.3\text{-}124)$$

and

$$\cos \theta_i = 0, \qquad \theta_i = 90°, \qquad (2.3\text{-}125)$$

substituting Eqs. (2.3-124) and (2.3-125) into Eq. (2.3-123) yields

$$\Phi = 0, \qquad \theta_i = \theta_c \qquad (2.3\text{-}126)$$

and

$$\Phi = \pi/2, \qquad \theta_i = 90°, \qquad (2.3\text{-}127)$$

respectively. Therefore, substituting Eqs. (2.3-126) and (2.3-127) into Eq. (2.3-122) yields

$$R_{12} = 1, \qquad c_2 > c_1, \qquad \theta_i = \theta_c \qquad (2.3\text{-}128)$$

and

$$R_{12} = \exp(+j\pi) = -1, \qquad c_2 > c_1, \qquad \theta_i = 90°, \qquad (2.3\text{-}129)$$

respectively. Equations (2.3-128) and (2.3-129) indicate that at the critical angle of incidence and at grazing incidence, the boundary acts like an ideal rigid and a pressure-release surface, respectively. In addition, by comparing Eqs. (2.3-106) and (2.3-129), it can be seen that the reflection coefficient is always equal to -1 at grazing incidence, regardless of the properties of the two fluid media.

   For the special case when the angle of incidence is between critical and grazing, the plane-wave velocity-potential transmission coefficient is

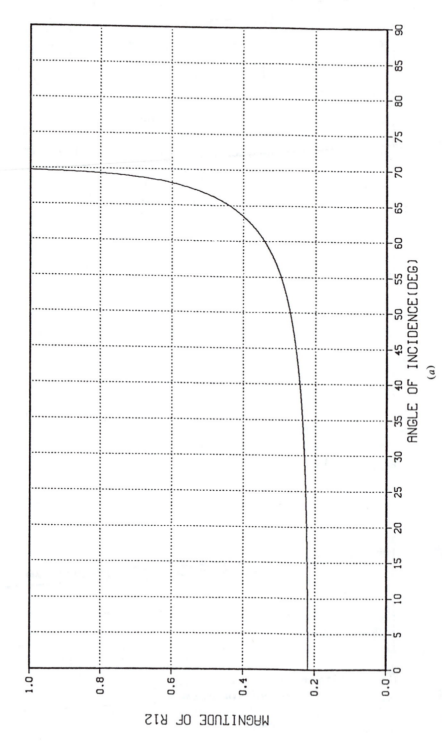

(a)

Figure 2.3-3. (a) Magnitude and (b) phase of the reflection coefficient R_{12}. Medium I is sea water with $\rho_1 = 1026$ kg/m^3 and $c_1 = 1500$ m, Medium II is the sea bottom with $\rho_2 = 1500$ kg/m^3 and $c_2 = 1600$ m/sec. Note that a critical angle of incidence exists at $69.6°$.

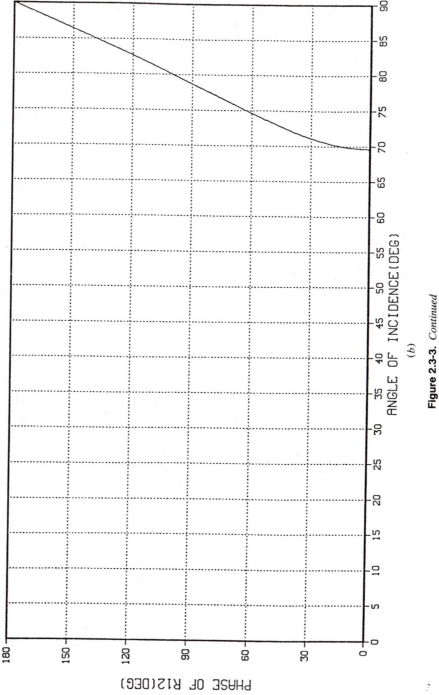

(b)

Figure 2.3-3. *Continued*

not equal to zero, although the magnitude of the reflection coefficient is equal to one. The transmission coefficient is complex, and can be obtained by substituting Eqs. (2.3-115) and (2.3-118) into Eq. (2.3-49). Doing so yields

$$T_{12} = \frac{2\rho_1 c_2 \cos\theta_i}{\sqrt{Z_2^2 \cos^2\theta_i + Z_1^2 b^2}}\exp(+j\Phi), \qquad c_2 > c_1, \quad \theta_c \le \theta_i \le 90°,$$

(2.3-130)

where b is given by Eq. (2.3-118) and Φ is given by Eq. (2.3-123). The nonzero plane-wave *pressure* transmission coefficient can also be obtained by substituting Eq. (2.3-130) into Eq. (2.3-66). And upon substituting Eqs. (2.3-33), (2.3-48), (2.3-123), and (2.3-124) into Eq. (2.3-44), we obtain the following expression for the transmitted acoustic field:

$$\varphi_t(t, x, z) = A_i T_{12} \exp(-jk_1 x \sin\theta_i)\exp(-k_2 b z)\exp(+j2\pi ft),$$

$$c_2 > c_1, \quad \theta_c \le \theta_i \le 90°$$

(2.3-131)

where T_{12} is given by Eq. (2.3-130), k_1 is given by Eq. (2.3-74), k_2 is given by Eq. (2.3-75), and b is given by Eq. (2.3-118). Note that the transmitted field propagates along the boundary in the positive X direction. Also note that it decays very rapidly to zero as it tries to enter into medium II because of the presence of the evanescent wave (decaying exponential) in the Z direction. Recall that since the magnitude of the plane-wave velocity-potential reflection coefficient is equal to one, the plane-wave time-average power transmission coefficient given by Eq. (2.3-80) is equal to zero, and as a result, no time-average acoustic power propagates into medium II.

Skin Depth. On the boundary, that is, at $z = 0$, the magnitude of the transmitted acoustic field given by Eq. (2.3-131) is $|A_i||T_{12}|$. At the depth

$$z_{SD} = \frac{1}{k_2 b}, \qquad c_2 > c_1, \quad \theta_c < \theta_i < 90°, \qquad (2.3\text{-}132)$$

the magnitude of φ_t is reduced to $|A_i||T_{12}|e^{-1}$. The depth (in meters) given by Eq. (2.3-132) is called the *skin depth*, and it is a monotonically decreasing function of both f and θ_i, that is, z_{SD} decreases as either f

or θ_i or both increase. Note that the skin depth is not defined at the critical angle of incidence, since $b = 0$ at $\theta_i = \theta_c$ [see Eq. (2.3-124)]. Also note that the skin depth is not defined at grazing incidence, since $T_{12} = 0$ at $\theta_i = 90°$ [see Eq. (2.3-130)].

Example 2.3-4 (Angle of intromission)

When the angle of incidence θ_i is equal to the *angle of intromission* θ_I, the reflection coefficient is equal to *zero*, and as a result, there is *total transmission* of time-average acoustic power into medium II. In order to solve for θ_I, we refer back to Eq. (2.3-53) and note that the reflection coefficient R_{12} is equal to zero when $\theta_i = \theta_I$, where

$$Z_2 \cos \theta_I = Z_1 \cos \theta_t, \qquad (2.3\text{-}133)$$

or

$$\cos^2 \theta_I = (Z_1/Z_2)^2 \cos^2 \theta_t, \qquad (2.3\text{-}134)$$

and since $\sin^2 \alpha + \cos^2 \alpha = 1$,

$$\sin^2 \theta_I - (Z_1/Z_2)^2 \sin^2 \theta_t = 1 - (Z_1/Z_2)^2. \qquad (2.3\text{-}135)$$

Substituting Snell's law given by Eq. (2.3-33) into Eq. (2.3-135) yields

$$\sin \theta_I = \sqrt{\frac{(\rho_2/\rho_1)^2 - (c_1/c_2)^2}{(\rho_2/\rho_1)^2 - 1}}, \qquad (2.3\text{-}136)$$

where use has been made of Eqs. (2.3-50) and (2.3-51). Since the angle of intromission θ_I is a particular angle of incidence, it must be real. As a result, θ_I will exist only if $\rho_2/\rho_1 > c_1/c_2 > 1$, or if $\rho_2/\rho_1 < c_1/c_2 < 1$ (see Fig. 2.3-4). Note that if both θ_I and θ_c exist, then $\theta_I < \theta_c$ (see Fig. 2.3-5).

Example 2.3-5 (Reflection coefficient at the boundary between two viscous fluid media)

In this example we shall derive an expression for the plane-wave velocity-potential reflection coefficient at the boundary between two viscous fluid media. The results obtained in this example are very important because all real fluids possess viscosity. For example, in ocean acoustics, the ocean medium is correctly modeled as a viscous fluid with a frequency-dependent attenuation coefficient. And the ocean bottom is sometimes modeled as a viscous, fluidlike medium with a

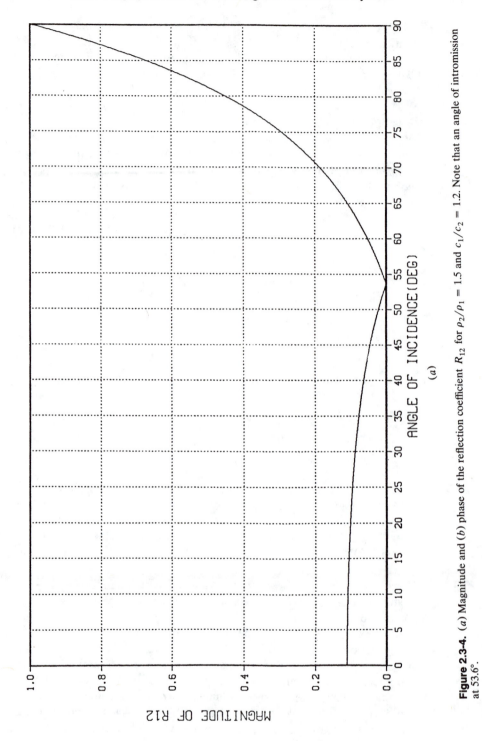

Figure 2.3-4. (*a*) Magnitude and (*b*) phase of the reflection coefficient R_{12} for $\rho_2/\rho_1 = 1.5$ and $c_1/c_2 = 1.2$. Note that an angle of intromission at 53.6°.

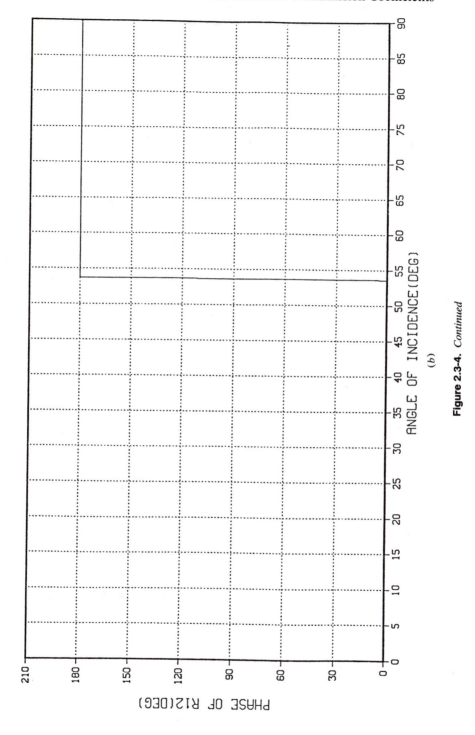

Figure 2-3-4. *Continued*

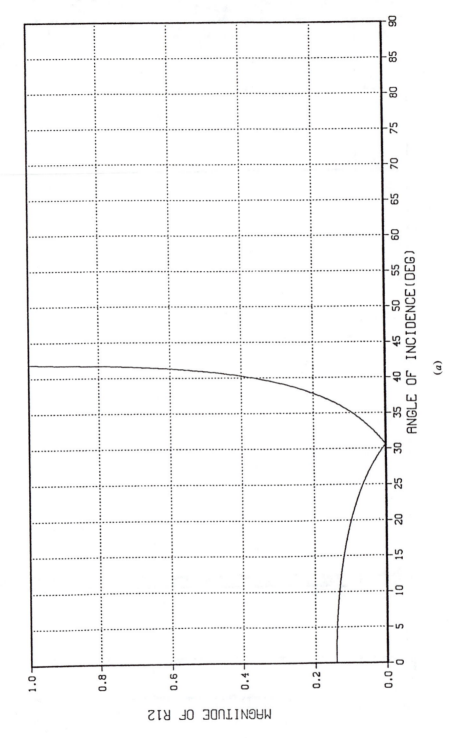

(a)

Figure 2.3-5. (a) Magnitude and (b) phase of the reflection coefficient R_{12} for $\rho_2/\rho_1 = \frac{1}{2}$ and $c_1/c_2 = \frac{2}{3}$. Note that an angle of intromission exi 30.6° and a critical angle of incidence exists at 41.8°.

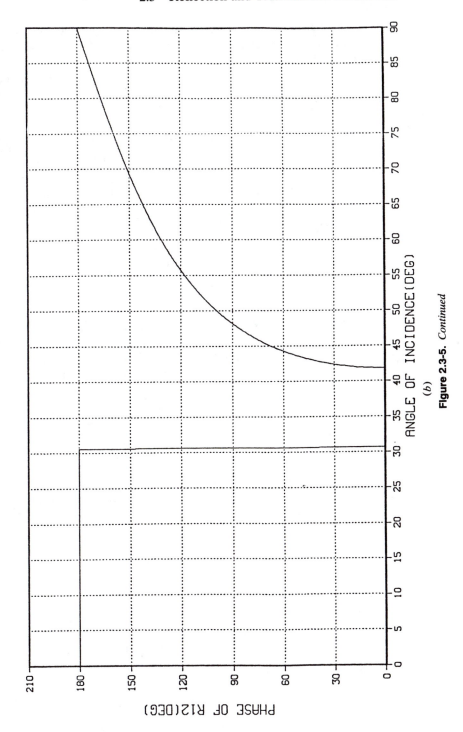

Figure 2.3-5. *Continued*

(b)

frequency-dependent attenuation coefficient as well. Recall that the plane-wave velocity-potential and pressure reflection coefficients are equal [see Eq. (2.3-65)].

We begin by rewriting the expression for the reflection coefficient given by Eq. (2.3-57). Substituting Eqs. (2.3-58) and (2.3-59) into Eq. (2.3-57) yields

$$R_{12} = \frac{\rho_2 k_1 \cos \theta_i - \rho_1 k_2 \cos \theta_t}{\rho_2 k_1 \cos \theta_i + \rho_1 k_2 \cos \theta_t}, \qquad (2.3\text{-}137)$$

where the wave numbers k_1 and k_2 are given by Eqs. (2.3-74) and (2.3-75), respectively. Since the incident wave is in medium I, the *real index of refraction* n_{12} is defined as follows:

$$\boxed{n_{12} \triangleq \frac{c_1}{c_2} = \frac{k_2}{k_1}.} \qquad (2.3\text{-}138)$$

Note that the index of refraction is *dimensionless*. Also note that if $c_2 > c_1$, then $n_{12} = \sin \theta_c$ [see Eq. (2.3-111)]. If we divide both the numerator and denominator of the right-hand side of Eq. (2.3-137) by k_1 and make use of Eqs. (2.3-55) and (2.3-138), then Eq. (2.3-137) can be rewritten as

$$R_{12} = \frac{\rho_2 \cos \theta_i - \rho_1 \sqrt{n_{12}^2 - \sin^2 \theta_i}}{\rho_2 \cos \theta_i + \rho_1 \sqrt{n_{12}^2 - \sin^2 \theta_i}}. \qquad (2.3\text{-}139)$$

In order to allow for *viscous* fluid media, the following more general definition of the *complex index of refraction* is used instead of Eq. (2.3-138):

$$\boxed{N_{12} \triangleq \frac{K_2}{K_1},} \qquad (2.3\text{-}140)$$

where (see Example 2.2-3)

$$K_1 = k_1 - j\alpha_1 \qquad (2.3\text{-}141)$$

and

$$K_2 = k_2 - j\alpha_2 \qquad (2.3\text{-}142)$$

are the *complex* wave numbers in media I and II, respectively; k_1 and

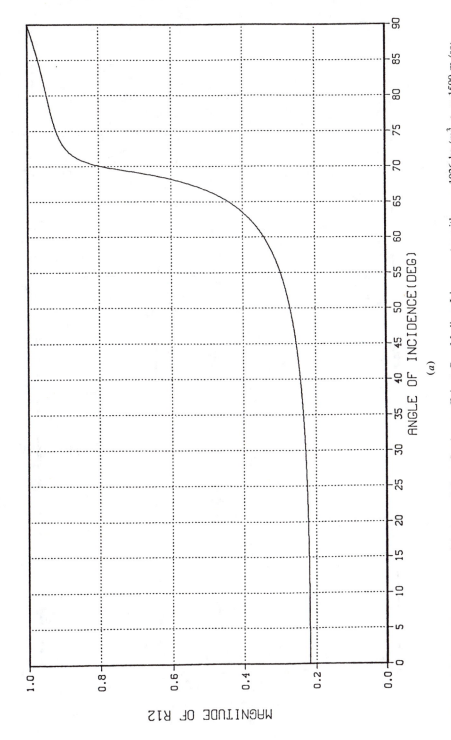

Figure 2.3-6. (*a*) Magnitude and (*b*) phase of the reflection coefficient R_{12}. Medium I is sea water with $\rho_1 = 1026$ kg/m^3, $c_1 = 1500$ m/sec $\alpha'_1 = 2.6 \times 10^{-7}$ dB/m at $f = 50$ Hz. Medium II is the sea bottom with $\rho_2 = 1500$ kg/m^3, $c_2 = 1600$ m/sec, and $\alpha'_2 = 1 \times 10^{-2}$ dB/m at $f = 5$ Note that a critical angle of incidence exists at 69.6°.

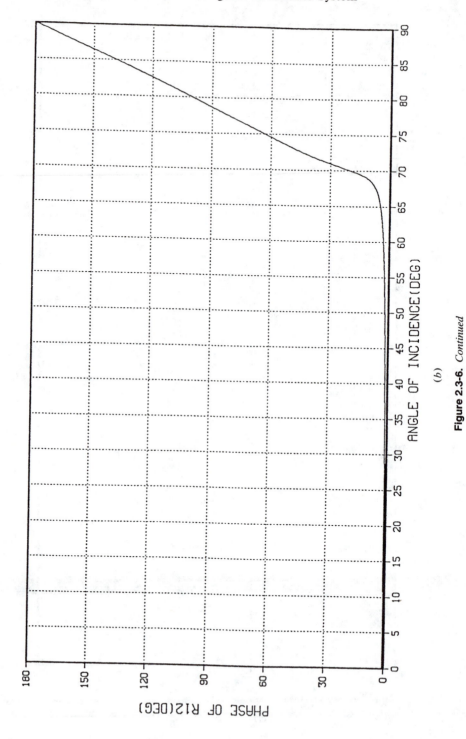

Figure 2.3-6. *Continued*

k_2 are the real wave numbers in radians per meter in media I and II, respectively; and α_1 and α_2 are the real attenuation coefficients in nepers per meter in media I and II, respectively. Note that for ideal, nonviscous fluids (i.e., $\alpha_1 = 0$ and $\alpha_2 = 0$), Eq. (2.3-140) reduces to Eq. (2.3-138). Therefore, in order to evaluate Eq. (2.3-139) for viscous fluid media, we need to rewrite it as follows:

$$R_{12} = \frac{\rho_2 \cos \theta_i - \rho_1 B}{\rho_2 \cos \theta_i + \rho_1 B},$$ (2.3-143)

where

$$B = \begin{cases} \sqrt{N_{12}^2 - \sin^2 \theta_i}, & \mathrm{Re}\{N_{12}^2\} \geq \sin^2 \theta_i, \\ -j\sqrt{\sin^2 \theta_i - N_{12}^2}, & \mathrm{Re}\{N_{12}^2\} < \sin^2 \theta_i, \end{cases}$$ (2.3-144)

and N_{12} is given by Eq. (2.3-140). Note that Eqs. (2.3-143) and (2.3-144) are also valid for ideal, nonviscous fluids where $N_{12} = n_{12}$. In order to clearly see the effects of viscosity on the process of reflection, compare Figs. 2.3-3 and 2.3-6: both figures are based on the same parameter values, with the exception that Fig. 2.3-6 includes viscosity, whereas Fig. 2.3-3 does not. Figure 2.3-6 shows that because of viscosity, the magnitude of R_{12} is now less than one for $\theta_c \leq \theta_i < 90°$, and as a result, time-average acoustic power can propagate into medium II.

2.4 The Rectangular Cavity

In this section we shall derive expressions for the velocity potential and the acoustic pressure inside a rectangular cavity of volume $L_X L_Y L_Z$ cubic meters, as illustrated in Fig. 2.4-1. Let us assume that all of the walls inside the cavity are *ideal rigid boundaries*. The solution of this problem can be used, for example, to represent the acoustic field inside of a closed room. Recall from Example 2.3-1 that the plane-wave velocity-potential reflection coefficient of an ideal rigid boundary is equal to one, and as a result, no time-average acoustic power can escape from inside the cavity. The solution of the Helmholtz equation is given by *standing waves* in the X, Y, and Z directions due to reflections, that is (see Section 2.1),

$$\varphi_f(x, y, z) = X(x)Y(y)Z(z),$$ (2.4-1)

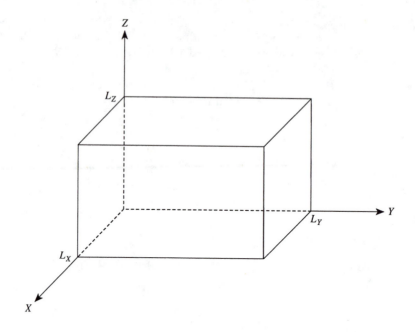

Figure 2.4-1. A rectangular cavity.

where

$$X(x) = A_X \exp(-jk_X x) + B_X \exp(+jk_X x), \qquad (2.4\text{-}2)$$

$$Y(y) = A_Y \exp(-jk_Y y) + B_Y \exp(+jk_Y y), \qquad (2.4\text{-}3)$$

$$Z(z) = A_Z \exp(-jk_Z z) + B_Z \exp(+jk_Z z), \qquad (2.4\text{-}4)$$

and

$$k_X^2 + k_Y^2 + k_Z^2 = k^2. \qquad (2.4\text{-}5)$$

Also recall from Example 2.3-1 that the normal component of the acoustic fluid velocity vector must be equal to zero at all points on an ideal rigid boundary. This problem is an example of a *Neumann boundary-value problem*. The values of the constants appearing in Eqs. (2.4-2) through (2.4-4) are obtained by satisfying this boundary condition.

First, let us obtain a general expression for the normal component of the acoustic fluid velocity vector. Since we are trying to solve the Helmholtz equation, the acoustic fields are assumed to be time harmonic, that is,

$$\varphi(t, \mathbf{r}) = \varphi_f(\mathbf{r}) \exp(+j2\pi ft) \qquad (2.4\text{-}6)$$

and

$$\mathbf{u}(t,\mathbf{r}) = \mathbf{u}_f(\mathbf{r}) \exp(+j2\pi ft), \qquad (2.4\text{-}7)$$

where

$$\mathbf{u}(t,\mathbf{r}) = \nabla\varphi(t,\mathbf{r}) \qquad (2.4\text{-}8)$$

and

$$\nabla = \frac{\partial}{\partial x}\hat{x} + \frac{\partial}{\partial y}\hat{y} + \frac{\partial}{\partial z}\hat{z} \qquad (2.4\text{-}9)$$

is the gradient expressed in the rectangular coordinates (x, y, z). Substituting Eq. (2.4-6) into Eq. (2.4-8) yields

$$\mathbf{u}(t,\mathbf{r}) = \nabla\varphi_f(\mathbf{r}) \exp(+j2\pi ft). \qquad (2.4\text{-}10)$$

The normal component of the acoustic fluid velocity vector can be obtained from Eq. (2.4-10) as follows:

$$u_n(t,\mathbf{r}) = \hat{n} \cdot \mathbf{u}(t,\mathbf{r}) = u_{f,n}(\mathbf{r}) \exp(+j2\pi ft) \qquad (2.4\text{-}11)$$

where

$$u_{f,n}(\mathbf{r}) = \hat{n} \cdot \nabla\varphi_f(\mathbf{r}) \qquad (2.4\text{-}12)$$

and $\hat{n}$ is a unit vector normal (perpendicular) to the boundary. Upon substituting Eqs. (2.4-1) and (2.4-9) into Eq. (2.4-12), we obtain

$$u_{f,n}(x,y,z) = \hat{n} \cdot \left[Y(y)Z(z)\frac{d}{dx}X(x)\hat{x} + X(x)Z(z)\frac{d}{dy}Y(y)\hat{y} \right.$$
$$\left. + X(x)Y(y)\frac{d}{dz}Z(z)\hat{z} \right], \qquad (2.4\text{-}13)$$

where the partial derivatives with respect to x, y, and z have been replaced by ordinary derivatives, since the functions $X(x)$, $Y(y)$, and $Z(z)$ depend on only one of the independent variables x, y, and z, respectively. In order to maintain consistency in the analysis that follows, we shall choose $\hat{n}$ such that it points in the outward direction, away from the inside of the cavity. However, it would be equally valid to decide, for example, to choose $\hat{n}$ such that it always points in the positive X, Y, or Z direction, whichever is appropriate for the boundary under consideration.

The first boundary condition is at $x = 0$, which corresponds to the YZ plane (see Fig. 2.4-1). The unit vector that is normal to the boundary at $x = 0$ and points in the outward direction away from the inside of the cavity is

$$\hat{n} = -\hat{x}.$$ (2.4-14)

With the use of Eq. (2.4-14), and since the normal component of the acoustic fluid velocity vector must be equal to zero on the rigid boundary at $x = 0$, Eq. (2.4-13) reduces to

$$u_{f,n}(0, y, z) = -Y(y)Z(z)\frac{d}{dx}X(x)\Big|_{x=0} = 0.$$ (2.4-15)

Substituting Eq. (2.4-2) into Eq. (2.4-15) yields

$$jk_X Y(y)Z(z)(A_X - B_X) = 0,$$ (2.4-16)

and since Eq. (2.4-16) must hold for all allowed values of y and z at $x = 0$,

$$\boxed{B_X = A_X.}$$ (2.4-17)

Since the surface at $x = 0$ is an ideal rigid boundary, Eq. (2.4-17) correctly predicts that the amplitudes A_X and B_X of the waves traveling in the positive and negative X directions, respectively, must be equal.

The second boundary condition is at $x = L_X$, which corresponds to a plane parallel to the YZ plane (see Fig. 2.4-1). The unit vector that is normal to the boundary at $x = L_X$ and points in the outward direction away from the inside of the cavity is

$$\hat{n} = \hat{x}.$$ (2.4-18)

With the use of Eq. (2.4-18), and since the normal component of the acoustic fluid velocity vector must be equal to zero on the rigid boundary at $x = L_X$, Eq. (2.4-13) reduces to

$$u_{f,n}(L_X, y, z) = Y(y)Z(z)\frac{d}{dx}X(x)\Big|_{x=L_X} = 0.$$ (2.4-19)

Substituting Eqs. (2.4-2) and (2.4-17) into Eq. (2.4-19) yields

$$2A_X k_X Y(y)Z(z)\sin(k_X L_X) = 0,$$ (2.4-20)

and since Eq. (2.4-20) must hold for all allowed values of y and z at $x = L_X$,

$$\sin(k_X L_X) = 0,$$ (2.4-21)

which implies that

$$k_X L_X = l\pi, \qquad l = 0, 1, 2, \ldots, \qquad (2.4\text{-}22)$$

or

$$\boxed{k_X = k_{X_l} = l\pi/L_X, \qquad l = 0, 1, 2, \ldots .} \qquad (2.4\text{-}23)$$

Therefore, in order to satisfy the boundary condition at $x = L_X$, the propagation-vector component in the X direction is only allowed certain *discrete values* rather than a continuum of values. Note that only positive integer values for l are used, in order to keep $k_X = k_{X_l}$ positive. Because of Eq. (2.4-2), a positive k_X value allows for wave propagation in both the positive and negative X directions.

The third boundary condition is at $y = 0$, which corresponds to the XZ plane (see Fig. 2.4-1). The unit vector that is normal to the boundary at $y = 0$ and points in the outward direction away from the inside of the cavity is

$$\hat{n} = -\hat{y}. \qquad (2.4\text{-}24)$$

With the use of Eq. (2.4-24), and since the normal component of the acoustic fluid velocity vector must be equal to zero on the rigid boundary at $y = 0$, Eq. (2.4-13) reduces to

$$u_{f,n}(x, 0, z) = -X(x)Z(z)\frac{d}{dy}Y(y)\bigg|_{y=0} = 0. \qquad (2.4\text{-}25)$$

Substituting Eq. (2.4-3) into Eq. (2.4-25) yields

$$jk_Y X(x)Z(z)(A_Y - B_Y) = 0, \qquad (2.4\text{-}26)$$

and since Eq. (2.4-26) must hold for all allowed values of x and z at $y = 0$,

$$\boxed{B_Y = A_Y.} \qquad (2.4\text{-}27)$$

Since the surface at $y = 0$ is an ideal rigid boundary, Eq. (2.4-27) correctly predicts that the amplitudes A_Y and B_Y of the waves traveling in the positive and negative Y directions, respectively, must be equal.

The fourth boundary condition is at $y = L_Y$, which corresponds to a plane parallel to the XZ plane (see Fig. 2.4-1). The unit vector that is normal to the boundary at $y = L_Y$ and points in the outward direction away from the inside of the cavity is

$$\hat{n} = \hat{y}. \qquad (2.4\text{-}28)$$

With the use of Eq. (2.4-28), and since the normal component of the acoustic fluid velocity vector must be equal to zero on the rigid boundary at $y = L_Y$, Eq. (2.4-13) reduces to

$$u_{f,n}(x, L_Y, z) = X(x)Z(z)\frac{d}{dy}Y(y)\Big|_{y=L_Y} = 0. \qquad (2.4\text{-}29)$$

Substituting Eqs. (2.4-3) and (2.4-27) into Eq. (2.4-29) yields

$$2A_Y k_Y X(x)Z(z)\sin(k_Y L_Y) = 0, \qquad (2.4\text{-}30)$$

and since Eq. (2.4-30) must hold for all allowed values of x and z at $y = L_Y$,

$$\sin(k_Y L_Y) = 0, \qquad (2.4\text{-}31)$$

which implies that

$$k_Y L_Y = m\pi, \qquad m = 0, 1, 2, \ldots, \qquad (2.4\text{-}32)$$

or

$$\boxed{k_Y = k_{Y_m} = m\pi/L_Y, \qquad m = 0, 1, 2, \ldots .} \qquad (2.4\text{-}33)$$

Therefore, in order to satisfy the boundary condition at $y = L_Y$, the propagation-vector component in the Y direction is only allowed certain discrete values rather than a continuum of values. Note that only positive integer values for m are used in order to keep $k_Y = k_{Y_m}$ positive. Because of Eq. (2.4-3), a positive k_Y value allows for wave propagation in both the positive and negative Y directions.

The fifth boundary condition is at $z = 0$, which corresponds to the XY plane (see Fig. 2.4-1). The unit vector that is normal to the boundary at $z = 0$ and points in the outward direction away from the inside of the cavity is

$$\hat{n} = -\hat{z}. \qquad (2.4\text{-}34)$$

With the use of Eq. (2.4-34), and since the normal component of the acoustic fluid velocity vector must be equal to zero on the rigid boundary at $z = 0$, Eq. (2.4-13) reduces to

$$u_{f,n}(x, y, 0) = -X(x)Y(y)\frac{d}{dz}Z(z)\Big|_{z=0} = 0. \qquad (2.4\text{-}35)$$

Substituting Eq. (2.4-4) into Eq. (2.4-35) yields

$$jk_z X(x)Y(y)(A_Z - B_Z) = 0, \qquad (2.4\text{-}36)$$

and since Eq. (2.4-36) must hold for all allowed values of x and y at $z = 0$,

$$\boxed{B_Z = A_Z.}$$

(2.4-37)

Since the surface at $z = 0$ is an ideal rigid boundary, Eq. (2.4-37) correctly predicts that the amplitudes A_Z and B_Z of the waves traveling in the positive and negative Z directions, respectively, must be equal.

The sixth boundary condition is at $z = L_Z$, which corresponds to a plane parallel to the XY plane (see Fig. 2.4-1). The unit vector that is normal to the boundary at $z = L_Z$ and points in the outward direction away from the inside of the cavity is

$$\hat{n} = \hat{z}.$$

(2.4-38)

With the use of Eq. (2.4-38), and since the normal component of the acoustic fluid velocity vector must be equal to zero on the rigid boundary at $z = L_Z$, Eq. (2.4-13) reduces to

$$u_{f,n}(x, y, L_Z) = X(x)Y(y)\frac{d}{dz}Z(z)\Big|_{z=L_Z} = 0.$$

(2.4-39)

Substituting Eqs. (2.4-4) and (2.4-37) into Eq. (2.4-39) yields

$$2A_Z k_Z X(x)Y(y)\sin(k_Z L_Z) = 0,$$

(2.4-40)

and since Eq. (2.4-40) must hold for all allowed values of x and y at $z = L_Z$,

$$\sin(k_Z L_Z) = 0,$$

(2.4-41)

which implies that

$$k_Z L_Z = n\pi, \qquad n = 0, 1, 2, \ldots,$$

(2.4-42)

or

$$\boxed{k_Z = k_{Z_n} = n\pi/L_Z, \qquad n = 0, 1, 2, \ldots.}$$

(2.4-43)

Therefore, in order to satisfy the boundary condition at $z = L_Z$, the propagation-vector component in the Z direction is only allowed certain discrete values rather than a continuum of values. Note that only positive integer values for n are used in order to keep $k_Z = k_{Z_n}$ positive. Because of Eq. (2.4-4), a positive k_Z value allows for wave propagation in both the positive and negative Z directions.

Now that we have satisfied all the boundary conditions at all the surfaces inside the cavity, we are in a position to obtain expressions for the velocity potential and the acoustic pressure inside the rectangular cavity. Substituting Eqs. (2.4-2) through (2.4-4), (2.4-17), (2.4-27), and (2.4-37) into Eq. (2.4-1) yields

$$\varphi_f(x, y, z) = A \cos(k_X x) \cos(k_Y y) \cos(k_Z z), \qquad (2.4\text{-}44)$$

where

$$A = 8 A_X A_Y A_Z. \qquad (2.4\text{-}45)$$

And upon substituting Eqs. (2.4-23), (2.4-33), and (2.4-43) into Eq. (2.4-44), we obtain

$$\varphi_{f,lmn}(x, y, z) = A_{lmn} \cos\left(\frac{l\pi}{L_X}x\right) \cos\left(\frac{m\pi}{L_Y}y\right) \cos\left(\frac{n\pi}{L_Z}z\right), \qquad l, m, n = 0, 1, 2, \ldots,$$

$$(2.4\text{-}46)$$

where the subscript *lmn* has been added to the velocity potential and the amplitude term because their values depend on the discrete values of the propaga-tion-vector components. Equation (2.4-46) is the spatial-dependent part of the velocity potential of the acoustic field inside a rectangular cavity with rigid walls corresponding to the (l, m, n) *eigenfunction* or *normal mode* (natural mode of vibration).

Since the propagation-vector components are only allowed certain discrete values, we shall show next that the frequencies of vibration of the acoustic field inside the cavity are also only allowed certain discrete values. Recalling that $k = 2\pi f/c$ and solving for the frequency f in Eq. (2.4-5) yields

$$f = \frac{c}{2\pi}\sqrt{k_X^2 + k_Y^2 + k_Z^2}. \qquad (2.4\text{-}47)$$

And upon substituting Eqs. (2.4-23), (2.4-33), and (2.4-43) into Eq. (2.4-47), we obtain

$$f = f_{lmn} = \frac{c}{2}\sqrt{\left(\frac{l}{L_X}\right)^2 + \left(\frac{m}{L_Y}\right)^2 + \left(\frac{n}{L_Z}\right)^2}, \qquad l, m, n = 0, 1, 2, \ldots.$$

$$(2.4\text{-}48)$$

Equation (2.4-48) is the (l, m, n) *eigenfrequency* or natural frequency of vibration (in hertz) of the acoustic field inside a rectangular cavity with rigid walls.

Therefore, with the use of Eq. (2.4-6), the time-harmonic velocity potential of the acoustic field inside a rectangular cavity with rigid walls corresponding to the

(l, m, n) normal mode is given by

$$\varphi_{lmn}(t, \mathbf{r}) = \varphi_{f, lmn}(\mathbf{r}) \exp(+j2\pi f_{lmn}t), \qquad l, m, n = 0, 1, 2, \ldots . \quad (2.4\text{-}49)$$

The complete time-harmonic normal-mode solution for the velocity potential is obtained by summing the contributions from *all* the normal modes, that is,

$$\varphi(t, \mathbf{r}) = \sum_{l=0}^{\infty} \sum_{m=0}^{\infty} \sum_{n=0}^{\infty} \varphi_{f, lmn}(\mathbf{r}) \exp(+j2\pi f_{lmn}t), \qquad (2.4\text{-}50)$$

and upon substituting Eq. (2.4-46) into Eq. (2.4-50), we obtain

$$\boxed{\begin{aligned} \varphi(t, x, y, z) = &\sum_{l=0}^{\infty} \sum_{m=0}^{\infty} \sum_{n=0}^{\infty} A_{lmn} \\ &\times \cos\left(\frac{l\pi}{L_X}x\right) \cos\left(\frac{m\pi}{L_Y}y\right) \cos\left(\frac{n\pi}{L_Z}z\right) \exp(+j2\pi f_{lmn}t), \end{aligned}}$$

$$(2.4\text{-}51)$$

where f_{lmn} is given by Eq. (2.4-48).

With the use of Eq. (2.2-28), the time-harmonic acoustic pressure inside a rectangular cavity with rigid walls corresponding to the (l, m, n) normal mode can be expressed as

$$p_{lmn}(t, \mathbf{r}) = -j2\pi f_{lmn}\rho_0 \varphi_{lmn}(t, \mathbf{r}), \qquad l, m, n = 0, 1, 2, \ldots . \quad (2.4\text{-}52)$$

Therefore, the complete time-harmonic normal-mode solution for the acoustic pressure is given by

$$p(t, \mathbf{r}) = -j2\pi\rho_0 \sum_{l=0}^{\infty} \sum_{m=0}^{\infty} \sum_{n=0}^{\infty} f_{lmn}\varphi_{f, lmn}(\mathbf{r}) \exp(+j2\pi f_{lmn}t), \quad (2.4\text{-}53)$$

and upon substituting Eq. (2.4-46) into Eq. (2.4-53), we obtain

$$\boxed{\begin{aligned} p(t, x, y, z) = &-j2\pi\rho_0 \sum_{l=0}^{\infty} \sum_{m=0}^{\infty} \sum_{n=0}^{\infty} A_{lmn} f_{lmn} \\ &\times \cos\left(\frac{l\pi}{L_X}x\right) \cos\left(\frac{m\pi}{L_Y}y\right) \cos\left(\frac{n\pi}{L_Z}z\right) \exp(+j2\pi f_{lmn}t), \end{aligned}}$$

$$(2.4\text{-}54)$$

where f_{lmn} is given by Eq. (2.4-48).

Although we have satisfied all the boundary conditions at all the surfaces inside the cavity, the amplitude term A_{lmn} is still unknown. In order to determine its value, the boundary condition at the location of a sound source inside the cavity must be satisfied. Although we shall not demonstrate how to satisfy the boundary condition at a source in this section, later, in Section 2.5, we will.

If *none* of the integers (l, m, n) are zero, then the propagation vector has three nonzero components and the mode is termed *oblique*. If *one* of the integers (l, m, n) is zero, then the propagation vector has two nonzero components. As a result, the propagation vector lies in a plane that is parallel to one of the three planes in the rectangular coordinate system, and the mode is termed *tangential*. And finally, if *two* of the integers (l, m, n) are zero, then the propagation vector has one nonzero component. As a result, the propagation vector is parallel to one of the three axes in the rectangular coordinate system, and the mode is termed *axial*.

2.5 Waveguide with Rectangular Cross-Sectional Area

In this section we shall derive expressions for the velocity potential and the acoustic pressure inside a waveguide with rectangular cross-sectional area $L_X L_Y$ square meters, as illustrated in Fig. 2.5-1. Let us assume that all of the walls inside the waveguide are *ideal rigid boundaries* with the exception of a vibrating surface (i.e., a sound source) located at $z = 0$. Also assume that the fluid media inside and outside the waveguide are the same, so that there is no impedance mismatch (i.e., no reflections) at the opened end of the waveguide.

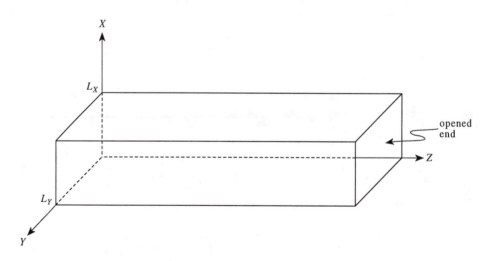

Figure 2.5-1. Waveguide with rectangular cross-sectional area.

The waveguide illustrated in Fig. 2.5-1 is nothing more than a rectangular cavity with one wall removed. Therefore, there will be *standing waves* in the X and Y directions and a *traveling wave* in the positive Z direction. Based on our discussion of the rectangular cavity with ideal rigid boundaries in Section 2.4, we can write immediately that the velocity potential of the time-harmonic eigenfunctions (normal modes) inside the rigid-walled waveguide is given by

$$\varphi_{lm}(t, \mathbf{r}) = \varphi_{f,lm}(\mathbf{r}) \exp(+j2\pi ft), \qquad l, m = 0, 1, 2, \ldots, \qquad (2.5\text{-}1)$$

where

$$\varphi_{f,lm}(x, y, z) = A_{lm} \cos(k_{X_l} x) \cos(k_{Y_m} y) \exp(-jk_Z z), \qquad l, m = 0, 1, 2, \ldots, \qquad (2.5\text{-}2)$$

or

$$\varphi_{f,lm}(x, y, z) = A_{lm} \cos\left(\frac{l\pi}{L_X} x\right) \cos\left(\frac{m\pi}{L_Y} y\right) \exp(-jk_Z z), \qquad l, m = 0, 1, 2, \ldots, \qquad (2.5\text{-}3)$$

since

$$k_X = k_{X_l} = l\pi/L_X, \qquad l = 0, 1, 2, \ldots, \qquad (2.5\text{-}4)$$

and

$$k_Y = k_{Y_m} = m\pi/L_Y, \qquad m = 0, 1, 2, \ldots, \qquad (2.5\text{-}5)$$

for ideal rigid boundaries in the X and Y directions.

Since

$$k_X^2 + k_Y^2 + k_Z^2 = k^2, \qquad (2.5\text{-}6)$$

we have

$$k_Z = \sqrt{k^2 - (k_X^2 + k_Y^2)}, \qquad (2.5\text{-}7)$$

and upon substituting Eqs. (2.5-4) and (2.5-5) into Eq. (2.5-7), we obtain

$$k_Z = k_{Z_{lm}} = k\sqrt{1 - (f_{lm}/f)^2}, \qquad f \geq f_{lm} \qquad (2.5\text{-}8)$$

where

$$k = 2\pi f/c = 2\pi/\lambda \qquad (2.5\text{-}9)$$

is the wave number and

$$f_{lm} = \frac{c}{2\pi}\sqrt{k_{X_l}^2 + k_{Y_m}^2} = \frac{c}{2}\sqrt{\left(\frac{l}{L_X}\right)^2 + \left(\frac{m}{L_Y}\right)^2}, \qquad l, m = 0, 1, 2, \ldots,$$

(2.5-10)

is the *cutoff frequency* in hertz for mode (l, m). By referring to Eq. (2.5-8), it can be seen that if $f = f_{lm}$, then $k_{Z_{lm}} = 0$, and as a result, there is no wave propagation in the positive Z direction. Hence the term "cutoff frequency." On the other hand, if $f > f_{lm}$, then $k_{Z_{lm}}$ is real and positive, and mode (l, m) is called a *propagating mode*. However, if $f < f_{lm}$, then $k_{Z_{lm}}$ is imaginary, and mode (l, m) is called an *evanescent mode*.

An evanescent mode is a decaying exponential, as we shall demonstrate next. If $f < f_{lm}$, then Eq. (2.5-8) can be rewritten as

$$k_Z = k_{Z_{lm}} = k\sqrt{-\left[(f_{lm}/f)^2 - 1\right]}, \qquad f < f_{lm},$$

(2.5-11)

or

$$k_Z = k_{Z_{lm}} = \pm jk\sqrt{(f_{lm}/f)^2 - 1}, \qquad f < f_{lm}.$$

(2.5-12)

The question now is which sign do we choose in Eq. (2.5-12). If we choose the minus sign and then substitute Eq. (2.5-12) into the complex exponential $\exp(-jk_Z z)$, we obtain

$$\exp(-jk_Z z) = \exp\left\{\left[-k\sqrt{(f_{lm}/f)^2 - 1}\right]z\right\}, \qquad f < f_{lm},$$

(2.5-13)

which approaches zero as z approaches positive infinity. If we had chosen the plus sign in Eq. (2.5-12), then Eq. (2.5-13) would have been a growing exponential (which does not make physical sense) instead of a decaying exponential. Therefore,

$$k_Z = k_{Z_{lm}} = -jk\sqrt{(f_{lm}/f)^2 - 1}, \qquad f < f_{lm}.$$

(2.5-14)

With the use of Eqs. (2.5-3), (2.5-8), and (2.5-14), the spatial-dependent part of the velocity potential inside a rigid-walled waveguide with rectangular cross-sectional area corresponding to the (l, m) normal mode is given by

$$\varphi_{f, lm}(x, y, z) = A_{lm}\cos\left(\frac{l\pi}{L_X}x\right)\cos\left(\frac{m\pi}{L_Y}y\right)\exp(-jk_{Z_{lm}}z), \quad l, m = 0, 1, 2, \ldots,$$

(2.5-15)

where

$$
k_{Z_{lm}} = \begin{cases} k\sqrt{1 - (f_{lm}/f)^2}, & f \geq f_{lm}, \\ -jk\sqrt{(f_{lm}/f)^2 - 1}, & f < f_{lm}, \end{cases} \tag{2.5-16}
$$

is the propagation-vector component in the Z direction for mode (l, m), k is the wave number given by Eq. (2.5-9), and f_{lm} is the cutoff frequency for mode (l, m) given by Eq. (2.5-10). The complete time-harmonic normal-mode solution for the velocity potential is given by

$$
\varphi(t, \mathbf{r}) = \sum_{l=0}^{\infty} \sum_{m=0}^{\infty} \varphi_{f,lm}(\mathbf{r}) \exp(+j2\pi ft), \tag{2.5-17}
$$

and upon substituting Eq. (2.5-15) into Eq. (2.5-17), we obtain

$$
\varphi(t, x, y, z) = \sum_{l=0}^{\infty} \sum_{m=0}^{\infty} A_{lm} \cos\left(\frac{l\pi}{L_X}x\right) \cos\left(\frac{m\pi}{L_Y}y\right) \exp(-jk_{Z_{lm}}z) \exp(+j2\pi ft), \tag{2.5-18}
$$

where $k_{Z_{lm}}$ is given by Eq. (2.5-16).

With the use of Eq. (2.2-28), the time-harmonic acoustic pressure inside a rigid-walled waveguide with rectangular cross-sectional area corresponding to the (l, m) normal mode can be expressed as

$$
p_{lm}(t, \mathbf{r}) = -j2\pi f\rho_0\varphi_{lm}(t, \mathbf{r}), \qquad l, m = 0, 1, 2, \ldots . \tag{2.5-19}
$$

Therefore, the complete time-harmonic normal-mode solution for the acoustic pressure is given by

$$
p(t, \mathbf{r}) = -j2\pi f\rho_0 \sum_{l=0}^{\infty} \sum_{m=0}^{\infty} \varphi_{f,lm}(\mathbf{r}) \exp(+j2\pi ft), \tag{2.5-20}
$$

and upon substituting Eq. (2.5-15) into Eq. (2.5-20), we obtain

$$
\begin{aligned}
p(t, x, y, z) = &-j2\pi f\rho_0 \sum_{l=0}^{\infty} \sum_{m=0}^{\infty} A_{lm} \cos\left(\frac{l\pi}{L_X}x\right) \cos\left(\frac{m\pi}{L_Y}y\right) \exp(-jk_{Z_{lm}}z) \\
&\times \exp(+j2\pi ft)
\end{aligned} \tag{2.5-21}
$$

where $k_{Z_{lm}}$ is given by Eq. (2.5-16).

Note that the upper limits on the summations in Eqs. (2.5-18) and (2.5-21) are infinity. This is a consequence of the fact that from a theoretical point of view, the complete normal-mode solution is obtained by summing the contributions from all the propagating and evanescent modes. However, since the evanescent modes are decaying exponentials, their contributions to the total acoustic field are only significant at short ranges from the source. Therefore, from a practical point of view, unless we are interested in the acoustic field near the sound source, we only need to sum the contributions from all the propagating modes. The finite number of propagating modes depends on the physical dimensions of the waveguide (i.e., L_X and L_Y) and the frequency components of the source. Also note that the amplitude term A_{lm} is still unknown. In order to determine its value, the boundary condition at the location of the sound source inside the waveguide must be satisfied. We shall satisfy the boundary condition at the source later in this section.

Group Speed, Phase Speed, and Angles of Propagation

The next topics of discussion are the *group speed* and the *phase speed* in the Z direction and the *angles of propagation* for the (l, m) propagating mode. Substituting Eq. (2.5-8) into Eqs. (2.2-49) and (2.2-54) yields

$$c_{g_{Z_{lm}}} = c\sqrt{1 - (f_{lm}/f)^2}, \qquad f \geq f_{lm}, \tag{2.5-22}$$

$$c_{p_{Z_{lm}}} = \frac{c}{\sqrt{1 - (f_{lm}/f)^2}}, \qquad f \geq f_{lm}, \tag{2.5-23}$$

respectively, where the cutoff frequency f_{lm} for mode (l, m) is given by Eq. (2.5-10). Note that both the group speed and the phase speed are functions of frequency and mode number. Recall from Section 2.2.2 that energy propagates at the group speed. Since the group speed given by Eq. (2.5-22) is a function of frequency and mode number, for a given mode (l, m) with cutoff frequency f_{lm} the energy associated with the high-frequency components from a transmitted pulse will arrive at a receiver before the energy associated with the low-frequency components. As a result, the shape of the transmitted pulse will be *distorted* by the time it reaches the receiver, due to the dispersion of the transmitted signal's energy. And since the phase speed given by Eq. (2.5-23) is a function of frequency as a result of satisfying the boundary conditions, this is known as *geometrical dispersion*, as opposed to *material* or *intrinsic dispersion* which was discussed in Section 1.4.

Next, let us compute the angles of propagation (θ_{lm}, ψ_{lm}) for the (l, m) propagating mode. Since

$$k_Z = kw = k \cos \theta, \tag{2.5-24}$$

substituting Eq. (2.5-8) into Eq. (2.5-24) yields

$$\theta_{lm} = \cos^{-1}\left(\sqrt{1 - (f_{lm}/f)^2}\right), \qquad f \geq f_{lm}, \qquad (2.5\text{-}25)$$

where θ_{lm} is the angle that the propagation vector

$$\mathbf{k}_{lm} = k_{X_l}\hat{x} + k_{Y_m}\hat{y} + k_{Z_{lm}}\hat{z} \qquad (2.5\text{-}26)$$

makes with the positive Z axis for the (l, m) propagating mode at frequency f.

In order to compute ψ_{lm}, we proceed as follows. First, note that

$$\sin^2 \theta_{lm} = 1 - \cos^2 \theta_{lm}, \qquad (2.5\text{-}27)$$

and upon substituting Eq. (2.5-25) into Eq. (2.5-27), we obtain

$$\sin \theta_{lm} = f_{lm}/f, \qquad f \geq f_{lm}, \qquad (2.5\text{-}28)$$

or

$$\theta_{lm} = \sin^{-1}(f_{lm}/f), \qquad f \geq f_{lm}. \qquad (2.5\text{-}29)$$

Next, since

$$k_X = ku = k \sin \theta \cos \psi, \qquad (2.5\text{-}30)$$

substituting Eqs. (2.5-4) and (2.5-28) into Eq. (2.5-30) yields

$$\psi_{lm} = \cos^{-1}\left(\frac{lc}{2L_X f_{lm}}\right), \qquad f \geq f_{lm}. \qquad (2.5\text{-}31)$$

Similarly, since

$$k_Y = kv = k \sin \theta \sin \psi, \qquad (2.5\text{-}32)$$

substituting Eqs. (2.5-5) and (2.5-28) into Eq. (2.5-32) yields

$$\psi_{lm} = \sin^{-1}\left(\frac{mc}{2L_Y f_{lm}}\right), \qquad f \geq f_{lm}. \qquad (2.5\text{-}33)$$

Note that if $l = m$, then it can be shown that

$$f_{mm} = \frac{mc}{2 L_X L_Y} \sqrt{L_X^2 + L_Y^2} \qquad (2.5\text{-}34)$$

and

$$\psi_{mm} = \cos^{-1}\left(\frac{L_Y}{\sqrt{L_X^2 + L_Y^2}} \right), \qquad f \geq f_{mm}, \qquad (2.5\text{-}35)$$

or

$$\psi_{mm} = \sin^{-1}\left(\frac{L_X}{\sqrt{L_X^2 + L_Y^2}} \right), \qquad f \geq f_{mm}. \qquad (2.5\text{-}36)$$

The lowest propagating mode for a rigid-walled waveguide with rectangular cross-sectional area is the $(0,0)$ mode. With $l = 0$ and $m = 0$, we have $k_{X_0} = 0$, $k_{Y_0} = 0$, $k_{Z_{00}} = k$, $f_{00} = 0$, $c_{g_{Z_{00}}} = c$, $c_{p_{Z_{00}}} = c$, $\theta_{00} = 0°$; and ψ_{00} is given by either Eq. (2.5-35) or Eq. (2.5-36). Actually, whenever $\theta_{lm} = 0°$, the value of ψ_{lm} is irrelevant. As can be seen from this set of values, the $(0,0)$ mode is a plane wave traveling in the positive Z direction, *parallel* to the Z axis. Therefore, the $(0,0)$ mode is referred to as the *longitudinal mode* or as the *plane-wave mode*.

Whenever $f = f_{lm}$, then according to Eqs. (2.5-22) and (2.5-23), the group speed and the phase speed in the Z direction are equal to zero and infinity, respectively, for mode (l, m). Also, according to either Eq. (2.5-25) or Eq. (2.5-29), $\theta_{lm} = 90°$. As a result, there is no propagation of energy in the positive Z direction, since the propagation vector for mode (l, m) is perpendicular to the Z axis.

Whenever $f \gg f_{lm}$, then according to Eqs. (2.5-22) and (2.5-23), both the group speed and the phase speed in the Z direction for mode (l, m) approach the speed of sound c. Also, according to either Eq. (2.5-25) or Eq. (2.5-29), $\theta_{lm} \rightarrow 0°$. As a result, mode (l, m) approaches being a longitudinal mode, since the propagation vector approaches being parallel to the Z axis.

Boundary Condition at the Source

As was mentioned previously, the amplitude terms A_{lm}, $l, m = 0, 1, 2, \ldots$, associated with the different modes are still unknown. In order to determine their values, the boundary condition at the location of the sound source inside the waveguide must be satisfied, which we shall do next. Recall that we assumed that there was a vibrating surface (i.e., a sound source) located at $z = 0$. In practice, we may know either the acoustic pressure at the source (a Dirichlet boundary value problem), or the normal component of the acoustic fluid velocity vector at the source (a Neumann boundary value problem), or both. Let us assume, for example purposes, that we are *given* the time-harmonic acoustic pressure distribution of the source at $z = 0$, that is,

$$p_S(t, x, y, 0) = p_{f,S}(x, y) \exp(+j2\pi f_S t), \qquad (2.5\text{-}37)$$

where the spatial acoustic pressure distribution $p_{f,S}(x, y)$ and the source frequency f_S in hertz are known. Recall that the complete time-harmonic normal-mode solution for the acoustic pressure inside the waveguide is given by Eq. (2.5-21). Evaluating Eq. (2.5-21) at $z = 0$ yields

$$p(t, x, y, 0) = -j2\pi f \rho_0 \sum_{l=0}^{\infty} \sum_{m=0}^{\infty} A_{lm} \cos\left(\frac{l\pi}{L_X}x\right) \cos\left(\frac{m\pi}{L_Y}y\right) \exp(+j2\pi ft).$$

$$(2.5\text{-}38)$$

If the boundary condition at the source is to be satisfied, then

$$p(t, x, y, 0) = p_S(t, x, y, 0). \qquad (2.5\text{-}39)$$

Therefore, substituting Eqs. (2.5-37) and (2.5-38) into Eq. (2.5-39) yields

$$-j2\pi f \rho_0 \sum_{l=0}^{\infty} \sum_{m=0}^{\infty} A_{lm} \cos\left(\frac{l\pi}{L_X}x\right) \cos\left(\frac{m\pi}{L_Y}y\right) \exp(+j2\pi ft)$$

$$= p_{f,S}(x, y) \exp(+j2\pi f_S t). \qquad (2.5\text{-}40)$$

Since Eq. (2.5-40) must hold for all allowed values of time t, that is, since Eq. (2.5-40) must be independent of time, this implies that

$$f = f_S. \qquad (2.5\text{-}41)$$

With the use of Eq. (2.5-41), Eq. (2.5-40) reduces to the following time-independent boundary condition:

$$\sum_{l=0}^{\infty} \sum_{m=0}^{\infty} A_{lm} \cos\left(\frac{l\pi}{L_X}x\right) \cos\left(\frac{m\pi}{L_Y}y\right) = j\frac{1}{2\pi f_S \rho_0} p_{f,S}(x, y). \quad (2.5\text{-}42)$$

In order to proceed further, we take advantage of the following orthogonality relationship:

$$\boxed{\int_0^L \cos\left(\frac{l\pi}{L}\zeta\right) \cos\left(\frac{l'\pi}{L}\zeta\right) d\zeta = C\,\delta_{ll'},} \qquad (2.5\text{-}43)$$

where

$$C = \begin{cases} L, & l = l' = 0, \\ L/2, & l = l', \quad l' \neq 0, \end{cases} \qquad (2.5\text{-}44)$$

and

$$\delta_{ll'} = \begin{cases} 1, & l = l', \\ 0, & l \neq l', \end{cases} \qquad (2.5\text{-}45)$$

is the *Kronecker delta*. Therefore, multiplying both sides of Eq. (2.5-42) by

$$\cos\left(\frac{l'\pi}{L_X}x\right)\cos\left(\frac{m'\pi}{L_Y}y\right),$$

integrating both sides of the resulting expression over x and y from 0 to L_X and from 0 to L_Y, respectively, and using Eqs. (2.5-43) through (2.5-45) yields

$$A_{00} = j\frac{1}{2\pi f\rho_0}\frac{1}{L_X L_Y}\int_0^{L_X}\int_0^{L_Y} p_{f,S}(x,y)\,dx\,dy, \qquad (2.5\text{-}46)$$

which is the amplitude of the longitudinal or plane-wave mode, and

$$A_{lm} = j\frac{1}{2\pi f\rho_0}\frac{4}{L_X L_Y}\int_0^{L_X}\int_0^{L_Y} p_{f,S}(x,y)\cos\left(\frac{l\pi}{L_X}x\right)\cos\left(\frac{m\pi}{L_Y}y\right)dx\,dy,$$
$$l, m = 1, 2, 3, \ldots,$$
$$(2.5\text{-}47)$$

where f_S has been replaced by f [see Eq. (2.5-41)].

Example 2.5-1 (Uniform radiation)

In this example we shall consider the case when the spatial acoustic pressure distribution of the source is uniform over the entire surface at $z = 0$, that is,

$$p_{f,S}(x,y) = P_0, \qquad 0 \leq x \leq L_X, \quad 0 \leq y \leq L_Y. \qquad (2.5\text{-}48)$$

Substituting Eq. (2.5-48) into Eqs. (2.5-46) and (2.5-47) yields

$$A_{00} = j\frac{P_0}{2\pi f\rho_0} \qquad (2.5\text{-}49)$$

and

$$A_{lm} = 0, \qquad l, m = 1, 2, 3, \ldots, \qquad \text{(2.5-50)}$$

since

$$\int_0^{L_X} \cos\left(\frac{l\pi}{L_X}x\right) dx = 0, \qquad l = 1, 2, 3, \ldots, \qquad \text{(2.5-51)}$$

and

$$\int_0^{L_Y} \cos\left(\frac{m\pi}{L_Y}y\right) dy = 0, \qquad m = 1, 2, 3, \ldots. \qquad \text{(2.5-52)}$$

Therefore, when the spatial acoustic pressure distribution of the source is uniform over the entire surface at $z = 0$, only the longitudinal or plane-wave mode is excited regardless of the source frequency.

2.6 The Time-Independent Free-Space Green's Function

2.6.1 Solution of the Homogeneous Helmholtz Equation

The *time-independent free-space Green's function* $g_f(\mathbf{r}|\mathbf{r}_0)$ is a very important function that appears often in acoustic wave propagation problems involving ideal (nonviscous), homogeneous, fluid media. In Section 2.6.2 we shall show that $g_f(\mathbf{r}|\mathbf{r}_0)$ is the free-space spatial impulse response of an ideal, homogeneous, fluid medium. Also, in Section 4.2.2, we shall show that the time-harmonic acoustic field produced by a spherical sound source in the monopole mode of vibration is directly proportional to $g_f(\mathbf{r}|\mathbf{r}_0)$. The time-independent free-space Green's function is defined as follows:

$$g_f(\mathbf{r}|\mathbf{r}_0) \triangleq -\frac{\exp(-jk|\mathbf{r} - \mathbf{r}_0|)}{4\pi|\mathbf{r} - \mathbf{r}_0|}, \qquad \text{(2.6-1)}$$

where

$$k = 2\pi f/c = 2\pi/\lambda \qquad \text{(2.6-2)}$$

is the wave number,

$$\mathbf{r} = x\hat{x} + y\hat{y} + z\hat{z} \qquad \text{(2.6-3)}$$

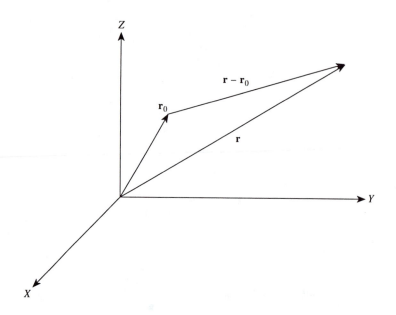

Figure 2.6-1. Position vectors $\mathbf{r}_0$ and $\mathbf{r}$ to source and field points, respectively.

is the position vector to a field point, and

$$\mathbf{r}_0 = x_0\hat{x} + y_0\hat{y} + z_0\hat{z} \tag{2.6-4}$$

is the position vector to a source point (see Fig. 2.6-1). An alternative expression for $g_f(\mathbf{r}|\mathbf{r}_0)$ is given by

$$g_f(\mathbf{r}|\mathbf{r}_0) = -\frac{\exp(-jkR)}{4\pi R}, \tag{2.6-5}$$

where

$$R = |\mathbf{r} - \mathbf{r}_0| = \sqrt{(x - x_0)^2 + (y - y_0)^2 + (z - z_0)^2} \tag{2.6-6}$$

is the range or distance between source and field points. Equation (2.6-6) can also be thought of as the radial distance measured from the origin of a spherical coordinate system centered at (x_0, y_0, z_0).

In this section we shall show that $g_f(\mathbf{r}|\mathbf{r}_0)$ is a solution of the following lossless, homogeneous Helmholtz equation:

$$\nabla^2 g_f(\mathbf{r}|\mathbf{r}_0) + k^2 g_f(\mathbf{r}|\mathbf{r}_0) = 0, \qquad \mathbf{r} \neq \mathbf{r}_0, \tag{2.6-7}$$

or

$$\nabla^2 g_f(\mathbf{r}|\mathbf{r}_0) = -k^2 g_f(\mathbf{r}|\mathbf{r}_0), \qquad \mathbf{r} \neq \mathbf{r}_0, \qquad (2.6\text{-}8)$$

where

$$\nabla^2 = \frac{\partial^2}{\partial x^2} + \frac{\partial^2}{\partial y^2} + \frac{\partial^2}{\partial z^2} \qquad (2.6\text{-}9)$$

is the Laplacian expressed in the rectangular coordinates (x, y, z). Note that as $\mathbf{r} \to \mathbf{r}_0$, $g_f(\mathbf{r}|\mathbf{r}_0) \to -\infty$.

In order to verify Eq. (2.6-8), we must compute the Laplacian of $g_f(\mathbf{r}|\mathbf{r}_0)$. We begin by noting that

$$\nabla^2 g_f(\mathbf{r}|\mathbf{r}_0) = \nabla \cdot \nabla g_f(\mathbf{r}|\mathbf{r}_0). \qquad (2.6\text{-}10)$$

If we use Eq. (2.6-5) to represent $g_f(\mathbf{r}|\mathbf{r}_0)$ instead of Eq. (2.6-1), then the gradient can be expressed in terms of the *spherical coordinate R* as follows (see Section 4.2.1):

$$\nabla = \frac{\partial}{\partial R}\hat{R} + \frac{1}{R}\frac{\partial}{\partial \theta}\hat{\theta} + \frac{1}{R \sin\theta}\frac{\partial}{\partial \psi}\hat{\psi} \qquad (2.6\text{-}11)$$

where $\hat{R}$ is the unit vector in the radial direction R. Therefore, taking the gradient of Eq. (2.6-5) using Eq. (2.6-11) yields

$$\nabla g_f(\mathbf{r}|\mathbf{r}_0) = \frac{\partial}{\partial R}g_f(\mathbf{r}|\mathbf{r}_0)\hat{R}, \qquad (2.6\text{-}12)$$

or

$$\nabla g_f(\mathbf{r}|\mathbf{r}_0) = -\left(\frac{1}{R} + jk\right)g_f(\mathbf{r}|\mathbf{r}_0)\hat{R}. \qquad (2.6\text{-}13)$$

Since

$$\nabla \cdot (a\mathbf{A}) = a\nabla \cdot \mathbf{A} + \mathbf{A} \cdot \nabla a, \qquad (2.6\text{-}14)$$

where a and $\mathbf{A}$ are arbitrary scalar and vector functions, respectively, substituting Eq. (2.6-13) into Eq. (2.6-10) yields

$$\nabla^2 g_f(\mathbf{r}|\mathbf{r}_0) = -\left(\frac{1}{R} + jk\right)\nabla \cdot g_f(\mathbf{r}|\mathbf{r}_0)\hat{R} - g_f(\mathbf{r}|\mathbf{r}_0)\hat{R} \cdot \nabla\left(\frac{1}{R} + jk\right). \qquad (2.6\text{-}15)$$

And since

$$\nabla \cdot g_f(\mathbf{r}|\mathbf{r}_0)\hat{R} = \frac{1}{R^2}\frac{\partial}{\partial R}\left[R^2 g_f(\mathbf{r}|\mathbf{r}_0)\right], \qquad (2.6\text{-}16)$$

substituting Eq. (2.6-16) into Eq. (2.6-15), and performing the indicated gradient operation using Eq. (2.6-11) yields

$$\nabla^2 g_f(\mathbf{r}|\mathbf{r}_0) = -\left(\frac{1}{R} + jk\right)\frac{1}{R^2}\frac{\partial}{\partial R}\left[R^2 g_f(\mathbf{r}|\mathbf{r}_0)\right] + \frac{1}{R^2}g_f(\mathbf{r}|\mathbf{r}_0), \quad (2.6\text{-}17)$$

and after further simplification,

$$\nabla^2 g_f(\mathbf{r}|\mathbf{r}_0) = -k^2 g_f(\mathbf{r}|\mathbf{r}_0), \qquad \mathbf{r} \neq \mathbf{r}_0. \qquad (2.6\text{-}18)$$

Note that Eq. (2.6-18) is identical to Eq. (2.6-8). Therefore, $g_f(\mathbf{r}|\mathbf{r}_0)$ given by Eq. (2.6-1) or, equivalently, by Eqs. (2.6-5) and (2.6-6), is a solution of the lossless, homogeneous Helmholtz equation given by Eq. (2.6-7).

2.6.2 Solution of the Inhomogeneous Helmholtz Equation

In this section we shall show that $g_f(\mathbf{r}|\mathbf{r}_0)$ is also a solution of the following lossless, inhomogeneous Helmholtz equation:

$$\boxed{\nabla^2 g_f(\mathbf{r}|\mathbf{r}_0) + k^2 g_f(\mathbf{r}|\mathbf{r}_0) = \delta(\mathbf{r} - \mathbf{r}_0)} \qquad (2.6\text{-}19)$$

where ∇^2 is the Laplacian in rectangular coordinates given by Eq. (2.6-9), and the impulse function $\delta(\mathbf{r} - \mathbf{r}_0)$ represents a unit-amplitude, omnidirectional point source at $\mathbf{r}_0$. Note that $\delta(\mathbf{r} - \mathbf{r}_0)$ only exists at $\mathbf{r} = \mathbf{r}_0$, that is, $\delta(\mathbf{r} - \mathbf{r}_0) = 0$ for $\mathbf{r} \neq \mathbf{r}_0$ [see Eq. (2.6-7)]. Substituting Eq. (2.6-10) into Eq. (2.6-19) and then integrating both sides of the resulting equation over the volume V of a sphere of radius R centered at $\mathbf{r}_0$ yields

$$\int_V \nabla \cdot \nabla g_f(\mathbf{r}|\mathbf{r}_0)\, dV + k^2 \int_V g_f(\mathbf{r}|\mathbf{r}_0)\, dV = 1, \qquad (2.6\text{-}20)$$

where

$$\int_V \delta(\mathbf{r} - \mathbf{r}_0)\, dV = 1, \qquad (2.6\text{-}21)$$

since the volume V includes the point source at $\mathbf{r}_0$. Therefore, the problem of verifying Eq. (2.6-19) has been transformed into the equivalent problem of verifying Eq. (2.6-20).

Using the divergence theorem, the first integral on the left-hand side of Eq. (2.6-20) can be expressed as

$$\int_V \nabla \cdot \nabla g_f(\mathbf{r}|\mathbf{r}_0)\, dV = \oint_S \nabla g_f(\mathbf{r}|\mathbf{r}_0) \cdot d\mathbf{S}, \qquad (2.6\text{-}22)$$

where S is the surface of a sphere of radius R in this development,

$$d\mathbf{S} = R^2 \sin\theta\, d\theta\, d\psi\, \hat{R}, \qquad (2.6\text{-}23)$$

and $\hat{R}$ is the unit vector in the outward, radial direction, normal to the surface of the sphere. Substituting Eqs. (2.6-13), (2.6-23), and (2.6-5) into Eq. (2.6-22) yields

$$\int_V \nabla \cdot \nabla g_f(\mathbf{r}|\mathbf{r}_0)\, dV = (1 + jkR)\frac{\exp(-jkR)}{4\pi} \int_0^{2\pi}\int_0^{\pi} \sin\theta\, d\theta\, d\psi \quad (2.6\text{-}24)$$

$$= (1 + jkR)\exp(-jkR), \qquad (2.6\text{-}25)$$

where R is given by Eq. (2.6-6).

With the use of Eq. (2.6-5), and since $dV = R^2 \sin\theta\, dR\, d\theta\, d\psi$ for a sphere of radius R, the second integral on the left-hand side of Eq. (2.6-20) reduces as follows:

$$\int_V g_f(\mathbf{r}|\mathbf{r}_0)\, dV = -\frac{1}{4\pi}\int_0^{2\pi}\int_0^{\pi}\int_0^R \exp(-jk\zeta)\,\zeta \sin\theta\, d\zeta\, d\theta\, d\psi \quad (2.6\text{-}26)$$

$$= -\int_0^R \exp(-jk\zeta)\,\zeta\, d\zeta \qquad (2.6\text{-}27)$$

$$= -(1 + jk\zeta)\frac{\exp(-jk\zeta)}{k^2}\Big|_0^R, \qquad (2.6\text{-}28)$$

and finally,

$$\int_V g_f(\mathbf{r}|\mathbf{r}_0)\, dV = -(1 + jkR)\frac{\exp(-jkR)}{k^2} + \frac{1}{k^2}. \qquad (2.6\text{-}29)$$

Substituting Eqs. (2.6-25) and (2.6-29) into Eq. (2.6-20) yields

$$(1 + jkR)\exp(-jkR) - (1 + jkR)\exp(-jkR) + 1 = 1, \qquad (2.6\text{-}30)$$

or $1 = 1$, which indicates that Eq. (2.6-20) has been verified. Therefore, $g_f(\mathbf{r}|\mathbf{r}_0)$ given by Eq. (2.6-1) or, equivalently, by Eqs. (2.6-5) and (2.6-6), is a solution of the lossless, inhomogeneous Helmholtz equation given by Eq. (2.6-19). Since $g_f(\mathbf{r}|\mathbf{r}_0)$

is the solution of the Helmholtz equation due to the application of a unit-amplitude impulse at spatial location $\mathbf{r}_0$, $g_f(\mathbf{r}|\mathbf{r}_0)$ is, by definition, the *time-independent free-space spatial impulse response of an ideal, homogeneous, fluid medium*. It is the response of the fluid medium (i.e., it is equal to the magnitude and phase of the spatial-dependent part of the velocity potential) at frequency f and spatial location $\mathbf{r}$ due to the application of a unit-amplitude impulse at $\mathbf{r}_0$. Note that the value of $g_f(\mathbf{r}|\mathbf{r}_0)$ does not change if the position vectors $\mathbf{r}$ and $\mathbf{r}_0$ are interchanged. That is, the response of the fluid medium is the same even if the source and field points are reversed. This is known as the *principle of reciprocity*. Also note that $g_f(\mathbf{r}|\mathbf{r}_0)$ is related to, but is not the same as, the *time-dependent* free-space impulse response, or Green's function, of an ideal, homogeneous, fluid medium (see Section 2.7.1). For the solution of the linear, three-dimensional, lossless, inhomogeneous wave equation with an *arbitrary* source distribution, see Section 2.7.

2.7 Solution of the Linear Three-Dimensional Inhomogeneous Wave Equation with Arbitrary Source Distribution

2.7.1 Free-Space Solution

The propagation of *small-amplitude* acoustic signals in ideal (nonviscous), homogeneous, fluid media can be described by the following *linear*, three-dimensional, lossless, inhomogeneous wave equation:

$$\nabla^2 \varphi(t, \mathbf{r}) - \frac{1}{c^2} \frac{\partial^2}{\partial t^2} \varphi(t, \mathbf{r}) = x_M(t, \mathbf{r}), \qquad (2.7\text{-}1)$$

where $\varphi(t, \mathbf{r})$ is the velocity potential at time t and position $\mathbf{r} = (x, y, z)$ with units of square meters per second; $x_M(t, \mathbf{r})$ is the input acoustic signal to the fluid medium (or source distribution) in inverse seconds, representing the volume flow rate per unit volume of fluid at time t and position $\mathbf{r}$; and c is the constant speed of sound in meters per second.

The solution of Eq. (2.7-1) can be obtained by treating the fluid medium as a *linear filter*. This treatment is valid because we are trying to solve the *linear wave equation*. Therefore, by using the principles of linear, time-variant, space-variant filter theory and time-domain and spatial-domain Fourier transforms, the free-space solution of the linear, three-dimensional, lossless, inhomogeneous wave equation will be obtained. The linear systems theory approach presented in this section demonstrates the consistency and relationships between linear systems theory and the physics of propagation of small-amplitude acoustic signals in fluid media. It provides an alternative to the traditional approach taken to solve this problem, which is discussed in Section 2.7.2. For a more detailed discussion on the relationships between linear systems theory and the physics of propagation of small-amplitude acoustic signals in fluid media, see Chapter 9.

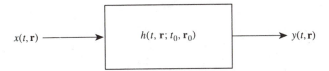

Figure 2.7-1. Illustration of a linear, time-variant, space-variant filter.

Figure 2.7-1 is an illustration of a linear, time-variant, space-variant filter with input-output relationship given by

$$y(t,\mathbf{r}) = \int_{-\infty}^{\infty} \int_{-\infty}^{\infty} x(t_0,\mathbf{r}_0) h(t,\mathbf{r};t_0,\mathbf{r}_0)\, dt_0\, d\mathbf{r}_0, \qquad (2.7\text{-}2)$$

where $h(t,\mathbf{r};t_0,\mathbf{r}_0)$ is the *impulse response* of the filter, that is, it is the response of the filter at time t and position $\mathbf{r} = (x, y, z)$ due to the application of a unit-amplitude impulse at time t_0 and position $\mathbf{r}_0 = (x_0, y_0, z_0)$, and $d\mathbf{r}_0 = dx_0\, dy_0\, dz_0$. Therefore, the right-hand side of Eq. (2.7-2) is shorthand notation for a fourfold integral. If the filter is *causal*, then $h(t,\mathbf{r};t_0,\mathbf{r}_0) = 0$ for $t < t_0$, that is, the filter cannot respond before the application of an input. The input-output relationship given by Eq. (2.7-2) is applicable to any linear, time-variant, space-variant filter, whether the linear filter is meant to model small-amplitude wave propagation in fluid media or any other physical system whose input and output are related by a linear partial differential equation. With this interpretation in mind, we identify the source distribution $x_M(t,\mathbf{r})$ as the *input signal* $x(t,\mathbf{r})$ to the filter and the velocity potential $\varphi(t,\mathbf{r})$ as the *output signal* $y(t,\mathbf{r})$ from the filter. Therefore, Eq. (2.7-2) can be rewritten as

$$\varphi(t,\mathbf{r}) = \int_{-\infty}^{\infty} \int_{-\infty}^{\infty} x_M(t_0,\mathbf{r}_0) h_M(t,\mathbf{r};t_0,\mathbf{r}_0)\, dt_0\, d\mathbf{r}_0, \qquad (2.7\text{-}3)$$

where $h_M(t,\mathbf{r};t_0,\mathbf{r}_0)$ is the *impulse response*, or *Green's function*, of the fluid medium. It is important to note that even if the speed of sound in the linear wave equation given by Eq. (2.7-1) is a function of position, that is, even if $c = c(\mathbf{r})$, Eq. (2.7-3) is still a valid representation for the solution of the linear wave equation (see Chapter 9).

In order to solve Eq. (2.7-1) for an unbounded, homogeneous medium, we shall first obtain the impulse response of the fluid medium, and then substitute the result into Eq. (2.7-3). We begin by taking the Fourier transform of both sides of Eq. (2.7-1) with respect to t. Doing so yields

$$\nabla^2 \Phi(\eta,\mathbf{r}) + (2\pi\eta/c)^2 \Phi(\eta,\mathbf{r}) = X_M(\eta,\mathbf{r}) \qquad (2.7\text{-}4)$$

where

$$\Phi(\eta, \mathbf{r}) = F_t\{\varphi(t, \mathbf{r})\} = \int_{-\infty}^{\infty} \varphi(t, \mathbf{r}) \exp(-j2\pi\eta t)\, dt, \qquad (2.7\text{-}5)$$

$$X_M(\eta, \mathbf{r}) = F_t\{x_M(t, \mathbf{r})\} = \int_{-\infty}^{\infty} x_M(t, \mathbf{r}) \exp(-j2\pi\eta t)\, dt, \qquad (2.7\text{-}6)$$

and from Appendix 2B,

$$F_t\left\{\frac{\partial^2}{\partial t^2}\varphi(t, \mathbf{r})\right\} = -(2\pi\eta)^2 \Phi(\eta, \mathbf{r}), \qquad (2.7\text{-}7)$$

where η corresponds to output frequencies in hertz.

The next step is to take the spatial Fourier transform of both sides of Eq. (2.7-4) with respect to $\mathbf{r}$. Doing so yields

$$F_\mathbf{r}\{\nabla^2 \Phi(\eta, \mathbf{r})\} + (2\pi\eta/c)^2 \Phi(\eta, \boldsymbol{\beta}) = X_M(\eta, \boldsymbol{\beta}) \qquad (2.7\text{-}8)$$

where

$$\Phi(\eta, \boldsymbol{\beta}) = F_\mathbf{r}\{\Phi(\eta, \mathbf{r})\} \triangleq \int_{-\infty}^{\infty} \Phi(\eta, \mathbf{r}) \exp(+j2\pi\boldsymbol{\beta} \cdot \mathbf{r})\, d\mathbf{r}, \qquad (2.7\text{-}9)$$

$$X_M(\eta, \boldsymbol{\beta}) = F_\mathbf{r}\{X_M(\eta, \mathbf{r})\} \triangleq \int_{-\infty}^{\infty} X_M(\eta, \mathbf{r}) \exp(+j2\pi\boldsymbol{\beta} \cdot \mathbf{r})\, d\mathbf{r}, \quad (2.7\text{-}10)$$

$F_\mathbf{r}\{\cdot\} = F_x F_y F_z\{\cdot\}$, $\boldsymbol{\beta} = (\beta_X, \beta_Y, \beta_Z)$ is a three-dimensional vector whose components β_X, β_Y, and β_Z are the output spatial frequencies in cycles per meter in the X, Y, and Z directions, respectively, and $d\mathbf{r} = dx\, dy\, dz$. Therefore, the right-hand sides of Eqs. (2.7-9) and (2.7-10) are shorthand notation for *three-dimensional spatial Fourier transforms*. Also, since (see Appendix 2C)

$$F_\mathbf{r}\{\nabla^2 \Phi(\eta, \mathbf{r})\} = -\left[(2\pi\beta_X)^2 + (2\pi\beta_Y)^2 + (2\pi\beta_Z)^2\right]\Phi(\eta, \boldsymbol{\beta}), \quad (2.7\text{-}11)$$

substituting Eq. (2.7-11) into Eq. (2.7-8) yields

$$\Phi(\eta, \boldsymbol{\beta}) = H_M(\eta, \boldsymbol{\beta}) X_M(\eta, \boldsymbol{\beta}), \qquad (2.7\text{-}12)$$

where

$$H_M(\eta, \boldsymbol{\beta}) = \frac{1}{(2\pi\eta/c)^2 - \left[(2\pi\beta_X)^2 + (2\pi\beta_Y)^2 + (2\pi\beta_Z)^2\right]} \qquad (2.7\text{-}13)$$

is the *transfer function* of an ideal, unbounded, homogeneous, fluid medium,

$\Phi(\eta, \boldsymbol{\beta})$ is the *frequency and angular spectrum* of the velocity potential, and $X_M(\eta, \boldsymbol{\beta})$ is the *frequency and angular spectrum* of the source distribution. The word "angular" is used because the spatial frequencies in the X, Y, and Z directions are related to the direction cosines in the X, Y, and Z directions, respectively, which can be expressed in terms of the spherical angles θ and ψ (see Section 2.2.1). Also recall from Section 2.2.1 that propagation-vector components and spatial frequencies are related by a factor of 2π, and as a result, spatial frequencies describe the direction of wave propagation.

The form of Eq. (2.7-12), that is, the output frequency and angular spectrum being equal to the *product* of the transfer function and the input frequency and angular spectrum, indicates that the filter is both *time-invariant* and *space-invariant*. This agrees with the physics of the problem, since the sound source is not in motion, and no other motion is being considered (time-invariant problem). Also, since the speed of sound c was assumed to be constant, the medium is unbounded, and no discrete point scatterers are being considered in this analysis, there is no refraction, no reflections, and no scattering, and hence, no angular spread or change in the direction of wave propagation (space-invariant problem).

In order to proceed further with the solution of the wave equation, we set the source distribution $x_M(t, \mathbf{r})$ equal to a unit-amplitude impulse applied at time t_0 and position $\mathbf{r}_0 = (x_0, y_0, z_0)$, that is,

$$x_M(t, \mathbf{r}) = \delta(t - t_0, \mathbf{r} - \mathbf{r}_0) = \delta(t - t_0)\,\delta(\mathbf{r} - \mathbf{r}_0). \qquad (2.7\text{-}14)$$

Substituting Eq. (2.7-14) into Eq. (2.7-6) yields the frequency spectrum of the source distribution:

$$X_M(\eta, \mathbf{r}) = \exp(-j2\pi\eta t_0)\,\delta(\mathbf{r} - \mathbf{r}_0), \qquad (2.7\text{-}15)$$

and upon substituting Eq. (2.7-15) into Eq. (2.7-10), we obtain the frequency and angular spectrum of the source distribution:

$$X_M(\eta, \boldsymbol{\beta}) = \exp(-j2\pi\eta t_0)\,\exp(+j2\pi\boldsymbol{\beta} \cdot \mathbf{r}_0), \qquad (2.7\text{-}16)$$

which is a time-harmonic plane wave evaluated at time t_0 and position $\mathbf{r}_0$. Since there is no motion and no change in the direction of wave propagation, there is no Doppler shift, so the output frequencies η are identical to the input or transmitted frequencies f, and the output spatial frequency vector $\boldsymbol{\beta}$ is identical to the input spatial frequency vector $\boldsymbol{\nu} = (f_X, f_Y, f_Z)$, since sound rays travel in straight lines in an unbounded, homogeneous medium. Therefore, substituting Eqs. (2.7-13) and (2.7-16) into Eq. (2.7-12), and replacing η with f and $\boldsymbol{\beta}$ with $\boldsymbol{\nu}$, we obtain the frequency and angular spectrum of the velocity potential:

$$\Phi(f, \boldsymbol{\nu}) = \frac{\exp(-j2\pi f t_0)\,\exp(+j2\pi\boldsymbol{\nu} \cdot \mathbf{r}_0)}{(2\pi f/c)^2 - \left[(2\pi f_X)^2 + (2\pi f_Y)^2 + (2\pi f_Z)^2\right]}. \qquad (2.7\text{-}17)$$

Next, compute the frequency spectrum of the velocity potential

$$\Phi(f,\mathbf{r}) = F_\nu^{-1}\{\Phi(f,\boldsymbol{\nu})\} \triangleq \int_{-\infty}^{\infty} \Phi(f,\boldsymbol{\nu}) \exp(-j2\pi\boldsymbol{\nu}\cdot\mathbf{r})\, d\boldsymbol{\nu}, \quad (2.7\text{-}18)$$

where $F_\nu^{-1}\{\cdot\} = F_{f_X}^{-1}F_{f_Y}^{-1}F_{f_Z}^{-1}\{\cdot\}$ and $d\boldsymbol{\nu} = df_X\, df_Y\, df_Z$. The right-hand side of Eq. (2.7-18) is shorthand notation for an *inverse* three-dimensional spatial Fourier transform. Substituting Eq. (2.7-17) into Eq. (2.7-18) yields

$$\Phi(f,\mathbf{r}) = g_f(\mathbf{r}|\mathbf{r}_0) \exp(-j2\pi f t_0) \qquad (2.7\text{-}19)$$

since it is shown in Section 2.8 that the time-independent free-space Green's function can be expressed as

$$g_f(\mathbf{r}|\mathbf{r}_0) = \int_{-\infty}^{\infty} \frac{\exp[-j2\pi\boldsymbol{\nu}\cdot(\mathbf{r}-\mathbf{r}_0)]}{(2\pi f/c)^2 - \left[(2\pi f_X)^2 + (2\pi f_Y)^2 + (2\pi f_Z)^2\right]}\, d\boldsymbol{\nu} \quad (2.7\text{-}20)$$

or, from Section 2.6.1,

$$g_f(\mathbf{r}|\mathbf{r}_0) = -\frac{\exp(-jk|\mathbf{r}-\mathbf{r}_0|)}{4\pi|\mathbf{r}-\mathbf{r}_0|} \qquad (2.7\text{-}21)$$

where

$$k = 2\pi f/c = 2\pi/\lambda \qquad (2.7\text{-}22)$$

is the wave number.

Next, compute the velocity potential

$$\varphi(t,\mathbf{r}) = F_f^{-1}\{\Phi(f,\mathbf{r})\} = \int_{-\infty}^{\infty} \Phi(f,\mathbf{r})\exp(+j2\pi ft)\, df. \quad (2.7\text{-}23)$$

Substituting Eqs. (2.7-21) and (2.7-22) into Eq. (2.7-19) yields

$$\Phi(f,\mathbf{r}) = -\frac{1}{4\pi|\mathbf{r}-\mathbf{r}_0|}\exp\left[-j2\pi f\left(t_0 + \frac{|\mathbf{r}-\mathbf{r}_0|}{c}\right)\right], \qquad (2.7\text{-}24)$$

and upon substituting Eq. (2.7-24) into Eq. (2.7-23), we obtain

$$\varphi(t,\mathbf{r}) = -\frac{1}{4\pi|\mathbf{r}-\mathbf{r}_0|}\int_{-\infty}^{\infty} \exp\left\{+j2\pi f\left[t - \left(t_0 + \frac{|\mathbf{r}-\mathbf{r}_0|}{c}\right)\right]\right\} df. \quad (2.7\text{-}25)$$

And since

$$\delta(t - t') = F_f^{-1}\{\exp(-j2\pi ft')\} = \int_{-\infty}^{\infty} \exp[+j2\pi f(t - t')] \, df, \quad (2.7\text{-}26)$$

Eq. (2.7-25) reduces to

$$\varphi(t, \mathbf{r}) = -\frac{1}{4\pi|\mathbf{r} - \mathbf{r}_0|}\delta\left[t - \left(t_0 + \frac{|\mathbf{r} - \mathbf{r}_0|}{c}\right)\right]. \qquad (2.7\text{-}27)$$

Since the solution of the wave equation due to the application of a unit amplitude impulse at time t_0 and position $\mathbf{r}_0$ is given by Eq. (2.7-27), the *time-dependent free-space impulse response of an ideal, homogeneous, fluid medium* is, by definition, equal to Eq. (2.7-27), that is,

$$h_M(t, \mathbf{r}; t_0, \mathbf{r}_0) = -\frac{1}{4\pi|\mathbf{r} - \mathbf{r}_0|}\delta\left[t - \left(t_0 + \frac{|\mathbf{r} - \mathbf{r}_0|}{c}\right)\right]. \qquad (2.7\text{-}28)$$

By inspecting Eq. (2.7-28), it can be seen that the impulse response is a function of the *time difference* $t - t_0$ and the *vector spatial difference* $\mathbf{r} - \mathbf{r}_0$, that is, $h_M(t, \mathbf{r}; t_0, \mathbf{r}_0) = h_M(t - t_0, \mathbf{r} - \mathbf{r}_0)$, which is the correct result for a linear, time-invariant (no motion), space-invariant (unbounded, homogeneous medium) filter as we are dealing with here. Also note that Eq. (2.7-28) satisfies the *principle of reciprocity*, that is, the acoustic field measured at $\mathbf{r}$ due to a source located at $\mathbf{r}_0$ is equal to the acoustic field that would be measured at $\mathbf{r}_0$ if the source were located at $\mathbf{r}$. Furthermore, the *time-dependent* free-space impulse response or Green's function given by Eq. (2.7-28) is related to the *time-independent* free-space Green's function given by Eq. (2.7-21) as follows:

$$\begin{aligned} g_f(\mathbf{r}|\mathbf{r}_0) &= H_M(t, \mathbf{r}; f, \mathbf{r}_0) \\ &\triangleq \int_{-\infty}^{\infty} h_M(t, \mathbf{r}; t_0, \mathbf{r}_0) \exp[-j2\pi f(t - t_0)] \, dt_0, \end{aligned} \qquad (2.7\text{-}29)$$

which can easily be verified by substituting Eq. (2.7-28) into Eq. (2.7-29) and then using the sifting property of impulse functions. Note that the right-hand side of Eq. (2.7-29) is *not* the Fourier transform of $h_M(t, \mathbf{r}; t_0, \mathbf{r}_0)$ with respect to t_0.

Finally, substituting Eq. (2.7-28) into Eq. (2.7-3) yields the desired result

$$\varphi(t,\mathbf{r}) = -\frac{1}{4\pi}\int_{V_0}\frac{x_M\big[t-(|\mathbf{r}-\mathbf{r}_0|/c),\mathbf{r}_0\big]}{|\mathbf{r}-\mathbf{r}_0|}\,dV_0, \qquad (2.7\text{-}30)$$

where $dV_0 = d\mathbf{r}_0 = dx_0\,dy_0\,dz_0$. Equation (2.7-30) is the solution of the linear, three-dimensional, lossless, inhomogeneous wave equation given by Eq. (2.7-1), where the speed of sound c is a constant (see Fig. 2.7-2). Equation (2.7-30) represents the solution of the wave equation as the linear superposition or "summing" of the contributions from all of the individual source points that make up the source distribution. Note that the amplitude decreases as the reciprocal of the distance between source and field points. The quantity $t - (|\mathbf{r} - \mathbf{r}_0|/c)$ is known as the *retarded time*. It is a measure of the amount of time that has elapsed since the signal transmitted by a source first appears at a receiver.

Equation (2.7-30) can also be used to solve problems involving source distributions that lie along the X, Y, or Z axes or in the XY, XZ, or YZ planes. In these cases, simply perform single or double integrations, respectively, instead of the triple integrations for volume source distributions. Also, with the appropriate change of variables, Eq. (2.7-30) can be expressed in terms of rectangular, cylindrical, or spherical coordinates. Equation (2.7-30) will be used in Section 6.2 for the

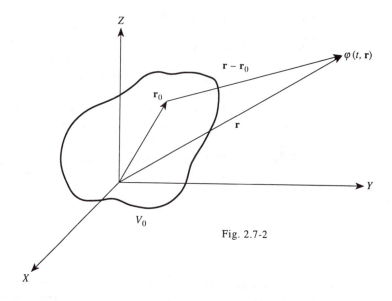

Fig. 2.7-2

Figure 2.7-2. Velocity potential $\varphi(t,\mathbf{r})$ due to a source distribution shown occupying a volume V_0.

derivation of the near-field and far-field directivity functions (beam patterns) of a volume aperture.

Example 2.7-1 (Omnidirectional point source)

In this example we shall evaluate Eq. (2.7-30) for the special case of an omnidirectional point source located at $\mathbf{r} = \mathbf{r}_S$ with arbitrary time dependence $a(t)$. The source distribution for this case is given by

$$x_M(t, \mathbf{r}) = a(t)\, \delta(\mathbf{r} - \mathbf{r}_S). \qquad (2.7\text{-}31)$$

Substituting Eq. (2.7-31) into Eq. (2.7-30) yields

$$\varphi(t, \mathbf{r}) = -\frac{1}{4\pi} \int_{V_0} \frac{a\big[t - (|\mathbf{r} - \mathbf{r}_0|/c)\big]\, \delta(\mathbf{r}_0 - \mathbf{r}_S)}{|\mathbf{r} - \mathbf{r}_0|}\, dV_0, \qquad (2.7\text{-}32)$$

and upon using the sifting property of impulse functions, we obtain the following expression for the velocity potential of a *spherical wave with arbitrary time dependence*:

$$\boxed{\varphi(t, \mathbf{r}) = -\frac{1}{4\pi R}\, a(t - \tau),} \qquad (2.7\text{-}33)$$

where

$$R = |\mathbf{r} - \mathbf{r}_S| \qquad (2.7\text{-}34)$$

is the distance in meters between source and field points, and

$$\tau = R/c \qquad (2.7\text{-}35)$$

is the travel time or time delay in seconds. Compare Eq. (2.7-33) with the solutions of the *homogeneous* wave equation given by Eq. (2.2-98) and in Problem 2-4.

If we let $a(t) = A \exp(+j2\pi ft)$, then Eq. (2.7-33) reduces to

$$\varphi(t, \mathbf{r}) = -\frac{A}{4\pi R}\exp\big[+j2\pi f(t - \tau)\big], \qquad (2.7\text{-}36)$$

and upon substituting Eq. (2.7-35) into Eq. (2.7-36), we obtain the following expression for the velocity potential of a *time-harmonic spherical wave*:

$$\varphi(t,\mathbf{r}) = -\frac{A}{4\pi R}\exp(-jkR)\exp(+j2\pi ft),\qquad (2.7\text{-}37)$$

or

$$\varphi(t,\mathbf{r}) = Ag_f(\mathbf{r}|\mathbf{r}_S)\exp(+j2\pi ft),\qquad (2.7\text{-}38)$$

where

$$g_f(\mathbf{r}|\mathbf{r}_S) = -\frac{\exp(-jkR)}{4\pi R} = -\frac{\exp(-jk|\mathbf{r}-\mathbf{r}_S|)}{4\pi|\mathbf{r}-\mathbf{r}_S|}\qquad (2.7\text{-}39)$$

is the time-independent free-space Green's function.

Example 2.7-2 (Time-harmonic source)

In this example we shall evaluate Eq. (2.7-30) for the special case of a time-harmonic source. The source distribution for this case is given by

$$x_M(t,\mathbf{r}) = x_{f,M}(\mathbf{r})\exp(+j2\pi ft).\qquad (2.7\text{-}40)$$

Substituting Eq. (2.7-40) into Eq. (2.7-30) yields

$$\varphi(t,\mathbf{r}) = -\frac{1}{4\pi}\int_{V_0} x_{f,M}(\mathbf{r}_0)\frac{\exp(-jk|\mathbf{r}-\mathbf{r}_0|)}{|\mathbf{r}-\mathbf{r}_0|}\,dV_0\,\exp(+j2\pi ft)$$

$$(2.7\text{-}41)$$

or

$$\varphi(t,\mathbf{r}) = \varphi_f(\mathbf{r})\exp(+j2\pi ft),\qquad (2.7\text{-}42)$$

where

$$\varphi_f(\mathbf{r}) = \int_{V_0} x_{f,M}(\mathbf{r}_0)g_f(\mathbf{r}|\mathbf{r}_0)\,dV_0\qquad (2.7\text{-}43)$$

is the spatial-dependent part of the velocity potential,

$$g_f(\mathbf{r}|\mathbf{r}_0) = -\frac{\exp(-jk|\mathbf{r}-\mathbf{r}_0|)}{4\pi|\mathbf{r}-\mathbf{r}_0|}\qquad (2.7\text{-}44)$$

is the time-independent free-space Green's function, and

$$R = |\mathbf{r} - \mathbf{r}_0| \qquad (2.7\text{-}45)$$

is the range or distance between source and field points. Since the time-independent free-space Green's function given by Eq. (2.7-44) is the spatial impulse response of an unbounded, homogeneous, fluid medium, Eq. (2.7-43) is, in fact, a *three-dimensional spatial convolution integral*.

2.7.2 Solution in the Presence of Boundaries

In this section we shall take what is considered to be the traditional approach of obtaining the solution of the linear, three-dimensional, lossless, inhomogeneous wave equation given by Eq. (2.7-1). We begin the solution of Eq. (2.7-1) by considering a time-harmonic source distribution, that is,

$$x_M(t,\mathbf{r}) = x_{f,M}(\mathbf{r}) \exp(+j2\pi ft), \qquad (2.7\text{-}46)$$

and as a result, the velocity potential

$$\varphi(t,\mathbf{r}) = \varphi_f(\mathbf{r}) \exp(+j2\pi ft) \qquad (2.7\text{-}47)$$

is also time-harmonic. Substituting Eqs. (2.7-46) and (2.7-47) into Eq. (2.7-1) yields the following lossless, inhomogeneous Helmholtz equation:

$$\nabla^2\varphi_f(\mathbf{r}) + k^2\varphi_f(\mathbf{r}) = x_{f,M}(\mathbf{r}), \qquad (2.7\text{-}48)$$

where

$$k = 2\pi f/c = 2\pi/\lambda \qquad (2.7\text{-}49)$$

is the wave number in radians per meter.

Next, recall from Section 2.6.2 that the time-independent free-space Green's function $g_f(\mathbf{r}|\mathbf{r}_0)$ is a solution of the following lossless, inhomogeneous Helmholtz equation:

$$\nabla^2 g_f(\mathbf{r}|\mathbf{r}_0) + k^2 g_f(\mathbf{r}|\mathbf{r}_0) = \delta(\mathbf{r} - \mathbf{r}_0), \qquad (2.7\text{-}50)$$

where the impulse function $\delta(\mathbf{r} - \mathbf{r}_0)$ represents a unit amplitude, omnidirectional point source at $\mathbf{r}_0$. Multiplying Eq. (2.7-48) by $g_f(\mathbf{r}|\mathbf{r}_0)$ and Eq. (2.7-50) by $\varphi_f(\mathbf{r})$ yields

$$g_f(\mathbf{r}|\mathbf{r}_0)\nabla^2\varphi_f(\mathbf{r}) + k^2\varphi_f(\mathbf{r})g_f(\mathbf{r}|\mathbf{r}_0) = x_{f,M}(\mathbf{r})g_f(\mathbf{r}|\mathbf{r}_0) \qquad (2.7\text{-}51)$$

and

$$\varphi_f(\mathbf{r})\nabla^2 g_f(\mathbf{r}|\mathbf{r}_0) + k^2\varphi_f(\mathbf{r})g_f(\mathbf{r}|\mathbf{r}_0) = \varphi_f(\mathbf{r})\delta(\mathbf{r} - \mathbf{r}_0), \qquad (2.7\text{-}52)$$

respectively, and upon subtracting Eq. (2.7-52) from Eq. (2.7-51), we obtain

$$g_f(\mathbf{r}|\mathbf{r}_0)\,\nabla^2\varphi_f(\mathbf{r}) - \varphi_f(\mathbf{r})\,\nabla^2 g_f(\mathbf{r}|\mathbf{r}_0) = x_{f,M}(\mathbf{r})g_f(\mathbf{r}|\mathbf{r}_0) - \varphi_f(\mathbf{r})\,\delta(\mathbf{r}-\mathbf{r}_0).$$

$$(2.7\text{-}53)$$

In order to proceed further, the position vectors $\mathbf{r}$ and $\mathbf{r}_0$ are interchanged in Eq. (2.7-53), noting that

$$g_f(\mathbf{r}_0|\mathbf{r}) = g_f(\mathbf{r}|\mathbf{r}_0) \tag{2.7-54}$$

and

$$\delta(\mathbf{r}_0 - \mathbf{r}) = \delta(\mathbf{r} - \mathbf{r}_0). \tag{2.7-55}$$

Equation (2.7-53) can then be rewritten as

$$g_f(\mathbf{r}|\mathbf{r}_0)\,\nabla_0^2\varphi_f(\mathbf{r}_0) - \varphi_f(\mathbf{r}_0)\,\nabla_0^2 g_f(\mathbf{r}|\mathbf{r}_0) = x_{f,M}(\mathbf{r}_0)g_f(\mathbf{r}|\mathbf{r}_0) - \varphi_f(\mathbf{r}_0)\,\delta(\mathbf{r}-\mathbf{r}_0),$$

$$(2.7\text{-}56)$$

where

$$\nabla_0^2 = \frac{\partial^2}{\partial x_0^2} + \frac{\partial^2}{\partial y_0^2} + \frac{\partial^2}{\partial z_0^2} \tag{2.7-57}$$

is the Laplacian expressed in the rectangular coordinates (x_0, y_0, z_0).
Integrating both sides of Eq. (2.7-56) over the volume V_0 yields

$$\varphi_f(\mathbf{r}) = \int_{V_0} x_{f,M}(\mathbf{r}_0)g_f(\mathbf{r}|\mathbf{r}_0)\,dV_0$$

$$- \int_{V_0} \nabla_0 \cdot \left[g_f(\mathbf{r}|\mathbf{r}_0)\,\nabla_0\varphi_f(\mathbf{r}_0) - \varphi_f(\mathbf{r}_0)\,\nabla_0 g_f(\mathbf{r}|\mathbf{r}_0) \right] dV_0, \quad (2.7\text{-}58)$$

where

$$\int_{V_0} \varphi_f(\mathbf{r}_0)\,\delta(\mathbf{r} - \mathbf{r}_0)\,dV_0 = \varphi_f(\mathbf{r}), \tag{2.7-59}$$

$$\nabla_0 \cdot \left[g_f(\mathbf{r}|\mathbf{r}_0)\,\nabla_0\varphi_f(\mathbf{r}_0) - \varphi_f(\mathbf{r}_0)\,\nabla_0 g_f(\mathbf{r}|\mathbf{r}_0) \right]$$

$$= g_f(\mathbf{r}|\mathbf{r}_0)\,\nabla_0^2\varphi_f(\mathbf{r}_0) - \varphi_f(\mathbf{r}_0)\,\nabla_0^2 g_f(\mathbf{r}|\mathbf{r}_0), \tag{2.7-60}$$

$$\nabla_0 = \frac{\partial}{\partial x_0}\hat{x} + \frac{\partial}{\partial y_0}\hat{y} + \frac{\partial}{\partial z_0}\hat{z} \tag{2.7-61}$$

is the gradient expressed in the rectangular coordinates (x_0, y_0, z_0), and $dV_0 = dx_0\, dy_0\, dz_0$.

With the use of the divergence theorem, Eq. (2.7-58) can be rewritten as

$$\varphi_f(\mathbf{r}) = \int_{V_0} x_{f,M}(\mathbf{r}_0)\, g_f(\mathbf{r}|\mathbf{r}_0)\, dV_0$$

$$-\oint_{S_0} \left[g_f(\mathbf{r}|\mathbf{r}_0)\, \nabla_0 \varphi_f(\mathbf{r}_0) - \varphi_f(\mathbf{r}_0)\, \nabla_0 g_f(\mathbf{r}|\mathbf{r}_0) \right] \cdot d\mathbf{S}_0, \quad (2.7\text{-}62)$$

where S_0 is any closed surface enclosing the source within the volume V_0,

$$d\mathbf{S}_0 = dS_0\, \hat{n}_0, \qquad (2.7\text{-}63)$$

and $\hat{n}_0$ is a unit vector normal to the surface S_0, pointing in the conventional outward direction away from the enclosed volume V_0, and hence, the source. Since

$$\hat{n}_0 \cdot \nabla_0 \varphi_f(\mathbf{r}_0) = \frac{\partial}{\partial n_0} \varphi_f(\mathbf{r}_0) \qquad (2.7\text{-}64)$$

and

$$\hat{n}_0 \cdot \nabla_0 g_f(\mathbf{r}|\mathbf{r}_0) = \frac{\partial}{\partial n_0} g_f(\mathbf{r}|\mathbf{r}_0), \qquad (2.7\text{-}65)$$

substituting Eqs. (2.7-63) through (2.7-65) into Eq. (2.7-62) yields the following desired result:

$$\boxed{\begin{aligned}
\varphi_f(\mathbf{r}) = &\int_{V_0} x_{f,M}(\mathbf{r}_0)\, g_f(\mathbf{r}|\mathbf{r}_0)\, dV_0 \\
&-\oint_{S_0} \left[g_f(\mathbf{r}|\mathbf{r}_0)\, \frac{\partial}{\partial n_0} \varphi_f(\mathbf{r}_0) - \varphi_f(\mathbf{r}_0)\, \frac{\partial}{\partial n_0} g_f(\mathbf{r}|\mathbf{r}_0) \right] dS_0,
\end{aligned}} \qquad (2.7\text{-}66)$$

where $\varphi_f(\mathbf{r}_0)$ is the velocity potential of the *total* acoustic field (incident plus reflected) at each point on the closed surface S_0, and $\partial/\partial n_0$ is the partial derivative in the outward direction away from the enclosed volume V_0, normal to the closed surface S_0, and evaluated at each point on S_0. Equation (2.7-66) is valid for computing the velocity potential $\varphi_f(\mathbf{r})$ at field points $\mathbf{r}$ inside V_0 and on S_0, but is invalid for field points $\mathbf{r}$ outside V_0.

As can be seen from Eq. (2.7-66), the velocity potential $\varphi_f(\mathbf{r})$ is equal to the sum of the acoustic field due to the source distribution $x_{f,M}(\mathbf{r}_0)$ within the volume V_0 and the acoustic field due to reflections from the boundary S_0. In the absence of boundaries, that is, for free-space propagation, Eq. (2.7-66) reduces to (see Example 2.7-2)

$$\varphi_f(\mathbf{r}) = \int_{V_0} x_{f,M}(\mathbf{r}_0) g_f(\mathbf{r}|\mathbf{r}_0) \, dV_0. \qquad (2.7\text{-}67)$$

2.8 Integral Representations of the Time-Independent Free-Space Green's Function in Rectangular Coordinates

In this section we shall represent the time-independent free-space Green's function $g_f(\mathbf{r}|\mathbf{r}_0)$ by two different integral expressions in the rectangular coordinates (x, y, z). These expressions represent expansions of $g_f(\mathbf{r}|\mathbf{r}_0)$ into an infinite number of plane waves and are very useful for solving problems involving the reflection of spherical waves from planar boundaries.

In Section 2.6.2 we showed that $g_f(\mathbf{r}|\mathbf{r}_0)$ is a solution of the following lossless, inhomogeneous Helmholtz equation:

$$\nabla^2 g_f(\mathbf{r}|\mathbf{r}_0) + k^2 g_f(\mathbf{r}|\mathbf{r}_0) = \delta(\mathbf{r} - \mathbf{r}_0), \qquad (2.8\text{-}1)$$

where $\delta(\mathbf{r} - \mathbf{r}_0)$ represents a unit-amplitude, omnidirectional point source at $\mathbf{r}_0$. Taking the spatial Fourier transform of both sides of Eq. (2.8-1) with respect to $\mathbf{r}$ yields

$$F_{\mathbf{r}}\{\nabla^2 g_f(\mathbf{r}|\mathbf{r}_0)\} + k^2 G_f(\boldsymbol{\nu}|\mathbf{r}_0) = \exp(+j2\pi\boldsymbol{\nu}\cdot\mathbf{r}_0), \qquad (2.8\text{-}2)$$

where

$$G_f(\boldsymbol{\nu}|\mathbf{r}_0) = F_{\mathbf{r}}\{g_f(\mathbf{r}|\mathbf{r}_0)\} \triangleq \int_{-\infty}^{\infty} g_f(\mathbf{r}|\mathbf{r}_0) \exp(+j2\pi\boldsymbol{\nu}\cdot\mathbf{r}) \, d\mathbf{r}, \qquad (2.8\text{-}3)$$

$$\boldsymbol{\nu} = (f_X, f_Y, f_Z) \qquad (2.8\text{-}4)$$

is a three-dimensional vector whose components are the spatial frequencies in cycles per meter in the X, Y, and Z directions (see Section 2.2.1), and

$$F_{\mathbf{r}}\{\delta(\mathbf{r} - \mathbf{r}_0)\} = \int_{-\infty}^{\infty} \delta(\mathbf{r} - \mathbf{r}_0) \exp(+j2\pi\boldsymbol{\nu}\cdot\mathbf{r}) \, d\mathbf{r} = \exp(+j2\pi\boldsymbol{\nu}\cdot\mathbf{r}_0). \quad (2.8\text{-}5)$$

Note that $F_{\mathbf{r}}\{\cdot\} = F_x F_y F_z\{\cdot\}$ and $d\mathbf{r} = dx\,dy\,dz$, and as a result, the right-hand side of Eq. (2.8-3), for example, is shorthand notation for a *three-dimensional spatial Fourier transform*. Since (see Appendix 2C)

$$F_{\mathbf{r}}\{\nabla^2 g_f(\mathbf{r}|\mathbf{r}_0)\} = -\left[(2\pi f_X)^2 + (2\pi f_Y)^2 + (2\pi f_Z)^2\right] G_f(\boldsymbol{\nu}|\mathbf{r}_0), \quad (2.8\text{-}6)$$

substituting Eq. (2.8-6) into Eq. (2.8-2) yields

$$G_f(\boldsymbol{\nu}|\mathbf{r}_0) = \frac{\exp(+j2\pi\boldsymbol{\nu}\cdot\mathbf{r}_0)}{(2\pi f/c)^2 - \left[(2\pi f_X)^2 + (2\pi f_Y)^2 + (2\pi f_Z)^2\right]}, \quad (2.8\text{-}7)$$

and since [see Eq. (2.8-3)]

$$g_f(\mathbf{r}|\mathbf{r}_0) = F_{\boldsymbol{\nu}}^{-1}\{G_f(\boldsymbol{\nu}|\mathbf{r}_0)\} \triangleq \int_{-\infty}^{\infty} G_f(\boldsymbol{\nu}|\mathbf{r}_0)\exp(-j2\pi\boldsymbol{\nu}\cdot\mathbf{r})\,d\boldsymbol{\nu}, \quad (2.8\text{-}8)$$

substituting Eq. (2.8-7) into Eq. (2.8-8) yields

$$\boxed{g_f(\mathbf{r}|\mathbf{r}_0) = \int_{-\infty}^{\infty} \frac{\exp[-j2\pi\boldsymbol{\nu}\cdot(\mathbf{r}-\mathbf{r}_0)]}{(2\pi f/c)^2 - \left[(2\pi f_X)^2 + (2\pi f_Y)^2 + (2\pi f_Z)^2\right]}\,d\boldsymbol{\nu},} \quad (2.8\text{-}9)$$

which is the first desired integral expression that represents an expansion of $g_f(\mathbf{r}|\mathbf{r}_0)$ into an infinite number of plane waves. The numerator of the integrand in Eq. (2.8-9) is the plane-wave term, whereas one over the denominator represents the amplitude of the plane wave. Note that $F_{\boldsymbol{\nu}}^{-1}\{\cdot\} = F_{f_X}^{-1}F_{f_Y}^{-1}F_{f_Z}^{-1}\{\cdot\}$ and $d\boldsymbol{\nu} = df_X\,df_Y\,df_Z$, and as a result, the right-hand side of Eq. (2.8-8) is shorthand notation for an *inverse three-dimensional spatial Fourier transform*.

The second and more common integral representation of $g_f(\mathbf{r}|\mathbf{r}_0)$ can be obtained by first rewriting Eq. (2.8-9) as follows:

$$g_f(\mathbf{r}|\mathbf{r}_0) = -\frac{1}{(2\pi)^2}\int_{-\infty}^{\infty}\int_{-\infty}^{\infty}\int_{-\infty}^{\infty} G(f_Z)\exp(-j2\pi f_Z z)\,df_Z$$

$$\times\exp\{-j2\pi[f_X(x-x_0)+f_Y(y-y_0)]\}\,df_X\,df_Y, \quad (2.8\text{-}10)$$

where

$$G(f_Z) = \frac{\exp(+j2\pi f_Z z_0)}{f_Z^2 - a^2} \quad (2.8\text{-}11)$$

and

$$a = \sqrt{(f/c)^2 - (f_X^2 + f_Y^2)}. \quad (2.8\text{-}12)$$

Note that the integral of $G(f_Z)$ is, by definition, the inverse spatial Fourier transform of $G(f_Z)$, that is,

$$\int_{-\infty}^{\infty} G(f_Z)\exp(-j2\pi f_Z z)\,df_Z \triangleq F_{f_Z}^{-1}\{G(f_Z)\}. \quad (2.8\text{-}13)$$

Since it can be shown that (see Problem 2-21)

$$\int_{-\infty}^{\infty} G(f_Z) \exp(-j2\pi f_Z z) \, df_Z = -j\frac{\pi}{a} \exp[-j2\pi a|z - z_0|], \quad (2.8\text{-}14)$$

substituting Eq. (2.8-14) into Eq. (2.8-10) yields

$$
\begin{aligned}
g_f(\mathbf{r}|\mathbf{r}_0) = {}&+j\frac{1}{4\pi}\int_{-\infty}^{\infty}\int_{-\infty}^{\infty} \exp\{-j2\pi[f_X(x - x_0) + f_Y(y - y_0)]\} \\
&\times \frac{\exp(-j2\pi f_Z|z - z_0|)}{f_Z} \, df_X \, df_Y,
\end{aligned}
\qquad (2.8\text{-}15)
$$

where, from Eq. (2.8-12),

$$
f_Z = a =
\begin{cases}
\sqrt{(f/c)^2 - (f_X^2 + f_Y^2)}, & f_X^2 + f_Y^2 \le (f/c)^2, \\
-j\sqrt{f_X^2 + f_Y^2 - (f/c)^2}, & f_X^2 + f_Y^2 > (f/c)^2.
\end{cases}
\qquad (2.8\text{-}16)
$$

Note that the second expression for f_Z ensures *evanescent waves* (i.e., decaying exponentials) in either the positive or the negative Z direction when $f_X^2 + f_Y^2 > (f/c)^2$. Equation (2.8-15) can also be expressed in terms of the propagation-vector components as follows:

$$
\begin{aligned}
g_f(\mathbf{r}|\mathbf{r}_0) = {}&+j\frac{1}{8\pi^2}\int_{-\infty}^{\infty}\int_{-\infty}^{\infty} \exp\{-j[k_X(x - x_0) + k_Y(y - y_0)]\} \\
&\times \frac{\exp(-jk_Z|z - z_0|)}{k_Z} \, dk_X \, dk_Y,
\end{aligned}
\qquad (2.8\text{-}17)
$$

where

$$
k_Z =
\begin{cases}
\sqrt{k^2 - (k_X^2 + k_Y^2)}, & k_X^2 + k_Y^2 \le k^2, \\
-j\sqrt{k_X^2 + k_Y^2 - k^2}, & k_X^2 + k_Y^2 > k^2,
\end{cases}
\qquad (2.8\text{-}18)
$$

$k = 2\pi f/c = 2\pi/\lambda$ is the wave number in radians per meter, and $k_X = 2\pi f_X$, $k_Y = 2\pi f_Y$, and $k_Z = 2\pi f_Z$ are the propagation-vector components in radians per meter in the X, Y, and Z directions, respectively (see Section 2.2.1). Therefore, either Eq. (2.8-15) or Eq. (2.8-17) is the second desired integral expression that represents an expansion of $g_f(\mathbf{r}|\mathbf{r}_0)$ into an infinite number of plane waves.

Later, in Section 3.6, we shall derive two integral expressions for the time-independent free-space Green's function in cylindrical coordinates by using Eq. (2.8-17) as
the starting point. These expressions are also very useful for solving problems
involving the reflection of spherical waves from planar boundaries (see Section 3.7)
and for solving ocean waveguide problems (see Section 3.9).

Problems

2-1   With the use of Fig. 2.2-2, verify Eqs. (2.2-11) through (2.2-13).

2-2   Show that

$$\nabla \exp(-j\mathbf{k} \cdot \mathbf{r}) = -j\mathbf{k} \exp(-j\mathbf{k} \cdot \mathbf{r}).$$

2-3   Consider a time-harmonic plane wave propagating in an unbounded fluid
medium with angles of propagation $\theta = 40°$ and $\psi = 108°$. The magnitude of
the acoustic pressure is 2 Pa, and $f = 250$ Hz. The fluid medium is air with
$\rho_0 = 1.21$ kg/m^3 and $c = 343$ m/sec. Compute
(a) the propagation vector,
(b) the group speeds in the X, Y, and Z directions,
(c) the phase speeds in the X, Y, and Z directions,
(d) the wavelengths in the X, Y, and Z directions, and
(e) the time-average intensity vector.

2-4   Both the time-harmonic solution of the homogeneous wave equation and the
solution for arbitrary time dependence given by Eqs. (2.2-92) and (2.2-98),
respectively, can be generalized by replacing the travel time or time delay τ
given by Eq. (2.2-93) with

$$\tau = \frac{\hat{n} \cdot (\mathbf{r} - \mathbf{r}_0)}{c},$$

where $\mathbf{r}_0 = (x_0, y_0, z_0)$ is the position vector to a source point that is not
located at the origin.
(a) Verify that

$$\varphi(t,\mathbf{r}) = a(t - \tau)$$

is a solution of the homogeneous wave equation given by Eq. (2.1-1) with
τ as given in this problem.
(b) If the angles of propagation of a plane wave are $\theta = 20°$ and $\psi = 285°$,
and if the position vectors to source and field points are $\mathbf{r}_0 = 10\hat{x} + 20\hat{y} - 30\hat{z}$ and $\mathbf{r} = 100\hat{x} - 50\hat{y} + 1000\hat{z}$, respectively, then find the
distance (range) and travel time between source and field points. Use
$c = 1500$ m/sec.

2-5   Consider a plane wave propagating in an unbounded fluid medium. The plane wave has arbitrary time dependence given by

$$\varphi(t,\mathbf{r}) = a(t - \tau),$$

where

$$\tau = \frac{\hat{n} \cdot (\mathbf{r} - \mathbf{r}_0)}{c}.$$

Show that the specific acoustic impedance is equal to the characteristic impedance. **Hint:** $Z = p(t,\mathbf{r})/u(t,\mathbf{r})$.

2-6   Verify that Eq. (2.2-101) is a solution of the lossy Helmholtz equation given by Eq. (2.2-99).

2-7   Consider a time-harmonic plane wave propagating in an unbounded viscous fluid medium with an attenuation coefficient equal to 2×10^{-4} Np/m.
(a) If the plane wave travels 1.5 km, then find the amount of attenuation in decibels that the plane wave will experience.
(b) If the amplitude of the plane wave at the source is A, then at what range will the amplitude of the plane wave be $0.1A$?

2-8   In this problem we shall investigate an approximate ideal pressure-release boundary that occurs in nature. The incident wave is in medium I. Compute the ratios ρ_2/ρ_1 and Z_2/Z_1, and the reflection coefficient R_{12} at normal incidence, for the following combination of parameters: Medium I is sea water with $\rho_1 = 1026$ kg/m^3 and $c_1 = 1500$ m/sec. Medium II is air with $\rho_2 = 1.21$ kg/m^3 and $c_2 = 343$ m/sec.

2-9   In this problem we shall investigate an approximate ideal rigid boundary that occurs in nature. The incident wave is in medium I. Compute the ratios ρ_2/ρ_1 and Z_2/Z_1, and the reflection coefficient R_{12} at normal incidence, using the parameters given in Problem 2-8, but with medium I as air and medium II as sea water.

2-10 If medium I is oxygen with $\rho_1 = 1.43$ kg/m^3 and $c_1 = 317.2$ m/sec, and medium II is hydrogen with $\rho_2 = 0.09$ kg/m^3 and $c_2 = 1269.5$ m/sec, compute:
(a) The ratios ρ_2/ρ_1 and Z_2/Z_1.
(b) The critical angle of incidence θ_c. Why does θ_c exist in this problem?
(c) The angle of intromission θ_I. Why does θ_I exist in this problem?

2-11 Using the combination of parameters given in Problem 2-10, compute the skin depth z_{SD} for
(a) $f = 100$ Hz and $\theta_i = 15°$, $45°$, and $85°$,
(b) $f = 1$ kHz and $\theta_i = 15°$, $45°$, and $85°$.

2-12 Using the combination of parameters given in Problem 2-10, compute the reflection coefficient R_{12} for $\theta_i = 0°$ and $10°$.

2-13 If medium I is sea water with $\rho_1 = 1026$ kg/m^3 and $c_1 = 1500$ m/sec, and medium II is the fluidlike sea bottom of quartz sand with $\rho_2 = 2070$ kg/m^3 and $c_2 = 1730$ m/sec, compute
(a) the critical angle of incidence θ_c,
(b) the skin depth z_{SD} for $f = 220$ Hz and $\theta_i = 65°$, $75°$, and $85°$,
(c) the reflection coefficient R_{12} for $\theta_i = 45°$ and $75°$.

2-14 Consider a time-harmonic plane wave incident upon a boundary between two different fluid media. If the plane-wave pressure reflection coefficient is equal to -0.45, and the time-average power of the incident acoustic field is 15 W, then find the value of the time-average power of the transmitted acoustic field.

2-15 Assume that all of the surfaces inside a rectangular cavity are pressure-release. This is an example of a Dirichlet boundary-value problem. Derive expressions for the time-harmonic eigenfunctions (natural modes) and corresponding eigenfrequencies (natural frequencies) inside the cavity.

2-16 **Water tank**. Figure P2-16 illustrates a water tank with rigid walls. Assume that the tank is filled with water so that the surface at $y = 0$ is pressure-release. This is an example of a mixed boundary-value problem, also known

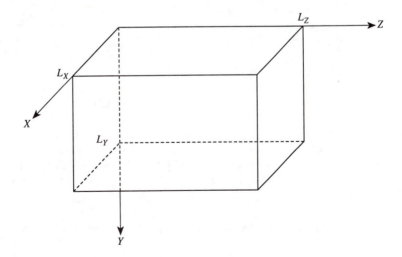

Figure P2-16.

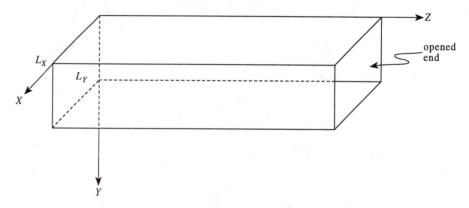

Figure P2-17.

as a Robin boundary-value problem. Derive expressions for the time-harmonic eigenfunctions (natural modes) and corresponding eigenfrequencies (natural frequencies) inside the water tank.

2-17 **Monterey Bay**. A very approximate model of Monterey Bay, California, is a waveguide with rectangular cross-sectional area as illustrated in Fig. P2-17. Assume that all surfaces inside the waveguide are rigid except the ocean surface at $y = 0$, which is pressure-release. Also assume that a time-harmonic sound source is located in the XY plane at $z = 0$. This is an example of a mixed boundary-value problem, also known as a Robin boundary-value problem. Derive expressions for the normal modes inside the waveguide. **Note:** This problem reveals that by making one surface of the waveguide pressure-release, there is no longitudinal or plane-wave mode, since $f_{00} \neq 0$ Hz. Therefore, if $f < f_{00}$, then no modes propagate.

2-18 Verify Eq. (2.6-13).

2-19 Verify that Eq. (2.6-17) reduces to Eq. (2.6-18).

2-20 Consider a unit-amplitude, time-harmonic, omnidirectional point source in air located at $\mathbf{r}_0 = 10\hat{x} + 5\hat{y} + 10\hat{z}$ with $c = 343$ m/sec and $f = 250$ Hz.
   (a) Find the magnitude and phase (in radians) of the spatial-dependent part of the velocity potential of the acoustic field at $\mathbf{r} = 100\hat{x} + 50\hat{y} + 200\hat{z}$.
   (b) Find the travel time in seconds between source and field points.

2-21 Let $G(f_Z)$ be the spatial Fourier transform of $g(z)$, that is,

$$G(f_Z) = F_z\{g(z)\} \triangleq \int_{-\infty}^{\infty} g(z)\exp(+j2\pi f_Z z)\, dz.$$

(a) Show that

$$F_z\{g(z - z_0)\} = G(f_Z)\exp(+j2\pi f_Z z_0).$$

(b) If

$$g(z) = A\exp(-b|z|), \qquad \mathrm{Re}\{b\} > 0,$$

where A is an arbitrary complex constant, $b = b_1 + jb_2$, and $\mathrm{Re}\{b\} = b_1 > 0$, show that

$$G(f_Z) = \frac{2Ab}{(2\pi f_Z)^2 + b^2}, \qquad \mathrm{Re}\{b\} > 0.$$

(c) If $b_1 = 0$, then $b = jb_2$, and as a result,

$$g(z) = A\exp(-jb_2|z|).$$

Show that

$$G(f_Z) = j\frac{2Ab_2}{(2\pi f_Z)^2 - b_2^2}.$$

Note that if $A = -j\pi/a$ and $b_2 = 2\pi a$, then

$$G(f_Z) = \frac{1}{f_Z^2 - a^2}.$$

Appendix 2A

In this appendix we shall derive Eq. (2.3-79), and in addition, we shall define and discuss the fourth set of coefficients, namely the *plane-wave time-average power reflection and transmission coefficients*. Conservation of energy requires that

$$P_{\mathrm{avg}_r} + P_{\mathrm{avg}_t} = P_{\mathrm{avg}_i}. \tag{2A-1}$$

In order to express Eq. (2A-1) in terms of the plane-wave time-average intensity reflection and transmission coefficients R_{12}'' and T_{12}'', respectively, we need to calculate the time-average power of the incident, reflected, and transmitted fields.

By referring to Fig. 2A-1, it can be seen that the time-average power flowing through surfaces S_i, S_r, and S_t can be expressed as

$$P_{\mathrm{avg}_i} = I_{\mathrm{avg}_i}S_i = I_{\mathrm{avg}_i}S\cos\theta_i, \tag{2A-2}$$

$$P_{\mathrm{avg}_r} = I_{\mathrm{avg}_r}S_r = I_{\mathrm{avg}_r}S\cos\theta_r = I_{\mathrm{avg}_r}S\cos\theta_i, \tag{2A-3}$$

and

$$P_{\mathrm{avg}_t} = I_{\mathrm{avg}_t}S_t = I_{\mathrm{avg}_t}S\cos\theta_t. \tag{2A-4}$$

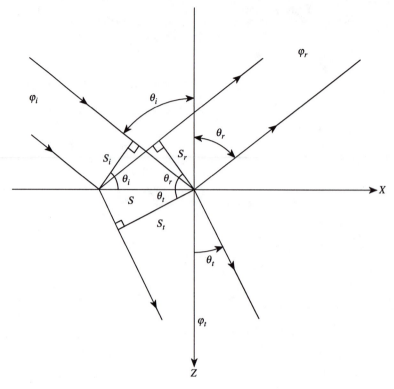

Figure 2A-1. Incident, reflected, and transmitted plane waves at the boundary between two different fluid media.

Substituting Eqs. (2A-2) through (2A-4) into Eq. (2A-1) yields

$$I_{\text{avg}_r} + \frac{\cos \theta_t}{\cos \theta_i} I_{\text{avg}_t} = I_{\text{avg}_i}, \tag{2A-5}$$

and upon dividing both sides of Eq. (2A-5) by I_{avg_i}, we obtain

$$R''_{12} + \frac{\cos \theta_t}{\cos \theta_i} T''_{12} = 1, \tag{2A-6}$$

which is identical to Eq. (2.3-79).

Next, let us define the plane-wave time-average power reflection and transmission coefficients:

$$R'''_{12} \triangleq \frac{P_{\text{avg}_r}}{P_{\text{avg}_i}} \tag{2A-7}$$

and

$$T_{12}''' \triangleq \frac{P_{\mathrm{avg}_t}}{P_{\mathrm{avg}_i}}. \qquad (2A-8)$$

Substituting Eqs. (2A-2) and (2A-3) into Eq. (2A-7) yields

$$R_{12}''' = \frac{I_{\mathrm{avg}_r}}{I_{\mathrm{avg}_i}} = R_{12}'', \qquad (2A-9)$$

and with the use of Eq. (2.3-76),

$$R_{12}''' = R_{12}'' = |R_{12}'|^2 = |R_{12}|^2. \qquad (2A-10)$$

Similarly, substituting Eqs. (2A-2) and (2A-4) into Eq. (2A-8) yields

$$T_{12}''' = \frac{I_{\mathrm{avg}_t}}{I_{\mathrm{avg}_i}} \frac{\cos \theta_t}{\cos \theta_i} = \frac{\cos \theta_t}{\cos \theta_i} T_{12}''. \qquad (2A-11)$$

And finally, dividing both sides of Eq. (2A-1) by P_{avg_i} yields the following alternative way of expressing conservation of energy:

$$R_{12}''' + T_{12}''' = 1. \qquad (2A-12)$$

Appendix 2B

In this appendix we shall compute the time-domain Fourier transform of $\partial^2 \varphi(t,\mathbf{r})/\partial t^2$. From Eq. (2.7-5) we can write that

$$\varphi(t,\mathbf{r}) = F_\eta^{-1}\{\Phi(\eta,\mathbf{r})\} = \int_{-\infty}^{\infty} \Phi(\eta,\mathbf{r}) \exp(+j2\pi\eta t)\, d\eta, \qquad (2B-1)$$

and as a result,

$$\frac{\partial^2}{\partial t^2}\varphi(t,\mathbf{r}) = \int_{-\infty}^{\infty} \Phi(\eta,\mathbf{r}) \frac{\partial^2}{\partial t^2}\exp(+j2\pi\eta t)\, d\eta. \qquad (2B-2)$$

Since

$$\frac{\partial^2}{\partial t^2} \exp(+j2\pi\eta t) = \frac{\partial}{\partial t}\left[\frac{\partial}{\partial t}\exp(+j2\pi\eta t)\right] \tag{2B-3}$$

$$= +j2\pi\eta\frac{\partial}{\partial t}\exp(+j2\pi\eta t) \tag{2B-4}$$

$$= (+j2\pi\eta)^2 \exp(+j2\pi\eta t), \tag{2B-5}$$

substituting Eq. (2B-5) into Eq. (2B-2) yields

$$\frac{\partial^2}{\partial t^2}\varphi(t,\mathbf{r}) = \int_{-\infty}^{\infty} (+j2\pi\eta)^2\Phi(\eta,\mathbf{r})\exp(+j2\pi\eta t)\,d\eta. \tag{2B-6}$$

Equation (2B-6) can be expressed as

$$\frac{\partial^2}{\partial t^2}\varphi(t,\mathbf{r}) = F_\eta^{-1}\{(+j2\pi\eta)^2\Phi(\eta,\mathbf{r})\}, \tag{2B-7}$$

and as a result,

$$\boxed{F_t\left\{\frac{\partial^2}{\partial t^2}\varphi(t,\mathbf{r})\right\} = -(2\pi\eta)^2\Phi(\eta,\mathbf{r}),} \tag{2B-8}$$

where $\Phi(\eta,\mathbf{r})$ is given by Eq. (2.7-5).

Appendix 2C

In this appendix we shall compute the spatial-domain Fourier transform of $\nabla^2\Phi(\eta,\mathbf{r})$. Since

$$\nabla^2\Phi(\eta,\mathbf{r}) = \frac{\partial^2}{\partial x^2}\Phi(\eta,\mathbf{r}) + \frac{\partial^2}{\partial y^2}\Phi(\eta,\mathbf{r}) + \frac{\partial^2}{\partial z^2}\Phi(\eta,\mathbf{r}), \tag{2C-1}$$

we have

$$F_\mathbf{r}\{\nabla^2\Phi(\eta,\mathbf{r})\} = F_\mathbf{r}\left\{\frac{\partial^2}{\partial x^2}\Phi(\eta,\mathbf{r})\right\} + F_\mathbf{r}\left\{\frac{\partial^2}{\partial y^2}\Phi(\eta,\mathbf{r})\right\} + F_\mathbf{r}\left\{\frac{\partial^2}{\partial z^2}\Phi(\eta,\mathbf{r})\right\}.$$

$$\tag{2C-2}$$

From Eq. (2.7-9) we can write that

$$\Phi(\eta, \mathbf{r}) = F_{\boldsymbol{\beta}}^{-1}\{\Phi(\eta, \boldsymbol{\beta})\} \triangleq \int_{-\infty}^{\infty} \Phi(\eta, \boldsymbol{\beta}) \exp(-j2\pi\boldsymbol{\beta} \cdot \mathbf{r}) \, d\boldsymbol{\beta}, \quad (2C\text{-}3)$$

and as a result,

$$\frac{\partial^2}{\partial x^2} \Phi(\eta, \mathbf{r}) = \int_{-\infty}^{\infty} \Phi(\eta, \boldsymbol{\beta}) \frac{\partial^2}{\partial x^2} \exp(-j2\pi\boldsymbol{\beta} \cdot \mathbf{r}) \, d\boldsymbol{\beta}, \quad (2C\text{-}4)$$

where $F_{\boldsymbol{\beta}}^{-1}\{\cdot\} = F_{\beta_X}^{-1} F_{\beta_Y}^{-1} F_{\beta_Z}^{-1}\{\cdot\}$ and $d\boldsymbol{\beta} = d\beta_X \, d\beta_Y \, d\beta_Z$. Therefore, the right-hand side of Eq. (2C-3) is shorthand notation for an *inverse three-dimensional spatial Fourier transform*. Since $\boldsymbol{\beta} = (\beta_X, \beta_Y, \beta_Z)$ and $\mathbf{r} = (x, y, z)$,

$$\frac{\partial^2}{\partial x^2} \exp(-j2\pi\boldsymbol{\beta} \cdot \mathbf{r}) = \frac{\partial}{\partial x}\left[\frac{\partial}{\partial x} \exp\left[-j2\pi(\beta_X x + \beta_Y y + \beta_Z z)\right]\right] \quad (2C\text{-}5)$$

$$= -j2\pi\beta_X \frac{\partial}{\partial x} \exp\left[-j2\pi(\beta_X x + \beta_Y y + \beta_Z z)\right] \quad (2C\text{-}6)$$

$$= (-j2\pi\beta_X)^2 \exp(-j2\pi\boldsymbol{\beta} \cdot \mathbf{r}). \quad (2C\text{-}7)$$

Substituting Eq. (2C-7) into Eq. (2C-4) yields

$$\frac{\partial^2}{\partial x^2} \Phi(\eta, \mathbf{r}) = \int_{-\infty}^{\infty} (-j2\pi\beta_X)^2 \Phi(\eta, \boldsymbol{\beta}) \exp(-j2\pi\boldsymbol{\beta} \cdot \mathbf{r}) \, d\boldsymbol{\beta}, \quad (2C\text{-}8)$$

which can also be expressed as

$$\frac{\partial^2}{\partial x^2} \Phi(\eta, \mathbf{r}) = F_{\boldsymbol{\beta}}^{-1}\{(-j2\pi\beta_X)^2 \Phi(\eta, \boldsymbol{\beta})\}, \quad (2C\text{-}9)$$

and as a result,

$$F_{\mathbf{r}}\left\{\frac{\partial^2}{\partial x^2} \Phi(\eta, \mathbf{r})\right\} = (-j2\pi\beta_X)^2 \Phi(\eta, \boldsymbol{\beta}), \quad (2C\text{-}10)$$

where $\Phi(\eta, \boldsymbol{\beta})$ is given by Eq. (2.7-9). Following the same procedure, it can also be shown that

$$F_{\mathbf{r}}\left\{\frac{\partial^2}{\partial y^2} \Phi(\eta, \mathbf{r})\right\} = (-j2\pi\beta_Y)^2 \Phi(\eta, \boldsymbol{\beta}) \quad (2C\text{-}11)$$

and

$$F_{\mathbf{r}}\left\{\frac{\partial^2}{\partial z^2}\Phi(\eta,\mathbf{r})\right\} = (-j2\pi\beta_Z)^2\Phi(\eta,\boldsymbol{\beta}). \qquad \text{(2C-12)}$$

Substituting Eqs. (2C-10) through (2C-12) into Eq. (2C-2) yields

$$F_{\mathbf{r}}\{\nabla^2\Phi(\eta,\mathbf{r})\} = -\left[(2\pi\beta_X)^2 + (2\pi\beta_Y)^2 + (2\pi\beta_Z)^2\right]\Phi(\eta,\boldsymbol{\beta}). \qquad \text{(2C-13)}$$

Bibliography

L. Brekhovskikh and Yu. Lysanov, *Fundamentals of Ocean Acoustics*, 2nd ed., Springer-Verlag, New York, 1991.

C. S. Clay and H. Medwin, *Acoustical Oceanography*, Wiley, New York, 1977.

L. E. Kinsler, A. R. Frey, A. B. Coppens, and J. V. Sanders, *Fundamentals of Acoustics*, 3rd ed., Wiley, New York, 1982.

P. M. Morse and K. U. Ingard, *Theoretical Acoustics*, Princeton University Press, Princeton, New Jersey, 1987.

C. B. Officer, *Introduction to the Theory of Sound Transmission*, McGraw-Hill, New York, 1958.

S. Temkin, *Elements of Acoustics*, Wiley, New York, 1981.

Chapter 3

Wave Propagation in the Cylindrical Coordinate System

3.1 Solution of the Linear Three-Dimensional Homogeneous Wave Equation

At the beginning of Chapter 2 we mentioned that the symmetry of a wave-propagation problem dictates which coordinate system is the most appropriate to express a solution in. The main focus of this chapter is to solve several well-known wave-propagation problems that are best suited for the cylindrical coordinate system. As we shall soon discover, solutions of the wave equation for a homogeneous medium will involve Bessel and Hankel functions in addition to complex exponentials.

In this section we shall solve the following linear, three-dimensional, lossless, homogeneous wave equation in the cylindrical coordinate system:

$$\nabla^2 \varphi(t, \mathbf{r}) - \frac{1}{c^2} \frac{\partial^2}{\partial t^2} \varphi(t, \mathbf{r}) = 0, \tag{3.1-1}$$

where

$$\nabla^2 = \frac{\partial^2}{\partial r^2} + \frac{1}{r} \frac{\partial}{\partial r} + \frac{1}{r^2} \frac{\partial^2}{\partial \psi^2} + \frac{\partial^2}{\partial z^2} \tag{3.1-2}$$

is the *Laplacian* expressed in the cylindrical coordinates (r, ψ, z) (see Fig. 3.1-1), $\varphi(t, \mathbf{r})$ is the *velocity potential* at time t and position $\mathbf{r} = (r, \psi, z)$ with units of square meters per second, and c is the *constant* speed of sound in meters per

151

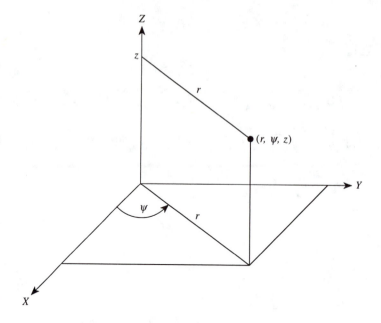

Figure 3.1-1. The cylindrical coordinates (r, ψ, z).

second. When the speed of sound is constant, the fluid medium is referred to as a *homogeneous medium*.

The wave equation given by Eq. (3.1-1) is referred to as being *homogeneous* since there is no source distribution (i.e., no input or forcing function). Solutions of the wave equation for several different source distribution models will be discussed in Section 3.2. Equation (3.1-1) describes the propagation of small-amplitude acoustic signals in ideal (nonviscous), homogeneous fluids (see Section 1.4).

We begin the solution of Eq. (3.1-1) by assuming that the velocity potential $\varphi(t, \mathbf{r})$ has a time-harmonic dependence, that is,

$$\varphi(t, \mathbf{r}) = \varphi_f(\mathbf{r}) \exp(+j2\pi ft), \qquad (3.1\text{-}3)$$

where $\varphi_f(\mathbf{r})$ is the spatial-dependent part of the velocity potential and f is the frequency in hertz. A time-harmonic field is one whose value at any time and point in space depends on a single frequency component. Substituting Eq. (3.1-3) into the time-dependent wave equation given by Eq. (3.1-1) yields

$$\nabla^2 \varphi_f(\mathbf{r}) + k^2 \varphi_f(\mathbf{r}) = 0, \qquad (3.1\text{-}4)$$

which is the time-independent *lossless Helmholtz equation*, where

$$k = 2\pi f/c = 2\pi/\lambda \qquad (3.1\text{-}5)$$

is the *wave number* in radians per meter and

$$c = f\lambda, \tag{3.1-6}$$

where λ is the *wavelength* in meters. If we can solve the Helmholtz equation given by Eq. (3.1-4), then substituting the solution $\varphi_f(\mathbf{r})$ into Eq. (3.1-3) will yield a time-harmonic solution of the wave equation given by Eq. (3.1-1), which is the main objective in this section. Time-harmonic solutions are very important, because solutions of the wave equation with arbitrary time dependence can be obtained from time-harmonic solutions by using Fourier transform techniques, as will be discussed in Sections 3.8 and 3.9.

The solution of the Helmholtz equation given by Eq. (3.1-4) will be obtained by using the *method of separation of variables*, that is, we assume a solution of the form

$$\boxed{\varphi_f(\mathbf{r}) = \varphi_f(r, \psi, z) = R(r)\Psi(\psi)Z(z).} \tag{3.1-7}$$

Substituting Eq. (3.1-7) into Eq. (3.1-4) and making use of Eq. (3.1-2) yields

$$\frac{1}{R(r)}\frac{d^2}{dr^2}R(r) + \frac{1}{rR(r)}\frac{d}{dr}R(r) + \frac{1}{r^2\Psi(\psi)}\frac{d^2}{d\psi^2}\Psi(\psi) + k^2$$

$$= -\frac{1}{Z(z)}\frac{d^2}{dz^2}Z(z), \tag{3.1-8}$$

where the second-order partial derivatives with respect to r, ψ, and z have been replaced by ordinary second-order derivatives, since the functions $R(r)$, $\Psi(\psi)$, and $Z(z)$ depend on only one of the independent variables r, ψ, and z, respectively. Since the left-hand side of Eq. (3.1-8) is a function of both r and ψ, and the right-hand side is a function of z, equality is possible only if both sides of Eq. (3.1-8) are equal to a constant; call it k_Z^2. The constant k_Z^2 is referred to as a *separation constant*. Therefore, since

$$-\frac{1}{Z(z)}\frac{d^2}{dz^2}Z(z) = k_Z^2, \tag{3.1-9}$$

Eq. (3.1-9) can be rewritten as

$$\frac{d^2}{dz^2}Z(z) + k_Z^2 Z(z) = 0. \tag{3.1-10}$$

Equation (3.1-10) is a second-order, homogeneous (no input or forcing function), ordinary differential equation (ODE) with *exact* solution

$$Z(z) = A_Z \exp(-jk_Z z) + B_Z \exp(+jk_Z z), \qquad (3.1\text{-}11)$$

where A_Z and B_Z are complex constants in general, whose values are determined by satisfying boundary conditions. If k_Z is *positive*, then the first term on the right-hand side of Eq. (3.1-11) represents a plane wave traveling in the *positive Z* direction, whereas the second term represents a plane wave traveling in the *negative Z* direction (see Section 2.1). The constant k_Z is the propagation-vector component in the Z direction with units of radians per meter (see Section 2.2.1).

Returning to the solution of Eq. (3.1-8), recall that the left-hand side of Eq. (3.1-8) must also be equal to the separation constant k_Z^2. Therefore,

$$\frac{1}{R(r)}\frac{d^2}{dr^2}R(r) + \frac{1}{rR(r)}\frac{d}{dr}R(r) + \frac{1}{r^2\Psi(\psi)}\frac{d^2}{d\psi^2}\Psi(\psi) + k^2 = k_Z^2, \quad (3.1\text{-}12)$$

or, multiplying both sides of Eq. (3.1-12) by r^2,

$$\frac{r^2}{R(r)}\frac{d^2}{dr^2}R(r) + \frac{r}{R(r)}\frac{d}{dr}R(r) + k_r^2 r^2 = -\frac{1}{\Psi(\psi)}\frac{d^2}{d\psi^2}\Psi(\psi), \quad (3.1\text{-}13)$$

where

$$k_r^2 = k^2 - k_Z^2. \qquad (3.1\text{-}14)$$

We shall demonstrate later in this section that k_r corresponds to the propagation-vector component in the horizontal radial direction r. Since the left-hand side of Eq. (3.1-13) is a function of r and the right-hand side is a function of ψ, equality is possible only if both sides of Eq. (3.1-13) are equal to a constant; call it n^2. The constant n^2 is also referred to as a separation constant. Therefore, since

$$-\frac{1}{\Psi(\psi)}\frac{d^2}{d\psi^2}\Psi(\psi) = n^2, \qquad (3.1\text{-}15)$$

Eq. (3.1-15) can be rewritten as

$$\frac{d^2}{d\psi^2}\Psi(\psi) + n^2\Psi(\psi) = 0. \qquad (3.1\text{-}16)$$

Equation (3.1-16) is a second-order, homogeneous ODE with *exact* solution

$$\Psi(\psi) = A'_\psi \exp(-jn\psi) + B'_\psi \exp(+jn\psi), \qquad (3.1\text{-}17)$$

or, with the use of Euler's formula,

$$\boxed{\Psi(\psi) = A_\psi \cos(n\psi) + B_\psi \sin(n\psi),} \qquad (3.1\text{-}18)$$

where

$$A_\psi = A'_\psi + B'_\psi \qquad (3.1\text{-}19)$$

and

$$B_\psi = j(B'_\psi - A'_\psi) \qquad (3.1\text{-}20)$$

are complex constants in general, whose values are determined by satisfying boundary conditions.

Finally, recall that the left-hand side of Eq. (3.1-13) must also be equal to n^2. Therefore,

$$\frac{r^2}{R(r)} \frac{d^2}{dr^2} R(r) + \frac{r}{R(r)} \frac{d}{dr} R(r) + k_r^2 r^2 = n^2, \qquad (3.1\text{-}21)$$

or, multiplying both sides of Eq. (3.1-21) by $R(r)/r^2$,

$$\frac{d^2}{dr^2} R(r) + \frac{1}{r} \frac{d}{dr} R(r) + \left[k_r^2 - \frac{n^2}{r^2} \right] R(r) = 0, \qquad (3.1\text{-}22)$$

where k_r^2 is given by Eq. (3.1-14). The next step is to transform Eq. (3.1-22) into *Bessel's differential equation*, which has known, exact solutions.

We begin the transformation by letting

$$R(r) = g(k_r r). \qquad (3.1\text{-}23)$$

Therefore,

$$\frac{d}{dr} R(r) = \frac{d}{dr} g(k_r r) = \frac{d}{d(k_r r)} g(k_r r) \frac{d}{dr} k_r r, \qquad (3.1\text{-}24)$$

or

$$\frac{d}{dr} R(r) = k_r \frac{d}{d(k_r r)} g(k_r r). \qquad (3.1\text{-}25)$$

Also,

$$\frac{d^2}{dr^2}R(r) = \frac{d}{dr}\frac{d}{dr}R(r) = \frac{d}{dr}\left[k_r\frac{d}{d(k_rr)}g(k_rr)\right] \tag{3.1-26}$$

$$= k_r\frac{d}{d(k_rr)}\frac{d}{dr}g(k_rr) \tag{3.1-27}$$

$$= k_r\frac{d}{d(k_rr)}\left[k_r\frac{d}{d(k_rr)}g(k_rr)\right], \tag{3.1-28}$$

or

$$\frac{d^2}{dr^2}R(r) = k_r^2\frac{d^2}{d(k_rr)^2}g(k_rr). \tag{3.1-29}$$

Substituting Eqs. (3.1-23), (3.1-25), and (3.1-29) into Eq. (3.1-22) yields *Bessel's differential equation*

$$\frac{d^2}{d\zeta^2}g(\zeta) + \frac{1}{\zeta}\frac{d}{d\zeta}g(\zeta) + \left[1 - \frac{n^2}{\zeta^2}\right]g(\zeta) = 0, \tag{3.1-30}$$

where

$$\zeta = k_rr \tag{3.1-31}$$

and

$$R(r) = g(\zeta) = g(k_rr). \tag{3.1-32}$$

The *exact* solution of Bessel's differential equation for *arbitrary* values of n (i.e., n can be real or complex, integer or noninteger, and positive or negative in value) is given by either

$$g(\zeta) = AJ_n(\zeta) + BY_n(\zeta) \tag{3.1-33}$$

or

$$g(\zeta) = AH_n^{(1)}(\zeta) + BH_n^{(2)}(\zeta), \tag{3.1-34}$$

where $J_n(\zeta)$ and $Y_n(\zeta)$ are the *nth-order Bessel functions of the first and second kind*, respectively, and $H_n^{(1)}(\zeta)$ and $H_n^{(2)}(\zeta)$ are the *nth-order Hankel functions of the first and second kind*, respectively, also known as *Bessel functions of the third kind*. The function $Y_n(\zeta)$ is also known as the *Neumann function* when n is an integer. By referring to Eqs. (3.1-31) through (3.1-34), we can finally write that the *exact* solution of Eq. (3.1-22) for *arbitrary* values of n is given by either

$$R(r) = A_r J_n(k_r r) + B_r Y_n(k_r r) \tag{3.1-35}$$

or

$$R(r) = A_r H_n^{(1)}(k_r r) + B_r H_n^{(2)}(k_r r), \tag{3.1-36}$$

where A_r and B_r are complex constants in general, whose values are determined by satisfying boundary conditions.

Therefore, upon substituting Eqs. (3.1-11), (3.1-18), and (3.1-35) into Eq. (3.1-7), we obtain

$$\varphi_f(r, \psi, z) = \left[A_r J_n(k_r r) + B_r Y_n(k_r r) \right]\left[A_\psi \cos(n\psi) + B_\psi \sin(n\psi) \right]$$
$$\times \left[A_Z \exp(-jk_Z z) + B_Z \exp(+jk_Z z) \right],$$

$$\tag{3.1-37}$$

and, upon substituting Eqs. (3.1-11), (3.1-18), and (3.1-36) into Eq. (3.1-7), we obtain

$$\varphi_f(r, \psi, z) = \left[A_r H_n^{(1)}(k_r r) + B_r H_n^{(2)}(k_r r) \right]\left[A_\psi \cos(n\psi) + B_\psi \sin(n\psi) \right]$$
$$\times \left[A_Z \exp(-jk_Z z) + B_Z \exp(+jk_Z z) \right],$$

$$\tag{3.1-38}$$

where [see Eq. (3.1-14)]

$$k_r^2 + k_Z^2 = k^2. \tag{3.1-39}$$

Both Eqs. (3.1-37) and (3.1-38) are solutions of the Helmholtz equation given by Eq. (3.1-4). Equation (3.1-39) relates the wave number k, given by Eq. (3.1-5), to the propagation-vector components k_r and k_Z. And, upon substituting either Eq. (3.1-37) or Eq. (3.1-38) into Eq. (3.1-3), we obtain the time-harmonic solution (in cylindrical coordinates) of the linear, three-dimensional, lossless, homogeneous wave equation given by Eq. (3.1-1).

We shall end our discussion of the solution of the wave equation in this section by making the following important comments concerning Eqs. (3.1-37) and (3.1-38). If the fluid medium occupies an angular sector $\psi_1 \leq \psi \leq \psi_2$ around the Z axis that is less than $360°$, then the value of n is determined by satisfying the boundary conditions at the ends of the angular sector at $\psi = \psi_1$ and $\psi = \psi_2$. As a result, the value of n can be arbitrary. If, for example, n is complex, then the argument $n\psi$ of the cosine and sine functions appearing in Eqs. (3.1-37) and (3.1-38) is also complex. Therefore, if

$$n = a + jb \tag{3.1-40}$$

where a and b are real numbers, then

$$\cos(n\psi) = \cos(a\psi + jb\psi) = \cos(a\psi)\cosh(b\psi) - j\sin(a\psi)\sinh(b\psi) \tag{3.1-41}$$

and

$$\sin(n\psi) = \sin(a\psi + jb\psi) = \sin(a\psi)\cosh(b\psi) + j\cos(a\psi)\sinh(b\psi), \tag{3.1-42}$$

where

$$\cosh\alpha = \tfrac{1}{2}(e^{\alpha} + e^{-\alpha}) \tag{3.1-43}$$

and

$$\sinh\alpha = \tfrac{1}{2}(e^{\alpha} - e^{-\alpha}) \tag{3.1-44}$$

are the hyperbolic cosine and hyperbolic sine functions, respectively.

However, if the fluid medium completely surrounds the Z axis, then $\Psi(\psi)$ given by Eq. (3.1-18) must be periodic with period 2π, so that the velocity potential $\varphi_f(r, \psi, z)$ will be a single-valued function of position. Therefore, in order to ensure that

$$\varphi_f(r, \psi, z) = \varphi_f(r, \psi + 2\pi, z), \tag{3.1-45}$$

the values of n must be integers, that is, $n = 0, \pm 1, \pm 2, \ldots$ (see Problem 3-1). Also, if $n = 0$, then [see Eq. (3.1-18)]

$$\Psi(\psi) = A_{\psi}, \qquad n = 0, \tag{3.1-46}$$

and as a result, the velocity potential $\varphi_f(r, \psi, z)$ given by either Eq. (3.1-37) or Eq. (3.1-38) is *not* a function of the azimuthal angle ψ. This is known as the *axisymmetric case*.

Also note that

$$J_0(0) = 1, \tag{3.1-47}$$

$$J_n(0) = 0, \qquad n = 1, 2, 3, \ldots, \tag{3.1-48}$$

and

$$Y_n(k_r r) \rightarrow -\infty \quad \text{as} \quad k_r r \rightarrow 0, \qquad n = 0, 1, 2, \dots . \qquad (3.1\text{-}49)$$

Therefore, because of Eq. (3.1-49), if the value of the velocity potential is required for $r \geq 0$, then we neglect $Y_n(k_r r)$ by setting the constant $B_r = 0$ in Eq. (3.1-37) so that the solution will be finite at $r = 0$ [see Eqs. (3.1-47) and (3.1-48)]. Equation (3.1-37), with $B_r = 0$, is best suited for cavity and waveguide problems which require solutions at $r = 0$. However, if the value of the velocity potential is *not* required at $r = 0$, then $Y_n(k_r r)$ should be retained in the solution given by Eq. (3.1-37). It should also be noted that

$$J_{-n}(k_r r) = (-1)^n J_n(k_r r), \qquad n = 1, 2, 3, \dots , \qquad (3.1\text{-}50)$$

and

$$Y_{-n}(k_r r) = (-1)^n Y_n(k_r r), \qquad n = 1, 2, 3, \dots . \qquad (3.1\text{-}51)$$

Next, if $k_r r \gg |n^2|/2$, then

$$H_n^{(1)}(k_r r) \approx \sqrt{\frac{2}{\pi k_r r}} \, \exp\left[+j\left(k_r r - \frac{n\pi}{2} - \frac{\pi}{4}\right)\right], \qquad k_r r \gg \frac{|n^2|}{2}, \qquad (3.1\text{-}52)$$

and

$$H_n^{(2)}(k_r r) \approx \sqrt{\frac{2}{\pi k_r r}} \, \exp\left[-j\left(k_r r - \frac{n\pi}{2} - \frac{\pi}{4}\right)\right], \qquad k_r r \gg \frac{|n^2|}{2}, \qquad (3.1\text{-}53)$$

for *arbitrary values* of n. Equations (3.1-52) and (3.1-53) are *asymptotic (far-field) approximations*. Since we chose $\exp(+j2\pi ft)$ for our time-harmonic dependence, Eq. (3.1-52) represents an *incoming wave* traveling in the direction of *decreasing r*, whereas Eq. (3.1-53) represents an *outgoing wave* traveling in the direction of *increasing r*, where k_r is the propagation-vector component in the radial direction. However, if we had chosen $\exp(-j2\pi ft)$ instead, then Eqs. (3.1-52) and (3.1-53) would represent *outgoing* and *incoming* waves, respectively. Therefore, Eq. (3.1-38), in conjunction with Eq. (3.1-52) and/or Eq. (3.1-53), is best suited for radiation and scattering problems, which typically require far-field solutions that do not include $r = 0$. In addition, by inspecting Eqs. (3.1-52) and (3.1-53), it can be seen that for a given value of n,

$$k_r r = \text{constant} \qquad (3.1\text{-}54)$$

defines a surface of constant phase or, in other words, a wavefront. Since r is the horizontal range measured from the Z axis (see Fig. 3.1-1), the wavefront defined by Eq. (3.1-54) is the surface of a cylinder. Hence the term "cylindrical wave." Also note that both Eqs. (3.1-52) and (3.1-53) indicate that the amplitude of a cylindrical wave decreases as $1/\sqrt{r}$. Although the following asymptotic (far-field) approximations are not used as often as Eqs. (3.1-52) and (3.1-53), they are worth being familiar with: if $k_r r \gg |n^2|/2$, then

$$J_n(k_r r) \approx \sqrt{\frac{2}{\pi k_r r}} \cos\left(k_r r - \frac{n\pi}{2} - \frac{\pi}{4}\right), \qquad k_r r \gg \frac{|n^2|}{2}, \qquad (3.1\text{-}55)$$

and

$$Y_n(k_r r) \approx \sqrt{\frac{2}{\pi k_r r}} \sin\left(k_r r - \frac{n\pi}{2} - \frac{\pi}{4}\right), \qquad k_r r \gg \frac{|n^2|}{2}, \qquad (3.1\text{-}56)$$

for arbitrary values of n.

The Bessel functions of the first, second, and third kind are related as follows:

$$H_n^{(1)}(\zeta) = J_n(\zeta) + jY_n(\zeta), \qquad (3.1\text{-}57)$$

$$H_n^{(2)}(\zeta) = J_n(\zeta) - jY_n(\zeta), \qquad (3.1\text{-}58)$$

$$W[J_n(\zeta), Y_n(\zeta)] = J_{n+1}(\zeta)Y_n(\zeta) - J_n(\zeta)Y_{n+1}(\zeta) = \frac{2}{\pi\zeta}, \qquad (3.1\text{-}59)$$

and

$$W[H_n^{(1)}(\zeta), H_n^{(2)}(\zeta)] = H_{n+1}^{(1)}(\zeta)H_n^{(2)}(\zeta) - H_n^{(1)}(\zeta)H_{n+1}^{(2)}(\zeta) = -j\frac{4}{\pi\zeta} \qquad (3.1\text{-}60)$$

for arbitrary values of n. Equations (3.1-59) and (3.1-60) are referred to as the *Wronskians* of $J_n(\zeta)$ and $Y_n(\zeta)$ and of $H_n^{(1)}(\zeta)$ and $H_n^{(2)}(\zeta)$, respectively, and are extremely useful when calculating time-average intensity vectors. The Bessel functions also obey the following recurrence relationships, where the generic function $f_n(\zeta)$ can represent $J_n(\zeta)$, $Y_n(\zeta)$, $H_n^{(1)}(\zeta)$, and $H_n^{(2)}(\zeta)$:

$$f_{n+1}(\zeta) = \frac{2n}{\zeta}f_n(\zeta) - f_{n-1}(\zeta) \qquad (3.1\text{-}61)$$

and

$$f_n'(\zeta) = f_{n-1}(\zeta) - \frac{n}{\zeta}f_n(\zeta) \qquad (3.1\text{-}62)$$

for arbitrary values of n, where

$$f_n'(\zeta) = \frac{d}{d\zeta}f_n(\zeta). \qquad (3.1\text{-}63)$$

3.2 Free-Space Propagation

3.2.1 Radiation from a Vibrating Cylinder

In this section we shall compute the radiated acoustic field produced by a vibrating cylinder. The cylinder has a radius of *a* meters and is *L* meters long (see Fig. 3.2-1). The radiated acoustic field is assumed to propagate in free space (i.e., in an unbounded medium), and as a result, there will be *no reflected waves*. It is important to note that we shall *not* consider radiation from the ends of the cylinder in our analysis. Therefore, the expressions we shall obtain for the radiated acoustic field are approximate. To date, there is no known exact analytic expression for the acoustic field produced by a vibrating cylinder of finite length. However, the techniques developed in Section 3.2 can be used, for example, to model, approximately, the radiation from a finite-length electroacoustic transducer that is cylindrical in shape (i.e., a continuous line source) or to model, very roughly, the radiation from a vibrating submarine hull.

We begin our analysis by making the following observations. First, since the radius of the cylinder is *a* meters, the velocity potential of the radiated acoustic

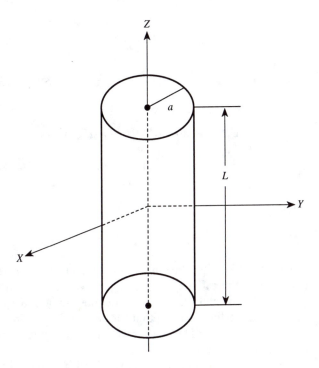

Figure 3.2-1. Vibrating cylinder of radius *a* meters and length *L* meters.

field will never have to be evaluated at $r = 0$. Therefore, we shall work with the solution of the Helmholtz equation given by Eq. (3.1-38). Second, since the fluid medium completely surrounds the cylinder, and hence the Z axis, the value of n in Eq. (3.1-38) must be an integer. Third, since we are dealing with free-space propagation, there are no reflected waves, and as a result, we set $A_r = 0$ in Eq. (3.1-38). Therefore, with the use of these three observations, Eq. (3.1-38) reduces to

$$\varphi_{f,n}(r,\psi,z) = H_n^{(2)}(k_r r)\left[A_n \cos(n\psi) + B_n \sin(n\psi)\right]$$

$$\times\left[A_Z \exp(-jk_Z z) + B_Z \exp(+jk_Z z)\right], \qquad (3.2\text{-}1)$$

where

$$A_n = B_r A_\psi, \qquad (3.2\text{-}2)$$

$$B_n = B_r B_\psi, \qquad (3.2\text{-}3)$$

and the subscript n has been added to $\varphi_f(r,\psi,z)$ in order to emphasize the dependence of the velocity potential on the value of the integer n. Since the sum of all possible solutions of the Helmholtz equation is also a solution, if we let

$$\varphi_f(r,\psi,z) = \sum_{n=0}^{\infty} \varphi_{f,n}(r,\psi,z), \qquad (3.2\text{-}4)$$

then substituting Eq. (3.2-1) into Eq. (3.2-4) yields

$$\varphi_f(r,\psi,z) = \sum_{n=0}^{\infty} H_n^{(2)}(k_r r)\left[A_n \cos(n\psi) + B_n \sin(n\psi)\right]$$
$$\times\left[A_Z \exp(-jk_Z z) + B_Z \exp(+jk_Z z)\right], \qquad (3.2\text{-}5)$$

where

$$k_r^2 + k_Z^2 = k^2. \qquad (3.2\text{-}6)$$

Equation (3.2-5) is the spatial-dependent part of the time-harmonic velocity potential of the radiated acoustic field where the values of the constants A_n, B_n, A_Z, and B_Z are determined by matching the boundary condition at the *surface* of the vibrating cylinder. Note that the summation in Eq. (3.2-5) only involves positive integer values of n. The reason why negative integer values were not included will be explained in Section 3.2.2.

In order to proceed further, we shall assume that the velocity vector of the surface of the vibrating cylinder has only a time-harmonic *radial* component given by

$$v_r(t, a, \psi, z) = v_{f,r}(a, \psi, z) \exp(+j2\pi ft) \tag{3.2-7}$$

where $v_{f,r}(a, \psi, z)$ is a known, real function of ψ and z. This is equivalent to knowing the radial component of the acoustic fluid velocity vector at the surface of the cylinder. The next step is to compute the radial component of the acoustic fluid velocity vector.

The acoustic fluid velocity vector in meters per second is given by

$$\mathbf{u}(t, \mathbf{r}) = \nabla \varphi(t, \mathbf{r}), \tag{3.2-8}$$

where

$$\nabla = \frac{\partial}{\partial r} \hat{r} + \frac{1}{r} \frac{\partial}{\partial \psi} \hat{\psi} + \frac{\partial}{\partial z} \hat{z} \tag{3.2-9}$$

is the *gradient* expressed in the cylindrical coordinates (r, ψ, z), and where $\hat{r}$, $\hat{\psi}$, and $\hat{z}$ are unit vectors in the r, ψ, and z directions, respectively. The radial component $u_r(t, \mathbf{r})$ is given by

$$u_r(t, \mathbf{r}) = \hat{r} \cdot \mathbf{u}(t, \mathbf{r}) = \hat{r} \cdot \nabla \varphi(t, \mathbf{r}), \tag{3.2-10}$$

or

$$u_r(t, \mathbf{r}) = \frac{\partial}{\partial r} \varphi(t, \mathbf{r}). \tag{3.2-11}$$

Therefore, at $r = a$,

$$u_r(t, a, \psi, z) = v_r(t, a, \psi, z) = \frac{\partial}{\partial r} \varphi(t, r, \psi, z) \Big|_{r=a}. \tag{3.2-12}$$

Substituting Eqs. (3.1-3), (3.2-5), and (3.2-7) into Eq. (3.2-12) yields

$$\boxed{\begin{aligned} v_{f,r}(a, \psi, z) = \; & k_r \sum_{n=0}^{\infty} H_n^{(2)\prime}(k_r a)[A_n \cos(n\psi) + B_n \sin(n\psi)] \\ & \times [A_Z \exp(-jk_Z z) + B_Z \exp(+jk_Z z)], \end{aligned}} \tag{3.2-13}$$

where

$$H_n^{(2)\prime}(k_r a) = \frac{d}{d(k_r r)} H_n^{(2)}(k_r r) \Big|_{r=a}. \tag{3.2-14}$$

The constants A_n, B_n, A_Z, and B_Z are determined from Eq. (3.2-13), which represents the boundary condition at the surface of the cylinder. Since the radial component is normal to the surface of the cylinder, this is an example of a Neumann boundary-value problem. Once the values of these constants are known, they are substituted into Eq. (3.2-5), which is then substituted into Eq. (3.1-3), which yields an approximate expression for the time-harmonic velocity potential of the radiated acoustic field produced by a vibrating cylinder of radius a meters and length L meters. Recall that we did not consider radiation from the ends of the cylinder in our analysis.

3.2.2 Monopole and Dipole Modes of Vibration

In the previous section, the radial component of the velocity vector of the surface of the vibrating cylinder was given by Eq. (3.2-7), where $v_{f,r}(a, \psi, z)$ was assumed to be a known, real function of ψ and z. Now, in this section, we shall investigate the simpler problem where the vibration of the surface of the cylinder is only a function of ψ, and is the same along the entire length of the cylinder, that is,

$$v_{f,r}(a, \psi, z) = V_0 x(\psi), \tag{3.2-15}$$

where V_0 is a real constant with units of meters per second, and $x(\psi)$ is a real, periodic function of ψ with period 2π. Both V_0 and $x(\psi)$ are assumed to be known. Since $x(\psi)$ is a real function, it can be represented by the following real, trigonometric form of the Fourier series:

$$x(\psi) = a_0 + \sum_{n=1}^{\infty} \left[a_n \cos(n\psi) + b_n \sin(n\psi) \right], \tag{3.2-16}$$

where the Fourier series coefficients are given by

$$a_0 = \frac{1}{2\pi} \int_0^{2\pi} x(\psi)\, d\psi, \tag{3.2-17}$$

which is the average value of $x(\psi)$;

$$a_n = \frac{1}{\pi} \int_0^{2\pi} x(\psi) \cos(n\psi)\, d\psi, \qquad n = 1, 2, 3, \ldots, \tag{3.2-18}$$

and

$$b_n = \frac{1}{\pi} \int_0^{2\pi} x(\psi) \sin(n\psi)\, d\psi, \qquad n = 1, 2, 3, \ldots. \tag{3.2-19}$$

Note that $b_0 = 0$. Also note that both $\cos(n\psi)$ and $\sin(n\psi)$ form an *orthogonal set of functions* in the closed interval $0 \le \psi \le 2\pi$, that is,

$$\int_0^{2\pi} \cos(m\psi) \cos(n\psi) \, d\psi = \pi\delta_{mn}, \qquad m, n = 1, 2, 3, \ldots, \qquad (3.2\text{-}20)$$

and

$$\int_0^{2\pi} \sin(m\psi) \sin(n\psi) \, d\psi = \pi\delta_{mn}, \qquad m, n = 1, 2, 3, \ldots, \qquad (3.2\text{-}21)$$

where

$$\delta_{mn} = \begin{cases} 1, & m = n, \\ 0, & m \ne n, \end{cases} \qquad (3.2\text{-}22)$$

is the *Kronecker delta*. In addition, $\cos(n\psi)$ and $\sin(n\psi)$ are *orthogonal* to each other in the closed interval $0 \le \psi \le 2\pi$, that is,

$$\int_0^{2\pi} \cos(m\psi) \sin(n\psi) \, d\psi = 0, \qquad m, n = 0, 1, 2, \ldots. \qquad (3.2\text{-}23)$$

Finally, $\cos(n\psi)$ and $\sin(n\psi)$ for $n = 1, 2, 3, \ldots$ have zero average values, since

$$\int_0^{2\pi} \cos(n\psi) \, d\psi = 0, \qquad n = 1, 2, 3, \ldots, \qquad (3.2\text{-}24)$$

and

$$\int_0^{2\pi} \sin(n\psi) \, d\psi = 0, \qquad n = 1, 2, 3, \ldots. \qquad (3.2\text{-}25)$$

By referring back to Eq. (3.2-16), it can be seen that only positive integer values of n are used in the real, trigonometric form of the Fourier series. This is the reason why only positive integer values of n were used in Eq. (3.2-5).

In order to determine the values of the unknown constants A_n, B_n, A_Z, and B_Z, we substitute Eqs. (3.2-15) and (3.2-16) into Eq. (3.2-13). Doing so yields

$$a_0 V_0 + V_0 \sum_{n=1}^{\infty} \left[a_n \cos(n\psi) + b_n \sin(n\psi) \right]$$

$$= k_r H_0^{(2)\prime}(k_r a) A_0 \left[A_Z \exp(-jk_Z z) + B_Z \exp(+jk_Z z) \right]$$

$$+ k_r \sum_{n=1}^{\infty} H_n^{(2)\prime}(k_r a) \left[A_n \cos(n\psi) + B_n \sin(n\psi) \right]$$

$$\times \left[A_Z \exp(-jk_Z z) + B_Z \exp(+jk_Z z) \right]. \qquad (3.2\text{-}26)$$

Since the left-hand side of Eq. (3.2-26) is not a function of z, in order to get rid of the z dependence on the right-hand side, we first set

$$k_z = 0 \qquad\qquad (3.2\text{-}27)$$

and then we set

$$A_z = B_z = 0.5. \qquad\qquad (3.2\text{-}28)$$

Substituting Eq. (3.2-27) into Eq. (3.2-6) yields

$$k_r = k. \qquad\qquad (3.2\text{-}29)$$

Therefore, substituting Eqs. (3.2-27) through (3.2-29) into Eq. (3.2-26) yields

$$a_0 V_0 + V_0 \sum_{n=1}^{\infty} a_n \cos(n\psi) + V_0 \sum_{n=1}^{\infty} b_n \sin(n\psi)$$

$$= k H_0^{(2)\prime}(ka) A_0 + k \sum_{n=1}^{\infty} H_n^{(2)\prime}(ka) A_n \cos(n\psi) + k \sum_{n=1}^{\infty} H_n^{(2)\prime}(ka) B_n \sin(n\psi).$$

$$(3.2\text{-}30)$$

By matching terms on both sides of Eq. (3.2-30), we obtain:

$$A_0 = \frac{V_0}{k H_0^{(2)\prime}(ka)} a_0, \qquad\qquad (3.2\text{-}31)$$

$$A_n = \frac{V_0}{k H_n^{(2)\prime}(ka)} a_n, \qquad n = 1,2,3,\ldots, \qquad\qquad (3.2\text{-}32)$$

$$B_n = \frac{V_0}{k H_n^{(2)\prime}(ka)} b_n, \qquad n = 1,2,3,\ldots. \qquad\qquad (3.2\text{-}33)$$

It should also be mentioned for completeness that the constants given by Eqs. (3.2-31) through (3.2-33) could have been obtained otherwise than by matching terms. For example, in order to solve for A_0, integrate both sides of Eq. (3.2-30) over ψ from 0 to 2π and use Eqs. (3.2-24) and (3.2-25). Next, in order to solve for A_n, multiply both sides of Eq. (3.2-30) by $\cos(m\psi)$, integrate both sides over ψ from 0 to 2π, and use Eqs. (3.2-20), (3.2-23), and (3.2-24). And finally, in order to solve for B_n, multiply both sides of Eq. (3.2-30) by $\sin(m\psi)$, integrate both sides over ψ from 0 to 2π, and use Eqs. (3.2-21), (3.2-23), and (3.2-25).

Substituting Eqs. (3.2-27) through (3.2-29) and Eqs. (3.2-31) through (3.2-33) into Eq. (3.2-5) yields

$$\varphi_f(r,\psi,z) = \frac{a_0 V_0}{k} \frac{H_0^{(2)}(kr)}{H_0^{(2)'}(ka)}$$

$$+ \frac{V_0}{k} \sum_{n=1}^{\infty} \frac{H_n^{(2)}(kr)}{H_n^{(2)'}(ka)} \left[a_n \cos(n\psi) + b_n \sin(n\psi) \right], \qquad r \geq a,$$

$$(3.2\text{-}34)$$

where the Fourier series coefficients a_0, a_n, and b_n are given by Eqs. (3.2-17) through (3.2-19), respectively. Equation (3.2-34) represents the spatial-dependent part of the time-harmonic velocity potential of the radiated acoustic field. The first term on the right-hand side of Eq. (3.2-34) represents the axisymmetric (no ψ dependence) term. Note that the radiated field does not depend on z, since the vibration of the surface of the cylinder was assumed not to depend on z [see Eq. (3.2-15)].

The time-harmonic velocity potential of the radiated acoustic field is obtained by substituting Eq. (3.2-34) into Eq. (3.1-3).

Monopole Mode of Vibration
If the real, periodic function

$$x(\psi) = 1, \qquad (3.2\text{-}35)$$

then substituting Eqs. (3.2-15) and (3.2-35) into Eq. (3.2-7) yields

$$v_r(t,a,\psi,z) = V_0 \exp(+j2\pi ft). \qquad (3.2\text{-}36)$$

Equation (3.2-36) defines the special case of *uniform radiation*, that is, the surface of the cylinder is vibrating (expanding and contracting) uniformly in the radial direction with an amplitude of V_0 meters per second, at a frequency of f hertz. This special case is also known as the *monopole mode* or *breathing mode* of vibration. Substituting Eq. (3.2-35) into Eqs. (3.2-17) through (3.2-19) yields the following values for the Fourier series coefficients [see Eqs. (3.2-24) and (3.2-25)]:

$$a_0 = 1, \qquad (3.2\text{-}37)$$

$$a_n = 0, \qquad n = 1,2,3,\ldots, \qquad (3.2\text{-}38)$$

$$b_n = 0, \qquad n = 1,2,3,\ldots. \qquad (3.2\text{-}39)$$

Therefore, upon substituting Eqs. (3.2-37) through (3.2-39) into Eq. (3.2-34), we obtain

$$\varphi_f(r, \psi, z) = \frac{V_0}{k} \frac{H_0^{(2)}(kr)}{H_0^{(2)\prime}(ka)}, \qquad r \geq a, \tag{3.2-40}$$

or, since [let $f_n(\zeta) = H_n^{(2)}(\zeta)$ and $n = 0$ in Eqs. (3.1-61) and (3.1-62)]

$$H_0^{(2)\prime}(ka) = -H_1^{(2)}(ka), \tag{3.2-41}$$

substituting Eq. (3.2-41) into Eq. (3.2-40) yields

$$\varphi_f(r, \psi, z) = -\frac{V_0}{k} \frac{H_0^{(2)}(kr)}{H_1^{(2)}(ka)}, \qquad r \geq a. \tag{3.2-42}$$

Note that for the monopole mode of vibration, the velocity potential of the radiated acoustic field given by either Eq. (3.2-40) or Eq. (3.2-42) is only a function of horizontal range r. With the use of Eq. (3.2-42), we shall compute the acoustic pressure, the acoustic fluid velocity vector, the time-average intensity vector, and the time-average power produced by the vibrating cylinder.

The time-harmonic radiated acoustic pressure can be expressed as

$$p(t, \mathbf{r}) = p_f(\mathbf{r}) \exp(+j2\pi ft), \tag{3.2-43}$$

where (see Example 2.2-1)

$$p_f(\mathbf{r}) = -jk\rho_0 c \varphi_f(\mathbf{r}), \tag{3.2-44}$$

and upon substituting Eq. (3.2-42) into Eq. (3.2-44), we obtain

$$p_f(r, \psi, z) = j\rho_0 c V_0 \frac{H_0^{(2)}(kr)}{H_1^{(2)}(ka)}, \qquad r \geq a, \tag{3.2-45}$$

for the monopole mode of vibration. With the use of Eq. (3.1-53), the far-field approximation of Eq. (3.2-45) is given by

$$p_f(r, \psi, z) \approx j \frac{\rho_0 c V_0}{H_1^{(2)}(ka)} \sqrt{\frac{2}{\pi kr}} \exp\left[-j\left(kr - \frac{\pi}{4}\right)\right], \qquad kr \gg 1. \tag{3.2-46}$$

Similarly, the time-harmonic radiated acoustic fluid velocity vector can be expressed as

$$\mathbf{u}(t,\mathbf{r}) = \mathbf{u}_f(\mathbf{r})\exp(+j2\pi ft), \qquad (3.2\text{-}47)$$

where

$$\mathbf{u}_f(\mathbf{r}) = \nabla\varphi_f(\mathbf{r}). \qquad (3.2\text{-}48)$$

Substituting Eq. (3.2-42) into Eq. (3.2-48) and using the gradient given by Eq. (3.2-9) yields

$$\mathbf{u}_f(r,\psi,z) = -\frac{V_0}{kH_1^{(2)}(ka)}\frac{d}{dr}H_0^{(2)}(kr)\hat{r}, \qquad r \geq a. \qquad (3.2\text{-}49)$$

Since

$$\frac{d}{dr}H_0^{(2)}(kr) = \frac{d}{d(kr)}H_0^{(2)}(kr)\frac{d}{dr}kr = kH_0^{(2)\prime}(kr), \qquad (3.2\text{-}50)$$

and since [see Eq. (3.2-41)]

$$H_0^{(2)\prime}(kr) = -H_1^{(2)}(kr), \qquad (3.2\text{-}51)$$

substituting Eqs. (3.2-50) and (3.2-51) into Eq. (3.2-49) yields

$$\boxed{\mathbf{u}_f(r,\psi,z) = V_0\frac{H_1^{(2)}(kr)}{H_1^{(2)}(ka)}\hat{r}, \qquad r \geq a,} \qquad (3.2\text{-}52)$$

for the monopole mode of vibration. Note that at the surface of the cylinder, at $r = a$,

$$\mathbf{u}(t,a,\psi,z) = \mathbf{u}_f(a,\psi,z)\exp(+j2\pi ft) = V_0\exp(+j2\pi ft)\,\hat{r}, \qquad (3.2\text{-}53)$$

and as a result, the radial component is given by

$$u_r(t,a,\psi,z) = \hat{r}\cdot\mathbf{u}(t,a,\psi,z) = V_0\exp(+j2\pi ft), \qquad (3.2\text{-}54)$$

which is identical to the boundary condition given by Eq. (3.2-36). With the use of Eq. (3.1-53), the far-field approximation of Eq. (3.2-52) is given by

$$\boxed{\mathbf{u}_f(r,\psi,z) \approx j\frac{V_0}{H_1^{(2)}(ka)}\sqrt{\frac{2}{\pi kr}}\exp\left[-j\left(kr-\frac{\pi}{4}\right)\right]\hat{r}, \qquad kr \gg 1,} \qquad (3.2\text{-}55)$$

where use has been made of the fact that $\exp(+j\pi/2) = j$. If Eq. (3.2-55) is rewritten as

$$\mathbf{u}_f(r, \psi, z) = u_f(r, \psi, z)\hat{r}, \tag{3.2-56}$$

where

$$u_f(r, \psi, z) \approx j\frac{V_0}{H_1^{(2)}(ka)}\sqrt{\frac{2}{\pi kr}}\exp\left[-j\left(kr - \frac{\pi}{4}\right)\right], \qquad kr \gg 1, \tag{3.2-57}$$

is the *complex speed*, then by comparing Eqs. (3.2-46) and (3.2-57), it can be seen that

$$p_f(r, \psi, z) = \rho_0 c u_f(r, \psi, z), \qquad kr \gg 1, \tag{3.2-58}$$

which is a plane-wave relationship (see Example 2.2-1). Therefore, the radiated acoustic field behaves like a plane wave in the far-field region of the cylinder.

Now that we have expressions for both the acoustic pressure and the acoustic fluid velocity vector, we can compute the time-average intensity vector. Recall that the time-average intensity vector for time-harmonic acoustic fields is given by (see Section 1.5)

$$\mathbf{I}_{avg}(\mathbf{r}) = \tfrac{1}{2}\,\mathrm{Re}\{p_f(\mathbf{r})\mathbf{u}_f^*(\mathbf{r})\}. \tag{3.2-59}$$

Substituting Eqs. (3.2-45) and (3.2-52) into Eq. (3.2-59) yields

$$\mathbf{I}_{avg}(r, \psi, z) = \frac{\rho_0 c}{2}\frac{V_0^2}{\left|H_1^{(2)}(ka)\right|^2}\,\mathrm{Re}\{jH_0^{(2)}(kr)H_1^{(2)*}(kr)\}\hat{r}, \qquad r \geq a. \tag{3.2-60}$$

By referring to Eq. (3.1-58), we can write that

$$H_0^{(2)}(kr) = J_0(kr) - jY_0(kr) \tag{3.2-61}$$

and

$$H_1^{(2)*}(kr) = J_1(kr) + jY_1(kr), \tag{3.2-62}$$

since $J_n(\cdot)$ and $Y_n(\cdot)$ are real functions for integer values of n. And with the use of Eqs. (3.2-61), (3.2-62), and (3.1-59), it can be shown that

$$\mathrm{Re}\{jH_0^{(2)}(kr)H_1^{(2)*}(kr)\} = J_1(kr)Y_0(kr) - J_0(kr)Y_1(kr) = \frac{2}{\pi kr}. \tag{3.2-63}$$

Therefore, upon substituting Eq. (3.2-63) into Eq. (3.2-60), we finally obtain

$$\mathbf{I}_{\text{avg}}(r, \psi, z) = \frac{\rho_0 c}{\pi k r} \frac{V_0^2}{\left| H_1^{(2)}(ka) \right|^2} \hat{r}, \qquad r \geq a, \tag{3.2-64}$$

which is the time-average intensity vector for the monopole mode of vibration.

The last calculation to be performed is that of the time-average radiated power. If we enclose the vibrating cylinder with another cylinder of radius r meters and length L meters, then the time-average power is given by (see Section 1.5)

$$P_{\text{avg}} = \oint_S \mathbf{I}_{\text{avg}}(r, \psi, z) \cdot \mathbf{dS}, \tag{3.2-65}$$

where

$$\mathbf{dS} = r \, d\psi \, dz \, \hat{r}. \tag{3.2-66}$$

Note that since the time-average intensity vector given by Eq. (3.2-64) only has a radial component, we need only concern ourselves with the infinitesimal vector surface area element $\mathbf{dS}$ in the $\hat{r}$ direction given by Eq. (3.2-66), since there is no power flow in the $\pm \hat{z}$ direction. Recall that we ignored radiation at the ends of the cylinder in our analysis. Therefore, substituting Eqs. (3.2-64) and (3.2-66) into Eq. (3.2-65) yields

$$P_{\text{avg}} = \frac{\rho_0 c}{\pi k} \frac{V_0^2}{\left| H_1^{(2)}(ka) \right|^2} \int_{-L/2}^{L/2} \int_0^{2\pi} d\psi \, dz \tag{3.2-67}$$

or

$$P_{\text{avg}} = 2L \frac{\rho_0 c}{k} \frac{V_0^2}{\left| H_1^{(2)}(ka) \right|^2}, \tag{3.2-68}$$

which is the time-average power radiated by the cylinder in the monopole mode of vibration.

Dipole Mode of Vibration

If the real, periodic function

$$x(\psi) = \cos \psi, \tag{3.2-69}$$

then substituting Eqs. (3.2-15) and (3.2-69) into Eq. (3.2-7) yields

$$v_r(t, a, \psi, z) = V_0 \cos \psi \exp(+j2\pi ft). \tag{3.2-70}$$

Equation (3.2-70) defines the special case of the *dipole mode of vibration*. Evaluating Eq. (3.2-70) at $\psi = 0$, $\pi/2$, π, $3\pi/2$, and 2π yields

$$v_r(t, a, 0, z) = V_0 \exp(+j2\pi ft), \tag{3.2-71}$$

$$v_r(t, a, \pi/2, z) = 0, \tag{3.2-72}$$

$$v_r(t, a, \pi, z) = -V_0 \exp(+j2\pi ft), \tag{3.2-73}$$

$$v_r(t, a, 3\pi/2, z) = 0, \tag{3.2-74}$$

and

$$v_r(t, a, 2\pi, z) = V_0 \exp(+j2\pi ft). \tag{3.2-75}$$

By comparing Eqs. (3.2-71) and (3.2-73), it can be seen that when the surface of the cylinder at $\psi = 0$ is expanding in the positive radial direction, the surface of the cylinder at $\psi = \pi$ is contracting in the negative radial direction, and vice versa. Hence the term "dipole mode of vibration." Equations (3.2-72) and (3.2-74) indicate that the surface of the cylinder at $\psi = \pi/2$ and $\psi = 3\pi/2$ are *nodes*, that is, there is no vibration at these points.

Substituting Eq. (3.2-69) into Eqs. (3.2-17) through (3.2-19) and making use of Eqs. (3.2-20), (3.2-23), and (3.2-24) yields the following values for the Fourier series coefficients:

$$a_0 = 0, \tag{3.2-76}$$

$$a_1 = 1, \tag{3.2-77}$$

$$a_n = 0, \quad n = 2, 3, 4, \ldots, \tag{3.2-78}$$

$$b_n = 0, \quad n = 1, 2, 3, \ldots. \tag{3.2-79}$$

Therefore, upon substituting Eqs. (3.2-76) through (3.2-79) into Eq. (3.2-34), we obtain

$$\boxed{\varphi_f(r, \psi, z) = \frac{V_0}{k} \frac{H_1^{(2)}(kr)}{H_1^{(2)\prime}(ka)} \cos \psi, \quad r \geq a,} \tag{3.2-80}$$

which is the spatial-dependent part of the time-harmonic velocity potential of the radiated acoustic field for the dipole mode of vibration.

3.3 The Cylindrical Cavity

In this section we shall derive expressions for the velocity potential and the acoustic pressure inside a cylindrical cavity of radius a meters and length L meters as shown in Fig. 3.3-1. This problem corresponds to solving for the acoustic field inside of a *closed* tube or pipe. Let us assume that all of the surfaces inside the cavity are *ideal rigid boundaries*. Therefore, the normal component of the acoustic fluid velocity vector must be equal to zero at all points on the ideal rigid boundaries (see Example 2.3-1). This problem is an example of a *Neumann boundary-value problem*. Also recall from Example 2.3-1 that the reflection coefficient of an ideal rigid boundary is equal to one, and as a result, time-average acoustic power cannot escape from inside the cavity. The solution of the Helmholtz equation is given by *standing waves* in the r and Z directions due to reflections.

Since values of the acoustic field inside the cavity at $r = 0$ are required, we shall use the solution of the Helmholtz equation given by Eq. (3.1-37) with $B_r = 0$ as the starting point. Also, since the fluid medium inside the cavity completely surrounds the Z axis, the values of n must be integers. Therefore, with the use of these two

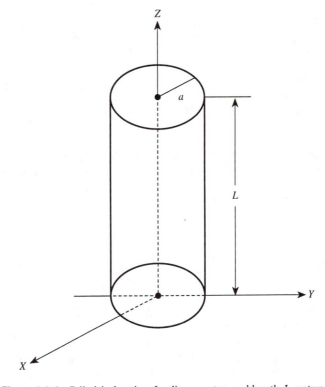

Figure 3.3-1. Cylindrical cavity of radius a meters and length L meters.

observations, Eq. (3.1-37) reduces to

$$\varphi_f(r, \psi, z) = J_n(k_r r)\left[A_n \cos(n\psi) + B_n \sin(n\psi)\right]$$

$$\times \left[A_Z \exp(-jk_Z z) + B_Z \exp(+jk_Z z)\right], \qquad (3.3\text{-}1)$$

where

$$A_n = A_r A_\psi, \qquad (3.3\text{-}2)$$

$$B_n = A_r B_\psi, \qquad (3.3\text{-}3)$$

and

$$k_r^2 + k_Z^2 = k^2. \qquad (3.3\text{-}4)$$

The values of the unknown constants A_n, B_n, A_Z, and B_Z are determined by satisfying the boundary conditions.

The time-harmonic acoustic fluid velocity vector is given by

$$\mathbf{u}(t, \mathbf{r}) = \mathbf{u}_f(\mathbf{r}) \exp(+j2\pi ft), \qquad (3.3\text{-}5)$$

where

$$\mathbf{u}_f(\mathbf{r}) = \nabla \varphi_f(\mathbf{r}), \qquad (3.3\text{-}6)$$

and

$$\nabla = \frac{\partial}{\partial r}\hat{r} + \frac{1}{r}\frac{\partial}{\partial \psi}\hat{\psi} + \frac{\partial}{\partial z}\hat{z} \qquad (3.3\text{-}7)$$

is the gradient expressed in the cylindrical coordinates (r, ψ, z). The normal component of the acoustic fluid velocity vector can be obtained from Eq. (3.3-5) as follows:

$$u_n(t, \mathbf{r}) = \hat{n} \cdot \mathbf{u}(t, \mathbf{r}) = u_{f,n}(\mathbf{r}) \exp(+j2\pi ft), \qquad (3.3\text{-}8)$$

where

$$u_{f,n}(\mathbf{r}) = \hat{n} \cdot \nabla \varphi_f(\mathbf{r}), \qquad (3.3\text{-}9)$$

and $\hat{n}$ is a unit vector normal (perpendicular) to the boundary. In order to maintain consistency in the analysis that follows, we shall choose $\hat{n}$ such that it points in the outward direction, away from the inside of the cavity. However, it would be equally valid to decide, for example, to choose $\hat{n}$ such that it always points in the positive r or Z direction, whichever is appropriate for the boundary under consideration.

The first boundary condition is at $z = 0$, which corresponds to the end of the closed cylinder lying in the XY plane (see Fig. 3.3-1). The unit vector that is normal to the boundary at $z = 0$ and points in the outward direction away from the inside of the cavity is

$$\hat{n} = -\hat{z}. \tag{3.3-10}$$

With the use of Eq. (3.3-10), and since the normal component of the acoustic fluid velocity vector must be equal to zero on the rigid boundary at $z = 0$, Eq. (3.3-9) becomes

$$u_{f,n}(r,\psi,0) = -\frac{\partial}{\partial z}\varphi_f(r,\psi,z)\bigg|_{z=0} = 0. \tag{3.3-11}$$

Substituting Eq. (3.3-1) into Eq. (3.3-11) yields

$$-jk_z J_n(k_r r)\big[A_n \cos(n\psi) + B_n \sin(n\psi)\big](A_Z - B_Z) = 0, \tag{3.3-12}$$

and since Eq. (3.3-12) must hold for all allowed values of r and ψ at $z = 0$,

$$\boxed{B_Z = A_Z.} \tag{3.3-13}$$

The second boundary condition is at $z = L$, which corresponds to the end of the closed cylinder lying in a plane parallel to the XY plane (see Fig. 3.3-1). The unit vector that is normal to the boundary at $z = L$ and points in the outward direction away from the inside of the cavity is

$$\hat{n} = \hat{z}. \tag{3.3-14}$$

With the use of Eq. (3.3-14), and since the normal component of the acoustic fluid velocity vector must be equal to zero on the rigid boundary at $z = L$, Eq. (3.3-9) becomes

$$u_{f,n}(r,\psi,L) = \frac{\partial}{\partial z}\varphi_f(r,\psi,z)\bigg|_{z=L} = 0. \tag{3.3-15}$$

Substituting Eqs. (3.3-1) and (3.3-13) into Eq. (3.3-15) yields

$$-2A_Z k_z \sin(k_z L)\, J_n(k_r r)\big[A_n \cos(n\psi) + B_n \sin(n\psi)\big] = 0, \tag{3.3-16}$$

and since Eq. (3.3-16) must hold for all allowed values of r and ψ at $z = L$,

$$\sin(k_z L) = 0, \tag{3.3-17}$$

which implies that

$$k_z L = m\pi, \qquad m = 0, 1, 2, \ldots, \qquad (3.3\text{-}18)$$

or

$$\boxed{k_z = k_{z_m} = m\pi/L, \qquad m = 0, 1, 2, \ldots .} \qquad (3.3\text{-}19)$$

Therefore, in order to satisfy the boundary condition at $z = L$, the propagation-vector component in the Z direction is only allowed certain *discrete values* rather than a continuum of values. Note that only positive integer values for m are used, in order to keep $k_z = k_{z_m}$ positive. Because of Eq. (3.3-1), a positive k_z value allows for wave propagation in both the positive and negative Z directions.

The third boundary condition is at $r = a$. The unit vector that is normal to the boundary at $r = a$ and points in the outward direction away from the inside of the cavity is

$$\hat{n} = \hat{r}. \qquad (3.3\text{-}20)$$

With the use of Eq. (3.3-20), and since the normal component of the acoustic fluid velocity vector must be equal to zero on the rigid boundary at $r = a$, Eq. (3.3-9) becomes

$$u_{f,n}(a, \psi, z) = \left. \frac{\partial}{\partial r} \varphi_f(r, \psi, z) \right|_{r=a} = 0. \qquad (3.3\text{-}21)$$

Substituting Eqs. (3.3-1), (3.3-13), and (3.3-19) into Eq. (3.3-21) yields

$$2 A_z k_r \cos(k_{z_m} z) J_n'(k_r a) \left[A_n \cos(n\psi) + B_n \sin(n\psi) \right] = 0, \qquad (3.3\text{-}22)$$

where

$$J_n'(k_r a) = \left. \frac{d}{d(k_r r)} J_n(k_r r) \right|_{r=a}. \qquad (3.3\text{-}23)$$

Since Eq. (3.3-22) must hold for all allowed values of ψ and z at $r = a$,

$$J_n'(k_r a) = 0. \qquad (3.3\text{-}24)$$

Note that for a given value of integer n, there are infinitely many values of $k_r a$ that will satisfy Eq. (3.3-24). If we let $\alpha_{n,l}$ represent the roots of the equation

$$J_n'(\alpha_{n,l}) = 0, \qquad n, l = 0, 1, 2, \ldots, \qquad (3.3\text{-}25)$$

where the index l counts the number of roots for a given value of n, then

$$k_r a = \alpha_{n,l}, \qquad n, l = 0, 1, 2, \ldots,$$ (3.3-26)

or

$$\boxed{k_r = k_{r_{n,l}} = \frac{\alpha_{n,l}}{a}, \qquad n, l = 0, 1, 2, \ldots.}$$ (3.3-27)

Therefore, in order to satisfy the boundary condition at $r = a$, the propagation-vector component in the horizontal radial direction is only allowed certain discrete values rather than a continuum of values.

Now that we have satisfied all the boundary conditions at all the surfaces inside the cylindrical cavity, we are in a position to obtain expressions for the velocity potential and the acoustic pressure inside the cylindrical cavity. Substituting Eqs. (3.3-13), (3.3-19), and (3.3-27) into Eq. (3.3-1) yields

$$\varphi_{f,mn,l}(r, \psi, z) = J_n(k_{r_{n,l}} r)[A_{mn,l} \cos(n\psi) + B_{mn,l} \sin(n\psi)] \cos(k_{z_m} z),$$

$$m, n, l = 0, 1, 2, \ldots, \quad (3.3\text{-}28)$$

where

$$A_{mn,l} = 2 A_n A_Z$$ (3.3-29)

and

$$B_{mn,l} = 2 B_n A_Z,$$ (3.3-30)

or

$$\boxed{\varphi_{f,mn,l}(r, \psi, z) = J_n\!\left(\alpha_{n,l} \frac{r}{a}\right)[A_{mn,l} \cos(n\psi) + B_{mn,l} \sin(n\psi)] \cos\!\left(\frac{m\pi}{L} z\right),}$$

$$m, n, l = 0, 1, 2, \ldots,$$

(3.3-31)

where the subscript mn, l has been added to the velocity potential and the unknown constants because their values depend on the discrete values of the propagation-vector components. Equation (3.3-31) is the spatial-dependent part of the velocity potential of the acoustic field inside a cylindrical cavity with rigid walls corresponding to the (m, n, l) *eigenfunction*, or *normal mode* (natural mode of vibration).

Since the propagation-vector components are only allowed certain discrete values, we shall show next that the frequencies of vibration of the acoustic field inside the cavity are also only allowed certain discrete values. Since $k = 2\pi f/c$, solving for the frequency f from Eq. (3.3-4) yields

$$f = \frac{c}{2\pi} \sqrt{k_r^2 + k_z^2}, \qquad (3.3\text{-}32)$$

and upon substituting Eqs. (3.3-19) and (3.3-27) into Eq. (3.3-32), we obtain

$$\boxed{f = f_{mn,l} = \frac{c}{2\pi} \sqrt{\left(\frac{\alpha_{n,l}}{a}\right)^2 + \left(\frac{m\pi}{L}\right)^2}, \qquad m,n,l = 0,1,2,\dots.} \qquad (3.3\text{-}33)$$

Equation (3.3-33) is the (m, n, l) *eigenfrequency*, or natural frequency of vibration (in hertz), of the acoustic field inside a cylindrical cavity with rigid walls.

Therefore, with the use of Eq. (3.1-3), the time-harmonic velocity potential inside a cylindrical cavity with rigid walls corresponding to the (m, n, l) normal mode is given by

$$\varphi_{mn,l}(t, \mathbf{r}) = \varphi_{f,mn,l}(\mathbf{r}) \exp(+j2\pi f_{mn,l}t), \qquad m,n,l = 0,1,2,\dots. \qquad (3.3\text{-}34)$$

The complete time-harmonic normal-mode solution for the velocity potential is obtained by summing the contributions from *all* the normal modes, that is,

$$\varphi(t, \mathbf{r}) = \sum_{m=0}^{\infty} \sum_{n=0}^{\infty} \sum_{l=0}^{\infty} \varphi_{f,mn,l}(\mathbf{r}) \exp(+j2\pi f_{mn,l}t), \qquad (3.3\text{-}35)$$

and upon substituting Eq. (3.3-31) into Eq. (3.3-35), we obtain

$$\boxed{\begin{aligned} \varphi(t, r, \psi, z) &= \sum_{m=0}^{\infty} \sum_{n=0}^{\infty} \sum_{l=0}^{\infty} J_n\left(\alpha_{n,l}\frac{r}{a}\right)\left[A_{mn,l}\cos(n\psi) + B_{mn,l}\sin(n\psi)\right] \\ &\quad \times \cos\left(\frac{m\pi}{L}z\right)\exp(+j2\pi f_{mn,l}t), \end{aligned}}$$

$$(3.3\text{-}36)$$

where $f_{mn,l}$ is given by Eq. (3.3-33).

With the use of Eq. (2.2-28), the time-harmonic acoustic pressure inside a cylindrical cavity with rigid walls corresponding to the (m, n, l) normal mode can be expressed as

$$p_{mn,l}(t, \mathbf{r}) = -j2\pi f_{mn,l}\rho_0 \varphi_{mn,l}(t, \mathbf{r}), \qquad m,n,l = 0,1,2,\dots. \qquad (3.3\text{-}37)$$

Therefore, the complete time-harmonic normal-mode solution for the acoustic pressure is given by

$$p(t, \mathbf{r}) = -j2\pi\rho_0 \sum_{m=0}^{\infty} \sum_{n=0}^{\infty} \sum_{l=0}^{\infty} f_{mn,l} \varphi_{f,mn,l}(\mathbf{r}) \exp(+j2\pi f_{mn,l}t), \quad (3.3\text{-}38)$$

and upon substituting Eq. (3.3-31) into Eq. (3.3-38), we obtain

$$
\begin{aligned}
p(t, r, \psi, z) = &-j2\pi\rho_0 \\
&\times \sum_{m=0}^{\infty} \sum_{n=0}^{\infty} \sum_{l=0}^{\infty} f_{mn,l} J_n\!\left(\alpha_{n,l}\frac{r}{a}\right)[A_{mn,l}\cos(n\psi) + B_{mn,l}\sin(n\psi)] \\
&\times \cos\!\left(\frac{m\pi}{L}z\right) \exp(+j2\pi f_{mn,l}t),
\end{aligned}
$$

$$(3.3\text{-}39)$$

where $f_{mn,l}$ is given by Eq. (3.3-33).

Let us conclude this section with the following additional remarks. Although we have satisfied all the boundary conditions at all the surfaces inside the cavity, the constants $A_{mn,l}$ and $B_{mn,l}$ are still unknown. In order to determine their values, the boundary condition at the location of a sound source inside the cavity must be satisfied.

3.4 Waveguide with Circular Cross-Sectional Area

In this section we shall derive expressions for the velocity potential and the acoustic pressure inside a waveguide with a circular cross-sectional area of πa^2 square meters, as illustrated in Fig. 3.4-1. Let us assume that the surface inside the waveguide is an *ideal rigid boundary*, with the exception of a vibrating surface (i.e., a sound source) located at $z = 0$. Also assume that the fluid media inside and outside the waveguide are the same, so that there is no impedance mismatch (i.e., no reflections) at the opened end of the waveguide.

The waveguide illustrated in Fig. 3.4-1 is nothing more than a cylindrical cavity with one end removed. Therefore, there will be *standing waves* in the horizontal, radial direction and a *traveling wave* in the positive Z direction. Based on our discussion of the cylindrical cavity with ideal rigid boundaries in Section 3.3, we can write immediately that the velocity potential of the time-harmonic eigenfunctions (normal modes) inside the rigid-walled waveguide is given by

$$\varphi_{n,l}(t, \mathbf{r}) = \varphi_{f,n,l}(\mathbf{r}) \exp(+j2\pi ft), \qquad n, l = 0, 1, 2, \ldots, \quad (3.4\text{-}1)$$

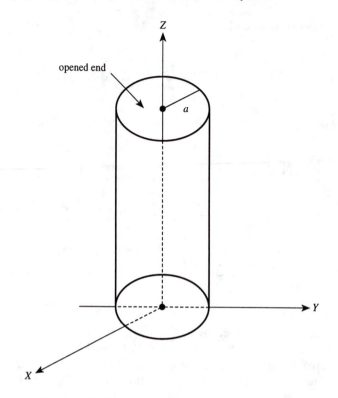

Figure 3.4-1. Waveguide with circular cross-sectional area.

where

$$\varphi_{f,n,l}(r,\psi,z) = J_n\left(\alpha_{n,l}\frac{r}{a}\right)\left[A_{n,l}\cos(n\psi) + B_{n,l}\sin(n\psi)\right]\exp(-jk_z z),$$

$$n,l = 0,1,2,\ldots, \quad (3.4\text{-}2)$$

and

$$J_n'(\alpha_{n,l}) = 0, \quad n,l = 0,1,2,\ldots, \quad (3.4\text{-}3)$$

where the index l counts the number of roots for a given value of n in Eq. (3.4-3). Since

$$k_r^2 + k_z^2 = k^2, \quad (3.4\text{-}4)$$

we have

$$k_z = \sqrt{k^2 - k_r^2}, \quad (3.4\text{-}5)$$

and upon substituting Eq. (3.3-27) into Eq. (3.4-5), we obtain

$$k_Z = k_{Z_{n,l}} = k\sqrt{1 - (f_{n,l}/f)^2}\,, \qquad f \geq f_{n,l}, \qquad (3.4\text{-}6)$$

where

$$k = 2\pi f/c = 2\pi/\lambda \qquad (3.4\text{-}7)$$

is the wave number and

$$\boxed{f_{n,l} = \frac{c}{2\pi a}\alpha_{n,l}\,, \qquad n, l = 0, 1, 2, \ldots,} \qquad (3.4\text{-}8)$$

is the *cutoff frequency* in hertz for mode (n, l). By referring to Eq. (3.4-6), it can be seen that if $f = f_{n,l}$, then $k_{Z_{n,l}} = 0$, and as a result, there is no wave propagation in the positive Z direction. Hence the term "cutoff frequency." In addition, if $f > f_{n,l}$, then $k_{Z_{n,l}}$ is real and positive and mode (n, l) is called a *propagating mode*. However, if $f < f_{n,l}$, then $k_{Z_{n,l}}$ is imaginary and mode (n, l) is called an *evanescent mode*.

An evanescent mode is a decaying exponential, as we shall demonstrate next. If $f < f_{n,l}$, then Eq. (3.4-6) can be rewritten as

$$k_Z = k_{Z_{n,l}} = k\sqrt{-\left[(f_{n,l}/f)^2 - 1\right]}\,, \qquad f < f_{n,l}, \qquad (3.4\text{-}9)$$

or

$$k_Z = k_{Z_{n,l}} = \pm jk\sqrt{(f_{n,l}/f)^2 - 1}\,, \qquad f < f_{n,l}. \qquad (3.4\text{-}10)$$

The question now is which sign do we choose in Eq. (3.4-10). If we choose the minus sign and then substitute Eq. (3.4-10) into the complex exponential $\exp(-jk_Z z)$, we obtain

$$\exp(-jk_Z z) = \exp\left\{\left[-k\sqrt{(f_{n,l}/f)^2 - 1}\right]z\right\}, \qquad f < f_{n,l}, \qquad (3.4\text{-}11)$$

which approaches zero as z approaches positive infinity. If we had chosen the plus sign in Eq. (3.4-10), then Eq. (3.4-11) would have been a growing exponential (which does not make physical sense) instead of a decaying exponential. Therefore,

$$k_Z = k_{Z_{n,l}} = -jk\sqrt{(f_{n,l}/f)^2 - 1}\,, \qquad f < f_{n,l}. \qquad (3.4\text{-}12)$$

With the use of Eqs. (3.4-2), (3.4-6), and (3.4-12), the spatial-dependent part of the velocity potential inside a rigid-walled waveguide with circular cross-sectional area corresponding to the (n, l) normal mode is given by

$$\varphi_{f,n,l}(r, \psi, z) = J_n\left(\alpha_{n,l}\frac{r}{a}\right)\left[A_{n,l}\cos(n\psi) + B_{n,l}\sin(n\psi)\right]\exp(-jk_{Z_{n,l}}z),$$

$$n, l = 0, 1, 2, \ldots,$$

(3.4-13)

where

$$k_{Z_{n,l}} = \begin{cases} k\sqrt{1 - (f_{n,l}/f)^2}, & f \geq f_{n,l}, \\ -jk\sqrt{(f_{n,l}/f)^2 - 1}, & f < f_{n,l}, \end{cases}$$

(3.4-14)

is the propagation-vector component in the Z direction of mode (n, l), k is the wave number given by Eq. (3.4-7), and $f_{n,l}$ is the cutoff frequency for mode (n, l) given by Eq. (3.4-8). The complete time-harmonic normal-mode solution for the velocity potential is given by

$$\varphi(t, \mathbf{r}) = \sum_{n=0}^{\infty}\sum_{l=0}^{\infty}\varphi_{f,n,l}(\mathbf{r})\exp(+j2\pi ft),$$

(3.4-15)

and upon substituting Eq. (3.4-13) into Eq. (3.4-15), we obtain

$$\varphi(t, r, \psi, z) = \sum_{n=0}^{\infty}\sum_{l=0}^{\infty}J_n\left(\alpha_{n,l}\frac{r}{a}\right)\left[A_{n,l}\cos(n\psi) + B_{n,l}\sin(n\psi)\right]$$

$$\times\exp(-jk_{Z_{n,l}}z)\exp(+j2\pi ft),$$

(3.4-16)

where $k_{Z_{n,l}}$ is given by Eq. (3.4-14).

With the use of Eq. (2.2-28), the time-harmonic acoustic pressure inside a rigid-walled waveguide with circular cross-sectional area corresponding to the (n, l) normal mode can be expressed as

$$p_{n,l}(t, \mathbf{r}) = -j2\pi f\rho_0\varphi_{n,l}(t, \mathbf{r}), \qquad n, l = 0, 1, 2, \ldots.$$

(3.4-17)

Therefore, the complete time-harmonic normal-mode solution for the acoustic pressure is given by

$$p(t, \mathbf{r}) = -j2\pi f\rho_0 \sum_{n=0}^{\infty} \sum_{l=0}^{\infty} \varphi_{f,n,l}(\mathbf{r}) \exp(+j2\pi ft), \qquad (3.4\text{-}18)$$

and upon substituting Eq. (3.4-13) into Eq. (3.4-18), we obtain

$$p(t, r, \psi, z) = -j2\pi f\rho_0 \sum_{n=0}^{\infty} \sum_{l=0}^{\infty} J_n\left(\alpha_{n,l} \frac{r}{a}\right) [A_{n,l} \cos(n\psi) + B_{n,l} \sin(n\psi)]$$

$$\times \exp\left(-jk_{z_{n,l}} z\right) \exp(+j2\pi ft),$$

$$(3.4\text{-}19)$$

where $k_{z_{n,l}}$ is given by Eq. (3.4-14).

Note that the upper limits on the summations in Eqs. (3.4-16) and (3.4-19) are infinity. This is a consequence of the fact that from a theoretical point of view, the complete normal-mode solution is obtained by summing the contributions from all the propagating and evanescent modes. However, since the evanescent modes are decaying exponentials, their contributions to the total acoustic field are only significant at short ranges from the source. Therefore, from a practical point of view, unless we are interested in the acoustic field near the sound source, we only need to sum the contributions from all the propagating modes. The finite number of propagating modes depends on the radius a of the waveguide and the frequency components of the source. Also note that the constants $A_{n,l}$ and $B_{n,l}$ are still unknown. In order to determine their values, the boundary condition at the location of the sound source inside the waveguide must be satisfied.

Group Speed, Phase Speed, and Angle of Propagation

The next topics of discussion are the *group speed* and *phase speed* in the Z direction, and the *angle of propagation* for the (n, l) propagating mode. Substituting Eq. (3.4-6) into Eqs. (2.2-49) and (2.2-54) yields

$$c_{g_{z_{n,l}}} = c\sqrt{1 - (f_{n,l}/f)^2}, \qquad f \geq f_{n,l}, \qquad (3.4\text{-}20)$$

$$c_{p_{z_{n,l}}} = \frac{c}{\sqrt{1 - (f_{n,l}/f)^2}}, \qquad f \geq f_{n,l}, \qquad (3.4\text{-}21)$$

respectively, where the cutoff frequency $f_{n,l}$ for mode (n, l) is given by Eq. (3.4-8). Note that both the group speed and the phase speed are functions of frequency and mode number. Recall from Section 2.2.2 that energy propagates at the group speed. Since the group speed given by Eq. (3.4-20) is a function of frequency and mode number, for a given mode (n, l) with cutoff frequency $f_{n,l}$ the energy associated with the high-frequency components from a transmitted pulse will arrive at a receiver before the energy associated with the low-frequency components. As a result, the shape of the transmitted pulse will be *distorted* by the time it reaches the receiver, due to the dispersion of the transmitted signal's energy. And since the phase speed given by Eq. (3.4-21) is a function of frequency as a result of satisfying the boundary conditions, this is known as *geometrical dispersion*, as opposed to *material* or *intrinsic dispersion* which was discussed in Section 1.4.

Next, let us compute the angle of propagation for the (n, l) propagating mode. Since

$$k_Z = kw = k \cos \theta, \qquad (3.4\text{-}22)$$

substituting Eq. (3.4-6) into Eq. (3.4-22) yields

$$\theta_{n,l} = \cos^{-1}\left(\sqrt{1 - (f_{n,l}/f)^2}\right), \qquad f \geq f_{n,l}, \qquad (3.4\text{-}23)$$

where $\theta_{n,l}$ is the angle that the propagation vector

$$\mathbf{k}_{n,l} = k_{r_{n,l}}\hat{r} + k_{Z_{n,l}}\hat{z} \qquad (3.4\text{-}24)$$

makes with the positive Z axis for the (n, l) propagating mode at frequency f. The radial component of the propagation vector $k_{r_{n,l}}$ is given by Eq. (3.3-27).

3.5 Scattering by a Cylinder

In this section we shall compute the velocity potential of the *scattered* acoustic field due to a time-harmonic plane wave incident upon a fixed (motionless) cylinder of radius a meters. The time-harmonic velocity potential of the incident plane wave can be expressed as follows:

$$\varphi_i(t, \mathbf{R}) = \varphi_{f,i}(\mathbf{R}) \exp(+j2\pi ft), \qquad (3.5\text{-}1)$$

where

$$\varphi_{f,i}(\mathbf{R}) = A_i \exp(-j\mathbf{k}_i \cdot \mathbf{R}) \qquad (3.5\text{-}2)$$

is the spatial-dependent part of the velocity potential; A_i is the amplitude of the incident plane wave;

$$\mathbf{k}_i = k\hat{n}_i \qquad (3.5\text{-}3)$$

is the incident propagation vector;

$$k = 2\pi f/c = 2\pi/\lambda \tag{3.5-4}$$

is the wave number in radians per meter;

$$\hat{n}_i = u_i\hat{x} + v_i\hat{y} + w_i\hat{z} \tag{3.5-5}$$

is the unit vector in the direction of the incident propagation vector, where

$$u_i = \sin\theta_i \cos\psi_i, \tag{3.5-6}$$

$$v_i = \sin\theta_i \sin\psi_i, \tag{3.5-7}$$

and

$$w_i = \cos\theta_i \tag{3.5-8}$$

are the dimensionless direction cosines with respect to the X, Y, and Z axes, respectively; and

$$\mathbf{R} = x\hat{x} + y\hat{y} + z\hat{z}, \tag{3.5-9}$$

or $\mathbf{R} = (x, y, z)$, is the position vector measured from the origin of the coordinate system to some arbitrary field point with rectangular coordinates (x, y, z) or, equivalently, with cylindrical coordinates (r, ψ, z) (see Fig. 3.5-1).

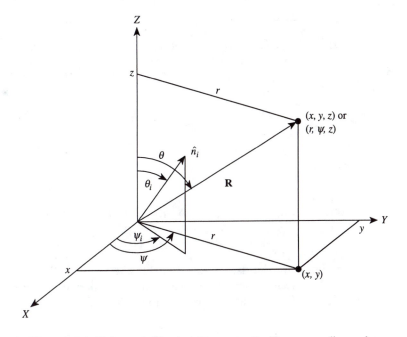

Figure 3.5-1. Unit vector $\hat{n}_i$ and position vector $\mathbf{R}$ with corresponding angles.

The first step in solving this scattering problem is to express $\varphi_{f,i}(\mathbf{R})$ given by Eq. (3.5-2) in terms of the cylindrical coordinates (r, ψ, z). Substituting Eq. (3.5-3) and Eqs. (3.5-5) through (3.5-9) into Eq. (3.5-2) yields

$$\varphi_{f,i}(x, y, z) = A_i \exp\left[-jk \sin \theta_i (x \cos \psi_i + y \sin \psi_i)\right] \exp(-jkz \cos \theta_i), \quad (3.5\text{-}10)$$

and since (see Fig. 3.5-1)

$$x = r \cos \psi \qquad (3.5\text{-}11)$$

and

$$y = r \sin \psi, \qquad (3.5\text{-}12)$$

substituting Eqs. (3.5-11) and (3.5-12) into Eq. (3.5-10) yields

$$\varphi_{f,i}(r, \psi, z) = A_i \exp\left[-jkr \sin \theta_i \cos(\psi - \psi_i)\right] \exp(-jkz \cos \theta_i), \quad (3.5\text{-}13)$$

where use has been made of the following trigonometric identity:

$$\cos(\alpha - \beta) = \cos \alpha \cos \beta + \sin \alpha \sin \beta. \qquad (3.5\text{-}14)$$

In order to proceed further, we shall make use of the following identity:

$$\exp\left[-jkr \sin \theta_i \cos(\psi - \psi_i)\right] = \sum_{n=-\infty}^{\infty} (-j)^n J_n(kr \sin \theta_i) \exp\left[+jn(\psi - \psi_i)\right]$$

$$(3.5\text{-}15)$$

or, equivalently,

$$\exp\left[-jkr \sin \theta_i \cos(\psi - \psi_i)\right]$$

$$= J_0(kr \sin \theta_i) + 2 \sum_{n=1}^{\infty} (-j)^n J_n(kr \sin \theta_i) \cos\left[n(\psi - \psi_i)\right]. \qquad (3.5\text{-}16)$$

Therefore, with the use of Eqs. (3.5-13) and (3.5-16), the velocity potential of the incident, time-harmonic plane wave can be expressed in terms of the cylindrical coordinates (r, ψ, z) as follows (see Fig. 3.5-2):

$$\varphi_i(t, \mathbf{r}) = \varphi_{f,i}(\mathbf{r}) \exp(+j2\pi ft), \qquad (3.5\text{-}17)$$

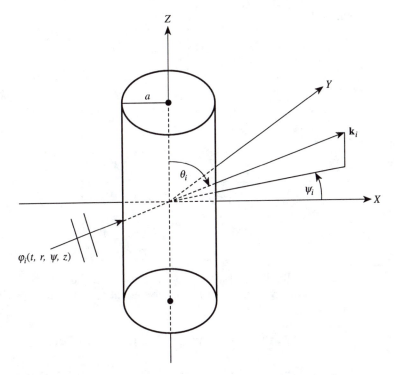

Figure 3.5-2. Time-harmonic plane wave incident upon a cylinder of radius a meters.

where the position vector $\mathbf{r} = (r, \psi, z)$, and

$$
\varphi_{f,i}(r, \psi, z) = A_i\left[J_0(kr \sin \theta_i) + 2 \sum_{n=1}^{\infty} (-j)^n J_n(kr \sin \theta_i) \cos\left[n(\psi - \psi_i)\right] \right]
$$
$$
\times \exp(-jkz \cos \theta_i).
$$

$$(3.5\text{-}18)$$

In addition to the incident plane wave, there is present a *scattered*, outgoing wave traveling away from the cylinder. Since the scattered acoustic field is analogous to the acoustic field radiated by a vibrating cylinder, by referring to Section 3.2.1, the velocity potential of the time-harmonic scattered acoustic field can be expressed as

$$
\varphi_s(t, \mathbf{r}) = \varphi_{f,s}(\mathbf{r}) \exp(+j2\pi ft), \tag{3.5-19}
$$

where

$$\varphi_{f,s}(r,\psi,z) = \sum_{n=0}^{\infty} H_n^{(2)}(k_r r)\left[A_n \cos(n\psi) + B_n \sin(n\psi)\right]$$

$$\times\left[A_Z \exp(-jk_Z z) + B_Z \exp(+jk_Z z)\right], \qquad (3.5\text{-}20)$$

or

$$\varphi_{f,s}(r,\psi,z) = \left[A_0 H_0^{(2)}(k_r r) + \sum_{n=1}^{\infty} H_n^{(2)}(k_r r)\left[A_n \cos(n\psi) + B_n \sin(n\psi)\right]\right]$$

$$\times\left[A_Z \exp(-jk_Z z) + B_Z \exp(+jk_Z z)\right], \qquad (3.5\text{-}21)$$

where

$$k_r^2 + k_Z^2 = k^2. \qquad (3.5\text{-}22)$$

If the cylinder is assumed to be *rigid*, then the normal component of the *total* acoustic fluid velocity vector must be equal to zero on the surface of the cylinder, that is,

$$u_n(t,r,\psi,z)\big|_{r=a} = \hat{n} \cdot \mathbf{u}(t,r,\psi,z)\big|_{r=a} = \hat{n} \cdot \nabla\varphi(t,r,\psi,z)\big|_{r=a} = 0, \quad (3.5\text{-}23)$$

where $\hat{n}$ is a unit vector normal to the surface of the cylinder and

$$\varphi(t,\mathbf{r}) = \varphi_i(t,\mathbf{r}) + \varphi_s(t,\mathbf{r}). \qquad (3.5\text{-}24)$$

Since

$$\hat{n} = \hat{r}, \qquad (3.5\text{-}25)$$

substituting Eqs. (3.2-9), (3.5-24), and (3.5-25) into Eq. (3.5-23) yields the following boundary condition:

$$\frac{\partial}{\partial r}\varphi_i(t,r,\psi,z)\bigg|_{r=a} = -\frac{\partial}{\partial r}\varphi_s(t,r,\psi,z)\bigg|_{r=a}. \qquad (3.5\text{-}26)$$

Substituting Eqs. (3.5-17) and (3.5-18) into the left-hand side of Eq. (3.5-26) yields

$$\frac{\partial}{\partial r}\varphi_i(t,r,\psi,z)\bigg|_{r=a} = A_i k \sin\theta_i\left[J_0'(ka\sin\theta_i) + 2\sum_{n=1}^{\infty}(-j)^n J_n'(ka\sin\theta_i)\right.$$

$$\times\left[\cos(n\psi)\cos(n\psi_i) + \sin(n\psi)\sin(n\psi_i)\right]$$

$$\times \exp(-jkz\cos\theta_i)\exp(+j2\pi ft), \qquad (3.5\text{-}27)$$

where

$$J_n'(ka \sin \theta_i) = \frac{d}{d(kr \sin \theta_i)} J_n(kr \sin \theta_i)\bigg|_{r=a}, \qquad n = 0, 1, 2, \dots, \quad (3.5\text{-}28)$$

and

$$\cos[n(\psi - \psi_i)] = \cos(n\psi)\cos(n\psi_i) + \sin(n\psi)\sin(n\psi_i). \quad (3.5\text{-}29)$$

And upon substituting Eqs. (3.5-19) and (3.5-21) into the right-hand side of Eq. (3.5-26), we obtain

$$-\frac{\partial}{\partial r}\varphi_s(t, r, \psi, z)\bigg|_{r=a}$$

$$= -k_r\left[A_0 H_0^{(2)\prime}(k_r a) + \sum_{n=1}^{\infty} H_n^{(2)\prime}(k_r a)\big[A_n \cos(n\psi) + B_n \sin(n\psi)\big]\right]$$

$$\times \big[A_Z \exp(-jk_Z z) + B_Z \exp(+jk_Z z)\big] \exp(+j2\pi ft), \qquad (3.5\text{-}30)$$

where

$$H_n^{(2)\prime}(k_r a) = \frac{d}{d(k_r r)} H_n^{(2)}(k_r r)\bigg|_{r=a}, \qquad n = 0, 1, 2, \dots. \quad (3.5\text{-}31)$$

Since Eqs. (3.5-27) and (3.5-30) must be equal,

$$A_Z = 1, \qquad (3.5\text{-}32)$$

$$B_Z = 0, \qquad (3.5\text{-}33)$$

$$k_Z = k \cos \theta_i, \qquad (3.5\text{-}34)$$

and upon substituting Eq. (3.5-34) into Eq. (3.5-22),

$$k_r = k \sin \theta_i. \qquad (3.5\text{-}35)$$

In addition,

$$\boxed{A_0 = -\frac{J_0'(k_r a)}{H_0^{(2)\prime}(k_r a)} A_i,} \qquad (3.5\text{-}36)$$

$$\boxed{A_n = -(-j)^n 2\frac{J_n'(k_r a)}{H_n^{(2)\prime}(k_r a)} \cos(n\psi_i) A_i, \qquad n = 1, 2, 3, \dots,} \qquad (3.5\text{-}37)$$

and

$$B_n = -(-j)^n 2 \frac{J_n'(k_r a)}{H_n^{(2)\prime}(k_r a)} \sin(n\psi_i) A_i, \qquad n = 1, 2, 3, \ldots . \qquad (3.5\text{-}38)$$

Therefore, upon substituting Eqs. (3.5-32) through (3.5-34) into Eq. (3.5-21), we obtain the following expression for the spatial-dependent part of the time-harmonic velocity potential of the scattered acoustic field due to a time-harmonic plane wave incident upon a fixed (motionless), *rigid* cylinder:

$$\varphi_{f,s}(r, \psi, z) = \left[A_0 H_0^{(2)}(k_r r) + \sum_{n=1}^{\infty} H_n^{(2)}(k_r r)\left[A_n \cos(n\psi) + B_n \sin(n\psi) \right] \right]$$
$$\times \exp(-jk_z z),$$

$$(3.5\text{-}39)$$

where A_0, A_n, and B_n are given by Eqs. (3.5-36) through (3.5-38), respectively, k_r is given by Eq. (3.5-35), and k_z is given by Eq. (3.5-34). The first term on the right-hand side of Eq. (3.5-39) is the *axisymmetric term* (i.e., no ψ dependence). However, this term does depend on the angle θ_i via k_r. It is important to note that Eq. (3.5-39) *does not take into account scattering from the ends of the cylinder*. Equation (3.5-39) could represent, for example, a *very rough solution* to the problem of scattering of sound by a submarine.

The corresponding time-harmonic scattered acoustic pressure field and scattered acoustic fluid velocity vector can be obtained from the following equations:

$$p_s(t, \mathbf{r}) = p_{f,s}(\mathbf{r}) \exp(+j2\pi ft), \qquad (3.5\text{-}40)$$

where

$$p_{f,s}(\mathbf{r}) = -jk\rho_0 c\varphi_{f,s}(\mathbf{r}), \qquad (3.5\text{-}41)$$

and

$$\mathbf{u}_s(t, \mathbf{r}) = \mathbf{u}_{f,s}(\mathbf{r}) \exp(+j2\pi ft), \qquad (3.5\text{-}42)$$

where

$$\mathbf{u}_{f,s}(\mathbf{r}) = \nabla \varphi_{f,s}(\mathbf{r}). \qquad (3.5\text{-}43)$$

Recall that the total velocity potential is equal to the sum of the incident and scattered velocity potentials [see Eq. (3.5-24)].

Scattering Cross Section

The last topic to be discussed in this section is the *scattering cross section* of the cylinder. The scattering cross section σ_s is defined as that *area* (in square meters) of the incident wavefront that transmits a power equal to the time-average scattered power, that is,

$$I_{\mathrm{avg}_i}\sigma_s = P_{\mathrm{avg}_s}, \tag{3.5-44}$$

or

$$\boxed{\sigma_s = \frac{P_{\mathrm{avg}_s}}{I_{\mathrm{avg}_i}},} \tag{3.5-45}$$

where

$$P_{\mathrm{avg}_s} = \oint_S \mathbf{I}_{\mathrm{avg}_s}(r,\psi,z) \cdot \mathbf{dS} \tag{3.5-46}$$

is the time-average scattered power,

$$\mathbf{I}_{\mathrm{avg}_s}(r,\psi,z) = \tfrac{1}{2}\,\mathrm{Re}\big\{p_{f,s}(r,\psi,z)\mathbf{u}_{f,s}^*(r,\psi,z)\big\} \tag{3.5-47}$$

is the time-average intensity vector of the scattered acoustic field,

$$\mathbf{dS} = r\,d\psi\,dz\,\hat{r}, \tag{3.5-48}$$

and

$$I_{\mathrm{avg}_i} = \tfrac{1}{2}k^2\rho_0 c\,|A_i|^2 \tag{3.5-49}$$

is the magnitude of the time-average intensity vector of the time-harmonic incident plane wave (see Example 2.2-1), where A_i is the amplitude of the velocity potential of the incident plane wave [see Eq. (3.5-2)].

Example 3.5-1 (Normal incidence)

In this example we shall obtain the expression for the time-harmonic velocity potential of the scattered acoustic field due to a time-harmonic plane wave traveling in the positive X direction, perpendicular to the Z axis, incident upon a fixed (motionless), rigid cylinder. The incident spherical angles that correspond to this problem are $\theta_i = \pi/2$ and

$\psi_i = 0$ (see Fig. 3.5-2). Therefore,

$$k_r = k, \qquad\qquad (3.5\text{-}50)$$

$$k_Z = 0, \qquad\qquad (3.5\text{-}51)$$

$$\boxed{A_0 = -\frac{J_0'(ka)}{H_0^{(2)'}(ka)}A_i,} \qquad\qquad (3.5\text{-}52)$$

$$\boxed{A_n = -(-j)^n 2\frac{J_n'(ka)}{H_n^{(2)'}(ka)}A_i, \qquad n = 1,2,3,\ldots,} \qquad (3.5\text{-}53)$$

and

$$\boxed{B_n = 0, \qquad n = 1,2,3,\ldots .} \qquad\qquad (3.5\text{-}54)$$

Upon substituting Eqs. (3.5-50), (3.5-51), and (3.5-54) into Eq. (3.5-39), we obtain

$$\boxed{\varphi_{f,s}(r,\psi,z) = A_0 H_0^{(2)}(kr) + \sum_{n=1}^{\infty} A_n H_n^{(2)}(kr)\cos(n\psi),} \quad (3.5\text{-}55)$$

where A_0 and A_n are given by Eqs. (3.5-52) and (3.5-53), respectively. Recall that scattering from the ends of the cylinder was not taken into account. Note that the first term on the right-hand side of Eq. (3.5-55) is the *omnidirectional term* (i.e., no angular dependence). Also note that for this problem, the acoustic field at $\theta = 90°$ and $\psi = 0°$ corresponds to *forward scattering*, whereas the acoustic field at $\theta = 90°$ and $\psi = 180°$ corresponds to *backscatter*. And finally, substituting Eq. (3.5-55) into Eq. (3.5-19) yields the time-harmonic velocity potential of the scattered acoustic field.

3.6 Integral Representations of the Time-Independent Free-Space Green's Function in Cylindrical Coordinates

In this section we shall represent the time-independent free-space Green's function $g_f(\mathbf{r}|\mathbf{r}_0)$ by two different integral expressions in the cylindrical coordinates r and z. These expressions are very useful for solving problems involving the reflection of spherical waves from planar boundaries (see Section 3.7) and for solving ocean waveguide problems (see Section 3.9).

Recall from Section 2.6.1 that

$$g_f(\mathbf{r}|\mathbf{r}_0) \triangleq -\frac{\exp(-jk|\mathbf{r}-\mathbf{r}_0|)}{4\pi|\mathbf{r}-\mathbf{r}_0|} = -\frac{\exp(-jkR)}{4\pi R},\qquad (3.6\text{-}1)$$

where

$$\mathbf{r} = x\hat{x} + y\hat{y} + z\hat{z} \qquad (3.6\text{-}2)$$

is the position vector to a field point,

$$\mathbf{r}_0 = x_0\hat{x} + y_0\hat{y} + z_0\hat{z} \qquad (3.6\text{-}3)$$

is the position vector to a source point, and the range (distance) between source and field points is given by

$$R = |\mathbf{r}-\mathbf{r}_0| = \left[(x-x_0)^2 + (y-y_0)^2 + (z-z_0)^2\right]^{1/2}. \qquad (3.6\text{-}4)$$

Also recall from Section 2.8 that

$$g_f(\mathbf{r}|\mathbf{r}_0) = +j\frac{1}{8\pi^2}\int_{-\infty}^{\infty}\int_{-\infty}^{\infty}\exp\{-j[k_X(x-x_0)+k_Y(y-y_0)]\}$$

$$\times\frac{\exp(-jk_Z|z-z_0|)}{k_Z}\,dk_X\,dk_Y, \qquad (3.6\text{-}5)$$

where

$$k_Z = \begin{cases} \sqrt{k^2-(k_X^2+k_Y^2)}, & k_X^2+k_Y^2 \le k^2, \\ -j\sqrt{k_X^2+k_Y^2-k^2}, & k_X^2+k_Y^2 > k^2, \end{cases} \qquad (3.6\text{-}6)$$

and

$$k = 2\pi f/c = 2\pi/\lambda. \qquad (3.6\text{-}7)$$

By referring to Fig. 3.6-1, it can be seen that

$$x - x_0 = r\cos\phi, \qquad (3.6\text{-}8)$$

$$y - y_0 = r\sin\phi, \qquad (3.6\text{-}9)$$

and

$$r = \sqrt{(x-x_0)^2 + (y-y_0)^2}. \qquad (3.6\text{-}10)$$

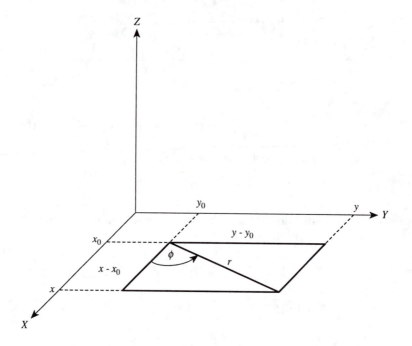

Figure 3.6-1. Horizontal range r.

Equation (3.6-10) is the horizontal range measured from the origin of a cylindrical coordinate system centered at (x_0, y_0, z_0). With the use of Eq. (3.6-10), Eq. (3.6-4) can be rewritten as

$$R = \sqrt{r^2 + (z - z_0)^2}, \tag{3.6-11}$$

and as a result,

$$g_f(\mathbf{r}|\mathbf{r}_0) = -\frac{\exp\left(-jk\sqrt{r^2 + (z - z_0)^2}\right)}{4\pi\sqrt{r^2 + (z - z_0)^2}}. \tag{3.6-12}$$

Equation (3.6-12) represents the time-independent free-space Green's function in terms of the cylindrical coordinates r and z.

Next, by referring to Fig. 3.6-2, it can be seen that

$$k_X = k_r \cos \psi, \tag{3.6-13}$$

$$k_Y = k_r \sin \psi, \tag{3.6-14}$$

$$k_r = \sqrt{k_X^2 + k_Y^2}, \tag{3.6-15}$$

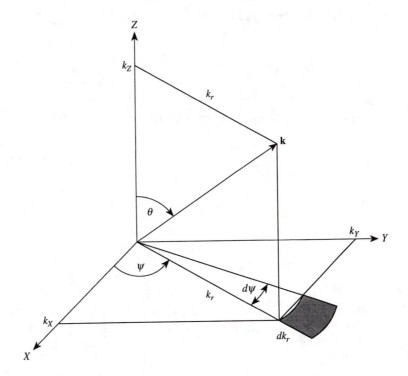

Figure 3.6-2. The propagation vector **k** and its components k_X, k_Y, and k_Z in the X, Y, and Z directions, respectively. Also shown are the radial component k_r, the spherical angles θ and ψ, and the infinitesimal area $k_r\, dk_r\, d\psi$.

and

$$dk_X\, dk_Y \rightarrow k_r\, dk_r\, d\psi. \tag{3.6-16}$$

Therefore, with the use of Eqs. (3.6-8), (3.6-9), (3.6-13), and (3.6-14),

$$k_X(x - x_0) + k_Y(y - y_0) = k_r r \cos(\psi - \phi), \tag{3.6-17}$$

since

$$\cos(\psi - \phi) = \cos\psi\cos\phi + \sin\psi\sin\phi. \tag{3.6-18}$$

We are now in a position to transform Eq. (3.6-5) from rectangular coordinates to cylindrical coordinates.

Upon substituting Eqs. (3.6-16) and (3.6-17) into Eq. (3.6-5), we obtain

$$g_f(\mathbf{r}|\mathbf{r}_0) = +j\frac{1}{4\pi}\int_0^\infty J_0(k_r r)\frac{\exp(-jk_z|z-z_0|)}{k_z}k_r\,dk_r, \qquad (3.6\text{-}19)$$

where $g_f(\mathbf{r}|\mathbf{r}_0)$ is given by Eq. (3.6-12),

$$J_0(k_r r) = \frac{1}{2\pi}\int_0^{2\pi}\exp[-jk_r r\cos(\psi-\phi)]\,d\psi \qquad (3.6\text{-}20)$$

is the zeroth-order Bessel function of the first kind,

$$k_z = \begin{cases} \sqrt{k^2-k_r^2}, & k_r^2 \le k^2, \\ -j\sqrt{k_r^2-k^2}, & k_r^2 > k^2, \end{cases} \qquad (3.6\text{-}21)$$

and r is given by Eq. (3.6-10). Equation (3.6-19) is the first desired integral representation of the time-independent free-space Green's function in terms of the cylindrical coordinates r and z. An alternative integral representation will be obtained next.

Since

$$J_0(k_r r) = \tfrac{1}{2}\big[H_0^{(1)}(k_r r) + H_0^{(2)}(k_r r)\big] \qquad (3.6\text{-}22)$$

and

$$H_0^{(1)}(k_r r) = -H_0^{(2)}(-k_r r), \qquad (3.6\text{-}23)$$

substituting Eqs. (3.6-22) and (3.6-23) into Eq. (3.6-19) yields

$$g_f(\mathbf{r}|\mathbf{r}_0) = +j\frac{1}{8\pi}\bigg[-\int_0^\infty H_0^{(2)}(-k_r r)\frac{\exp(-jk_z|z-z_0|)}{k_z}k_r\,dk_r$$

$$+\int_0^\infty H_0^{(2)}(k_r r)\frac{\exp(-jk_z|z-z_0|)}{k_z}k_r\,dk_r\bigg], \qquad (3.6\text{-}24)$$

or

$$g_f(\mathbf{r}|\mathbf{r}_0) = +j\frac{1}{8\pi}\int_{-\infty}^\infty H_0^{(2)}(k_r r)\frac{\exp(-jk_z|z-z_0|)}{k_z}k_r\,dk_r. \qquad (3.6\text{-}25)$$

Equation (3.6-25) is the second desired integral representation of the time-independent free-space Green's function in terms of the cylindrical coordinates r and z.

3.7 Reflection and Transmission of Spherical Waves at Planar Boundaries

In Section 3.6, we derived two integral representations of the time-independent free-space Green's function in terms of the cylindrical coordinates r and z. Now, in this section, we shall demonstrate how to use these expressions to solve problems involving the reflection and transmission of spherical waves at planar boundaries.

Figure 3.7-1 shows a time-harmonic, omnidirectional point source located in fluid medium I at $r = 0$ and $z = z_0$. Also shown is a boundary at $z = 0$ (i.e., the XY plane) between fluid media I and II, where $\rho_1 c_1$ and $\rho_2 c_2$ are the characteristic impedances of media I and II, respectively; where ρ_i, $i = 1, 2$, is the constant density and c_i, $i = 1, 2$, is the constant speed of sound in each of the two media.

As was discussed in Section 2.3, the following two boundary conditions must be satisfied at all times and at all spatial locations on the boundary between two fluid media: (1) *continuity of acoustic pressure*, and (2) *continuity of the normal component of the acoustic fluid velocity vector*. In order to apply the boundary conditions to the problem illustrated in Fig. 3.7-1, we need mathematical models for the incident, reflected, and transmitted time-harmonic acoustic fields. Since $g_f(\mathbf{r}|\mathbf{r}_0)$ is the time-independent free-space, spatial impulse response of an ideal, homogeneous, fluid medium (see Section 2.6.2), the incident acoustic field due to a time-harmonic point source can be expressed as [see Eq. (3.6-12)]

$$\varphi_i(t, r, \psi, z) = -\frac{\exp\left(-jk_1\sqrt{r^2 + (z - z_0)^2}\right)}{4\pi\sqrt{r^2 + (z - z_0)^2}}\exp(+j2\pi ft) \quad (3.7\text{-}1)$$

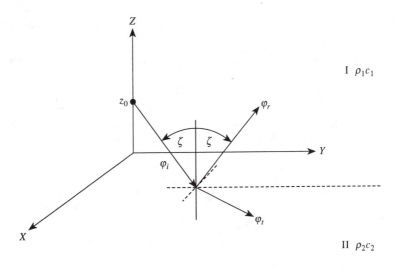

Figure 3.7-1. Incident, reflected, and transmitted velocity potentials.

or

$$\varphi_i(t, r, \psi, z) = +j\frac{1}{4\pi}\int_0^\infty J_0(k_{r_1}r)\frac{\exp(-jk_{z_1}|z - z_0|)}{k_{z_1}}k_{r_1}\,dk_{r_1}\exp(+j2\pi ft),$$

(3.7-2)

where

$$r = \sqrt{x^2 + y^2}$$

(3.7-3)

is the horizontal range to the field point, since the point source is located at $x_0 = 0$ and $y_0 = 0$;

$$k_{z_1} = \begin{cases} \sqrt{k_1^2 - k_{r_1}^2}, & k_{r_1}^2 \le k_1^2, \\ -j\sqrt{k_{r_1}^2 - k_1^2}, & k_{r_1}^2 > k_1^2, \end{cases}$$

(3.7-4)

is the propagation-vector component in the Z direction in medium I; and

$$k_1 = 2\pi f/c_1$$

(3.7-5)

is the wave number in medium I.

In view of our discussion of the solution of the wave equation in the cylindrical coordinate system in Section 3.1, referring to Fig. 3.7-1, and noting that the frequencies of the incident, reflected, and transmitted fields are equal (see Section 2.3), we can write that

$$\varphi_r(t, r, \psi, z) = AJ_0(k_{r_1}r)\exp(-jk_{z_1}z)\exp(+j2\pi ft)$$

(3.7-6)

is the reflected acoustic field traveling in the positive Z direction in medium I and

$$\varphi_t(t, r, \psi, z) = BJ_0(k_{r_2}r)\exp(+jk_{z_2}z)\exp(+j2\pi ft)$$

(3.7-7)

is the transmitted acoustic field traveling in the negative Z direction in medium II, where A and B are complex constants in general,

$$k_{z_2} = \begin{cases} \sqrt{k_2^2 - k_{r_2}^2}, & k_{r_2}^2 \le k_2^2, \\ -j\sqrt{k_{r_2}^2 - k_2^2}, & k_{r_2}^2 > k_2^2, \end{cases}$$

(3.7-8)

is the propagation-vector component in the Z direction in medium II, and

$$k_2 = 2\pi f/c_2$$

(3.7-9)

is the wave number in medium II. Note that the Neumann function was not included in Eqs. (3.7-6) and (3.7-7), since values for the acoustic fields are required at $r = 0$. Also note that Eqs. (3.7-6) and (3.7-7) are axisymmetric solutions, since the source is an omnidirectional point source and the boundary is a plane rather than arbitrarily shaped. Finally, note that Eqs. (3.7-6) and (3.7-7) can be general-ized by integrating over k_{r_1} and k_{r_2}, respectively, from zero to infinity as follows:

$$\varphi_r(t, r, \psi, z) = \int_0^\infty A(k_{r_1}) J_0(k_{r_1} r) \exp(-jk_{z_1} z) \, dk_{r_1} \exp(+j2\pi ft) \quad (3.7\text{-}10)$$

and

$$\varphi_t(t, r, \psi, z) = \int_0^\infty B(k_{r_2}) J_0(k_{r_2} r) \exp(+jk_{z_2} z) \, dk_{r_2} \exp(+j2\pi ft), \quad (3.7\text{-}11)$$

where $A(k_{r_1})$ and $B(k_{r_2})$ are determined by satisfying the boundary conditions. Note that both Eqs. (3.7-10) and (3.7-11) are solutions of the homogeneous wave equation in the cylindrical coordinate system. Because of the integrations over k_{r_1} and k_{r_2} from zero to infinity, Eqs. (3.7-10) and (3.7-11) are called *full-wave solutions* and will be used as our mathematical models of the reflected and transmitted fields, respectively.

The first boundary condition is continuity of acoustic pressure, that is,

$$p_1(t, r, \psi, z)\big|_{z=0} = p_2(t, r, \psi, z)\big|_{z=0}, \quad (3.7\text{-}12)$$

where

$$p_1(t, \mathbf{r}) = p_i(t, \mathbf{r}) + p_r(t, \mathbf{r}) \quad (3.7\text{-}13)$$

and

$$p_2(t, \mathbf{r}) = p_t(t, \mathbf{r}) \quad (3.7\text{-}14)$$

are the *total* acoustic pressures in media I and II, respectively, and where the position vector $\mathbf{r} = (r, \psi, z)$. Since for time-harmonic acoustic fields (see Section 2.3)

$$p_i(t, \mathbf{r}) = -j2\pi f\rho_1 \varphi_i(t, \mathbf{r}), \quad (3.7\text{-}15)$$

$$p_r(t, \mathbf{r}) = -j2\pi f\rho_1 \varphi_r(t, \mathbf{r}), \quad (3.7\text{-}16)$$

and

$$p_t(t, \mathbf{r}) = -j2\pi f\rho_2 \varphi_t(t, \mathbf{r}), \quad (3.7\text{-}17)$$

with the use of Eqs. (3.7-2), (3.7-10), and (3.7-11), Eq. (3.7-12) can be expressed as

$$\rho_1 \int_0^\infty \left[j\frac{1}{4\pi} \frac{\exp(-jk_{Z_1}z_0)}{k_{Z_1}} k_{r_1} + A(k_{r_1}) \right] J_0(k_{r_1}r)\, dk_{r_1}$$

$$= \rho_2 \int_0^\infty B(k_{r_2}) J_0(k_{r_2}r)\, dk_{r_2}. \qquad (3.7\text{-}18)$$

Since the boundary condition represented by Eq. (3.7-18) must hold for all allowed values of r, that is, since Eq. (3.7-18) must be *independent* of r, this can only be possible if

$$\boxed{k_{r_2} = k_{r_1} = k_r.} \qquad (3.7\text{-}19)$$

Therefore, upon substituting Eq. (3.7-19) into Eq. (3.7-18), we obtain

$$\int_0^\infty \left[j\rho_1 \frac{1}{4\pi} \frac{\exp(-jk_{Z_1}z_0)}{k_{Z_1}} k_r + \rho_1 A(k_r) - \rho_2 B(k_r) \right] J_0(k_r r)\, dk_r = 0, \quad (3.7\text{-}20)$$

or

$$\rho_1 A(k_r) - \rho_2 B(k_r) = -j\rho_1 \frac{1}{4\pi} \frac{\exp(-jk_{Z_1}z_0)}{k_{Z_1}} k_r, \qquad (3.7\text{-}21)$$

which is independent of r.

The second boundary condition is continuity of the normal component of the acoustic fluid velocity vector, that is,

$$u_{n_1}(t, r, \psi, z)\big|_{z=0} = u_{n_2}(t, r, \psi, z)\big|_{z=0}, \qquad (3.7\text{-}22)$$

where

$$u_{n_1}(t, \mathbf{r}) = u_{n_i}(t, \mathbf{r}) + u_{n_r}(t, \mathbf{r}) \qquad (3.7\text{-}23)$$

and

$$u_{n_2}(t, \mathbf{r}) = u_{n_t}(t, \mathbf{r}) \qquad (3.7\text{-}24)$$

are the *total* acoustic fluid *speeds* in media I and II, respectively, normal (perpendicular) to the boundary. Since the unit vector $\hat{z}$ is perpendicular to the planar boundary, the normal component of the acoustic fluid velocity vector is given by

$$u_n(t, r, \psi, z) = \hat{z} \cdot \mathbf{u}(t, r, \psi, z) = \hat{z} \cdot \nabla\varphi(t, r, \psi, z), \qquad (3.7\text{-}25)$$

or

$$u_n(t, r, \psi, z) = \frac{\partial}{\partial z} \varphi(t, r, \psi, z), \qquad (3.7\text{-}26)$$

where the gradient expressed in the cylindrical coordinates (r, ψ, z) is given by Eq. (3.2-9).

In order to calculate the normal component of the incident acoustic fluid velocity vector at $z = 0$ according to Eq. (3.7-26), Eq. (3.7-2) is rewritten as follows with the use of Eq. (3.7-19):

$$\varphi_i(t, r, \psi, z) = +j\frac{1}{4\pi} \int_0^\infty J_0(k_r r) \frac{\exp\left[-jk_{Z_1}(z_0 - z)\right]}{k_{Z_1}} k_r \, dk_r \exp(+j2\pi f t),$$

$$0 \le z \le z_0. \quad (3.7\text{-}27)$$

Therefore, with the use of Eqs. (3.7-27), (3.7-10), (3.7-11), (3.7-19), and (3.7-26), Eq. (3.7-22) can be expressed as

$$\int_0^\infty \left[+\frac{1}{4\pi} \exp(-jk_{Z_1}z_0)k_r + jk_{Z_1}A(k_r) + jk_{Z_2}B(k_r) \right] J_0(k_r r) \, dk_r = 0.$$

$$(3.7\text{-}28)$$

Since the boundary condition represented by Eq. (3.7-28) must hold for all allowed values of r, that is, since Eq. (3.7-28) must be independent of r, this can only be possible if

$$k_{Z_1}A(k_r) + k_{Z_2}B(k_r) = +j\frac{1}{4\pi} \exp(-jk_{Z_1}z_0)k_r, \qquad (3.7\text{-}29)$$

which is independent of r. We now have two equations, Eqs. (3.7-21) and (3.7-29), and two unknowns, $A(k_r)$ and $B(k_r)$.

In order to solve for $A(k_r)$, we multiply Eq. (3.7-21) by k_{Z_2} and Eq. (3.7-29) by ρ_2. Upon adding the resulting equations, we obtain

$$A(k_r) = +j\frac{1}{4\pi} R_{12} \frac{\exp(-jk_{Z_1}z_0)}{k_{Z_1}} k_r, \qquad (3.7\text{-}30)$$

where

$$R_{12} = \frac{\rho_2 k_{Z_1} - \rho_1 k_{Z_2}}{\rho_2 k_{Z_1} + \rho_1 k_{Z_2}} \qquad (3.7\text{-}31)$$

is the plane-wave velocity-potential reflection coefficient at the boundary between fluid media I and II (see Section 2.3).

Next, solving for $B(k_r)$ from Eq. (3.7-21) yields

$$B(k_r) = \frac{\rho_1}{\rho_2}\left[A(k_r) + j\frac{1}{4\pi}\frac{\exp(-jk_{Z_1}z_0)}{k_{Z_1}}k_r\right], \qquad (3.7\text{-}32)$$

and upon substituting Eqs. (3.7-30) and (3.7-31) into Eq. (3.7-32), we obtain

$$B(k_r) = +j\frac{1}{4\pi}T_{12}\frac{\exp(-jk_{Z_1}z_0)}{k_{Z_1}}k_r, \qquad (3.7\text{-}33)$$

where

$$T_{12} = \frac{2\rho_1 k_{Z_1}}{\rho_2 k_{Z_1} + \rho_1 k_{Z_2}} \qquad (3.7\text{-}34)$$

is the plane-wave velocity-potential transmission coefficient at the boundary between fluid media I and II (see Section 2.3).

Therefore, in summary,

$$\varphi_i(t,r,\psi,z) = +j\frac{1}{4\pi}\int_0^\infty J_0(k_r r)\frac{\exp(-jk_{Z_1}|z-z_0|)}{k_{Z_1}}k_r\,dk_r\,\exp(+j2\pi ft),$$

$$z \geq 0, \qquad (3.7\text{-}35)$$

$$\varphi_r(t,r,\psi,z) = +j\frac{1}{4\pi}\int_0^\infty R_{12}J_0(k_r r)\frac{\exp\left[-jk_{Z_1}(z+z_0)\right]}{k_{Z_1}}k_r\,dk_r$$

$$\times \exp(+j2\pi ft), \qquad z \geq 0, \qquad (3.7\text{-}36)$$

and

$$\varphi_t(t,r,\psi,z) = +j\frac{1}{4\pi}\int_0^\infty T_{12}J_0(k_r r)\frac{\exp\left[+j(k_{Z_2}z - k_{Z_1}z_0)\right]}{k_{Z_1}}k_r\,dk_r$$

$$\times \exp(+j2\pi ft), \qquad z \leq 0, \qquad (3.7\text{-}37)$$

where r is the horizontal range to the field point and is given by Eq. (3.7-3), R_{12} and T_{12} are the plane-wave velocity-potential reflection and transmission coefficients at the boundary between fluid media I and II and are given by Eqs. (3.7-31) and (3.7-34), respectively, and the propagation-vector components in the Z direction in media I and II are given by

$$
k_{Z_1} = \begin{cases} \sqrt{k_1^2 - k_r^2}, & k_r^2 \le k_1^2, \\ -j\sqrt{k_r^2 - k_1^2}, & k_r^2 > k_1^2, \end{cases}
\tag{3.7-38}
$$

and

$$
k_{Z_2} = \begin{cases} \sqrt{k_2^2 - k_r^2}, & k_r^2 \le k_2^2, \\ -j\sqrt{k_r^2 - k_2^2}, & k_r^2 > k_2^2, \end{cases}
\tag{3.7-39}
$$

respectively, where k_1 and k_2 are the wave numbers in media I and II and are given by Eqs. (3.7-5) and (3.7-9), respectively.

Example 3.7-1 (Ideal rigid and pressure-release boundaries)
Assume that the boundary at $z = 0$ (i.e., the XY plane) is an ideal rigid boundary. Therefore, the reflection coefficient $R_{12} = +1$ (see Example 2.3-1), and as a result, the reflected acoustic field given by Eq. (3.7-36) reduces to

$$
\varphi_r(t, r, \psi, z) = +j\frac{1}{4\pi} \int_0^\infty J_0(k_r r) \frac{\exp[-jk_{Z_1}(z + z_0)]}{k_{Z_1}} k_r \, dk_r
$$

$$
\times \exp(+j2\pi ft), \qquad z \ge 0.
\tag{3.7-40}
$$

Equation (3.7-40) represents the acoustic field from an *in-phase image point source* located at $z = -z_0$ (see Fig. 3.7-2). The image point source is in phase because the boundary is an ideal rigid boundary with reflection coefficient $R_{12} = +1$. With the use of Eqs. (3.7-35) and (3.7-40), it can be shown that on the ideal rigid boundary (i.e., at $z = 0$), the acoustic pressure in medium I is twice the incident acoustic pressure and the normal component of the acoustic fluid velocity vector in medium I is equal to zero (see Example 2.3-1).

Now assume that the boundary at $z = 0$ is an ideal pressure-release boundary. Therefore, the reflection coefficient $R_{12} = -1$ (see Example 2.3-1), and as a result, the reflected acoustic field given by Eq. (3.7-36) reduces to

$$
\varphi_r(t, r, \psi, z) = -j\frac{1}{4\pi} \int_0^\infty J_0(k_r r) \frac{\exp[-jk_{Z_1}(z + z_0)]}{k_{Z_1}} k_r \, dk_r
$$

$$
\times \exp(+j2\pi ft), \qquad z \ge 0.
\tag{3.7-41}
$$

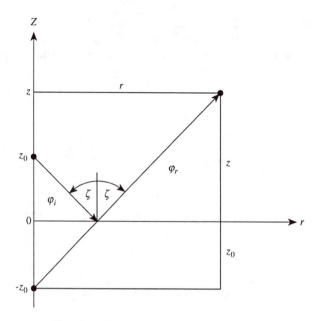

Figure 3.7-2. Image point source at $z = -z_0$.

Equation (3.7-41) represents the acoustic field from a 180°-*out-of-phase image point source* located at $z = -z_0$ (see Fig. 3.7-2). The image point source is 180° out of phase because the boundary is an ideal pressure-release boundary with reflection coefficient $R_{12} = -1$. With the use of Eqs. (3.7-35) and (3.7-41), it can be shown that on the ideal pressure-release boundary (i.e., at $z = 0$), the acoustic pressure in medium I is equal to zero and the normal component of the acoustic fluid velocity vector in medium I is twice the incident normal component of the acoustic fluid velocity vector (see Example 2.3-1).

3.8 Waveguide Models of the Ocean: Normal Modes

3.8.1 Pressure-Release Surface with a Rigid Bottom

One of the main theoretical problems of interest in underwater acoustics is the accurate prediction of sound propagation in the ocean. One classical approach that is used to solve this problem is to treat the ocean as a waveguide with plane, parallel boundaries, and to express the acoustic field in the ocean medium as a sum of normal modes. A simple waveguide model of the ocean is illustrated in Fig. 3.8-1, where $\rho_1 c_1$, $\rho_2 c_2$, and $\rho_3 c_3$ are the characteristic impedances of medium

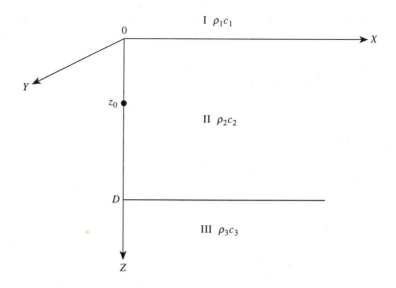

Figure 3.8-1. Simple waveguide model of the ocean where $\rho_1 c_1, \rho_2 c_2$, and $\rho_3 c_3$ are the characteristic impedances of medium I (air), medium II (ocean water), and medium III (ocean bottom), respectively. Also shown is an omnidirectional point source at $r = 0$ and $z = z_0$ meters.

I (air), medium II (ocean water), and medium III (ocean bottom), respectively, where ρ_i, $i = 1, 2, 3$, is the constant ambient (equilibrium) density, and c_i, $i = 1, 2, 3$, is the constant speed of sound in each of the three media. Both the ocean surface and ocean bottom are modeled as plane, parallel boundaries. An omnidirectional, time-harmonic point source is located in medium II at $r = 0$ and $z = z_0$, and the ocean is D meters deep.

In this section we shall model the ocean surface at $z = 0$ as an ideal pressure-release boundary (which is realistic) and the ocean bottom at $z = D$ as an ideal rigid boundary (which is not very realistic). Recall from Example 2.3-1 that ideal pressure-release and rigid boundaries are characterized by Rayleigh reflection coefficients of -1 and $+1$, respectively. Therefore, no time-average acoustic power will be present in media I and III due to the sound source located in medium II. As was mentioned previously, modeling the ocean bottom as an ideal rigid boundary is not very realistic. However, the main advantage of using this simplest of ocean-bottom models is that the resulting equations that describe the waveguide solution are relatively simple to evaluate.

The ocean waveguide problem can be stated as follows: derive an expression for the velocity potential in medium II (ocean water) that is a solution of the wave equation and satisfies all the boundary conditions, including the boundary condition at the source. Since both the ocean surface and ocean bottom are being modeled as plane, parallel boundaries, and since the sound source is assumed to be an omnidirectional point source, the solution will be *axisymmetric* (i.e., independent of the azimuthal angle ψ). Therefore, we can set $n = 0$ in Eq. (3.1-38). Also, since there are no reflected waves in the radial direction, we can neglect $H_n^{(1)}(k_r r)$ by setting $A_r = 0$ in Eq. (3.1-38). And upon taking into account the additional assumption of a time-harmonic source, the velocity potential in medium II can be expressed as

$$\varphi_2(t, r, \psi, z) = \varphi_{f,2}(r, \psi, z) \exp(+j2\pi ft), \qquad (3.8\text{-}1)$$

where

$$\varphi_{f,2}(r, \psi, z) = R_2(r) Z_2(z), \qquad (3.8\text{-}2)$$

$$R_2(r) = H_0^{(2)}(k_{r_2} r), \qquad (3.8\text{-}3)$$

$$Z_2(z) = A_2 \exp(-jk_{z_2} z) + B_2 \exp(+jk_{z_2} z), \qquad (3.8\text{-}4)$$

and [see Eq. (3.1-39)]

$$k_{r_2}^2 + k_{z_2}^2 = k_2^2 = (2\pi f/c_2)^2, \qquad (3.8\text{-}5)$$

where k_{r_2}, k_{z_2}, and k_2 are the propagation-vector components in the r and Z directions and the wave number, respectively, in medium II. Also, A_2 and B_2 are complex constants in general, whose values are determined by satisfying the boundary conditions. Note that Eq. (3.8-4) accounts for *standing waves* that exist in the Z direction in medium II due to reflections from the boundaries at the ocean surface and bottom. If k_{z_2} is positive, then the first term on the right-hand side of Eq. (3.8-4) represents a plane wave traveling in the positive Z direction, and the second term represents a plane wave traveling in the negative Z direction.

The first boundary condition is at $z = 0$. Since the boundary at $z = 0$ is assumed to be an ideal pressure-release surface, the acoustic pressure $p_2(t, r, \psi, z)$ must be equal to zero at $z = 0$ for all allowed values of t, r, and ψ. Since

$$p_2(t, r, \psi, z) = -\rho_2 \frac{\partial}{\partial t} \varphi_2(t, r, \psi, z), \qquad (3.8\text{-}6)$$

substituting Eq. (3.8-1) into Eq. (3.8-6) yields

$$p_2(t, r, \psi, z) = p_{f,2}(r, \psi, z) \exp(+j2\pi ft) \qquad (3.8\text{-}7)$$

where

$$p_{f,2}(r, \psi, z) = -jk_2 \rho_2 c_2 \varphi_{f,2}(r, \psi, z), \qquad (3.8\text{-}8)$$

and upon substituting Eq. (3.8-2) into Eq. (3.8-8), we obtain

$$p_{f,2}(r, \psi, z) = -jk_2 \rho_2 c_2 R_2(r) Z_2(z). \qquad (3.8\text{-}9)$$

Therefore,

$$p_2(t,r,\psi,z)\big|_{z=0} = 0 \tag{3.8-10}$$

is equivalent to

$$p_{f,2}(r,\psi,z)\big|_{z=0} = -jk_2\rho_2 c_2 R_2(r)Z_2(z)\big|_{z=0} = 0, \tag{3.8-11}$$

or

$$Z_2(z)\big|_{z=0} = 0. \tag{3.8-12}$$

Substituting Eq. (3.8-4) into Eq. (3.8-12) yields

$$B_2 = -A_2, \tag{3.8-13}$$

and upon substituting Eq. (3.8-13) into Eq. (3.8-4), we obtain

$$Z_2(z) = -j2A_2 \sin(k_{z_2}z). \tag{3.8-14}$$

The second boundary condition is at $z = D$. Since the boundary at $z = D$ is assumed to be an ideal rigid surface, and the unit vector normal to the boundary is $\hat{z}$, the normal component of the acoustic fluid velocity vector $u_{z_2}(t,r,\psi,z)$ must be equal to zero at $z = D$ for all allowed values of t, r, and ψ. Since

$$u_{z_2}(t,r,\psi,z) = \hat{z} \cdot \nabla\varphi_2(t,r,\psi,z) = \frac{\partial}{\partial z}\varphi_2(t,r,\psi,z), \tag{3.8-15}$$

substituting Eq. (3.8-1) into Eq. (3.8-15) yields

$$u_{z_2}(t,r,\psi,z) = u_{f,z_2}(r,\psi,z)\exp(+j2\pi ft), \tag{3.8-16}$$

where

$$u_{f,z_2}(r,\psi,z) = \frac{\partial}{\partial z}\varphi_{f,2}(r,\psi,z), \tag{3.8-17}$$

and upon substituting Eq. (3.8-2) into Eq. (3.8-17), we obtain

$$u_{f,z_2}(r,\psi,z) = R_2(r)\frac{d}{dz}Z_2(z). \tag{3.8-18}$$

Therefore,

$$u_{z_2}(t,r,\psi,z)\big|_{z=D} = 0 \tag{3.8-19}$$

is equivalent to

$$u_{f,Z_2}(r,\psi,z)\Big|_{z=D} = R_2(r)\frac{d}{dz}Z_2(z)\Big|_{z=D} = 0, \qquad (3.8\text{-}20)$$

or

$$\frac{d}{dz}Z_2(z)\Big|_{z=D} = 0. \qquad (3.8\text{-}21)$$

Substituting Eq. (3.8-14) into Eq. (3.8-21) yields

$$\cos(k_{Z_2}D) = 0, \qquad (3.8\text{-}22)$$

which implies that

$$k_{Z_2}D = (2n+1)\pi/2, \qquad n = 0,1,2,\ldots, \qquad (3.8\text{-}23)$$

or

$$\boxed{k_{Z_2} = k_{Z_{2,n}} = \frac{(2n+1)\pi}{2D}, \qquad n = 0,1,2,\ldots.} \qquad (3.8\text{-}24)$$

By inspecting Eq. (3.8-24), it can be seen that the propagation-vector component in the Z direction is only allowed certain discrete values in order to satisfy the boundary condition at $z = D$. Note that only positive integer values for n are used in order to keep $k_{Z_2} = k_{Z_{2,n}}$ positive. Because of Eq. (3.8-4), a positive k_{Z_2} value allows for wave propagation in both the positive and negative Z directions. Substituting Eq. (3.8-24) into Eq. (3.8-14) yields

$$\boxed{Z_2(z) = Z_{2,n}(z) = -j2A_{2,n}\sin(k_{Z_{2,n}}z), \qquad n = 0,1,2,\ldots,} \qquad (3.8\text{-}25)$$

where the set of functions $\sin(k_{Z_{2,n}}z)$, $n = 0,1,2,\ldots$, are referred to as the *eigenfunctions* or *normal modes*, since they describe the natural modes of vibration of the acoustic field in the ocean waveguide. The set of values $k_{Z_{2,n}}$, $n = 0,1,2,\ldots$, are referred to as the *eigenvalues*.

Next, let us solve for the propagation-vector component in the radial direction k_{r_2}. From Eq. (3.8-5)

$$k_{r_2} = \sqrt{k_2^2 - k_{Z_2}^2}, \qquad (3.8\text{-}26)$$

and upon substituting Eq. (3.8-24) into Eq. (3.8-26), we obtain

$$k_{r_2} = k_{r_{2,n}} = \sqrt{k_2^2\left[1 - \left(k_{z_{2,n}}/k_2\right)^2\right]}, \qquad (3.8\text{-}27)$$

or

$$k_{r_{2,n}} = k_2\sqrt{1 - \left(f_n/f\right)^2}, \quad f \ge f_n, \qquad n = 0,1,2,\ldots, \qquad (3.8\text{-}28)$$

where f is the source frequency in hertz and

$$f_n = \frac{(2n+1)c_2}{4D}, \qquad n = 0,1,2,\ldots, \qquad (3.8\text{-}29)$$

is the *cutoff frequency* in hertz of the nth mode. Note that if $f = f_n$, then $k_{r_{2,n}} = 0$, which means that the nth mode will *not* propagate in the radial direction. Also, if Eq. (3.8-28) is rewritten as

$$k_{r_{2,n}} = k_2\sqrt{-\left[(f_n/f)^2 - 1\right]}, \qquad (3.8\text{-}30)$$

then

$$k_{r_{2,n}} = -jk_2\sqrt{(f_n/f)^2 - 1}, \quad f < f_n, \qquad n = 0,1,2,\ldots, \qquad (3.8\text{-}31)$$

in order to ensure *evanescent modes* (i.e., decaying exponentials) in the radial direction. Evanescent modes are only important at short ranges from the source. Therefore, the nth *mode will be a propagating mode only if the source frequency f is greater than the cutoff frequency f_n of the nth mode.*

Before continuing further, several additional comments concerning the physical significance of the expressions for $k_{r_{2,n}}$ and f_n given by Eqs. (3.8-28) and (3.8-29), respectively, are in order. Since $f_0 \ne 0$ Hz, if the source frequency $f < f_0$, then *no* modes will propagate. Also, since $f_0 \ne 0$ Hz, we have $k_{r_{2,0}} \ne k_2$, and as a result, there is *no* longitudinal, or plane-wave mode. In order for the plane-wave mode to exist, which is conventionally mode $n = 0$, $k_{r_{2,0}}$ must be equal to the wave number k_2, implying that $k_{z_{2,0}} = 0$ [see Eq. (3.8-5)]. If $k_{z_{2,0}} = 0$, then the propagation vector for mode $n = 0$ is parallel to the boundaries, since it only has a radial component. This further implies that mode $n = 0$ propagates directly down the length of the waveguide without undergoing any surface and bottom reflections. However, if $k_{z_{2,0}} = 0$, then $Z_{2,0}(z) = 0$ [see Eq. (3.8-25)], which means that even if the plane-wave mode existed for the waveguide model in this section, it would be a *trivial mode*, since it would not contribute to the acoustic field in the waveguide. Also note that for any mode, if $f \gg f_n$, then $k_{r_{2,n}} \to k_2$, that is, the nth mode approaches being a plane wave that propagates directly down the length of the waveguide without undergoing any surface and bottom reflections. And as the depth D of the ocean increases, the value of f_n decreases.

Since $k_{r_2} = k_{r_{2,n}}$, Eq. (3.8-3) can be rewritten as

$$R_2(r) = R_{2,n}(r) = H_0^{(2)}(k_{r_{2,n}}r), \qquad n = 0, 1, 2, \ldots, \tag{3.8-32}$$

and upon substituting Eqs. (3.8-25) and (3.8-32) into Eq. (3.8-2), we obtain

$$\varphi_{f,2,n}(r, \psi, z) = -j2A_{2,n}H_0^{(2)}(k_{r_{2,n}}r)\sin(k_{z_{2,n}}z), \qquad n = 0, 1, 2, \ldots, \tag{3.8-33}$$

which is the spatial-dependent part of the velocity potential of the acoustic field in medium II corresponding to the nth normal mode. If we define the *complete normal-mode solution* as being equal to the sum (linear superposition) of the contributions from *all* the normal modes (propagating and evanescent), that is, if

$$\varphi_{f,2}(r, \psi, z) \triangleq \sum_{n=0}^{\infty} \varphi_{f,2,n}(r, \psi, z), \tag{3.8-34}$$

then substituting Eq. (3.8-33) into Eq. (3.8-34) yields

$$\varphi_{f,2}(r, \psi, z) = -j2 \sum_{n=0}^{\infty} A_{2,n}H_0^{(2)}(k_{r_{2,n}}r)\sin(k_{z_{2,n}}z). \tag{3.8-35}$$

And upon substituting Eq. (3.8-35) into Eq. (3.8-8), we obtain

$$p_{f,2}(r, \psi, z) = -2k_2\rho_2 c_2 \sum_{n=0}^{\infty} A_{2,n}H_0^{(2)}(k_{r_{2,n}}r)\sin(k_{z_{2,n}}z), \tag{3.8-36}$$

which is the complete normal-mode solution for the spatial-dependent part of the acoustic pressure in medium II. Note that the amplitude term $A_{2,n}$ is still unknown. In order to determine its value, the boundary condition at the location of the sound source inside the waveguide must be satisfied. We shall satisfy the boundary condition at the source later in this section.

In practice, only the contributions from the individual propagating modes are summed. However, if the acoustic field is being evaluated near the source, then the contributions from the evanescent modes must also be included. The *total number of propagating normal modes*, N_p, excited by a source frequency of f hertz can be obtained by setting $f_n = f$ and $n = N_p$ in Eq. (3.8-29) and solving for N_p. Doing so, with the following modifications, yields

$$N_p = \text{INT}\left(\frac{1}{2}\left[\frac{4Df}{c_2} - 1\right]\right) + 1 \qquad (3.8\text{-}37)$$

where $\text{INT}(\cdot)$ means form an integer by simply truncating the decimal portion of the real number inside the parentheses, and then add one to the result, since the first or lowest mode is mode $n = 0$. Therefore, if one is only interested in summing the contributions from all of the propagating modes and not the evanescent modes, then the upper limit of the summation in Eq. (3.8-36), for example, should be replaced by $N_p - 1$. Also, as can be seen from Eq. (3.8-37), *for a given value of source frequency f, the number of propagating modes increases as D increases, and for a given value of ocean depth D, the number of propagating modes increases as f increases*. As the number of propagating modes increases, more modes must be summed in order to produce the normal-mode solution. As a result, the normal-mode analysis approach is best suited for numerical solution of *low-frequency shallow-water* problems.

Group Speed, Phase Speed, Angle of Propagation, and Travel Time
The group and phase speeds in the horizontal radial direction, the angle of propagation, and the travel time of the nth propagating mode will be computed next. The *group speed* in the horizontal radial direction of the nth propagating mode is given by (see Section 2.2.2)

$$c_{g_{r2,n}} = c_2 k_{r_{2,n}}/k_2, \qquad (3.8\text{-}38)$$

and upon substituting Eq. (3.8-28) into Eq. (3.8-38), we obtain

$$c_{g_{r2,n}} = c_2\sqrt{1 - (f_n/f)^2}, \quad f \geq f_n, \qquad n = 0, 1, \ldots, N_p - 1, \qquad (3.8\text{-}39)$$

where f_n is given by Eq. (3.8-29). The group speed given by Eq. (3.8-39) is the speed at which *energy* propagates in the horizontal radial direction. Since the group speed given by Eq. (3.8-39) is a function of frequency and mode number, for a given mode n with cutoff frequency f_n, the energy associated with the high-frequency components from a transmitted pulse will arrive at a receiver before the energy associated with the low-frequency components. As a result, the shape of the transmitted pulse will be *distorted* by the time it reaches the receiver, due to the dispersion of the transmitted signal's energy. Note that when $f = f_n$, the group speed given by Eq. (3.8-39) is equal to zero, which means that no energy propagates in the radial direction.

The *phase speed* in the horizontal radial direction of the nth propagating mode is given by (see Section 2.2.2)

$$c_{p_{r_{2,n}}} = c_2 k_2 / k_{r_{2,n}},$$ (3.8-40)

and upon substituting Eq. (3.8-28) into Eq. (3.8-40), we obtain

$$c_{p_{r_{2,n}}} = \frac{c_2}{\sqrt{1 - (f_n/f)^2}}, \quad f \geq f_n, \quad n = 0, 1, \ldots, N_p - 1,$$ (3.8-41)

where f_n is given by Eq. (3.8-29). The phase speed given by Eq. (3.8-41) is the speed at which a surface of constant phase propagates in the horizontal radial direction. Note that the phase speed is also a function of frequency and mode number. Whenever the phase speed is a function of frequency, the effect is called *dispersion*. In our problem, the dispersion is more precisely known as *geometrical dispersion*, since it is the result of satisfying the boundary conditions at the ocean surface and bottom. Also, when $f = f_n$, the phase speed given by Eq. (3.8-41) is equal to infinity, which means that the propagation-vector component in the radial direction is zero [see Eq. (3.8-40)], which is consistent with no energy propagating in the radial direction.

The *propagation vector* of the nth propagating mode in medium II can be expressed as (see Fig. 3.8-2)

$$\mathbf{k}_{2,n} = k_{r_{2,n}} \hat{r} + k_{Z_{2,n}} \hat{z}, \quad n = 0, 1, \ldots, N_p - 1,$$ (3.8-42)

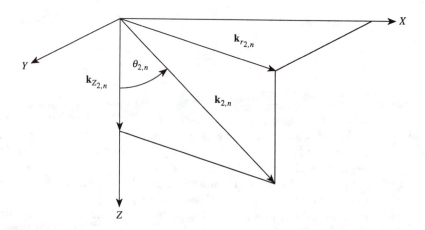

Figure 3.8-2. The propagation vector $\mathbf{k}_{2,n}$ of the nth propagating mode in medium II. Also shown are its r and Z components.

where

$$k_{r_{2,n}} = k_2 \sin \theta_{2,n}, \tag{3.8-43}$$

$$k_{z_{2,n}} = k_2 \cos \theta_{2,n}, \tag{3.8-44}$$

and

$$|\mathbf{k}_{2,n}| = k_2 = 2\pi f/c_2. \tag{3.8-45}$$

Therefore, from Eq. (3.8-43),

$$\theta_{2,n} = \sin^{-1}(k_{r_{2,n}}/k_2), \tag{3.8-46}$$

and upon substituting Eq. (3.8-28) into Eq. (3.8-46), we obtain the *angle of propagation* of the nth propagating mode,

$$\boxed{\theta_{2,n} = \sin^{-1}\left(\sqrt{1 - (f_n/f)^2}\right), \quad f \geq f_n, \quad n = 0, 1, \ldots, N_p - 1,} \tag{3.8-47}$$

where f_n is given by Eq. (3.8-29). Note that when $f = f_n$, the angle of propagation given by Eq. (3.8-47) is equal to zero, which means that the propagation-vector component in the radial direction is zero [see Eq. (3.8-43)], which is consistent with no energy propagating in the radial direction.

The *travel time*, or *time of arrival*, of the nth propagating mode at a horizontal range of r meters from the source is given by

$$\boxed{\tau_{r_{2,n}} = r/c_{g_{r_{2,n}}},} \tag{3.8-48}$$

where the group speed is given by Eq. (3.8-39).

Boundary Condition at the Source

The third and final boundary condition is at the source. Recall that the source was assumed to be an omnidirectional, time-harmonic, point source located at $r = 0$ and $z = z_0$. Therefore, if the acoustic pressure at the source at frequency f is P_0 pascals, then

$$p_S(t, r, \psi, z) = p_{f,S}(r, \psi, z) \exp(+j2\pi ft), \tag{3.8-49}$$

where

$$p_{f,S}(r, \psi, z) = P_0 \frac{\delta(r, z - z_0)}{2\pi r} = P_0 \frac{\delta(r)}{2\pi r} \delta(z - z_0). \tag{3.8-50}$$

Note that

$$\int_{-\infty}^{\infty} \int_{0}^{2\pi} \int_{0}^{\infty} \frac{\delta(r)}{2\pi r} \delta(z - z_0) r\, dr\, d\psi\, dz = 1, \qquad (3.8\text{-}51)$$

which indicates that the representation of the location of the point source at $r = 0$ and $z = z_0$ by the impulse functions on the right-hand side of Eq. (3.8-50) is correct, since the volume of an impulse function must be equal to one. Therefore, the acoustic pressure in medium II must also satisfy the following inhomogeneous Helmholtz equation:

$$\nabla^2 p_{f,2}(\mathbf{r}) + k_2^2 p_{f,2}(\mathbf{r}) = p_{f,s}(\mathbf{r}). \qquad (3.8\text{-}52)$$

Since the sound source is being modeled by an impulse function [see Eq. (3.8-50)], it is very important to realize that the solution $p_{f,2}(\mathbf{r})$ of Eq. (3.8-52) is the *spatial impulse response of the pressure-release-surface rigid-bottom ocean waveguide for acoustic pressure*. Substituting Eqs. (3.1-2), (3.8-36), and (3.8-50) into Eq. (3.8-52) yields

$$\sum_{n=0}^{\infty} A_{2,n} \left[\frac{d^2}{dr^2} H_0^{(2)}(k_{r_{2,n}} r) + \frac{1}{r} \frac{d}{dr} H_0^{(2)}(k_{r_{2,n}} r) + k_{r_{2,n}}^2 H_0^{(2)}(k_{r_{2,n}} r) \right] \sin(k_{Z_{2,n}} z)$$

$$= \frac{P_0}{A} \frac{\delta(r)}{2\pi r} \delta(z - z_0), \qquad (3.8\text{-}53)$$

where

$$A = -2k_2 \rho_2 c_2 \qquad (3.8\text{-}54)$$

and

$$k_{r_{2,n}}^2 = k_2^2 - k_{Z_{2,n}}^2. \qquad (3.8\text{-}55)$$

Since

$$\boxed{\frac{d^2}{dr^2} H_0^{(2)}(k_r r) + \frac{1}{r} \frac{d}{dr} H_0^{(2)}(k_r r) + k_r^2 H_0^{(2)}(k_r r) = j\frac{2}{\pi} \frac{\delta(r)}{r},} \qquad (3.8\text{-}56)$$

substituting Eq. (3.8-56) into Eq. (3.8-53) yields

$$\sum_{n=0}^{\infty} A_{2,n} \sin(k_{Z_{2,n}} z) = -j\frac{P_0}{4A} \delta(z - z_0). \qquad (3.8\text{-}57)$$

Next, multiplying both sides of Eq. (3.8-57) by the eigenfunction $\sin(k_{Z_{2,m}} z)$ and then integrating over z from 0 to D yields

$$\sum_{n=0}^{\infty} A_{2,n} \int_0^D \sin(k_{Z_{2,m}} z) \sin(k_{Z_{2,n}} z)\, dz = -j\frac{P_0}{4A} \int_0^D \sin(k_{Z_{2,m}} z)\, \delta(z - z_0)\, dz.$$

$$(3.8\text{-}58)$$

Since the eigenfunctions $\sin(k_{Z_{2,n}} z)$, $n = 0, 1, 2, \ldots$, are orthogonal in the interval $0 \le z \le D$ (see Appendix 3A), and using the sifting property of impulse functions, Eq. (3.8-58) reduces to

$$\sum_{n=0}^{\infty} A_{2,n} E_n \delta_{mn} = -j\frac{P_0}{4A} \sin(k_{Z_{2,m}} z_0),$$

$$(3.8\text{-}59)$$

$$A_{2,m} E_m = -j\frac{P_0}{4A} \sin(k_{Z_{2,m}} z_0),$$

$$(3.8\text{-}60)$$

and upon replacing the index m with the index n,

$$A_{2,n} = -j\frac{P_0}{4AE_n} \sin(k_{Z_{2,n}} z_0),$$

$$(3.8\text{-}61)$$

where

$$\boxed{E_n = D/2, \qquad n = 0, 1, 2, \ldots,}$$

$$(3.8\text{-}62)$$

is the *energy* contained in the eigenfunctions (see Appendix 3A). Finally, substituting Eqs. (3.8-54) and (3.8-62) into Eq. (3.8-61) yields the following expression for the *amplitude* term associated with the nth normal mode:

$$\boxed{A_{2,n} = j\frac{P_0}{4k_2\rho_2 c_2 D} \sin(k_{Z_{2,n}} z_0), \qquad n = 0, 1, 2, \ldots .}$$

$$(3.8\text{-}63)$$

Therefore, upon substituting Eqs. (3.8-35) and (3.8-63) into Eq. (3.8-1), and by only summing the contributions from the propagating modes, we obtain

$$\boxed{\begin{aligned} \varphi_2(t, r, \psi, z) &= \frac{P_0}{2k_2\rho_2 c_2 D} \sum_{n=0}^{N_p-1} \sin(k_{Z_{2,n}} z_0)\, H_0^{(2)}(k_{r_{2,n}} r) \sin(k_{Z_{2,n}} z) \\ &\quad \times \exp(+j2\pi ft), \end{aligned}}$$

$$(3.8\text{-}64)$$

which is the *time-harmonic normal-mode solution for the velocity potential* in medium II. And upon substituting Eqs. (3.8-36) and (3.8-63) into Eq. (3.8-7), and by only summing the contributions from the propagating modes, we obtain

$$p_2(t, r, \psi, z) = -j\frac{P_0}{2D} \sum_{n=0}^{N_p-1} \sin(k_{Z_{2,n}} z_0) H_0^{(2)}(k_{r_{2,n}} r) \sin(k_{Z_{2,n}} z) \exp(+j2\pi ft),$$

(3.8-65)

which is the *time-harmonic normal-mode solution for the acoustic pressure* in medium II, where, in summary,

$$k_{r_{2,n}} = \begin{cases} k_2\sqrt{1 - (f_n/f)^2}, & f \geq f_n, \\ -jk_2\sqrt{(f_n/f)^2 - 1}, & f < f_n, \end{cases}$$

(3.8-66)

is the propagation-vector component in the horizontal radial direction, in radians per meter, of the nth mode;

$$k_2 = 2\pi f/c_2 = 2\pi/\lambda_2$$

(3.8-67)

is the wave number in radians per meter;

$$f_n = \frac{(2n + 1)c_2}{4D}$$

(3.8-68)

is the cutoff frequency in hertz of the nth mode;

$$k_{Z_{2,n}} = \frac{(2n + 1)\pi}{2D}$$

(3.8-69)

is the propagation-vector component in the Z direction, in radians per meter, of the nth mode; and N_p is the total number of propagating modes excited by a source frequency of f hertz and is given by Eq. (3.8-37). Note that in the far field, that is, when $k_{r_{2,n}} r \gg 1$ (see Section 3.1),

$$H_0^{(2)}(k_{r_{2,n}} r) \approx \sqrt{\frac{2}{\pi k_{r_{2,n}} r}} \exp\left[-j\left(k_{r_{2,n}} r - \frac{\pi}{4}\right)\right], \qquad k_{r_{2,n}} r \gg 1. \quad (3.8\text{-}70)$$

As was mentioned previously, the spatial-dependent part of Eq. (3.8-65) is the spatial impulse response of the pressure-release-surface, rigid-bottom ocean waveguide for acoustic pressure.

Example 3.8-1 (**Pulse propagation**)

In this example we shall demonstrate a procedure that is commonly used to compute the acoustic field due to the transmission of a finite-duration, time-domain pulse or transient by a sound source. If we let $X(f)$ be the complex frequency spectrum of the transmitted time-domain pulse, then the resulting acoustic pressure field can be expressed as an inverse Fourier transform as follows:

$$p_2(t, \mathbf{r}) = F_f^{-1}\{X(f)p_{f,2}(\mathbf{r})\} = \int_{-\infty}^{\infty} X(f)p_{f,2}(\mathbf{r}) \exp(+j2\pi ft)\, df,$$

$$(3.8\text{-}71)$$

where $p_{f,2}(\mathbf{r})$ is the spatial-dependent part of the acoustic pressure obtained from the time-harmonic solution. By inspecting the right-hand side of Eq. (3.8-71), it can be seen that the pulse solution is obtained by simply multiplying the time-harmonic solution by $X(f)$ and then integrating over frequency f from $-\infty$ to ∞. In other words, we "sum" the contributions from all the frequency components contained in the transmitted pulse.

For example, with the use of Eq. (3.8-65) and by setting $P_0 = 1$ Pa, the acoustic pressure in an ocean waveguide with a pressure-release surface and a rigid bottom due to a transmitted pulse is given by

$$p_2(t, r, \psi, z) = -j\frac{1}{2D}\int_{-\infty}^{\infty} X(f) \sum_{n=0}^{N_p - 1} \sin(k_{Z_{2,n}} z_0)\, H_0^{(2)}(k_{r_{2,n}} r)$$

$$\times \sin(k_{Z_{2,n}} z)\exp(+j2\pi ft)\, df. \qquad (3.8\text{-}72)$$

Recall that the number of propagating modes, N_p, is *frequency dependent* [see Eq. (3.8-37)]. The parameter P_0 must be set equal to one, since $X(f)$ automatically assigns the correct amplitudes to the various frequency components.

Figure 3.8-3 illustrates a Hamming-envelope continuous-wave (CW) transmitted electrical pulse. Amplitude- and angle-modulated carriers will be discussed in detail in Chapter 8, Section 8.2. The pulse shown in Fig. 3.8-3 is composed of 101 frequency components; it has an amplitude of 10, a pulse length of 100 msec, a pulse-repetition frequency (fundamental frequency) of 0.4 Hz, and a carrier frequency of 250 Hz.

Figure 3.8-4 illustrates the corresponding received acoustic pulse obtained by evaluating Eq. (3.8-72) for $z_0 = 30$ m, $D = 100$ m, $\rho_2 = 1026$ kg/m^3, and $c_2 = 1500$ m/sec. Figure 3.8-4a is for a receiver located above the source at $x = 1$ km, $y = 0$ m, and $z = 20$ m.

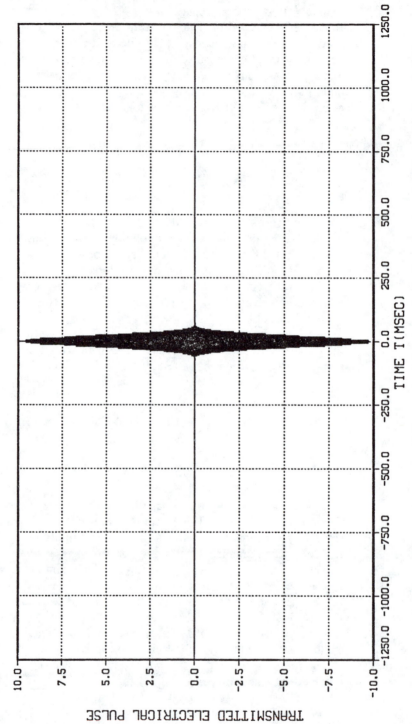

Figure 3.8-3. Hamming-envelope continuous-wave (CW) transmitted electrical pulse for the parameter values discussed in Example 3.8-1.

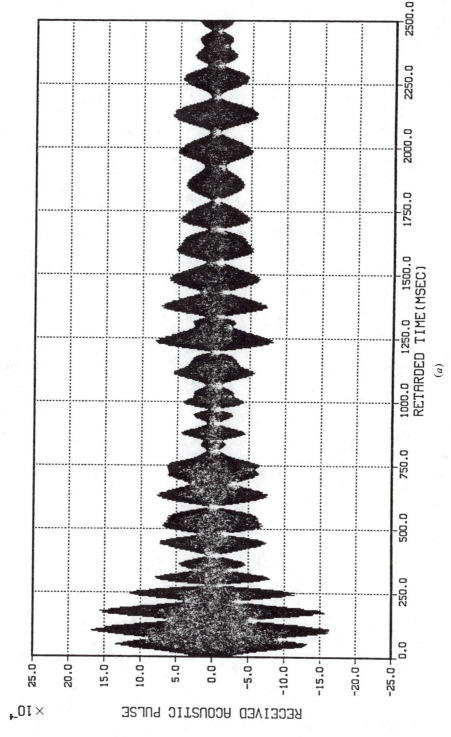

Figure 3.8-4. Received acoustic pulse for a receiver located (*a*) above the source and (*b*) below the source, using the parameter values discussed in Example 3.8-1.

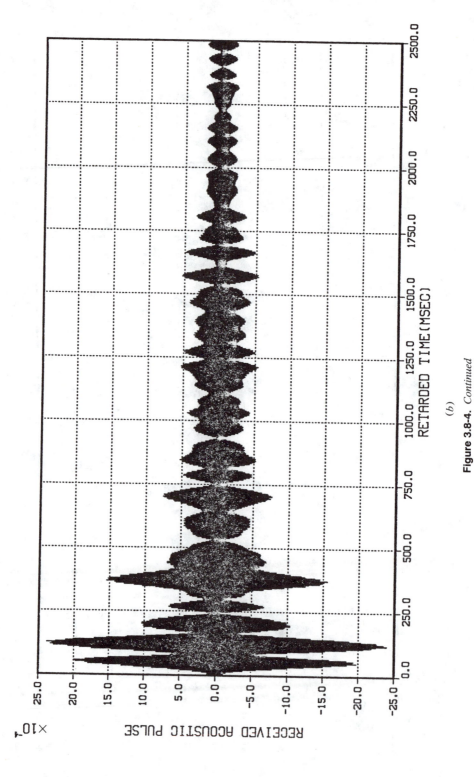

(b)

Figure 3.8-4. *Continued*

Figure 3.8-4b is for a receiver located below the source at $x = 1$ km, $y = 0$ m, and $z = 80$ m. By comparing Figs. 3.8-3 and 3.8-4, the effects of geometrical dispersion are evident. Also, by comparing Figs. 3.8-4a and b, it can be seen that for a constant horizontal range, the shape of the received acoustic pulse changes dramatically as a function of depth.

And finally, Table 3.8-1 lists the important corresponding parameter values for all of the propagating modes excited by the carrier frequency.

Table 3.8-1   Important parameter values for all of the propagating modes excited by the carrier frequency $f = 250$ Hz in Example 3.8-1. Note that the wave number in medium II is $k_2 = 1.04720$ rad/m at $f = 250$ Hz.

n	f_n (Hz)	$\theta_{2,n}$ (deg)	$k_{r_{2,n}}$ (rad/m)	$k_{Z_{2,n}}$ (rad/m)	$c_{g_{r_{2,n}}}$ (m/sec)	E_n	$\tau_{r_{2,n}}$ (sec)
0	3.750	89.141	1.04708	0.01571	1499.831	50.000	0.666742
1	11.250	87.421	1.04614	0.04712	1498.480	50.000	0.667343
2	18.750	85.699	1.04425	0.07854	1495.775	50.000	0.668550
3	26.250	83.973	1.04141	0.10996	1491.708	50.000	0.670372
4	33.750	82.241	1.03761	0.14137	1486.268	50.000	0.672826
5	41.250	80.503	1.03284	0.17279	1479.440	50.000	0.675931
6	48.750	78.755	1.02709	0.20420	1471.205	50.000	0.679715
7	56.250	76.997	1.02035	0.23562	1461.538	50.000	0.684211
8	63.750	75.226	1.01258	0.26704	1450.412	50.000	0.689459
9	71.250	73.441	1.00377	0.29845	1437.791	50.000	0.695511
10	78.750	71.639	0.99389	0.32987	1423.638	50.000	0.702426
11	86.250	69.818	0.98290	0.36128	1407.904	50.000	0.710276
12	93.750	67.976	0.97078	0.39270	1390.537	50.000	0.719147
13	101.250	66.109	0.95747	0.42412	1371.475	50.000	0.729142
14	108.750	64.215	0.94293	0.45553	1350.646	50.000	0.740387
15	116.250	62.290	0.92710	0.48695	1327.966	50.000	0.753031
16	123.750	60.330	0.90990	0.51836	1303.339	50.000	0.767260
17	131.250	58.332	0.89127	0.54978	1276.653	50.000	0.783298
18	138.750	56.289	0.87111	0.58119	1247.776	50.000	0.801426
19	146.250	54.197	0.84931	0.61261	1216.550	50.000	0.821997
20	153.750	52.048	0.82574	0.64403	1182.791	50.000	0.845458
21	161.250	49.834	0.80025	0.67544	1146.274	50.000	0.872392
22	168.750	47.546	0.77264	0.70686	1106.727	50.000	0.903566
23	176.250	45.170	0.74268	0.73827	1063.811	50.000	0.940017
24	183.750	42.693	0.71007	0.76969	1017.101	50.000	0.983187
25	191.250	40.093	0.67443	0.80111	966.045	50.000	1.035148
26	198.750	37.345	0.63524	0.83252	909.914	50.000	1.099005
27	206.250	34.412	0.59181	0.86394	847.699	50.000	1.179664
28	213.750	31.240	0.54311	0.89535	777.942	50.000	1.285443
29	221.250	27.748	0.48757	0.92677	698.387	50.000	1.431872
30	228.750	23.794	0.42250	0.95819	605.181	50.000	1.652399
31	236.250	19.091	0.34251	0.98960	490.605	50.000	2.038298
32	243.750	12.839	0.23269	1.02102	333.307	50.000	3.000234

3.8.2 Pressure-Release Surface with a Fluid Bottom

In this section we shall consider a slightly more realistic ocean waveguide model compared to the one discussed in Section 3.8.1. The ocean surface at $z = 0$ is once again modeled as an ideal pressure-release boundary, as was done in Section 3.8.1. However, in this section we shall model the ocean bottom at $z = D$ as a boundary between two different fluid media (which is more realistic) instead of as an ideal rigid boundary as was done in Section 3.8.1. This particular model for the ocean bottom will have several major consequences, one of them being that a transcendental equation must be solved in order to obtain a normal-mode solution. Note that this ocean waveguide model is sometimes referred to as the *Pekeris waveguide* in honor of C. L. Pekeris who first analyzed this model in detail. As in Section 3.8.1, the problem is to derive an expression for the velocity potential in medium II (ocean water) that is a solution of the wave equation and satisfies all the boundary conditions, including the boundary condition at the source.

The first boundary condition is at $z = 0$. Since the boundary at $z = 0$ is modeled as an ideal pressure-release surface, Eq. (3.8-14) is still valid. The second and third boundary conditions are at $z = D$. Since the surface at $z = D$ is now modeled as the boundary between two different fluid media, the second and third boundary conditions are *continuity of acoustic pressure* and *continuity of the normal component of the acoustic fluid velocity vector* at $z = D$, respectively. In order to satisfy the two boundary conditions at $z = D$, we introduce the following expression for the time-harmonic velocity potential in medium III:

$$\varphi_3(t, r, \psi, z) = \varphi_{f,3}(r, \psi, z) \exp(+j2\pi ft), \qquad (3.8\text{-}73)$$

where

$$\varphi_{f,3}(r, \psi, z) = R_3(r) Z_3(z), \qquad (3.8\text{-}74)$$

$$R_3(r) = H_0^{(2)}(k_{r_3} r), \qquad (3.8\text{-}75)$$

$$Z_3(z) = A_3 \exp(-jk_{Z_3} z), \qquad (3.8\text{-}76)$$

and

$$k_{r_3}^2 + k_{Z_3}^2 = k_3^2 = (2\pi f/c_3)^2, \qquad (3.8\text{-}77)$$

where k_{r_3}, k_{Z_3}, and k_3 are the propagation-vector components in the r and Z directions and the wave number, respectively, in medium III (see Fig. 3.8-5). Also, A_3 is a complex constant in general, whose value is determined by satisfying the boundary conditions. Equation (3.8-76) indicates that medium III is being modeled as an infinite half-space, since reflected waves traveling in the negative Z direction are not included.

The second boundary condition is continuity of acoustic pressure at $z = D$, that is,

$$p_{f,2}(r, \psi, z)\big|_{z=D} = p_{f,3}(r, \psi, z)\big|_{z=D}, \qquad (3.8\text{-}78)$$

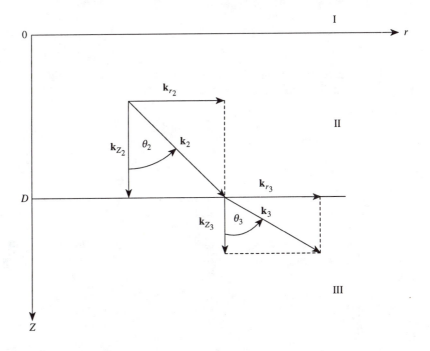

Figure 3.8-5. Propagation vectors k_2 and k_3 in media II and III, respectively. Also shown are the corresponding r and Z components and associated angles.

or [see Eq. (3.8-9)]

$$-j2\pi f\rho_2 R_2(r)Z_2(z)\big|_{z=D} = -j2\pi f\rho_3 R_3(r)Z_3(z)\big|_{z=D}. \qquad (3.8\text{-}79)$$

Substituting Eqs. (3.8-3), (3.8-14), (3.8-75), and (3.8-76) into Eq. (3.8-79) yields

$$-j2\rho_2 A_2 H_0^{(2)}(k_{r_2}r)\sin(k_{Z_2}D) = \rho_3 A_3 H_0^{(2)}(k_{r_3}r)\exp(-jk_{Z_3}D). \qquad (3.8\text{-}80)$$

Since Eq. (3.8-80) must hold for all allowed values of r, that is, since Eq. (3.8-80) must be *independent* of r, this requires that

$$\boxed{k_{r_3} = k_{r_2},} \qquad (3.8\text{-}81)$$

and as a result, Eq. (3.8-80) reduces to

$$A_2 = j\frac{\rho_3}{2\rho_2}\frac{\exp(-jk_{Z_3}D)}{\sin(k_{Z_2}D)}A_3 \qquad (3.8\text{-}82)$$

and Eq. (3.8-77) becomes

$$k_{r_2}^2 + k_{Z_3}^2 = k_3^2 = (2\pi f/c_3)^2. \tag{3.8-83}$$

The third boundary condition is continuity of the normal component of the acoustic fluid velocity vector at $z = D$, that is,

$$u_{f,Z_2}(r,\psi,z)\big|_{z=D} = u_{f,Z_3}(r,\psi,z)\big|_{z=D}, \tag{3.8-84}$$

or [see Eq. (3.8-18)]

$$R_2(r)\frac{d}{dz}Z_2(z)\bigg|_{z=D} = R_3(r)\frac{d}{dz}Z_3(z)\bigg|_{z=D}. \tag{3.8-85}$$

Substituting Eqs. (3.8-3), (3.8-14), (3.8-75), (3.8-76), and (3.8-81) into Eq. (3.8-85) yields

$$A_2 = \frac{k_{Z_3}}{2k_{Z_2}} \frac{\exp(-jk_{Z_3}D)}{\cos(k_{Z_2}D)} A_3. \tag{3.8-86}$$

By equating the right-hand sides of Eqs. (3.8-82) and (3.8-86), we obtain the following relationship:

$$\tan(k_{Z_2}D) = j\frac{\rho_3 k_{Z_2}}{\rho_2 k_{Z_3}}, \tag{3.8-87}$$

where the propagation-vector components in the Z direction can be expressed as (see Fig. 3.8-5)

$$k_{Z_2} = k_2 \cos\theta_2 \tag{3.8-88}$$

and

$$k_{Z_3} = k_3 \cos\theta_3. \tag{3.8-89}$$

Note that the angles θ_2 and θ_3 are the angles of incidence and transmission, respectively, with respect to the boundary between media II and III (see Fig. 3.8-5). Also note that normal incidence and grazing incidence correspond to $\theta_2 = 0$ and $\theta_2 = \pi/2$, respectively.

What we want to do next is to express Eq. (3.8-89), and hence Eq. (3.8-87), in terms of θ_2. Since

$$\cos\theta_3 = \sqrt{1 - \sin^2\theta_3}, \tag{3.8-90}$$

and from Snell's law (see Section 2.3)

$$\sin \theta_3 = \frac{c_3}{c_2} \sin \theta_2, \qquad (3.8\text{-}91)$$

substituting Eqs. (3.8-90) and (3.8-91) into Eq. (3.8-89) yields

$$k_{Z_3} = k_3 \sqrt{1 - \left(\frac{c_3}{c_2} \sin \theta_2\right)^2}. \qquad (3.8\text{-}92)$$

At this point in the analysis we have two possible situations. The first possibility is that $c_3 < c_2$, which is known as the *slow-bottom case*. If $c_3 < c_2$, then $(c_3/c_2) \sin \theta_2 < 1$ for all angles of incidence $0 \leq \theta_2 \leq \pi/2$. Therefore, both k_{Z_2} and k_{Z_3} given by Eqs. (3.8-88) and (3.8-92), respectively, are always real for $0 \leq \theta_2 \leq \pi/2$. As a result, the right-hand side of Eq. (3.8-87) is always imaginary, which does not make physical sense, since $\tan(k_{Z_2}D) = \tan(k_2 D \cos \theta_2)$ cannot be imaginary. Therefore, *we cannot solve the ocean waveguide problem when $c_3 < c_2$ with the set of assumptions and analysis approach used in this section.*

The second possibility is that $c_3 > c_2$, which is known as the *fast-bottom case*. From the slow-bottom case we learned that since k_{Z_2} is always real for $0 \leq \theta_2 \leq \pi/2$, if k_{Z_3} is also always real for $0 \leq \theta_2 \leq \pi/2$, then we cannot solve Eq. (3.8-87). Therefore, we need to determine the range of values for θ_2 that will make $(c_3/c_2) \sin \theta_2 > 1$, and hence k_{Z_3}, imaginary. If $c_3 > c_2$, then $(c_3/c_2) \sin \theta_2 > 1$ only if $\sin \theta_2 > c_2/c_3$ or, in other words,

$$\frac{c_3}{c_2} \sin \theta_2 > 1, \qquad \theta_c < \theta_2 \leq \frac{\pi}{2}, \qquad (3.8\text{-}93)$$

where

$$\theta_c = \sin^{-1}(c_2/c_3), \qquad c_3 > c_2, \qquad (3.8\text{-}94)$$

is known as the *critical angle of incidence* (see Example 2.3-3). Using Eqs. (3.8-93) and (3.8-94), Eq. (3.8-92) can be rewritten as

$$k_{Z_3} = -jk_3 \sqrt{\left(\frac{\sin \theta_2}{\sin \theta_c}\right)^2 - 1}, \qquad c_3 > c_2, \quad \theta_c < \theta_2 \leq \frac{\pi}{2}, \qquad (3.8\text{-}95)$$

where $-j$ is used in order to ensure evanescent waves (i.e., decaying exponentials) in the positive Z direction in medium III. Note that $\theta_2 = \theta_c$ is *not* included, since when $\theta_2 = \theta_c$, we have $k_{Z_3} = 0$, which is a real number. Therefore, upon substituting Eqs. (3.8-88) and (3.8-95) into Eq. (3.8-87), and since $k_2 = 2\pi f/c_2$ and

$k_3 = 2\pi f/c_3$, we obtain the following *transcendental equation* in terms of θ_2:

$$\tan\left(\frac{2\pi f}{c_2} D \cos\theta_2\right) + \frac{\rho_3 c_3 \cos\theta_2}{\rho_2 c_2 \sqrt{\left(\dfrac{\sin\theta_2}{\sin\theta_c}\right)^2 - 1}} = 0, \qquad c_3 > c_2, \quad \theta_c < \theta_2 < \pi/2,$$

(3.8-96)

where $\theta_2 \neq \pi/2$ shall be justified later. From Eq. (3.8-96) it can also be seen that we *cannot solve the ocean waveguide problem for angles of incidence less than* θ_c *with the set of assumptions and analysis approach used in this section*. This limitation is not considered to be a major drawback, since the magnitude of the ocean-bottom reflection coefficient is relatively small for angles of incidence less than critical (see Figs. 2.3-3 and 2.3-6). Therefore, at long horizontal ranges from the source, those normal modes with angles of propagation less than critical will have very small amplitudes by the time they reach the receiver because of the many bottom reflections they will undergo. As a result, their contributions to the total acoustic field will be negligible.

Before continuing further, let us digress for a moment. By referring to Example 2.3-3, we can express the Rayleigh reflection coefficient at the ocean bottom as follows:

$$R_{23} = \exp(+j2\Phi), \qquad c_3 > c_2, \quad \theta_c \leq \theta_2 \leq \pi/2, \tag{3.8-97}$$

where

$$\Phi = \tan^{-1}\left(\frac{Z_2 b}{Z_3 \cos\theta_2}\right), \tag{3.8-98}$$

$$b = \sqrt{\left(\frac{\sin\theta_2}{\sin\theta_c}\right)^2 - 1}, \tag{3.8-99}$$

$$\sin\theta_c = c_2/c_3, \qquad c_3 > c_2, \tag{3.8-100}$$

$$Z_2 = \rho_2 c_2, \tag{3.8-101}$$

and

$$Z_3 = \rho_3 c_3. \tag{3.8-102}$$

Therefore, with the use of Eqs. (3.8-98) through (3.8-102), Eq. (3.8-96) can be rewritten as

$$\tan\left(\frac{2\pi f}{c_2}D\cos\theta_2\right) + \frac{1}{\tan\Phi} = 0, \qquad c_3 > c_2, \quad \theta_c < \theta_2 < \frac{\pi}{2}. \qquad (3.8\text{-}103)$$

Equation (3.8-103) demonstrates the connection between the angle of incidence θ_2 and the reflection coefficient at the ocean bottom R_{23} via the phase term Φ.

For a given value of source frequency f, the allowed angles of propagation in medium II correspond to the *roots* $\theta_2 = \theta_{2,n}$, $n = 0, 1, \ldots, N_t - 1$, of Eq. (3.8-96), where N_t is the *total number of trapped* (or *propagating*) *normal modes*, to be derived later. The normal modes corresponding to the roots of Eq. (3.8-96) are known as trapped modes because for angles of incidence $\theta_c < \theta_{2,n} < \pi/2$, the magnitudes of the reflection coefficients at the ocean bottom [see Eq. (3.8-97)] and at the ideal pressure-release ocean surface are equal to unity. As a result, no time-average acoustic power is present in media I and III. All the time-average acoustic power is "trapped" in medium II.

In order to derive an equation for N_t, we first need to derive an equation for the *cutoff frequency* f_n in hertz for the nth trapped normal mode. Since $\theta_2 = \theta_{2,n}$, $n = 0, 1, \ldots, N_t - 1$, Eq. (3.8-96) can be rewritten as

$$\tan^2\left(\frac{2\pi f}{c_2}D\cos\theta_{2,n}\right) = \frac{Z_3^2\cos^2\theta_{2,n}}{Z_2^2\left[\left(\dfrac{\sin\theta_{2,n}}{\sin\theta_c}\right)^2 - 1\right]}, \qquad c_3 > c_2, \quad \theta_c < \theta_{2,n} < \frac{\pi}{2}.$$

$$(3.8\text{-}104)$$

If we let $\theta_{2,n} = \theta_c$, then Eq. (3.8-104) reduces to

$$\tan^2\left(\frac{2\pi f}{c_2}D\cos\theta_c\right) = \infty, \qquad (3.8\text{-}105)$$

which implies that

$$\frac{2\pi f}{c_2}D\cos\theta_c = \frac{(2n+1)\pi}{2}, \qquad n = 0, 1, 2, \ldots. \qquad (3.8\text{-}106)$$

By setting $f = f_n$ in Eq. (3.8-106) and then solving for f_n, we obtain

$$f_n = \frac{(2n+1)c_2}{4D\cos\theta_c}, \qquad n = 0, 1, 2, \ldots, \qquad (3.8\text{-}107)$$

where f_n is the frequency at which the nth trapped normal mode has an angle of propagation $\theta_{2,n} = \theta_c$. The frequency f_n is referred to as the *cutoff frequency* of the nth trapped normal mode, since for source frequencies $f \geq f_n$, one has $\theta_{2,n} \geq \theta_c$. The critical angle of incidence θ_c is given by Eq. (3.8-94). From Eq. (3.8-94) it can be seen that as $c_2/c_3 \to 0$, $\theta_c \to 0°$, and as a result, $\cos \theta_c \to 1$. As $\cos \theta_c \to 1$, Eq. (3.8-107) approaches Eq. (3.8-29), which is the cutoff frequency of the nth propagating normal mode for the ocean waveguide model that uses an ideal rigid bottom (i.e., $Z_3/Z_2 \to \infty$ or $\rho_3/\rho_2 \to \infty$).

The total number of trapped normal modes N_t excited by a source frequency of f hertz can be obtained by setting $f_n = f$ and $n = N_t$ in Eq. (3.8-107) and solving for N_t. Doing so, with the following modifications, yields

$$N_t = \text{INT}\left(\frac{1}{2}\left[\frac{4Df \cos \theta_c}{c_2} - 1\right]\right) + 1, \qquad (3.8\text{-}108)$$

where $\text{INT}(\cdot)$ means form an integer by simply truncating the decimal portion of the real number inside the parentheses, and then add one to the result, since the first (lowest) mode is mode $n = 0$. The number of trapped modes predicted by Eq. (3.8-108), which is also equal to the number of roots of Eq. (3.8-96), does *not* include the mode associated with the *trivial root* $\theta_2 = \pi/2$, as shall be explained later. Also note that the roots of Eq. (3.8-96) are separated *approximately* by the constant amount $\Delta\theta_2$ (in degrees) given by

$$\Delta\theta_2 = \frac{90° - \theta_c}{N_t + 1}. \qquad (3.8\text{-}109)$$

Therefore, $\theta_{2,0} \approx 90° - \Delta\theta_2$, $\theta_{2,1} \approx \theta_{2,0} - \Delta\theta_2$, $\theta_{2,2} \approx \theta_{2,1} - \Delta\theta_2$, and so on. Computer algorithms are available to accurately solve for the roots of Eq. (3.8-96). However, as N_t increases, the approximate values for $\theta_{2,n}$ calculated using $\Delta\theta_2$ become more accurate, that is, they are in better agreement with the values obtained from a root-finding computer algorithm (see Example 3.8-2).

Once the angles (roots) $\theta_{2,n}$ are computed and known, then the trapped (propagating) normal modes are also known, since the propagation vector for the nth propagating mode in medium II can be expressed as (see Fig. 3.8-2)

$$\mathbf{k}_{2,n} = k_{r_{2,n}}\hat{r} + k_{Z_{2,n}}\hat{z}, \qquad n = 0, 1, \ldots, N_t - 1, \qquad (3.8\text{-}110)$$

where

$$k_{r_{2,n}} = k_2 \sin \theta_{2,n}, \qquad (3.8\text{-}111)$$

$$k_{Z_{2,n}} = k_2 \cos \theta_{2,n}, \qquad (3.8\text{-}112)$$

and

$$|\mathbf{k}_{2,n}| = k_2 = 2\pi f/c_2. \qquad (3.8\text{-}113)$$

Therefore, with $k_{r_{2,n}}$ and $k_{z_{2,n}}$ known, the spatial-dependent part of the time-harmonic velocity potential and the acoustic pressure in medium II are given by

$$\varphi_{f,2}(r, \psi, z) = -j2 \sum_{n=0}^{N_t - 1} A_{2,n} H_0^{(2)}(k_{r_{2,n}} r) \sin(k_{z_{2,n}} z) \qquad (3.8\text{-}114)$$

and

$$p_{f,2}(r, \psi, z) = -2k_2 \rho_2 c_2 \sum_{n=0}^{N_t - 1} A_{2,n} H_0^{(2)}(k_{r_{2,n}} r) \sin(k_{z_{2,n}} z), \qquad (3.8\text{-}115)$$

respectively.

Next we shall justify $\theta_2 \neq \pi/2$ in Eq. (3.8-96). Although $\theta_2 = \pi/2$ is a valid mathematical root, it has no physical significance. For example, if we let the root $\pi/2$ correspond to mode $n = 0$, that is, if $\theta_{2,0} = \pi/2$, then mode $n = 0$ is the *longitudinal*, or *plane-wave*, *mode*, and modes $n = 1, 2, \ldots, N_t - 1$ correspond to angles of propagation $\theta_{2,n}$ in decreasing order from $\pi/2$. In reality, there is *no* plane-wave mode. If we substitute $\theta_{2,0} = \pi/2$ into Eq. (3.8-112), we obtain $k_{z_{2,0}} = 0$, as expected. However, with $k_{z_{2,0}} = 0$, the contribution of mode $n = 0$ to the velocity potential and acoustic pressure in medium II is zero, since $\sin 0 = 0$ [see Eqs. (3.8-114) and (3.8-115), respectively]. Therefore, $\theta_2 = \pi/2$ is considered a *trivial root* and is not part of the solution. *We shall still adopt the convention that the lowest mode is mode $n = 0$. However, $\theta_{2,0} < \pi/2$ and modes $n = 1, 2, \ldots, N_t - 1$ correspond to angles of propagation $\theta_{2,n}$ in decreasing order from the value of $\theta_{2,0}$.*

Group Speed, Phase Speed, and Travel Time

With $k_{r_{2,n}}$ known, the *group* and *phase speeds* in the horizontal radial direction are given by

$$c_{g_{r_{2,n}}} = c_2 k_{r_{2,n}}/k_2 = c_2 \sin \theta_{2,n} \qquad (3.8\text{-}116)$$

and

$$c_{p_{r_{2,n}}} = c_2 k_2/k_{r_{2,n}} = c_2/\sin \theta_{2,n} \qquad (3.8\text{-}117)$$

respectively, and the *travel time*, or *time of arrival*, of the nth trapped (propagating) mode at a horizontal range of r meters from the source is given by

$$T_{r_{2,n}} = r/c_{g_{r_{2,n}}}, \qquad (3.8\text{-}118)$$

where the group speed is given by Eq. (3.8-116).

Boundary Condition at the Source

The fourth and final boundary condition is at the source. In order to satisfy the boundary condition at the source, we must first rewrite the expression for the velocity potential in medium III. Substituting Eq. (3.8-81) into Eq. (3.8-75) and noting that $k_{r_2} = k_{r_{2,n}}$ (since $\theta_2 = \theta_{2,n}$) yields

$$R_3(r) = R_{3,n}(r) = H_0^{(2)}(k_{r_{2,n}} r), \qquad (3.8\text{-}119)$$

where $k_{r_{2,n}}$ is given by Eq. (3.8-111). Substituting Eqs. (3.8-82) and (3.8-95) into Eq. (3.8-76) and noting that $A_2 = A_{2,n}$ and $k_{z_2} = k_{z_{2,n}}$ (since $\theta_2 = \theta_{2,n}$) yields

$$Z_3(z) = Z_{3,n}(z) = -j2\frac{\rho_2}{\rho_3} A_{2,n} \sin(k_{z_{2,n}} D) \exp[-k_3 b_n (z - D)], \quad (3.8\text{-}120)$$

where

$$b_n = \sqrt{\left(\frac{\sin \theta_{2,n}}{\sin \theta_c}\right)^2 - 1} > 0, \qquad \theta_c < \theta_{2,n} < \frac{\pi}{2}, \qquad (3.8\text{-}121)$$

and $k_{z_{2,n}}$ is given by Eq. (3.8-112). Therefore, upon substituting Eqs. (3.8-119) and (3.8-120) into Eq. (3.8-74), we obtain

$$\varphi_{f,3,n}(r, \psi, z) = -j2\frac{\rho_2}{\rho_3} A_{2,n} \sin(k_{z_{2,n}} D) H_0^{(2)}(k_{r_{2,n}} r) \exp[-k_3 b_n (z - D)],$$

$$(3.8\text{-}122)$$

which is the spatial-dependent part of the time-harmonic velocity potential in medium III for the nth trapped normal mode. If we define the normal-mode solution in medium III as

$$\varphi_{f,3}(r, \psi, z) \triangleq \sum_{n=0}^{N_t-1} \varphi_{f,3,n}(r, \psi, z), \qquad (3.8\text{-}123)$$

then substituting Eq. (3.8-122) into Eq. (3.8-123) yields

$$\varphi_{f,3}(r,\psi,z) = -j2\frac{\rho_2}{\rho_3} \sum_{n=0}^{N_t-1} A_{2,n} \sin(k_{Z_{2,n}}D) H_0^{(2)}(k_{r_{2,n}}r) \exp[-k_3b_n(z-D)].$$

(3.8-124)

Since [see Eq. (3.8-8)]

$$p_{f,3}(r,\psi,z) = -j2\pi f\rho_3\varphi_{f,3}(r,\psi,z),$$ (3.8-125)

substituting Eq. (3.8-124) into Eq. (3.8-125) yields

$$p_{f,3}(r,\psi,z) = -2k_2\rho_2c_2 \sum_{n=0}^{N_t-1} A_{2,n} \sin(k_{Z_{2,n}}D) H_0^{(2)}(k_{r_{2,n}}r) \exp[-k_3b_n(z-D)],$$

(3.8-126)

which is the normal-mode solution for the spatial-dependent part of the time-harmonic acoustic pressure in medium III. Note that at $z = D$, $p_{f,2}(r,\psi,D)$ obtained from Eq. (3.8-115) is equal to $p_{f,3}(r,\psi,D)$ obtained from Eq. (3.8-126) as required (i.e., continuity of acoustic pressure at $z = D$).

Next, by comparing Eqs. (3.8-115) and (3.8-126), the acoustic pressure in media II and III can be represented by the following single expression:

$$p_f(r,\psi,z) = -2k_2\rho_2c_2 \sum_{n=0}^{N_t-1} A_{2,n} H_0^{(2)}(k_{r_{2,n}}r)Z_n(z), \qquad 0 \le z \le \infty,$$

(3.8-127)

where the *eigenfunction* $Z_n(z)$ is given by

$$Z_n(z) = \begin{cases} \sin(k_{Z_{2,n}}z), & 0 \le z \le D, \quad \theta_c \le \theta_{2,n} < \pi/2, & \text{(3.8-128a)} \\[2mm] \sin(k_{Z_{2,n}}D) & & \text{(3.8-128b)} \\ \times\exp[-k_3b_n(z-D)], & D < z \le \infty, \quad \theta_c < \theta_{2,n} < \pi/2, & \\[2mm] 0, & D < z \le \infty, \quad \theta_{2,n} = \theta_c; & \text{(3.8-128c)} \end{cases}$$

$k_{r_{2,n}}$ and the *eigenvalue* $k_{Z_{2,n}}$ are given by Eqs. (3.8-111) and (3.8-112), respectively, where $\theta_{2,n}$, $n = 0, 1, \ldots, N_t - 1$ are the roots of Eq. (3.8-96); and b_n is given by Eq. (3.8-121). Concerning Eq. (3.8-128c), note that when $\theta_{2,n} = \theta_c$, the reflection coefficient at the ocean bottom $R_{23} = +1$ [see Eqs. (3.8-97) through (3.8-99)], and as a result, the ocean bottom acts like an *ideal rigid boundary*. Therefore, when $\theta_{2,n} = \theta_c$, there is no acoustic pressure field present in medium III ($D < z \leq \infty$) due to the sound source located in medium II, since the plane-wave velocity-potential transmission coefficient is equal to zero for an ideal rigid boundary (see Example 2.3-1). Since the transmitted acoustic pressure is directly proportional to the transmitted velocity potential, which is equal to zero when $\theta_{2,n} = \theta_c$, the transmitted acoustic pressure is also equal to zero when $\theta_{2,n} = \theta_c$. In addition, since the ocean surface at $z = 0$ was modeled as an ideal pressure-release boundary, no acoustic pressure field will be present in medium I (at any time) due to the sound source located in medium II, since the plane-wave pressure transmission coefficient is equal to zero for an ideal pressure-release boundary (see Example 2.3-1).

Recall that the source was assumed to be an omnidirectional, time-harmonic point source located at $r = 0$ and $z = z_0$. Therefore, the acoustic pressure in media II and III, as given by Eqs. (3.8-127) and (3.8-128a) through (3.8-128c), must also satisfy the following inhomogeneous Helmholtz equations:

$$\nabla^2 p_f(\mathbf{r}) + k_2^2 p_f(\mathbf{r}) = p_{f,s}(\mathbf{r}), \qquad 0 \leq z \leq D, \qquad (3.8\text{-}129)$$

and

$$\nabla^2 p_f(\mathbf{r}) + k_3^2 p_f(\mathbf{r}) = p_{f,s}(\mathbf{r}), \qquad D < z \leq \infty, \qquad (3.8\text{-}130)$$

where $p_{f,s}(\mathbf{r})$ is given by Eq. (3.8-50). Substituting Eqs. (3.8-50), (3.8-127), and (3.8-128a) through (3.8-128c) into Eqs. (3.8-129) and (3.8-130) yields the following single equation:

$$\sum_{n=0}^{N_t-1} A_{2,n} \left[\frac{d^2}{dr^2} H_0^{(2)}(k_{r_{2,n}} r) + \frac{1}{r} \frac{d}{dr} H_0^{(2)}(k_{r_{2,n}} r) + k_{r_{2,n}}^2 H_0^{(2)}(k_{r_{2,n}} r) \right] Z_n(z)$$

$$= \frac{P_0}{A} \frac{\delta(r)}{2\pi r} \delta(z - z_0), \qquad 0 \leq z \leq \infty, \qquad (3.8\text{-}131)$$

where

$$A = -2 k_2 \rho_2 c_2, \qquad (3.8\text{-}132)$$

$$k_{r_{2,n}}^2 = k_2^2 - k_{z_{2,n}}^2, \qquad 0 \leq z \leq D, \qquad (3.8\text{-}133)$$

and [see Eqs. (3.8-81) and (3.8-83)]

$$k_{r_{2,n}}^2 = k_3^2 - k_{z_{3,n}}^2, \qquad D < z \leq \infty, \qquad (3.8\text{-}134)$$

where [see Eqs. (3.8-95) and (3.8-99)]

$$k_{Z_{3,n}}^2 = -k_3^2 b_n^2, \qquad \theta_c < \theta_{2,n} < \pi/2. \tag{3.8-135}$$

And upon substituting Eq. (3.8-56) into Eq. (3.8-131), we obtain

$$\sum_{n=0}^{N_t-1} A_{2,n} Z_n(z) = -j\frac{P_0}{4A}\delta(z - z_0), \qquad 0 \le z \le \infty. \tag{3.8-136}$$

Multiplying both sides of Eq. (3.8-136) by the real eigenfunction $Z_m(z)$ and then integrating over z from 0 to ∞ yields

$$\sum_{n=0}^{N_t-1} A_{2,n} \int_0^\infty Z_m(z) Z_n(z)\, dz = -j\frac{P_0}{4A} \int_0^\infty Z_m(z)\delta(z - z_0)\, dz. \tag{3.8-137}$$

Since the real eigenfunctions $Z_n(z)$ given by Eqs. (3.8-128a) through (3.8-128c) are orthogonal in the interval $0 \le z \le \infty$ (see Appendix 3B), and using the sifting property of impulse functions, Eq. (3.8-137) reduces to

$$\sum_{n=0}^{N_t-1} A_{2,n} E_n \delta_{mn} = -j\frac{P_0}{4A} Z_m(z_0), \tag{3.8-138}$$

$$A_{2,m} E_m = -j\frac{P_0}{4A} Z_m(z_0), \tag{3.8-139}$$

and upon replacing the index m with the index n,

$$A_{2,n} = -j\frac{P_0}{4AE_n} Z_n(z_0), \qquad n = 0, 1, \ldots, N_t - 1, \tag{3.8-140}$$

where

$$E_n = \begin{cases} D/2, & \theta_{2,n} = \theta_c, \quad n = 0, 1, \ldots, N_t - 1, \\[2mm] \dfrac{2Dk_{Z_{2,n}} - \sin(2k_{Z_{2,n}} D) - 2(\rho_2/\rho_3)\tan(k_{Z_{2,n}} D)\sin^2(k_{Z_{2,n}} D)}{4k_{Z_{2,n}}}, \\[3mm] \qquad\qquad\qquad \theta_c < \theta_{2,n} < \pi/2, \quad n = 0, 1, \ldots, N_t - 1, \end{cases} \tag{3.8-141}$$

is the *energy* contained in the eigenfunctions (see Appendix 3B). Finally, substituting Eqs. (3.8-128a) and (3.8-132) into Eq. (3.8-140) yields the following expression for the *amplitude* term associated with the nth trapped mode:

$$A_{2,n} = j \frac{P_0}{8k_2\rho_2 c_2 E_n} \sin(k_{Z_{2,n}} z_0), \qquad 0 \le z \le D, \quad n = 0, 1, \dots, N_t - 1.$$

(3.8-142)

Therefore, upon substituting Eqs. (3.8-114) and (3.8-142) into Eq. (3.8-1), we obtain

$$\varphi_2(t, r, \psi, z) = \frac{P_0}{4k_2\rho_2 c_2} \sum_{n=0}^{N_t - 1} E_n^{-1} \sin(k_{Z_{2,n}} z_0) H_0^{(2)}(k_{r_{2,n}} r) \sin(k_{Z_{2,n}} z)$$
$$\times \exp(+j2\pi ft),$$

(3.8-143)

which is the *time-harmonic normal-mode solution for the velocity potential* in medium II. And upon substituting Eqs. (3.8-115) and (3.8-142) into Eq. (3.8-7), we obtain

$$p_2(t, r, \psi, z) = -j \frac{P_0}{4} \sum_{n=0}^{N_t - 1} E_n^{-1} \sin(k_{Z_{2,n}} z_0) H_0^{(2)}(k_{r_{2,n}} r) \sin(k_{Z_{2,n}} z)$$
$$\times \exp(+j2\pi ft),$$

(3.8-144)

which is the *time-harmonic normal-mode solution for the acoustic pressure* in medium II, where E_n is given by Eq. (3.8-141), $k_{r_{2,n}}$ and $k_{Z_{2,n}}$ are given by Eqs. (3.8-111) and (3.8-112), respectively, where $\theta_{2,n}$, $n = 0, 1, \dots, N_t - 1$, are the roots of Eq. (3.8-96), and the far-field approximation of the zeroth-order Hankel function of the second kind is given by Eq. (3.8-70). Since the sound source was modeled by an impulse function [see Eq. (3.8-50)], the *spatial-dependent part of* Eq. (3.8-144) *is the spatial impulse response of the pressure-release-surface, fast-fluid-bottom ocean waveguide for acoustic pressure in medium II.*

Example 3.8-2 (Pulse propagation)

Setting $P_0 = 1$ Pa in Eq. (3.8-144) and using the procedure discussed in Example 3.8-1, the acoustic pressure in an ocean waveguide with a pressure-release surface and a *fast* fluid bottom due to a transmitted pulse is given by

$$p_2(t, r, \psi, z) = -j\frac{1}{4}\int_{-\infty}^{\infty} X(f) \sum_{n=0}^{N_t-1} E_n^{-1} \sin(k_{Z_{2,n}} z_0) H_0^{(2)}(k_{r_{2,n}} r)$$

$$\times \sin(k_{Z_{2,n}} z) \exp(+j2\pi f t)\, df, \qquad (3.8\text{-}145)$$

where $X(f)$ is the complex frequency spectrum of the transmitted time-domain pulse. Recall that the number of trapped (propagating) modes, N_t, is *frequency dependent* [see Eq. (3.8-108)]. Also recall that the parameter P_0 must be set equal to one, since $X(f)$ automatically assigns the correct amplitudes to the various frequency components.

Figure 3.8-6 illustrates a Hamming-envelope continuous-wave (CW) transmitted electrical pulse. As was mentioned in Example 3.8-1, amplitude- and angle-modulated carriers will be discussed in detail in Chapter 8, Section 8.2. The pulse shown in Fig. 3.8-6 is composed of 51 frequency components; it has an amplitude of 10, a pulse length of 100 msec, a pulse-repetition frequency (fundamental frequency) of 0.8 Hz, and a carrier frequency of 250 Hz.

Figure 3.8-7 illustrates the evaluation of the transcendental equation given by Eq. (3.8-96) for $f = 250$ Hz, $D = 100$ m, $\rho_2 = 1026$ kg/m^3, $c_2 = 1500$ m/sec, and for a fluidlike, quartz sand ocean bottom, $\rho_3 = 2070$ kg/m^3 and $c_3 = 1730$ m/sec. The critical angle of incidence for this example is $\theta_c = 60.1°$, and the total number of trapped modes excited by the carrier frequency $f = 250$ Hz is $N_t = 17$. Recall that the total number of trapped modes does *not* include the trivial mode propagating at $\theta_2 = 90°$.

Figure 3.8-8 illustrates the corresponding received acoustic pulse obtained by evaluating Eq. (3.8-145) for $z_0 = 30$ m and the previously mentioned parameter values. Figure 3.8-8a is for a receiver located above the source at $x = 10$ km, $y = 0$ m, and $z = 20$ m. Figure 3.8-8b is for a receiver located below the source at $x = 10$ km, $y = 0$ m, and $z = 80$ m. By comparing Figs. 3.8-6 and 3.8-8, the effects of geometrical dispersion are evident. Also, by comparing Figs. 3.8-8a and 3.8-8b, it can be seen that for a constant horizontal range, the shape of the received acoustic pulse changes dramatically as a function of depth.

Finally, Table 3.8-2 lists the important corresponding parameter values for all of the trapped modes excited by the carrier frequency. Included in Table 3.8-2 are the roots $\theta_{2,n}$ of Eq. (3.8-96) obtained from

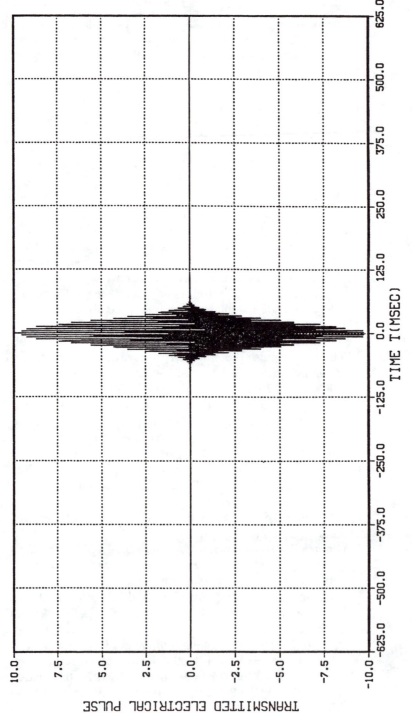

Figure 3.8-6. Hamming-envelope continuous-wave (CW) transmitted electrical pulse for the parameter values discussed in Example 3.8-2.

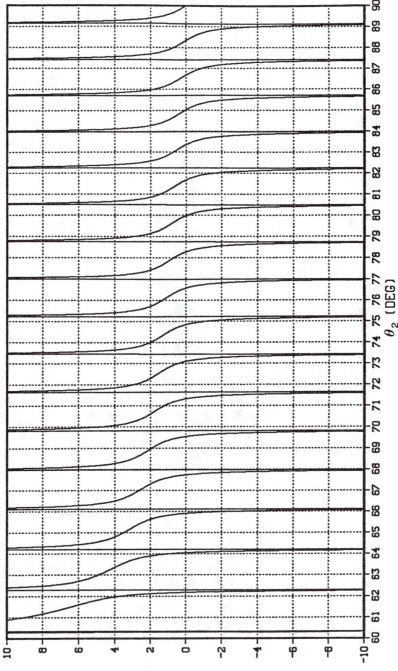

Figure 3.8-7. Evaluation of the transcendental equation given by Eq. (3.8-96) for the parameter values discussed in Example 3.8-2.

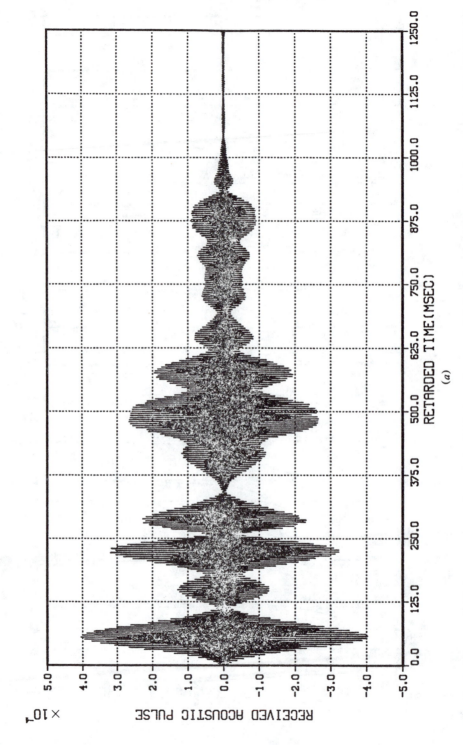

Figure 3.8-8. Received acoustic pulse for a receiver located (*a*) above the source and (*b*) below the source, using the parameter values discussed in Example 3.8-2.

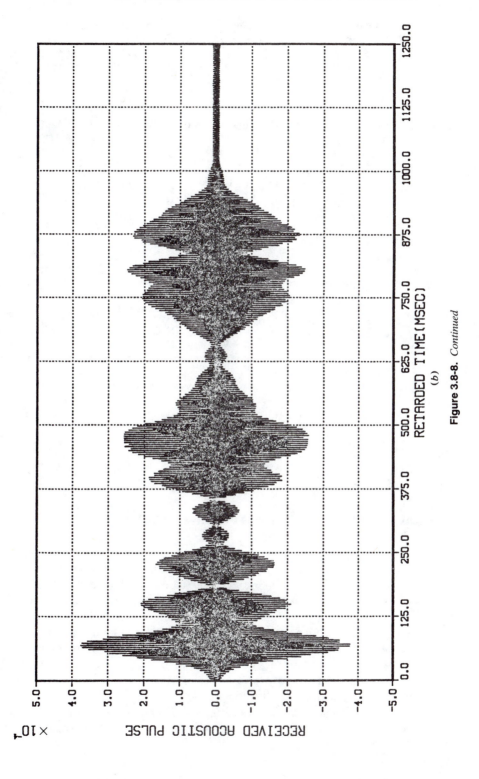

Figure 3.8-8. *Continued*

(b)

Table 3.8-2 Important parameter values for all of the trapped (propagating) modes excited by the carrier frequency $f = 250$ Hz in Example 3.8-2. Note that the wave number in medium II is $k_2 = 1.04720$ rad/m at $f = 250$ Hz.

n	f_n (Hz)	$\theta_{2,n}$ (deg)	$k_{r_{2,n}}$ (rad/m)	$k_{z_{2,n}}$ (rad/m)	$c_{g_{r_{2,n}}}$ (m/sec)	E_n	$\tau_{r_{2,n}}$ (sec)
0	7.527	88.345	1.04676	0.03025	1499.374	51.924	6.669450
1	22.580	86.687	1.04545	0.06052	1497.493	51.895	6.677827
2	37.634	85.025	1.04325	0.09082	1494.348	51.852	6.691882
3	52.688	83.355	1.04016	0.12117	1489.924	51.799	6.711751
4	67.741	81.677	1.03617	0.15158	1484.202	51.743	6.737627
5	82.795	79.988	1.03125	0.18205	1477.159	51.688	6.769753
6	97.849	78.287	1.02539	0.21258	1468.768	51.639	6.808428
7	112.902	76.573	1.01857	0.24317	1458.999	51.601	6.854015
8	127.956	74.843	1.01077	0.27381	1447.819	51.576	6.906941
9	143.010	73.096	1.00195	0.30449	1435.192	51.567	6.967711
10	158.063	71.331	0.99210	0.33521	1421.076	51.580	7.036919
11	173.117	69.546	0.98117	0.36595	1405.429	51.621	7.115263
12	188.171	67.739	0.96915	0.39670	1388.203	51.704	7.203558
13	203.224	65.909	0.95598	0.42745	1369.347	51.854	7.302753
14	218.278	64.054	0.94165	0.45817	1348.814	52.142	7.413919
15	233.332	62.177	0.92613	0.48878	1326.586	52.832	7.538146
16	248.385	60.298	0.90961	0.51888	1302.915	59.207	7.675097

a root-finding computer program. Note that the approximate values of the roots obtained by using Eq. (3.8-109) are in very good agreement with those listed in Table 3.8-2 (see Problem 3-12).

3.9 Waveguide Model of the Ocean: Full-Wave Solution

The waveguide model of the ocean that we shall use in this section is illustrated in Fig. 3.9-1, where $\rho_1 c_1$, $\rho_2 c_2$, and $\rho_3 c_3$ are the characteristic impedances of medium I (air), medium II (ocean water), and medium III (ocean bottom), respectively, where ρ_i, $i = 1, 2, 3$, is the constant ambient (equilibrium) density and c_i, $i = 1, 2, 3$, is the constant speed of sound in each of the three media, respectively. Both the ocean surface and ocean bottom are modeled as plane, parallel boundaries. An omnidirectional, time-harmonic point source is located in medium II at $r = 0$ and $z = z_0$, and the ocean is D meters deep. Note, however, that medium II has been divided into medium IIa and medium IIb, both with the same characteristic impedance $\rho_2 c_2$, by the introduction of another plane, parallel boundary at the depth of the source $z = z_0$ (see Fig. 3.9-1). This was not done in Section 3.8. The additional boundary is introduced in order to satisfy the boundary conditions at

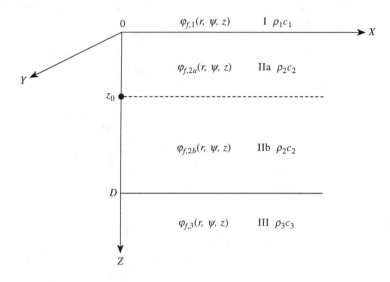

Figure 3.9-1. Ocean waveguide model.

the source, as will be discussed later, contrary to using the orthogonality property of eigenfunctions as was done in Section 3.8.

In this section we shall treat the ocean surface at $z = 0$ and the ocean bottom at $z = D$ as boundaries between two different fluid media whose properties are characterized by the Rayleigh reflection coefficients R_{21} and R_{23}, respectively. Since the surfaces at $z = 0$ and $z = D$ are no longer being modeled as ideal pressure-release and rigid boundaries, respectively, as was done in Section 3.8.1, time-average acoustic power will, in general, be present in both media I and III due to a sound source located in medium II.

The ocean waveguide problem in this section can be stated as follows: derive expressions for the velocity potentials in both media IIa and IIb that are solutions of the wave equation and satisfy all the boundary conditions, including the boundary conditions at the source. We shall obtain full-wave solutions that are equivalent to infinite summations of "ray integrals" as opposed to normal modes. Contrary to the normal-mode solution discussed in Section 3.8.2, our full-wave solutions will be valid for both slow and fast ocean bottoms and for all angles of incidence at the ocean bottom.

Since both the ocean surface and ocean bottom are being modeled as plane, parallel boundaries, and since the sound source is assumed to be an omnidirectional point source, the solution will be *axisymmetric* (i.e., independent of the azimuthal angle ψ). Therefore, we can set $n = 0$ in Eq. (3.1-37). Also, since the value of the velocity potential is required at $r = 0$, we can neglect $Y_n(k_r r)$ by setting $B_r = 0$ in Eq. (3.1-37). And upon taking into account the additional assumption of a time-harmonic source, the velocity potential in medium I can be expressed as

$$\varphi_1(t, r, \psi, z) = \varphi_{f,1}(r, \psi, z) \exp(+j2\pi f t), \tag{3.9-1}$$

where

$$\varphi_{f,1}(r, \psi, z) = R_1(r) Z_1(z), \tag{3.9-2}$$

$$R_1(r) = J_0(k_{r_1} r), \tag{3.9-3}$$

$$Z_1(z) = B_1 \exp(+jk_{Z_1} z), \tag{3.9-4}$$

and

$$k_{r_1}^2 + k_{Z_1}^2 = k_1^2 = (2\pi f/c_1)^2, \tag{3.9-5}$$

where k_{r_1}, k_{Z_1}, and k_1 are the propagation-vector components in the r and Z directions and the wave number, respectively, in medium I. According to Eq. (3.9-4) the acoustic field in medium I propagates only in the negative Z direction, implying that medium I is being modeled as an infinite half space.

In medium IIa and medium IIb,

$$\varphi_{2a}(t, r, \psi, z) = \varphi_{f,2a}(r, \psi, z) \exp(+j2\pi f t) \tag{3.9-6}$$

and

$$\varphi_{2b}(t, r, \psi, z) = \varphi_{f,2b}(r, \psi, z) \exp(+j2\pi f t), \tag{3.9-7}$$

respectively, where

$$\varphi_{f,2a}(r, \psi, z) = R_{2a}(r) Z_{2a}(z), \tag{3.9-8}$$

$$R_{2a}(r) = J_0(k_{r_2} r), \tag{3.9-9}$$

$$Z_{2a}(z) = A_{2a} \exp(-jk_{Z_2} z) + B_{2a} \exp(+jk_{Z_2} z); \tag{3.9-10}$$

$$\varphi_{f,2b}(r, \psi, z) = R_{2b}(r) Z_{2b}(z), \tag{3.9-11}$$

$$R_{2b}(r) = J_0(k_{r_2} r), \tag{3.9-12}$$

$$Z_{2b}(z) = A_{2b} \exp(-jk_{Z_2} z) + B_{2b} \exp(+jk_{Z_2} z); \tag{3.9-13}$$

and

$$k_{r_2}^2 + k_{Z_2}^2 = k_2^2 = (2\pi f/c_2)^2, \tag{3.9-14}$$

where k_{r_2}, k_{Z_2}, and k_2 are the propagation-vector components in the r and Z directions and the wave number, respectively, in medium II, and hence, in media IIa and IIb. Equations (3.9-10) and (3.9-13) take into account *standing waves* that exist in the Z direction in media IIa and IIb due to reflections from the boundaries.

Finally, in medium III,

$$\varphi_3(t, r, \psi, z) = \varphi_{f,3}(r, \psi, z) \exp(+j2\pi ft), \tag{3.9-15}$$

where

$$\varphi_{f,3}(r, \psi, z) = R_3(r)Z_3(z), \tag{3.9-16}$$

$$R_3(r) = J_0(k_{r_3}r), \tag{3.9-17}$$

$$Z_3(z) = A_3 \exp(-jk_{Z_3}z), \tag{3.9-18}$$

and

$$k_{r_3}^2 + k_{Z_3}^2 = k_3^2 = (2\pi f/c_3)^2, \tag{3.9-19}$$

where k_{r_3}, k_{Z_3}, and k_3 are the propagation-vector components in the r and Z directions and the wave number, respectively, in medium III. According to Eq. (3.9-18), the acoustic field in medium III propagates only in the positive Z direction, implying that medium III is also being modeled as an infinite half space. The values of the constants B_1, A_{2a}, B_{2a}, A_{2b}, B_{2b}, and A_3 are complex in general and are determined by satisfying the boundary conditions.

The first boundary condition is continuity of acoustic pressure at $z = 0$, that is,

$$p_{f,1}(r, \psi, z)\big|_{z=0} = p_{f,2a}(r, \psi, z)\big|_{z=0}, \tag{3.9-20}$$

where

$$p_{f,1}(r, \psi, z) = -jk_1\rho_1c_1\varphi_{f,1}(r, \psi, z) \tag{3.9-21}$$

and

$$p_{f,2a}(r, \psi, z) = -jk_2\rho_2c_2\varphi_{f,2a}(r, \psi, z). \tag{3.9-22}$$

Substituting Eqs. (3.9-21), (3.9-22), (3.9-2) through (3.9-4), and (3.9-8) through (3.9-10) into Eq. (3.9-20) yields

$$\rho_1B_1J_0(k_{r_1}r) = \rho_2J_0(k_{r_2}r)(A_{2a} + B_{2a}). \tag{3.9-23}$$

Since Eq. (3.9-23) must hold for all allowed values of r or, in other words, since Eq. (3.9-23) must be *independent* of r, this requires that

$$k_{r_2} = k_{r_1} = k_r,$$
(3.9-24)

and as a result, Eq. (3.9-23) reduces to

$$\rho_2 A_{2a} + \rho_2 B_{2a} - \rho_1 B_1 = 0.$$
(3.9-25)

The second boundary condition is continuity of the normal component of the acoustic fluid velocity vector at $z = 0$, that is,

$$u_{f,z_1}(r, \psi, z)\big|_{z=0} = u_{f,z_{2a}}(r, \psi, z)\big|_{z=0},$$
(3.9-26)

where

$$u_{f,z_1}(r, \psi, z) = \frac{\partial}{\partial z}\varphi_{f,1}(r, \psi, z) = R_1(r)\frac{d}{dz}Z_1(z)$$
(3.9-27)

and

$$u_{f,z_{2a}}(r, \psi, z) = \frac{\partial}{\partial z}\varphi_{f,2a}(r, \psi, z) = R_{2a}(r)\frac{d}{dz}Z_{2a}(z).$$
(3.9-28)

Note that because of Eq. (3.9-24),

$$R_{2a}(r) = R_1(r) = J_0(k_r r).$$
(3.9-29)

Substituting Eqs. (3.9-27) through (3.9-29), (3.9-4), and (3.9-10) into Eq. (3.9-26) yields

$$k_{z_2} A_{2a} - k_{z_2} B_{2a} + k_{z_1} B_1 = 0.$$
(3.9-30)

The third boundary condition is continuity of acoustic pressure at the depth of the point source $z = z_0$ (see Appendix 3C), that is,

$$p_{f,2a}(r, \psi, z)\big|_{z=z_0} = p_{f,2b}(r, \psi, z)\big|_{z=z_0},$$
(3.9-31)

where $p_{f,2a}(r, \psi, z)$ is given by Eq. (3.9-22) and

$$p_{f,2b}(r, \psi, z) = -jk_2\rho_2 c_2\varphi_{f,2b}(r, \psi, z).$$
(3.9-32)

With the use of Eq. (3.9-24),

$$R_{2b}(r) = R_{2a}(r) = J_0(k_r r).$$
(3.9-33)

Substituting Eqs. (3.9-22), (3.9-32), (3.9-33), and Eqs. (3.9-8) through (3.9-13) into Eq. (3.9-31) yields

$$\exp(-jk_{Z_2}z_0)\,A_{2a} + \exp(+jk_{Z_2}z_0)\,B_{2a}$$

$$- \exp(-jk_{Z_2}z_0)\,A_{2b} - \exp(+jk_{Z_2}z_0)\,B_{2b} = 0. \qquad (3.9\text{-}34)$$

The fourth boundary condition is *discontinuity* of the normal component of the acoustic fluid velocity vector at the depth of the point source $z = z_0$ (see Appendix 3C), that is,

$$u_{f,Z_{2b}}(r,\psi,z)\big|_{z=z_0} - u_{f,Z_{2a}}(r,\psi,z)\big|_{z=z_0} = \frac{1}{2\pi}k_r J_0(k_r r), \qquad (3.9\text{-}35)$$

where $u_{f,Z_{2a}}(r,\psi,z)$ is given by Eq. (3.9-28) and

$$u_{f,Z_{2b}}(r,\psi,z) = \frac{\partial}{\partial z}\varphi_{f,2b}(r,\psi,z) = R_{2b}(r)\frac{d}{dz}Z_{2b}(z). \qquad (3.9\text{-}36)$$

Substituting Eqs. (3.9-28), (3.9-36), and (3.9-33) into Eq. (3.9-35) yields

$$\frac{d}{dz}Z_{2b}(z)\bigg|_{z=z_0} - \frac{d}{dz}Z_{2a}(z)\bigg|_{z=z_0} = \frac{1}{2\pi}k_r. \qquad (3.9\text{-}37)$$

The discontinuity equal to $k_r J_0(k_r r)/(2\pi)$ in Eq. (3.9-35) or, equivalently, $k_r/(2\pi)$ in Eq. (3.9-37) ensures that in the absence of the boundaries at $z = 0$ and $z = D$, the final expressions for the velocity potentials in media IIa and IIb will reduce to the expression for the velocity potential of an acoustic field in free space produced by an omnidirectional point source. This shall be demonstrated later. Returning to the fourth boundary-condition problem, substituting Eqs. (3.9-10) and (3.9-13) into Eq. (3.9-37) yields

$$- \exp(-jk_{Z_2}z_0)\,A_{2a} + \exp(+jk_{Z_2}z_0)\,B_{2a} + \exp(-jk_{Z_2}z_0)\,A_{2b}$$

$$- \exp(+jk_{Z_2}z_0)\,B_{2b} = +j\frac{1}{2\pi}\frac{k_r}{k_{Z_2}}. \qquad (3.9\text{-}38)$$

The fifth boundary condition is continuity of acoustic pressure at $z = D$, that is,

$$p_{f,2b}(r,\psi,z)\big|_{z=D} = p_{f,3}(r,\psi,z)\big|_{z=D}, \qquad (3.9\text{-}39)$$

where $p_{f,2b}(r,\psi,z)$ is given by Eq. (3.9-32) and

$$p_{f,3}(r,\psi,z) = -jk_3\rho_3 c_3 \varphi_{f,3}(r,\psi,z). \qquad (3.9\text{-}40)$$

Substituting Eqs. (3.9-32), (3.9-40), (3.9-11), (3.9-12), (3.9-16), and (3.9-17) into Eq. (3.9-39), and making use of Eq. (3.9-24) yields

$$\rho_2 J_0(k_r r) Z_{2b}(z)\big|_{z=D} = \rho_3 A_3 J_0(k_{r_3} r) Z_3(z)\big|_{z=D}. \tag{3.9-41}$$

Since the boundary condition represented by Eq. (3.9-41) must be independent of r, this requires that

$$k_{r_3} = k_r, \tag{3.9-42}$$

and as a result, Eq. (3.9-41) reduces to

$$\rho_2 Z_{2b}(z)\big|_{z=D} = \rho_3 A_3 Z_3(z)\big|_{z=D}. \tag{3.9-43}$$

Substituting Eqs. (3.9-13) and (3.9-18) into Eq. (3.9-43) yields

$$\rho_2 \exp(-jk_{Z_2} D)\, A_{2b} + \rho_2 \exp(+jk_{Z_2} D)\, B_{2b} - \rho_3 \exp(-jk_{Z_3} D) A_3 = 0. \tag{3.9-44}$$

By combining Eqs. (3.9-24) and (3.9-42),

$$\boxed{k_{r_3} = k_{r_2} = k_{r_1} = k_r,} \tag{3.9-45}$$

and as a result,

$$R_3(r) = R_{2b}(r) = R_{2a}(r) = R_1(r) = J_0(k_r r). \tag{3.9-46}$$

The sixth and final boundary condition is continuity of the normal component of the acoustic fluid velocity vector at $z = D$, that is,

$$u_{f,Z_{2b}}(r,\psi,z)\big|_{z=D} = u_{f,Z_3}(r,\psi,z)\big|_{z=D} \tag{3.9-47}$$

where $u_{f,Z_{2b}}(r,\psi,z)$ is given by Eq. (3.9-36) and

$$u_{f,Z_3}(r,\psi,z) = \frac{\partial}{\partial z}\varphi_{f,3}(r,\psi,z) = R_3(r)\frac{d}{dz}Z_3(z). \tag{3.9-48}$$

Substituting Eqs. (3.9-36), (3.9-48), (3.9-13), (3.9-18), and (3.9-46) into Eq. (3.9-47) yields

$$k_{Z_2}\exp(-jk_{Z_2}D)\,A_{2b} - k_{Z_2}\exp(+jk_{Z_2}D)\,B_{2b} - k_{Z_3}\exp(-jk_{Z_3}D)\,A_3 = 0. \tag{3.9-49}$$

Equations (3.9-25), (3.9-30), (3.9-34), (3.9-38), (3.9-44), and (3.9-49) represent six equations in terms of the six unknown constants B_1, A_{2a}, B_{2a}, A_{2b}, B_{2b}, and A_3.

With the use of these equations it can be shown that

$$A_{2a} = +j\frac{1}{4\pi}R_{21}\left[\frac{1 + R_{23}\exp\left[-j2k_{Z_2}(D - z_0)\right]}{1 - R_{21}R_{23}\exp\left(-j2k_{Z_2}D\right)}\right]\frac{\exp(-jk_{Z_2}z_0)}{k_{Z_2}}k_r, \quad (3.9\text{-}50)$$

$$B_{2a} = A_{2a}/R_{21}, \quad (3.9\text{-}51)$$

$$A_{2b} = +j\frac{1}{4\pi}\left[\frac{R_{21}\exp(-jk_{Z_2}z_0) + \exp(+jk_{Z_2}z_0)}{1 - R_{21}R_{23}\exp\left(-j2k_{Z_2}D\right)}\right]\frac{k_r}{k_{Z_2}}, \quad (3.9\text{-}52)$$

and

$$B_{2b} = A_{2b}R_{23}\exp\left(-j2k_{Z_2}D\right), \quad (3.9\text{-}53)$$

where

$$R_{21} = \frac{\rho_1 k_{Z_2} - \rho_2 k_{Z_1}}{\rho_1 k_{Z_2} + \rho_2 k_{Z_1}} \quad (3.9\text{-}54)$$

and

$$R_{23} = \frac{\rho_3 k_{Z_2} - \rho_2 k_{Z_3}}{\rho_3 k_{Z_2} + \rho_2 k_{Z_3}} \quad (3.9\text{-}55)$$

are the plane-wave velocity-potential reflection coefficients, or Rayleigh reflection coefficients, at the ocean surface and bottom, respectively (see Section 2.3). In addition,

$$B_1 = B_{2a}T_{21} \quad (3.9\text{-}56)$$

and

$$A_3 = A_{2b}T_{23}\exp\left[-j(k_{Z_2} - k_{Z_3})D\right], \quad (3.9\text{-}57)$$

where

$$T_{21} = \frac{2\rho_2 k_{Z_2}}{\rho_1 k_{Z_2} + \rho_2 k_{Z_1}} \quad (3.9\text{-}58)$$

and

$$T_{23} = \frac{2\rho_2 k_{Z_2}}{\rho_3 k_{Z_2} + \rho_2 k_{Z_3}} \quad (3.9\text{-}59)$$

are the plane-wave velocity-potential transmission coefficients at the ocean surface and bottom, respectively (see Section 2.3). And with the use of Eqs. (3.9-5), (3.9-14), (3.9-19), and (3.9-45), the propagation-vector components in the Z direction in media I, II, and III can be expressed as follows:

$$
k_{Z_i} = \begin{cases} \sqrt{k_i^2 - k_r^2}, & k_r^2 \le k_i^2, \quad i = 1,2,3, \\ -j\sqrt{k_r^2 - k_i^2}, & k_r^2 > k_i^2, \quad i = 1,2,3, \end{cases} \tag{3.9-60}
$$

where

$$
k_i = 2\pi f/c_i, \quad i = 1,2,3, \tag{3.9-61}
$$

is the wave number in media I, II, and III, respectively. Note that for $k_r > k_1$, the sign convention chosen for k_{Z_1} will ensure *evanescent waves* (i.e., decaying exponentials) in the negative Z direction in medium I. Similarly, for $k_r > k_3$, the sign convention chosen for k_{Z_3} will ensure *evanescent waves* in the positive Z direction in medium III.

Now that all of the unknown constants have been determined, the velocity potentials in all three media can be expressed as follows:

$$
\begin{aligned}
\varphi_{f,1}(r,\psi,z) = {}&+j\frac{1}{4\pi}\int_0^\infty T_{21}\frac{F(k_r,z_0)}{k_{Z_2}} \exp(-jk_{Z_2}z_0)\exp(+jk_{Z_1}z) \\
&\times J_0(k_r r)k_r\, dk_r, \quad z \le 0,
\end{aligned} \tag{3.9-62}
$$

$$
\begin{aligned}
\varphi_{f,2a}(r,\psi,z) = {}&+j\frac{1}{4\pi}\int_0^\infty \frac{F(k_r,z_0)}{k_{Z_2}} \\
&\times \left\{ R_{21}\exp\left[-jk_{Z_2}(z+z_0)\right] + \exp\left[+jk_{Z_2}(z-z_0)\right] \right\} \\
&\times J_0(k_r r)k_r\, dk_r, \quad 0 \le z \le z_0,
\end{aligned} \tag{3.9-63}
$$

$$
\begin{aligned}
\varphi_{f,2b}(r,\psi,z) = {}&+j\frac{1}{4\pi}\int_0^\infty \frac{F(k_r,z)}{k_{Z_2}} \\
&\times \left\{ R_{21}\exp\left[-jk_{Z_2}(z+z_0)\right] + \exp\left[-jk_{Z_2}(z-z_0)\right] \right\} \\
&\times J_0(k_r r)k_r\, dk_r, \quad z_0 \le z \le D,
\end{aligned} \tag{3.9-64}
$$

and, finally,

$$
\boxed{
\begin{aligned}
\varphi_{f,3}(r,\psi,z) &= +j\frac{1}{4\pi}\int_0^\infty T_{23}\frac{G(k_r,z_0)}{k_{Z_2}}\exp(-jk_{Z_2}D)\exp\left[-jk_{Z_3}(z-D)\right] \\
&\quad \times J_0(k_r r)k_r\, dk_r, \qquad z \ge D,
\end{aligned}
}
$$

$$(3.9\text{-}65)$$

where

$$
F(k_r,z) = \frac{1 + R_{23}\exp\left[-j2k_{Z_2}(D-z)\right]}{1 - R_{21}R_{23}\exp(-j2k_{Z_2}D)},
\tag{3.9-66}
$$

$$
G(k_r,z) = \frac{R_{21}\exp(-jk_{Z_2}z) + \exp(+jk_{Z_2}z)}{1 - R_{21}R_{23}\exp(-j2k_{Z_2}D)},
\tag{3.9-67}
$$

$$
F(k_r,z_0) = F(k_r,z)|_{z=z_0},
\tag{3.9-68}
$$

and

$$
G(k_r,z_0) = G(k_r,z)|_{z=z_0}.
\tag{3.9-69}
$$

Note that the function $F(k_r,z)$ in the integrand of Eq. (3.9-64) is *not* evaluated at $z = z_0$. Also note that our solutions have been *generalized* by integrating over k_r from zero to infinity, as was discussed in Section 3.7. Because of these integrations, Eqs. (3.9-62) through (3.9-65) are called *full-wave solutions*. Recall that k_{Z_1}, k_{Z_2}, and k_{Z_3} can be expressed in terms of k_r [see Eq. (3.9-60)]. Also, since R_{21}, R_{23}, T_{21}, and T_{23} depend on k_{Z_1}, k_{Z_2}, and k_{Z_3} [see Eqs. (3.9-54), (3.9-55), (3.9-58), and (3.9-59), respectively], they too can be expressed in terms of k_r. Therefore, the velocity-potential expressions given by Eqs. (3.9-62) through (3.9-65) are really *inverse zeroth-order Hankel transforms*, that is, in terms of the generic functions $\varphi_f(\cdot)$ and $\Phi_f(\cdot)$,

$$
\varphi_f(r,\psi,z) = H_0^{-1}\{\Phi_f(k_r,\psi,z)\} \triangleq \int_0^\infty \Phi_f(k_r,\psi,z)J_0(k_r r)k_r\, dk_r.
\tag{3.9-70}
$$

Since the sound source was modeled as an omnidirectional point source [i.e., an impulse function (see Appendix 3C)], Eqs. (3.9-62) through (3.9-65) are the *spatial impulse responses of the ocean waveguide for the velocity potentials in media I, IIa, IIb, and III, respectively*.

In order to determine if our velocity-potential expressions satisfy the boundary conditions at the source, let us examine $\varphi_{f,2a}(r,\psi,z)$ and $\varphi_{f,2b}(r,\psi,z)$ given by Eqs. (3.9-63) and (3.9-64), respectively. If the boundaries at $z = 0$ and $z = D$ are *removed*, that is, if $R_{21} = R_{23} = 0$ and $T_{21} = T_{23} = 1$, then

$$\varphi_{f,2a}(r, \psi, z) = +j\frac{1}{4\pi} \int_0^\infty \frac{\exp\left[+jk_{z_2}(z - z_0)\right]}{k_{z_2}} J_0(k_r r) k_r \, dk_r, \qquad z \leq z_0,$$

$$(3.9\text{-}71)$$

and

$$\varphi_{f,2b}(r, \psi, z) = +j\frac{1}{4\pi} \int_0^\infty \frac{\exp\left[-jk_{z_2}(z - z_0)\right]}{k_{z_2}} J_0(k_r r) k_r \, dk_r, \qquad z \geq z_0.$$

$$(3.9\text{-}72)$$

Equations (3.9-71) and (3.9-72) can be combined into the single expression

$$\varphi_f(r, \psi, z) = +j\frac{1}{4\pi} \int_0^\infty \frac{\exp\left[-jk_z|z - z_0|\right]}{k_z} J_0(k_r r) k_r \, dk_r, \quad (3.9\text{-}73)$$

which is valid for all z, where the subscripts $2a$ and $2b$ on $\varphi_f(\cdot)$ and the subscript 2 on k_z have been removed, since there is only one fluid medium now; where

$$k_z = \begin{cases} \sqrt{k^2 - k_r^2}, & k_r^2 \leq k^2, \\ -j\sqrt{k_r^2 - k^2}, & k_r^2 > k^2, \end{cases} \qquad (3.9\text{-}74)$$

and

$$k = 2\pi f/c. \qquad (3.9\text{-}75)$$

It was shown in Section 3.6 that Eq. (3.9-73) is equal to

$$\varphi_f(r, \psi, z) = -\frac{\exp\left(-jk\sqrt{r^2 + (z - z_0)^2}\right)}{4\pi\sqrt{r^2 + (z - z_0)^2}} = g_f(\mathbf{r}|\mathbf{r}_0), \qquad (3.9\text{-}76)$$

where

$$g_f(\mathbf{r}|\mathbf{r}_0) = -\frac{\exp(-jkR)}{4\pi R} \qquad (3.9\text{-}77)$$

is the time-independent free-space Green's function,

$$R = |\mathbf{r} - \mathbf{r}_0| = \sqrt{r^2 + (z - z_0)^2} \qquad (3.9\text{-}78)$$

is the line-of-sight range between source and field points, and

$$r = \sqrt{(x - x_0)^2 + (y - y_0)^2} \qquad (3.9\text{-}79)$$

is the horizontal range between source and field points. Therefore, in the absence of any boundaries, the velocity potential in medium II (which includes media IIa and IIb) reduces to the velocity potential of an acoustic field in free space produced by an omnidirectional point source, as it should, and as a result, the boundary conditions at the source are satisfied.

Ray Integrals

The next topic to be discussed is the concept of *ray integrals*. We begin by noting that the expression

$$\frac{1}{1 - R_{21} R_{23} \exp(-j2k_{z_2}D)}$$

is common to all of the velocity-potential expressions given by Eqs. (3.9-62) through (3.9-65). Since

$$\frac{1}{1 - x} = \sum_{n=0}^{\infty} x^n, \qquad |x| < 1, \tag{3.9-80}$$

we can write that

$$\frac{1}{1 - R_{21} R_{23} \exp(-j2k_{z_2}D)} = \sum_{n=0}^{\infty} R_{21}^n R_{23}^n \exp(-j2nk_{z_2}D), \qquad |R_{21} R_{23}| < 1. \tag{3.9-81}$$

The series expansion given by Eq. (3.9-81) will *converge*, that is, $|R_{21} R_{23}| < 1$, since the ocean surface is not exactly an ideal pressure-release boundary, and as a result, $|R_{21}| < 1$ for all angles of incidence less than grazing (i.e., less than 90°). In addition, although $|R_{23}| = 1$ for ideal (nonviscous) fluids for angles of incidence equal to or greater than the critical angle of incidence θ_c when θ_c exists (i.e., when $c_3 > c_2$), $|R_{23}| < 1$ for angles of incidence less than θ_c (see Fig. 2.3-3). However, for real (viscous) fluids, $|R_{23}| < 1$ for all angles of incidence less than grazing (see Fig. 2.3-6). Finally, when θ_c does not exist (i.e., when $c_3 < c_2$), $|R_{23}| < 1$ for all angles of incidence less than grazing, since the ocean bottom is not an ideal rigid boundary.

Substituting Eq. (3.9-81) into Eq. (3.9-63), for example, and rewriting yields

$$\varphi_{f,2a}(r, \psi, z) = +j\frac{1}{4\pi} \int_0^\infty k_{z_2}^{-1} \{ \exp[-jk_{z_2}(z_0 - z)] + R_{21} \exp[-jk_{z_2}(z_0 + z)]$$

$$+ R_{23} \exp[-jk_{z_2}(2D - z_0 - z)] + R_{21} R_{23} \exp[-jk_{z_2}(2D - z_0 + z)] \}$$

$$\times \sum_{n=0}^{\infty} R_{21}^n R_{23}^n \exp(-j2nk_{z_2}D) J_0(k_r r) k_r \, dk_r, \qquad 0 \le z \le z_0,$$

$$\tag{3.9-82}$$

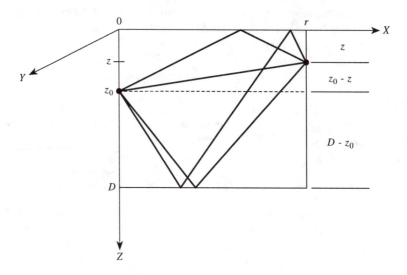

Figure 3.9-2. Four possible ray paths between source and field points.

where the first term inside the braces $\{\cdot\}$ corresponds to the direct ray path between source and field points, the second term corresponds to the ray path that has undergone one surface reflection, the third term corresponds to the ray path that has undergone one bottom reflection, and the fourth term corresponds to the ray path that has undergone one bottom reflection first, followed by one surface reflection, between source and field points. These four ray paths can be easily identified by considering the distance traveled in the Z direction (see Fig. 3.9-2). Multiplication of each of the four terms inside the braces by the infinite summation yields an integrand that contains terms of the form

$$R_{21}^i R_{23}^q, \qquad i, q = 0, 1, 2, \ldots,$$

that can be associated with the different possible ray paths between source and field points. For example, the values of the integers i and q indicate the number of surface and bottom reflections, respectively, that a particular ray path has undergone. All of the aforementioned ray paths are known as *eigenrays* or as *eigenray paths*. An eigenray exactly connects source and receiver.

Finally, if we move the summation in the integrand of Eq. (3.9-82) in front of the integral, then the velocity potential $\varphi_{f,2a}$, for example, can be thought of as being equal to an infinite sum of "ray integrals."

Example 3.9-1 (Pulse propagation)
Using the procedure discussed in Example 3.8-1 and the generic expression given by Eq. (3.9-70), the full-wave pulse solution can be expressed

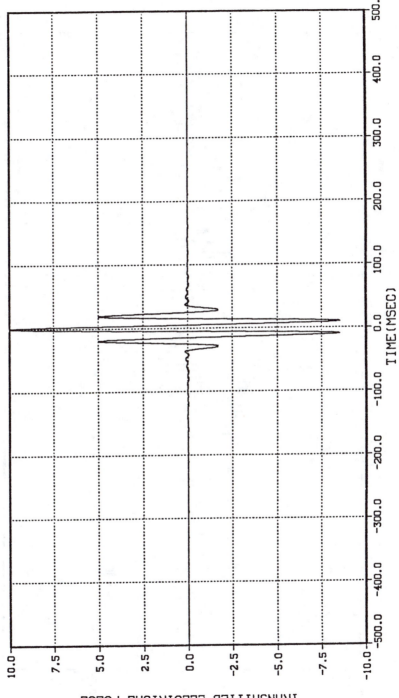

Figure 3.9-3. Hanning-envelope continuous-wave (CW) transmitted electrical pulse for the parameter values discussed in Example 3.9-1.

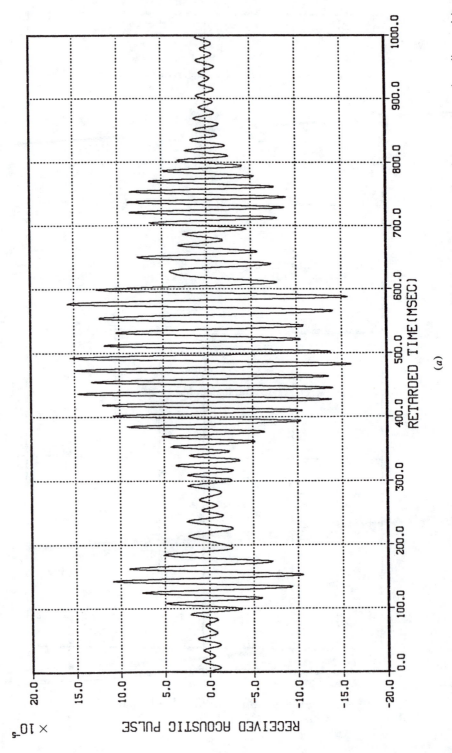

Figure 3.9-4. Received acoustic pulse for a receiver located (*a*) above the source and (*b*) below the source, using the parameter values discussed in Example 3.9-1.

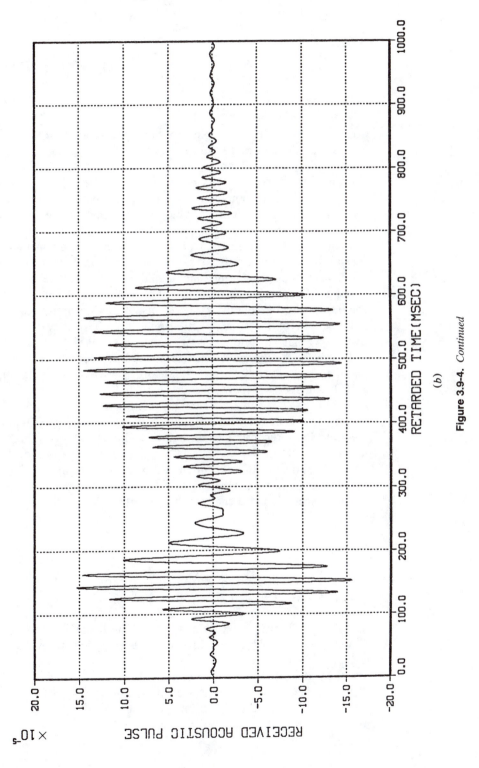

Figure 3.9-4. *Continued*

(b)

as

$$\varphi(t, r, \psi, z) = \int_{-\infty}^{\infty} X(f) \int_{0}^{\infty} \Phi_f(k_r, \psi, z) J_0(k_r r) k_r \, dk_r \, \exp(+j2\pi ft) \, df,$$

(3.9-83)

where $X(f)$ is the complex frequency spectrum of the trans-mitted time-domain pulse. Since we are mainly interested in the velocity potential in the ocean medium, that is, in media IIa and IIb, the appropriate expression to use for the generic integrand term $\Phi_f(k_r, \psi, z)$ can be obtained from Eq. (3.9-63) when the receiver is *above* the source, or from Eq. (3.9-64) when the receiver is *below* the source.

Figure 3.9-3 illustrates a Hanning-envelope continuous-wave (CW) transmitted electrical pulse. As was mentioned in Examples 3.8-1 and 3.8-2, amplitude- and angle-modulated carriers will be discussed in detail in Chapter 8, Section 8.2. The pulse shown in Fig. 3.9-3 is composed of 53 frequency components; it has an amplitude of 10, a pulse length of 80 msec, a pulse-repetition frequency (fundamental frequency) of 1 Hz, and a carrier frequency of 50 Hz.

Figure 3.9-4 illustrates the corresponding received acoustic pulse obtained by evaluating Eq. (3.9-83) for $z_0 = 25$ m, $D = 100$ m, $\rho_1 = 1.21$ kg/m^3, $c_1 = 343$ m/sec, $\rho_2 = 1026$ kg/m^3, $c_2 = 1500$ m/sec, $\rho_3 = 1500$ kg/m^3, and $c_3 = 1600$ m/sec. Figure 3.9-4a is for a receiver located above the source at $x = 30$ km, $y = 0$ m, and $z = 20$ m. Figure 3.9-4b is for a receiver located below the source at $x = 30$ km, $y = 0$ m, and $z = 80$ m. By comparing Figs. 3.9-3 and 3.9-4, the effects of geometrical dispersion are evident. Also, by comparing Figs. 3.9-4a and b, it can be seen that for a constant horizontal range, the shape of the received acoustic pulse changes dramatically as a function of depth.

Problems

3-1   Show that $\Psi(\psi)$ given by Eq. (3.1-18) is periodic with period 2π only for integer values of n, that is, $n = 0, \pm 1, \pm 2, \ldots$.

3-2   Consider a vibrating cylinder of length L meters and radius a meters in the dipole mode of vibration. Ignoring the radiation from the ends of the cylinder, compute
(a) the radiated acoustic pressure,
(b) the radiated acoustic fluid velocity vector,
(c) the time-average radiated intensity vector, and
(d) the time-average radiated power.

3-3 Consider a vibrating cylinder of length L meters and radius a meters whose radial component of surface velocity is given by

$$v_r(t, a, \psi, z) = V_0 x(\psi) \exp(+j2\pi ft),$$

where

$$x(\psi) = 1 + 5\sin(3\psi).$$

Ignoring the radiation from the ends of the cylinder, determine the far-field radiated acoustic pressure field $p(t, r, \psi, z)$.

3-4 Assume that all of the surfaces inside a cylindrical cavity are pressure-release (note that this is a Dirichlet boundary-value problem). Derive expressions for the time-harmonic velocity potential of the (l, m, n) eigenfunction and corresponding eigenfrequency.

3-5 Assume that the surface inside a cylindrical cavity at $r = a$ is rigid and that the surfaces at $z = 0$ and $z = L$ are pressure-release. This is an example of a mixed boundary-value problem, also known as a Robin boundary-value problem. Derive expressions for the time-harmonic velocity potential of the (l, m, n) eigenfunction and corresponding eigenfrequency.

3-6 Consider the simple pressure-release-surface, rigid-bottom waveguide model of the ocean. If $c_2 = 1500$ m/sec and the source frequency $f = 1$ kHz, how many modes will propagate in the ocean if
(a) $D = 10$ m?
(b) $D = 100$ m?
(c) $D = 1000$ m?

3-7 Consider the simple pressure-release-surface, rigid-bottom waveguide model of the ocean. If $c_2 = 1500$ m/sec and $D = 300$ m, how many modes will propagate in the ocean if
(a) $f = 100$ Hz?
(b) $f = 1$ kHz?
(c) $f = 10$ kHz?

3-8 Consider the simple pressure-release-surface, rigid-bottom waveguide model of the ocean. If $c_2 = 1500$ m/sec and $D = 300$ m, compute the group speed in the radial direction and the angle of propagation for modes $n = 16$ and $n = 32$ for the following source frequencies:
(a) $f = 100$ Hz,
(b) $f = 1$ kHz,
(c) $f = 10$ kHz.

3-9   Consider the simple pressure-release-surface, rigid-bottom waveguide model of the ocean. Let $c_2 = 1500$ m/sec, $D = 300$ m, and the sound source contain two frequency components at $f = 200$ Hz and $f = 300$ Hz.
(a) Is mode $n = 40$ a propagating mode or an evanescent mode? Why?
(b) If the horizontal range between source and receiver is 1 km, how long will it take the energy associated with the frequency components $f = 200$ Hz and $f = 300$ Hz for mode $n = 27$ to reach the receiver?
(c) Is mode $n = 100$ a propagating mode or an evanescent mode? Why?

3-10   Consider the pressure-release-surface, fluid-bottom waveguide model of the ocean. If $c_2 = 1500$ m/sec and medium III is the fluidlike ocean bottom of quartz sand with $c_3 = 1730$ m/sec, compute
(a) the critical angle of incidence θ_c,
(b) the total number of trapped modes excited by a source frequency of $f = 1$ kHz for $D = 10$, 100, and 1000 m, and
(c) the approximate angles of propagation for all of the trapped modes excited by a source frequency of $f = 1$ kHz for $D = 10$ m.

3-11   Show that Snell's law given by Eq. (3.8-91) can be obtained from Eq. (3.8-81). **Hint**: Refer to Fig. 3.8-5.

3-12   Use Eq. (3.8-109) to find the approximate values of the angles of propagation $\theta_{2,n}$ for $n = 0, 1, \ldots, 5$ for the carrier frequency and parameter values given in Example 3.8-2. Compare your results with the values obtained from a root-finding computer algorithm listed in Table 3.8-2.

Appendix 3A

In this appendix we shall show that the set of real eigenfunctions $\sin(k_{Z_{2,n}} z)$, $n = 0, 1, 2, \ldots$, are *orthogonal* in the interval $0 \le z \le D$. The depth interval $0 \le z \le D$ was chosen because the acoustic pressure field only exists in medium II, due to the assumptions of an ideal pressure-release ocean surface and an ideal rigid ocean bottom. A set of functions $g_n(t)$, $n = 0, 1, 2, \ldots$, are said to be *orthogonal* in the interval $a \le t \le b$ if the *inner product* between $g_m(t)$ and $g_n(t)$, $m \ne n$, is *zero*, that is, if

$$\langle g_m(t), g_n(t) \rangle \triangleq \int_a^b g_m(t) g_n^*(t)\, dt = 0, \qquad m \ne n. \qquad (3A\text{-}1)$$

Therefore, we must show that

$$\int_0^D \sin(k_{Z_{2,m}} z) \sin(k_{Z_{2,n}} z)\, dz = 0, \qquad m \ne n. \qquad (3A\text{-}2)$$

Since

$$\int_0^D \sin(k_{Z_{2,m}}z)\sin(k_{Z_{2,n}}z)\,dz = \frac{\sin\left[(k_{Z_{2,m}} - k_{Z_{2,n}})D\right]}{2(k_{Z_{2,m}} - k_{Z_{2,n}})} - \frac{\sin\left[(k_{Z_{2,m}} + k_{Z_{2,n}})D\right]}{2(k_{Z_{2,m}} + k_{Z_{2,n}})},$$

$$(3A\text{-}3)$$

substituting Eq. (3.8-24) into the right-hand side of Eq. (3A-3) yields

$$\int_0^D \sin(k_{Z_{2,m}}z)\sin(k_{Z_{2,n}}z)\,dz$$

$$= \frac{\sin[(m - n)\pi]}{2(m - n)\pi/D} - \frac{\sin[(m + n + 1)\pi]}{2(m + n + 1)\pi/D} = 0, \qquad m \neq n, \quad (3A\text{-}4)$$

which is the desired result.

Let us conclude by computing the *energy* E_n contained in the eigenfunctions in the interval $0 \leq z \leq D$, where

$$E_n \triangleq \int_0^D \sin^2(k_{Z_{2,n}}z)\,dz, \qquad n = 0, 1, 2, \ldots . \qquad (3A\text{-}5)$$

Since

$$\int_0^D \sin^2(k_{Z_{2,n}}z)\,dz = \frac{D}{2} - \frac{1}{4k_{Z_{2,n}}}\sin(2k_{Z_{2,n}}D), \qquad n = 0, 1, 2, \ldots, \quad (3A\text{-}6)$$

substituting Eq. (3.8-24) into Eq. (3A-6) yields

$$\boxed{E_n = D/2, \qquad n = 0, 1, 2, \ldots .} \qquad (3A\text{-}7)$$

Therefore, in summary

$$\boxed{\left\langle \sin(k_{Z_{2,m}}z), \sin(k_{Z_{2,n}}z) \right\rangle = \int_0^D \sin(k_{Z_{2,m}}z)\sin(k_{Z_{2,n}}z)\,dz = \frac{D}{2}\delta_{mn},} \qquad (3A\text{-}8)$$

where δ_{mn} is the Kronecker delta given by Eq. (3.2-22).

Appendix 3B

In this appendix we shall show that the set of real eigenfunctions $Z_n(z)$, $n = 0, 1, \ldots, N_t - 1$, given by Eqs. (3.8-128a) through (3.8-128c), are *orthogonal* in the interval $0 \le z \le \infty$, that is, we shall show that

$$\langle Z_m(z), Z_n(z) \rangle = \int_0^\infty Z_m(z) Z_n(z) \, dz = 0, \qquad m \ne n. \tag{3B-1}$$

The depth interval $0 \le z \le \infty$ was chosen because the acoustic pressure field only exists in media II and III, due to the assumptions of an ideal pressure-release ocean surface and a fluid ocean bottom. We begin by noting that $Z_n(z)$ satisfies the following equation:

$$\frac{d^2}{dz^2} Z_n(z) + k_{Z_n}^2 Z_n(z) = 0, \qquad 0 \le z \le \infty, \tag{3B-2}$$

where

$$k_{Z_n} = \begin{cases} k_{Z_{2,n}}, & 0 \le z \le D, & \text{(3B-3a)} \\ k_{Z_{3,n}}, & D < z \le \infty. & \text{(3B-3b)} \end{cases}$$

As a result, we can also write that

$$\frac{d^2}{dz^2} Z_m(z) + k_{Z_m}^2 Z_m(z) = 0, \qquad 0 \le z \le \infty. \tag{3B-4}$$

Next, multiplying Eq. (3B-2) by $Z_m(z)$ and Eq. (3B-4) by $Z_n(z)$ and subtracting the resulting equations yields

$$Z_m(z) \frac{d^2}{dz^2} Z_n(z) - Z_n(z) \frac{d^2}{dz^2} Z_m(z) = \left(k_{Z_m}^2 - k_{Z_n}^2 \right) Z_m(z) Z_n(z), \tag{3B-5}$$

or

$$\frac{d}{dz} \left[Z_m(z) \frac{d}{dz} Z_n(z) - Z_n(z) \frac{d}{dz} Z_m(z) \right] = \left(k_{Z_m}^2 - k_{Z_n}^2 \right) Z_m(z) Z_n(z), \tag{3B-6}$$

and upon integrating both sides of Eq. (3B-6) over z from 0 to ∞, we obtain

$$\left[Z_m(z) \frac{d}{dz} Z_n(z) - Z_n(z) \frac{d}{dz} Z_m(z) \right]_0^\infty = \left(k_{Z_m}^2 - k_{Z_n}^2 \right) \int_0^\infty Z_m(z) Z_n(z) \, dz. \tag{3B-7}$$

Case 1: $\theta_{2,n} = \theta_c$

Let us first evaluate Eq. (3B-7) for the case when $\theta_{2,n} = \theta_c$. With the use of Eq. (3.8-128c), Eq. (3B-7) reduces to

$$\left[Z_m(z)\frac{d}{dz}Z_n(z) - Z_n(z)\frac{d}{dz}Z_m(z) \right]_0^D = \left(k_{Z_m}^2 - k_{Z_n}^2 \right) \int_0^D Z_m(z)Z_n(z)\, dz.$$

$$(3B\text{-}8)$$

Since [see Eq. (3.8-128a)]

$$Z_n(z) = \sin\left(k_{Z_{2,n}} z \right), \quad 0 \le z \le D, \qquad \theta_{2,n} = \theta_c, \qquad (3B\text{-}9)$$

we can write that

$$\frac{d}{dz}Z_n(z) = k_{Z_{2,n}} \cos\left(k_{Z_{2,n}} z \right), \quad 0 \le z \le D, \qquad \theta_{2,n} = \theta_c, \qquad (3B\text{-}10)$$

and as a result,

$$Z_m(D) = \sin\left(k_{Z_{2,m}} D \right) \quad \text{and} \quad Z_n(D) = \sin\left(k_{Z_{2,n}} D \right), \qquad (3B\text{-}11)$$

$$\left. \frac{d}{dz}Z_m(z) \right|_{z=D} = k_{Z_{2,m}} \cos\left(k_{Z_{2,m}} D \right) \quad \text{and} \quad \left. \frac{d}{dz}Z_n(z) \right|_{z=D} = k_{Z_{2,n}} \cos\left(k_{Z_{2,n}} D \right),$$

$$(3B\text{-}12)$$

$$Z_m(0) = 0 \quad \text{and} \quad Z_n(0) = 0, \qquad (3B\text{-}13)$$

and

$$\left. \frac{d}{dz}Z_m(z) \right|_{z=0} = k_{Z_{2,m}} \quad \text{and} \quad \left. \frac{d}{dz}Z_n(z) \right|_{z=0} = k_{Z_{2,n}}. \qquad (3B\text{-}14)$$

Also [see Eq. (3B-3a)],

$$k_{Z_m} = k_{Z_{2,m}}, \qquad 0 \le z \le D, \qquad (3B\text{-}15)$$

and

$$k_{Z_n} = k_{Z_{2,n}}, \qquad 0 \le z \le D. \qquad (3B\text{-}16)$$

Therefore, substituting Eqs. (3B-11) through (3B-16) into Eq. (3B-8) yields

$$k_{Z_{2,n}} \sin\left(k_{Z_{2,m}} D \right) \cos\left(k_{Z_{2,n}} D \right) - k_{Z_{2,m}} \sin\left(k_{Z_{2,n}} D \right) \cos\left(k_{Z_{2,m}} D \right)$$

$$= \left(k_{Z_{2,m}}^2 - k_{Z_{2,n}}^2 \right) \int_0^D Z_m(z)Z_n(z)\, dz. \qquad (3B\text{-}17)$$

Next, we take advantage of the fact that when $\theta_{2,n} = \theta_c$, the ocean bottom acts like an ideal rigid boundary, since the reflection coefficient at the ocean bottom $R_{23} = +1$ [see Eqs. (3.8-97) through (3.8-99)]. Recall that [see Eq. (3.8-112)]

$$k_{Z_{2,n}} = k_2 \cos\theta_{2,n}. \tag{3B-18}$$

When $\theta_{2,n} = \theta_c$, we have $f = f_n$ [see Eq. (3.8-107)], so that Eq. (3B-18) becomes

$$k_{Z_{2,n}} = \frac{2\pi f_n}{c_2}\cos\theta_c, \qquad \theta_{2,n} = \theta_c, \tag{3B-19}$$

and upon substituting Eq. (3.8-107) into Eq. (3B-19), we obtain

$$k_{Z_{2,n}} = \frac{(2n+1)\pi}{2D}, \qquad \theta_{2,n} = \theta_c. \tag{3B-20}$$

Note that Eq. (3B-20) is identical to the propagation-vector component in the Z direction in medium II for the nth normal mode for an ideal pressure-release-surface, rigid-bottom ocean waveguide model [see Eq. (3.8-69)]. As a result, with the use of Eq. (3B-20),

$$\cos\left(k_{Z_{2,m}}D\right) = 0, \qquad \theta_{2,m} = \theta_c, \tag{3B-21}$$

and

$$\cos\left(k_{Z_{2,n}}D\right) = 0, \qquad \theta_{2,n} = \theta_c. \tag{3B-22}$$

Substituting Eqs. (3B-21) and (3B-22) into Eq. (3B-17) yields

$$\left(k_{Z_{2,m}}^2 - k_{Z_{2,n}}^2\right)\int_0^D Z_m(z)Z_n(z)\,dz = 0, \qquad \theta_{2,n} = \theta_c. \tag{3B-23}$$

Since $k_{Z_{2,m}} = k_{Z_{2,n}}$, $m \neq n$, yields a trivial solution, Eq. (3B-23) implies that

$$\int_0^D Z_m(z)Z_n(z)\,dz = 0, \quad m \neq n, \qquad \theta_{2,n} = \theta_c, \tag{3B-24}$$

and because of Eq. (3.8-128c),

$$\int_0^\infty Z_m(z)Z_n(z)\,dz = 0, \quad m \neq n, \qquad \theta_{2,n} = \theta_c, \tag{3B-25}$$

which proves that the set of real eigenfunctions $Z_n(z)$, $n = 0, 1, \ldots, N_t - 1$, are *orthogonal* in the interval $0 \leq z \leq \infty$ when $\theta_{2,n} = \theta_c$.

Case II: $\theta_c < \theta_{2,n} < \pi/2$

Next, let us evaluate Eq. (3B-7) for the case when $\theta_c < \theta_{2,n} < \pi/2$. Since [see Eq. (3.8-128a)]

$$Z_n(z) = \sin(k_{Z_{2,n}}z), \quad 0 \leq z \leq D, \qquad \theta_c < \theta_{2,n} < \pi/2, \qquad \text{(3B-26)}$$

Eqs. (3B-13) and (3B-14) are applicable in this case as well. Also, by referring to Eq. (3.8-128b),

$$Z_m(\infty) = 0 \quad \text{and} \quad Z_n(\infty) = 0, \qquad \text{(3B-27)}$$

and

$$\frac{d}{dz}Z_m(z)\bigg|_{z=\infty} = 0 \quad \text{and} \quad \frac{d}{dz}Z_n(z)\bigg|_{z=\infty} = 0. \qquad \text{(3B-28)}$$

Therefore, substituting Eqs. (3B-13), (3B-14), (3B-27), and (3B-28) into Eq. (3B-7) yields

$$\left(k_{Z_m}^2 - k_{Z_n}^2\right)\int_0^\infty Z_m(z)Z_n(z)\, dz = 0, \qquad \theta_c < \theta_{2,n} < \pi/2. \qquad \text{(3B-29)}$$

Since $k_{Z_m} = k_{Z_n}$, $m \neq n$, yields a trivial solution, Eq. (3B-29) implies that

$$\int_0^\infty Z_m(z)Z_n(z)\, dz = 0, \quad m \neq n, \qquad \theta_c < \theta_{2,n} < \pi/2, \qquad \text{(3B-30)}$$

which proves that the set of real eigenfunctions $Z_n(z)$, $n = 0, 1, \ldots, N_t - 1$, are *orthogonal* in the interval $0 \leq z \leq \infty$ when $\theta_c < \theta_{2,n} < \pi/2$.

Energy Calculations

Let us conclude by computing the energy E_n contained in the eigenfunction $Z_n(z)$ where

$$E_n \triangleq \int_0^\infty Z_n^2(z)\, dz. \qquad \text{(3B-31)}$$

When $\theta_{2,n} = \theta_c$, and with the use of Eqs. (3.8-128a) and (3.8-128c), Eq. (3B-31) reduces to

$$E_n = \int_0^D \sin^2(k_{Z_{2,n}}z)\, dz, \qquad \theta_{2,n} = \theta_c, \qquad \text{(3B-32)}$$

and upon substituting Eqs. (3A-6) and (3B-20) into Eq. (3B-32), we obtain

$$E_n = D/2, \quad \theta_{2,n} = \theta_c, \quad n = 0, 1, \ldots, N_t - 1, \qquad \text{(3B-33)}$$

which is identical to the energy contained in the eigenfunctions for an ideal pressure-release-surface, rigid-bottom ocean waveguide model [compare with Eq. (3A-7)].

When $\theta_c < \theta_{2,n} < \pi/2$, and with the use of Eqs. (3.8-128a) and (3.8-128b), Eq. (3B-31) can be rewritten as follows:

$$E_n = \int_0^D \sin^2(k_{Z_{2,n}} z)\, dz + \sin^2(k_{Z_{2,n}} D)\exp(+2k_3 b_n D)\int_D^\infty \exp(-2k_3 b_n z)\, dz,$$

$$\theta_c < \theta_{2,n} < \pi/2. \quad (3B\text{-}34)$$

Since

$$\int_D^\infty \exp(-2k_3 b_n z)\, dz = -\frac{1}{2k_3 b_n}\exp(-2k_3 b_n z)\Big|_D^\infty = \frac{\exp(-2k_3 b_n D)}{2k_3 b_n}, \quad (3B\text{-}35)$$

substituting Eqs. (3A-6) and (3B-35) into Eq. (3B-34) yields

$$E_n = \frac{D}{2} - \frac{1}{4k_{Z_{2,n}}}\sin(2k_{Z_{2,n}} D) + \frac{1}{2k_3 b_n}\sin^2(k_{Z_{2,n}} D), \qquad \theta_c < \theta_{2,n} < \frac{\pi}{2},$$

$$(3B\text{-}36)$$

or

$$E_n = \left[2Dk_{Z_{2,n}} k_3 b_n - k_3 b_n \sin(2k_{Z_{2,n}} D) + 2k_{Z_{2,n}}\sin^2(k_{Z_{2,n}} D)\right]\left(4k_{Z_{2,n}} k_3 b_n\right)^{-1},$$

$$\theta_c < \theta_{2,n} < \pi/2. \quad (3B\text{-}37)$$

Since $\theta_2 = \theta_{2,n}$, substituting Eq. (3.8-121) into Eq. (3.8-95) yields

$$k_{Z_{3,n}} = -jk_3 b_n, \qquad \theta_c < \theta_{2,n} < \pi/2. \qquad (3B\text{-}38)$$

If we next substitute $k_{Z_2} = k_{Z_{2,n}}$, $k_{Z_3} = k_{Z_{3,n}}$, and Eq. (3B-38) into Eq. (3.8-87), then

$$k_{Z_{2,n}} = -k_3 b_n \frac{\rho_2}{\rho_3}\tan(k_{Z_{2,n}} D), \qquad \theta_c < \theta_{2,n} < \frac{\pi}{2}. \qquad (3B\text{-}39)$$

Upon substituting Eq. (3B-39) into the coefficient of the $\sin^2$ term in the numerator of Eq. (3B-37), cancelling $k_3 b_n$ from the numerator and the denominator, and combining the resulting expression with Eq. (3B-33), we finally obtain the desired result

$$
E_n = \begin{cases} D/2, & \theta_{2,n} = \theta_c, \quad n = 0, 1, \ldots, N_t - 1, \\[2mm] \dfrac{2 D k_{Z_{2,n}} - \sin\left(2 k_{Z_{2,n}} D\right) - 2\left(\rho_2/\rho_3\right) \tan\left(k_{Z_{2,n}} D\right) \sin^2\left(k_{Z_{2,n}} D\right)}{4 k_{Z_{2,n}}}, \\[4mm] \qquad \theta_c < \theta_{2,n} < \pi/2, \quad n = 0, 1, \ldots, N_t - 1. \end{cases}
$$

(3B-40)

Therefore, in summary,

$$
\left\langle Z_m(z), Z_n(z) \right\rangle = \int_0^\infty Z_m(z) Z_n(z) \, dz = E_n \delta_{mn},
$$
(3B-41)

where $Z_n(z)$ is given by Eqs. (3.8-128a) through (3.8-128c), E_n is given by Eq. (3B-40), and δ_{mn} is the Kronecker delta given by Eq. (3.2-22).

Appendix 3C

In this appendix we shall show that at the depth $z = z_0$ of a point source, there is a *discontinuity* in the Z component of the acoustic fluid velocity vector and *continuity* of acoustic pressure. Consider the following inhomogeneous Helmholtz equation:

$$
\frac{\partial^2}{\partial r^2} \varphi_f(r, z) + \frac{1}{r} \frac{\partial}{\partial r} \varphi_f(r, z) + \frac{\partial^2}{\partial z^2} \varphi_f(r, z) + k^2(z) \varphi_f(r, z) = \frac{\delta(r)}{2\pi r} \delta(z - z_0),
$$

(3C-1)

where the velocity potential $\varphi_f(r, z)$ is axisymmetric (i.e., independent of the azimuthal angle ψ),

$$
k(z) = 2\pi f / c(z)
$$
(3C-2)

is the depth-dependent wave number, $c(z)$ is the depth-dependent speed of sound, and the impulse functions on the right-hand side of Eq. (3C-1) represent a unit-amplitude, omnidirectional point source at $r = 0$ and $z = z_0$. The solution $\varphi_f(r, z)$ of Eq. (3C-1) can be expressed as an *inverse zeroth-order Hankel transform* as follows:

$$\varphi_f(r, z) = H_0^{-1}\{\Phi_f(k_r, z)\} = \int_0^{\infty} \Phi_f(k_r, z) J_0(k_r r) k_r \, dk_r, \qquad (3\text{C-}3)$$

where

$$\Phi_f(k_r, z) = H_0\{\varphi_f(r, z)\} = \int_0^{\infty} \varphi_f(r, z) J_0(k_r r) r \, dr \qquad (3\text{C-}4)$$

is the *forward zeroth-order Hankel transform* (also known as the *Fourier-Bessel transform*) of $\varphi_f(r, z)$. Therefore, if we can find $\Phi_f(k_r, z)$, then substituting $\Phi_f(k_r, z)$ into Eq. (3C-3) will yield a solution of Eq. (3C-1).
Since

$$H_0\left\{ \frac{\partial^2}{\partial r^2}\varphi_f(r, z) + \frac{1}{r}\frac{\partial}{\partial r}\varphi_f(r, z) \right\} = -k_r^2 \Phi_f(k_r, z) \qquad (3\text{C-}5)$$

and

$$H_0\left\{ \frac{\delta(r)}{r} \right\} = 1, \qquad (3\text{C-}6)$$

taking the zeroth-order Hankel transform of Eq. (3C-1) yields

$$\frac{d^2}{dz^2}\Phi_f(k_r, z) + k_Z^2(z)\Phi_f(k_r, z) = \frac{\delta(z - z_0)}{2\pi}, \qquad (3\text{C-}7)$$

where

$$k_Z^2(z) = k^2(z) - k_r^2 \qquad (3\text{C-}8)$$

is the depth-dependent propagation-vector component in the Z direction. The solution of Eq. (3C-7) can be broken up into two separate problems.

The first problem is to determine the solution of Eq. (3C-7) when $z \neq z_0$. Since $\delta(z - z_0) = 0$ for $z \neq z_0$, Eq. (3C-7) reduces to

$$\frac{d^2}{dz^2}\Phi_f(k_r, z) + k_Z^2(z)\Phi_f(k_r, z) = 0, \qquad z \neq z_0. \qquad (3\text{C-}9)$$

Therefore, $\Phi_f(k_r, z)$ must be a solution of the homogeneous differential equation given by Eq. (3C-9) for $z \neq z_0$. It must also satisfy all the boundary conditions at both the ocean surface and ocean bottom.

The second problem is to determine the boundary conditions that must be satisfied by $\Phi_f(k_r, z)$ at $z = z_0$. Integrating both sides of Eq. (3C-7) over z from $z_0 - \varepsilon$ to $z_0 + \varepsilon$, where $\varepsilon > 0$, yields

$$\frac{d}{dz}\Phi_f(k_r, z)\Big|_{z_0-\varepsilon}^{z_0+\varepsilon} + \int_{z_0-\varepsilon}^{z_0+\varepsilon} k_Z^2(z)\Phi_f(k_r, z) \, dz = \frac{1}{2\pi}, \qquad (3\text{C-}10)$$

or

$$\frac{d}{dz}\Phi_f(k_r, z)\Big|_{z_0+\varepsilon} - \frac{d}{dz}\Phi_f(k_r, z)\Big|_{z_0-\varepsilon} + \int_{z_0-\varepsilon}^{z_0+\varepsilon} k_Z^2(z)\Phi_f(k_r, z)\, dz = \frac{1}{2\pi},$$

(3C-11)

since

$$\int_{z_0-\varepsilon}^{z_0+\varepsilon} \delta(z - z_0)\, dz = 1.$$

(3C-12)

In the limit as $\varepsilon \to 0$, Eq. (3C-11) reduces to the following expression:

$$\lim_{\varepsilon \to 0}\left[\frac{d}{dz}\Phi_f(k_r, z)\Big|_{z_0+\varepsilon} - \frac{d}{dz}\Phi_f(k_r, z)\Big|_{z_0-\varepsilon}\right] = \frac{1}{2\pi},$$

(3C-13)

which indicates a *discontinuity* in the Z component of the acoustic fluid velocity vector at $z = z_0$, since

$$U_{f,Z}(k_r, z) = \frac{d}{dz}\Phi_f(k_r, z).$$

(3C-14)

Next, if we let

$$g(z_0, \varepsilon) = \int_{z_0-\varepsilon}^{z_0+\varepsilon} k_Z^2(z)\Phi_f(k_r, z)\, dz,$$

(3C-15)

and if we integrate both sides of Eq. (3C-10) over z from $z_0 - \varepsilon$ to $z_0 + \varepsilon$, where $\varepsilon > 0$, then

$$\Phi_f(k_r, z)\Big|_{z_0-\varepsilon}^{z_0+\varepsilon} + g(z_0, \varepsilon)\int_{z_0-\varepsilon}^{z_0+\varepsilon} dz = \frac{1}{2\pi}\int_{z_0-\varepsilon}^{z_0+\varepsilon} dz,$$

(3C-16)

or

$$\Phi_f(k_r, z)\Big|_{z_0+\varepsilon} - \Phi_f(k_r, z)\Big|_{z_0-\varepsilon} + 2g(z_0, \varepsilon)\varepsilon = \frac{\varepsilon}{\pi}.$$

(3C-17)

In the limit as $\varepsilon \to 0$, Eq. (3C-17) reduces to the following expression:

$$\lim_{\varepsilon \to 0} \left[\Phi_f(k_r, z) \big|_{z_0 + \varepsilon} - \Phi_f(k_r, z) \big|_{z_0 - \varepsilon} \right] = 0. \qquad \text{(3C-18)}$$

Equation (3C-18) indicates *continuity* of the acoustic pressure at $z = z_0$, since the acoustic pressure is directly proportional to the velocity potential, that is,

$$P_f(k_r, z) = -j2\pi f \rho_0(z) \Phi_f(k_r, z). \qquad \text{(3C-19)}$$

Therefore, *two* boundary conditions must be satisfied at $z = z_0$: continuity of acoustic pressure and a discontinuity in the Z component of the acoustic fluid velocity vector equal in magnitude to $1/(2\pi)$.

Bibliography

M. Abramowitz and I. A. Stegun, editors, *Handbook of Mathematical Functions*, 9th printing, Dover, New York.

C. A. Boyles, *Acoustic Waveguides*, Wiley, New York, 1984.

L. Brekhovskikh, *Waves in Layered Media*, 2nd ed., Academic Press, New York, 1980.

L. Brekhovskikh and Yu. Lysanov, *Fundamentals of Ocean Acoustics*, 2nd ed., Springer-Verlag, New York, 1991.

C. S. Clay and H. Medwin, *Acoustical Oceanography*, Wiley, New York, 1977.

J. A. DeSanto, editor, *Ocean Acoustics*, Springer-Verlag, Berlin, 1979.

F. B. Jensen, W. A. Kuperman, M. B. Porter, and H. Schmidt, *Computational Ocean Acoustics*, AIP Press, Woodbury, New York, 1994.

J. B. Keller and J. S. Papadakis, editors, *Wave Propagation and Underwater Acoustics*, Springer-Verlag, Berlin, 1977.

L. E. Kinsler, A. R. Frey, A. B. Coppens, and J. V. Sanders, *Fundamentals of Acoustics*, 3rd ed., Wiley, New York, 1982.

P. M. Morse and K. U. Ingard, *Theoretical Acoustics*, Princeton University Press, Princeton, New Jersey, 1987.

T. Myint-U, *Partial Differential Equations of Mathematical Physics*, 2nd ed., North Holland, New York, 1980.

C. B. Officer, *Introduction to the Theory of Sound Transmission*, McGraw-Hill, New York, 1958.

J. A. Stratton, *Electromagnetic Theory*, McGraw-Hill, New York, 1941.

S. Temkin, *Elements of Acoustics*, Wiley, New York, 1981.

I. Tolstoy and C. S. Clay, *Ocean Acoustics*, Acoustical Society of America, Woodbury, New York, 1987.

Chapter 4

Wave Propagation in the Spherical Coordinate System

4.1 Solution of the Linear Three-Dimensional Homogeneous Wave Equation

Chapters 2 and 3 were devoted to expressing solutions of the wave equation in terms of rectangular and cylindrical coordinates, respectively. The main focus of this chapter is to solve several well-known wave-propagation problems for which the spherical coordinate system is best suited. Solutions of the wave equation for a homogeneous medium will involve *spherical* Bessel functions, *spherical* Hankel functions, and associated Legendre functions.

In this section we shall solve the following linear, three-dimensional, lossless, homogeneous wave equation in the spherical coordinate system:

$$\nabla^2 \varphi(t, \mathbf{r}) - \frac{1}{c^2} \frac{\partial^2}{\partial t^2} \varphi(t, \mathbf{r}) = 0, \tag{4.1-1}$$

where

$$\nabla^2 = \frac{\partial^2}{\partial r^2} + \frac{2}{r} \frac{\partial}{\partial r} + \frac{1}{r^2} \left[\frac{\partial^2}{\partial \theta^2} + \cot \theta \frac{\partial}{\partial \theta} \right] + \frac{1}{r^2 \sin^2 \theta} \frac{\partial^2}{\partial \psi^2} \tag{4.1-2}$$

is the *Laplacian* expressed in the spherical coordinates (r, θ, ψ) (see Fig. 4.1-1), $\varphi(t, \mathbf{r})$ is the *velocity potential* at time t and position $\mathbf{r} = (r, \theta, \psi)$ with units of square meters per second, and c is the *constant* speed of sound in meters per second. When the speed of sound is constant, the fluid medium is referred to as a *homogeneous medium*.

269

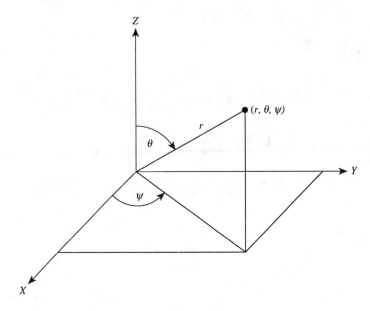

Figure 4.1-1. The spherical coordinates (r, θ, ψ).

The wave equation given by Eq. (4.1-1) is referred to as being *homogeneous* since there is no source distribution (i.e., no input or forcing function). Solutions of the wave equation for several different source distribution models will be discussed in Section 4.2. Equation (4.1-1) describes the propagation of small-amplitude acoustic signals in ideal (nonviscous), homogeneous fluids (see Section 1.4).

We begin the solution of Eq. (4.1-1) by assuming that the velocity potential $\varphi(t, \mathbf{r})$ has a time-harmonic dependence, that is,

$$\varphi(t, \mathbf{r}) = \varphi_f(\mathbf{r}) \exp(+j2\pi ft) \qquad (4.1\text{-}3)$$

where $\varphi_f(\mathbf{r})$ is the spatial-dependent part of the velocity potential and f is the frequency in hertz. A time-harmonic field is one whose value at any time and point in space depends on a single frequency component. Substituting Eq. (4.1-3) into the time-dependent wave equation given by Eq. (4.1-1) yields

$$\nabla^2 \varphi_f(\mathbf{r}) + k^2 \varphi_f(\mathbf{r}) = 0, \qquad (4.1\text{-}4)$$

which is the time-independent *lossless Helmholtz equation*, where

$$k = 2\pi f/c = 2\pi/\lambda \qquad (4.1\text{-}5)$$

is the *wave number* in radians per meter and

$$c = f\lambda, \tag{4.1-6}$$

where λ is the *wavelength* in meters. If we can solve the Helmholtz equation given by Eq. (4.1-4), then substituting the solution $\varphi_f(\mathbf{r})$ into Eq. (4.1-3) will yield a time-harmonic solution of the wave equation given by Eq. (4.1-1), which is the main objective in this section. Time-harmonic solutions are very important, because solutions of the wave equation with arbitrary time dependence can be obtained from time-harmonic solutions by using Fourier transform techniques, as was discussed in Section 2.2.3 and Examples 3.8-1, 3.8-2, and 3.9-1.

The solution of the Helmholtz equation given by Eq. (4.1-4) will be obtained by using the *method of separation of variables*, that is, we assume a solution of the form

$$\boxed{\varphi_f(\mathbf{r}) = \varphi_f(r, \theta, \psi) = R(r)\Theta(\theta)\Psi(\psi).} \tag{4.1-7}$$

Substituting Eq. (4.1-7) into Eq. (4.1-4) and making use of Eq. (4.1-2) yields

$$r^2 \sin^2 \theta \left[\frac{1}{R(r)} \frac{d^2}{dr^2} R(r) + \frac{2}{rR(r)} \frac{d}{dr} R(r) + k^2 \right]$$

$$+ \frac{\sin^2 \theta}{\Theta(\theta)} \left[\frac{d^2}{d\theta^2} \Theta(\theta) + \cot \theta \frac{d}{d\theta} \Theta(\theta) \right] = -\frac{1}{\Psi(\psi)} \frac{d^2}{d\psi^2} \Psi(\psi), \tag{4.1-8}$$

where the second-order partial derivatives with respect to r, θ, and ψ have been replaced by ordinary second-order derivatives, since the functions $R(r)$, $\Theta(\theta)$, and $\Psi(\psi)$ depend on only one of the independent variables r, θ, and ψ, respectively. Since the left-hand side of Eq. (4.1-8) is a function of both r and θ, and the right-hand side is a function of ψ, equality is possible only if both sides of Eq. (4.1-8) are equal to a constant; call it n^2. The constant n^2 is referred to as a *separation constant*. Therefore, since

$$-\frac{1}{\Psi(\psi)} \frac{d^2}{d\psi^2} \Psi(\psi) = n^2, \tag{4.1-9}$$

Eq. (4.1-9) can be rewritten as

$$\frac{d^2}{d\psi^2} \Psi(\psi) + n^2 \Psi(\psi) = 0. \tag{4.1-10}$$

Equation (4.1-10) is a second-order, homogeneous (no input or forcing function), ordinary differential equation (ODE) with *exact* solution

$$\Psi(\psi) = E_n \cos(n\psi) + F_n \sin(n\psi), \qquad n = 0, 1, 2, \ldots, \qquad (4.1\text{-}11)$$

where n is an *integer*, since we shall restrict ourselves to problems where the fluid medium completely surrounds the Z axis (see Section 3.1). The values of the constants E_n and F_n are complex, in general, and are determined by satisfying boundary conditions.

Returning to the solution of Eq. (4.1-8), recall that the left-hand side of Eq. (4.1-8) must also be equal to the separation constant n^2. Therefore,

$$r^2 \sin^2 \theta \left[\frac{1}{R(r)} \frac{d^2}{dr^2} R(r) + \frac{2}{rR(r)} \frac{d}{dr} R(r) + k^2 \right]$$

$$+ \frac{\sin^2 \theta}{\Theta(\theta)} \left[\frac{d^2}{d\theta^2} \Theta(\theta) + \cot \theta \frac{d}{d\theta} \Theta(\theta) \right] = n^2, \qquad (4.1\text{-}12)$$

or, dividing both sides of Eq. (4.1-12) by $\sin^2 \theta$,

$$\frac{r^2}{R(r)} \frac{d^2}{dr^2} R(r) + \frac{2r}{R(r)} \frac{d}{dr} R(r) + k^2 r^2$$

$$= \frac{n^2}{\sin^2 \theta} - \frac{1}{\Theta(\theta)} \left[\frac{d^2}{d\theta^2} \Theta(\theta) + \cot \theta \frac{d}{d\theta} \Theta(\theta) \right]. \qquad (4.1\text{-}13)$$

Since the left-hand side of Eq. (4.1-13) is a function of r and the right-hand side is a function of θ, equality is possible only if both sides of Eq. (4.1-13) are equal to a constant; call it M^2. The constant M^2 is also referred to as a separation constant. Therefore, since

$$\frac{n^2}{\sin^2 \theta} - \frac{1}{\Theta(\theta)} \left[\frac{d^2}{d\theta^2} \Theta(\theta) + \cot \theta \frac{d}{d\theta} \Theta(\theta) \right] = M^2, \qquad (4.1\text{-}14)$$

Eq. (4.1-14) can be rewritten as

$$\frac{d^2}{d\theta^2} \Theta(\theta) + \cot \theta \frac{d}{d\theta} \Theta(\theta) + \left[M^2 - \frac{n^2}{\sin^2 \theta} \right] \Theta(\theta) = 0. \qquad (4.1\text{-}15)$$

Equation (4.1-15) has solutions that are finite (bounded) in the closed interval $0 \le \theta \le \pi$ if

$$M^2 = m(m+1), \qquad m = 0, 1, 2, \ldots, \qquad (4.1\text{-}16)$$

where m is an *integer*. Therefore, substituting Eq. (4.1-16) into Eq. (4.1-15) yields

$$\frac{d^2}{d\theta^2}\Theta(\theta) + \cot\theta\,\frac{d}{d\theta}\Theta(\theta) + \left[m(m+1) - \frac{n^2}{\sin^2\theta}\right]\Theta(\theta) = 0,$$

$$m, n = 0, 1, 2, \ldots, \qquad (4.1\text{-}17)$$

which is known as *Legendre's associated equation*. The *exact* solution of Legendre's associated equation is given by

$$\Theta(\theta) = C_{mn} P_m^n(\cos\theta) + D_{mn} Q_m^n(\cos\theta), \qquad m, n = 0, 1, 2, \ldots, \qquad (4.1\text{-}18)$$

where $P_m^n(\cos\theta)$ and $Q_m^n(\cos\theta)$ are the *associated Legendre functions of degree m and order n of the first and second kind*, respectively. When $n = 0$, Eq. (4.1-17) reduces to

$$\frac{d^2}{d\theta^2}\Theta(\theta) + \cot\theta\,\frac{d}{d\theta}\Theta(\theta) + m(m+1)\Theta(\theta) = 0, \qquad m = 0, 1, 2, \ldots,$$

$$(4.1\text{-}19)$$

which is known as *Legendre's equation*. The *exact* solution of Legendre's equation is given by Eq. (4.1-18) with $n = 0$, that is,

$$\Theta(\theta) = C_m P_m(\cos\theta) + D_m Q_m(\cos\theta), \qquad m = 0, 1, 2, \ldots, \qquad (4.1\text{-}20)$$

where $C_m = C_{m0}$, $D_m = D_{m0}$, $P_m(\cos\theta) = P_m^0(\cos\theta)$, and $Q_m(\cos\theta) = Q_m^0(\cos\theta)$, where $P_m(\cos\theta)$ is the *Legendre polynomial of degree m* and $Q_m(\cos\theta)$ is the *Legendre function of the second kind of degree m*. The function $P_m(\cos\theta)$ is also known as a *zonal harmonic*. Note that when $n = 0$ [see Eq. (4.1-11)],

$$\Psi(\psi) = E_0, \qquad n = 0, \qquad (4.1\text{-}21)$$

and as a result, the solution of the wave equation does not depend on the azimuthal angle ψ. This is known as the *axisymmetric case*. The values of the constants C_{mn}, D_{mn}, C_m, and D_m are complex, in general, and are determined by satisfying boundary conditions.

Finally, recall that the left-hand side of Eq. (4.1-13) must also be equal to M^2. Therefore,

$$\frac{r^2}{R(r)} \frac{d^2}{dr^2} R(r) + \frac{2r}{R(r)} \frac{d}{dr} R(r) + k^2 r^2 = M^2, \tag{4.1-22}$$

and upon substituting Eq. (4.1-16) into Eq. (4.1-22),

$$\frac{d^2}{dr^2} R(r) + \frac{2}{r} \frac{d}{dr} R(r) + \left[k^2 - \frac{m(m+1)}{r^2} \right] R(r) = 0, \qquad m = 0, 1, 2, \ldots . \tag{4.1-23}$$

Next, let

$$R(r) = g(kr). \tag{4.1-24}$$

Therefore, since (see Section 3.1)

$$\frac{d}{dr} R(r) = k \frac{d}{d(kr)} g(kr) \tag{4.1-25}$$

and

$$\frac{d^2}{dr^2} R(r) = k^2 \frac{d^2}{d(kr)^2} g(kr), \tag{4.1-26}$$

substituting Eqs. (4.1-25) and (4.1-26) into Eq. (4.1-23) yields

$$\frac{d^2}{d\zeta^2} g(\zeta) + \frac{2}{\zeta} \frac{d}{d\zeta} g(\zeta) + \left[1 - \frac{m(m+1)}{\zeta^2} \right] g(\zeta) = 0, \qquad m = 0, 1, 2, \ldots, \tag{4.1-27}$$

where

$$\zeta = kr \tag{4.1-28}$$

and

$$R(r) = g(\zeta) = g(kr). \tag{4.1-29}$$

Note that Eq. (4.1-27) is similar to but is not the same as Bessel's differential equation [see Eq. (3.1-30)]. The *exact* solution of Eq. (4.1-27) is given by either

$$g(\zeta) = A_m j_m(\zeta) + B_m y_m(\zeta), \qquad m = 0, 1, 2, \ldots, \qquad (4.1\text{-}30)$$

or

$$g(\zeta) = A_m h_m^{(1)}(\zeta) + B_m h_m^{(2)}(\zeta), \qquad m = 0, 1, 2, \ldots, \qquad (4.1\text{-}31)$$

where $j_m(\zeta)$ and $y_m(\zeta)$ are the *mth-order spherical Bessel functions of the first and second kind*, respectively, and $h_m^{(1)}(\zeta)$ and $h_m^{(2)}(\zeta)$ are the *mth-order spherical Hankel functions of the first and second kind*, respectively, also known as *spherical Bessel functions of the third kind*. The function $y_m(\zeta)$ is also known as the *spherical Neumann function* when m is an integer. By referring to Eqs. (4.1-29) through (4.1-31), we can finally write that the *exact* solution of Eq. (4.1-23) is given by either

$$R(r) = A_m j_m(kr) + B_m y_m(kr), \qquad m = 0, 1, 2, \ldots, \qquad (4.1\text{-}32)$$

or

$$R(r) = A_m h_m^{(1)}(kr) + B_m h_m^{(2)}(kr), \qquad m = 0, 1, 2, \ldots, \qquad (4.1\text{-}33)$$

where A_m and B_m are complex constants in general, whose values are determined by satisfying boundary conditions.

Therefore, upon substituting Eqs. (4.1-11), (4.1-18), and (4.1-32) into Eq. (4.1-7), we obtain

$$\varphi_{f,mn}(r, \theta, \psi) = [A_m j_m(kr) + B_m y_m(kr)][C_{mn} P_m^n(\cos\theta) + D_{mn} Q_m^n(\cos\theta)]$$
$$\times [E_n \cos(n\psi) + F_n \sin(n\psi)], \qquad m, n = 0, 1, 2, \ldots,$$

$$(4.1\text{-}34)$$

and, upon substituting Eqs. (4.1-11), (4.1-18), and (4.1-33) into Eq. (4.1-7), we obtain

$$\varphi_{f,mn}(r, \theta, \psi) = [A_m h_m^{(1)}(kr) + B_m h_m^{(2)}(kr)][C_{mn} P_m^n(\cos\theta) + D_{mn} Q_m^n(\cos\theta)]$$
$$\times [E_n \cos(n\psi) + F_n \sin(n\psi)], \qquad m, n = 0, 1, 2, \ldots .$$

$$(4.1\text{-}35)$$

The subscript mn has been added to $\varphi_f(r, \theta, \psi)$ in order to emphasize the dependence of the velocity potential on the value of the integers m and n. Both Eqs. (4.1-34) and (4.1-35) are solutions of the Helmholtz equation given by Eq. (4.1-4) where k is the wave number given by Eq. (4.1-5). And, upon substituting either Eq. (4.1-34) or Eq. (4.1-35) into Eq. (4.1-3), we obtain the time-harmonic solution (in spherical coordinates) of the linear, three-dimensional, lossless, homogeneous wave equation given by Eq. (4.1-1). Note that in the spherical coordinate system, there is *no* interrelationship between the wave number k and the separation constants M^2 and n^2 as there is between the wave number and the separation constants in the rectangular and cylindrical coordinate systems. Also note that since we assumed that the fluid medium completely surrounds the Z axis, n must be an integer in order to ensure that $\Psi(\psi)$ will be periodic with period 2π, so that our solution $\varphi_{f, mn}(r, \theta, \psi)$ will be a single-valued function of position (see Section 3.1).

Next, let us discuss some of the properties of the spherical Bessel functions of the first, second, and third kind. Note that

$$j_m(\zeta) = \sqrt{\frac{\pi}{2\zeta}} \, J_{m+0.5}(\zeta), \tag{4.1-36}$$

$$y_m(\zeta) = \sqrt{\frac{\pi}{2\zeta}} \, Y_{m+0.5}(\zeta), \tag{4.1-37}$$

$$h_m^{(1)}(\zeta) = j_m(\zeta) + jy_m(\zeta) = \sqrt{\frac{\pi}{2\zeta}} \, H_{m+0.5}^{(1)}(\zeta), \tag{4.1-38}$$

and

$$h_m^{(2)}(\zeta) = j_m(\zeta) - jy_m(\zeta) = \sqrt{\frac{\pi}{2\zeta}} \, H_{m+0.5}^{(2)}(\zeta), \tag{4.1-39}$$

where $J_{m+0.5}(\zeta)$ and $Y_{m+0.5}(\zeta)$ are the Bessel functions of the first and second kind, respectively, of order $m + 0.5$, and $H_{m+0.5}^{(1)}(\zeta)$ and $H_{m+0.5}^{(2)}(\zeta)$ are the Hankel functions of the first and second kind, respectively, of order $m + 0.5$. Also,

$$j_0(0) = 1, \tag{4.1-40}$$

$$j_m(0) = 0, \quad m = 1, 2, 3, \ldots, \tag{4.1-41}$$

and

$$y_m(kr) \to -\infty \quad \text{as} \quad kr \to 0, \quad m = 0, 1, 2, \ldots. \tag{4.1-42}$$

Therefore, because of Eq. (4.1-42), if the value of the velocity potential is required for $r \geq 0$, then we neglect $y_m(kr)$ by setting the constant $B_m = 0$ in

Eq. (4.1-34) so that the solution will be finite at $r = 0$. Equation (4.1-34), with $B_m = 0$, is best suited for cavity problems which require solutions at $r = 0$. However, if the value of the velocity potential is *not* required at $r = 0$, then $y_m(kr)$ should be retained in the solution given by Eq. (4.1-34).

Next, if $kr \gg (m + 0.5)^2/2$, then for $m = 1, 2, 3, \ldots$,

$$h_m^{(1)}(kr) \approx (-j)^{m+1} \frac{\exp(+jkr)}{kr}, \qquad kr \gg \frac{(m + 0.5)^2}{2}, \qquad m = 1, 2, 3, \ldots,$$

$$(4.1\text{-}43)$$

and

$$h_m^{(2)}(kr) \approx (+j)^{m+1} \frac{\exp(-jkr)}{kr}, \qquad kr \gg \frac{(m + 0.5)^2}{2}, \qquad m = 1, 2, 3, \ldots.$$

$$(4.1\text{-}44)$$

For $m = 0$, the expressions obtained from Eqs. (4.1-43) and (4.1-44) are *exact* for all values of kr. Equations (4.1-43) and (4.1-44) are *asymptotic (far-field) approximations*. Since we chose $\exp(+j2\pi ft)$ for our time-harmonic dependence, Eq. (4.1-43) represents an *incoming wave* traveling in the direction of *decreasing r*, whereas Eq. (4.1-44) represents an *outgoing wave* traveling in the direction of *increasing r*. However, if we had chosen $\exp(-j2\pi ft)$ instead, then Eqs. (4.1-43) and (4.1-44) would represent *outgoing* and *incoming* waves, respectively. Therefore, Eq. (4.1-35), in conjunction with Eq. (4.1-43) and/or Eq. (4.1-44), is best suited for radiation and scattering problems, which typically require far-field solutions that do not include $r = 0$. In addition, by inspecting Eqs. (4.1-43) and (4.1-44), it can be seen that for a given value of m,

$$kr = \text{constant} \qquad\qquad\qquad (4.1\text{-}45)$$

defines a surface of constant phase or, in other words, a wavefront. Since r is the spherical range measured from the origin (see Fig. 4.1-1), the wavefront defined by Eq. (4.1-45) is the surface of a sphere. Hence the term "spherical wave." Also note that both Eq. (4.1-43) and Eq. (4.1-44) indicate that the amplitude of a spherical wave decreases as $1/r$. Although the following asymptotic expressions are not used as often as Eqs. (4.1-43) and (4.1-44), they are worth being familiar with: if $kr \gg (m + 0.5)^2/2$, then for $m = 1, 2, 3, \ldots$,

$$j_m(kr) \approx \frac{1}{kr} \cos\left(kr - \frac{m\pi}{2} - \frac{\pi}{2} \right), \qquad kr \gg \frac{(m + 0.5)^2}{2}, \qquad m = 1, 2, 3, \ldots,$$

$$(4.1\text{-}46)$$

and

$$y_m(kr) \approx \frac{1}{kr} \sin\left(kr - \frac{m\pi}{2} - \frac{\pi}{2}\right), \qquad kr \gg \frac{(m+0.5)^2}{2}, \qquad m = 1, 2, 3, \ldots .$$

$$(4.1\text{-}47)$$

For $m = 0$, the expressions obtained from Eqs. (4.1-46) and (4.1-47) are *exact* for all values of kr.

Finally, the spherical Bessel functions of the first, second, and third kind also obey the following relationships:

$$W[j_m(\zeta), y_m(\zeta)] = j_{m+1}(\zeta) y_m(\zeta) - j_m(\zeta) y_{m+1}(\zeta) = 1/\zeta^2, \qquad (4.1\text{-}48)$$

$$W[h_m^{(1)}(\zeta), h_m^{(2)}(\zeta)] = h_{m+1}^{(1)}(\zeta) h_m^{(2)}(\zeta) - h_m^{(1)}(\zeta) h_{m+1}^{(2)}(\zeta) = -j2/\zeta^2, \quad (4.1\text{-}49)$$

$$f_{m+1}(\zeta) = \frac{2m+1}{\zeta} f_m(\zeta) - f_{m-1}(\zeta), \qquad (4.1\text{-}50)$$

and

$$f_m'(\zeta) = f_{m-1}(\zeta) - \frac{m+1}{\zeta} f_m(\zeta), \qquad (4.1\text{-}51)$$

where

$$f_m'(\zeta) = \frac{d}{d\zeta} f_m(\zeta) \qquad (4.1\text{-}52)$$

and the generic function $f_m(\zeta)$ can represent $j_m(\zeta)$, $y_m(\zeta)$, $h_m^{(1)}(\zeta)$, and $h_m^{(2)}(\zeta)$ in the recurrence relationships given by Eqs. (4.1-50) and (4.1-51). Equations (4.1-48) and (4.1-49) are referred to as the *Wronskians* of $j_m(\zeta)$ and $y_m(\zeta)$ and of $h_m^{(1)}(\zeta)$ and $h_m^{(2)}(\zeta)$, respectively, and are extremely useful when calculating time-average intensity vectors.

Let us next discuss some of the properties of the associated Legendre functions of the first and second kind. Note that

$$P_m^n(\cos\theta) = \sin^n\theta \frac{d^n}{d\cos^n\theta} P_m(\cos\theta), \qquad m, n = 0, 1, 2, \ldots, \quad (4.1\text{-}53)$$

and

$$Q_m^n(\cos\theta) = \sin^n\theta \frac{d^n}{d\cos^n\theta} Q_m(\cos\theta), \qquad m, n = 0, 1, 2, \ldots, \quad (4.1\text{-}54)$$

where $P_m(\cos\theta)$ is the Legendre polynomial of degree m and $Q_m(\cos\theta)$ is the Legendre function of the second kind of degree m. Also note that

$$P_m^0(\cos\theta) = P_m(\cos\theta), \qquad m = 0,1,2,\dots, \tag{4.1-55}$$

$$Q_m^0(\cos\theta) = Q_m(\cos\theta), \qquad m = 0,1,2,\dots, \tag{4.1-56}$$

$$P_m^n(\cos\theta) = 0, \qquad n > m, \qquad m = 0,1,2,\dots, \qquad n = 1,2,3,\dots, \tag{4.1-57}$$

$$P_m^n(\cos\theta) = 0, \qquad \theta = 0,\pi, \qquad m = 0,1,2,\dots, \qquad n = 1,2,3,\dots, \tag{4.1-58}$$

and

$$\left| Q_m^n(\cos\theta) \right| = \infty, \qquad \theta = 0,\pi, \quad m,n = 0,1,2,\dots . \tag{4.1-59}$$

Therefore, because of Eq. (4.1-59), if the value of the velocity potential is required at $\theta = 0$ and $\theta = \pi$, then we neglect $Q_m^n(\cos\theta)$ by setting the constant $D_{mn} = 0$ in either Eq. (4.1-34) or Eq. (4.1-35), so that the solution will be finite at $\theta = 0$ and $\theta = \pi$. However, if the value of the velocity potential is *not* required at $\theta = 0$ and $\theta = \pi$, then $Q_m^n(\cos\theta)$ should be retained in the solution given by either Eq. (4.1-34) or Eq. (4.1-35).

The first few Legendre polynomials are given by the following expressions:

$$P_0(\cos\theta) = 1, \tag{4.1-60}$$

$$P_1(\cos\theta) = \cos\theta, \tag{4.1-61}$$

$$P_2(\cos\theta) = \tfrac{1}{2}(3\cos^2\theta - 1) = \tfrac{1}{4}[3\cos(2\theta) + 1], \tag{4.1-62}$$

$$P_3(\cos\theta) = \tfrac{1}{2}(5\cos^3\theta - 3\cos\theta) = \tfrac{1}{8}[5\cos(3\theta) + 3\cos\theta]. \tag{4.1-63}$$

The Legendre polynomials also obey the following recurrence relationships:

$$P_{m+1}(\cos\theta) = \frac{2m+1}{m+1}P_m(\cos\theta)\cos\theta - \frac{m}{m+1}P_{m-1}(\cos\theta), \qquad m \geq 1, \tag{4.1-64}$$

and

$$P'_{m+1}(\cos\theta) = P'_m(\cos\theta)\cos\theta + (m+1)P_m(\cos\theta), \qquad m \geq 0, \tag{4.1-65}$$

where

$$P'_m(\cos\theta) = \frac{d}{d\cos\theta}P_m(\cos\theta). \tag{4.1-66}$$

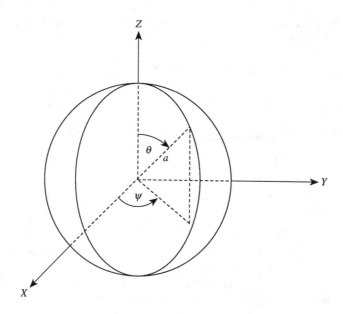

Figure 4.2-1. Vibrating sphere of radius a meters.

4.2 Free-Space Propagation

4.2.1 Radiation from a Vibrating Sphere

In this section we shall compute the velocity potential of the acoustic field produced by a vibrating sphere of radius a meters (see Fig. 4.2-1). The radiated acoustic field is assumed to propagate in free space (i.e., in an unbounded fluid medium), and as a result, there will be *no reflected waves*. After we obtain some general results in this section, we shall demonstrate in Section 4.2.2 the calculation of the time-average radiated power for the special case of the monopole mode of vibration of a sphere. This special case is very important, because if a sphere is in the monopole mode of vibration and if its circumference is small compared to a wavelength, then it radiates an acoustic field that is identical to the one produced by the theoretical omnidirectional point source that was modeled by an impulse function as discussed in Sections 2.6.2 and 2.7. The techniques developed here can be used to model the radiation from an electroacoustic transducer that is spherical in shape.

We begin our analysis by making the following observations. First, since the sphere is of radius a meters, the velocity potential of the radiated acoustic field will never have to be evaluated at $r = 0$. Therefore, we shall work with the solution of the Helmholtz equation given by Eq. (4.1-35). Second, since we are dealing with free-space propagation, there are no reflected waves, and as a result, we set $A_m = 0$ in Eq. (4.1-35). Third, since the value of the velocity potential is required at $\theta = 0$ and $\theta = \pi$, we set $D_{mn} = 0$ in Eq. (4.1-35). Therefore, with the use of these three observations, Eq. (4.1-35) reduces to

$$\varphi_{f,mn}(r,\theta,\psi) = h_m^{(2)}(kr)P_m^n(\cos\theta)\left[A_{mn}\cos(n\psi) + B_{mn}\sin(n\psi)\right],$$

$$m, n = 0, 1, 2, \ldots, \quad (4.2\text{-}1)$$

where

$$A_{mn} = B_m C_{mn} E_n \quad (4.2\text{-}2)$$

and

$$B_{mn} = B_m C_{mn} F_n. \quad (4.2\text{-}3)$$

Since the sum of all possible solutions of the Helmholtz equation is also a solution, if we let

$$\varphi_f(r,\theta,\psi) = \sum_{m=0}^{\infty}\sum_{n=0}^{m}\varphi_{f,mn}(r,\theta,\psi), \quad (4.2\text{-}4)$$

then substituting Eq. (4.2-1) into Eq. (4.2-4) yields

$$\varphi_f(r,\theta,\psi) = \sum_{m=0}^{\infty}\sum_{n=0}^{m} h_m^{(2)}(kr)P_m^n(\cos\theta)\left[A_{mn}\cos(n\psi) + B_{mn}\sin(n\psi)\right],$$

$$(4.2\text{-}5)$$

where

$$k = 2\pi f/c = 2\pi/\lambda. \quad (4.2\text{-}6)$$

Note that the upper limit on the summation with respect to n appearing in Eq. (4.2-5) is m because of Eq. (4.1-57). Equation (4.2-5) is the spatial-dependent part of the velocity potential of the radiated acoustic field where the values of the constants A_{mn} and B_{mn} are determined by matching the boundary condition at the surface of the vibrating sphere.

In order to proceed further, we shall assume that the velocity vector of the surface of the vibrating sphere has only a time-harmonic *radial* component given by

$$v_r(t, a, \theta, \psi) = v_{f,r}(a, \theta, \psi)\exp(+j2\pi ft), \quad (4.2\text{-}7)$$

where $v_{f,r}(a,\theta,\psi)$ is a known, real function of θ and ψ. This is equivalent to knowing the radial component of the acoustic fluid velocity vector at the surface of the sphere. Since the radial component is normal to the surface of the sphere, and since the surface of the sphere is the boundary in this problem, this problem is an example of a Neumann boundary-value problem. The next step is to compute the radial component of the acoustic fluid velocity vector.

The acoustic fluid velocity vector in meters per second is given by

$$\mathbf{u}(t,\mathbf{r}) = \nabla\varphi(t,\mathbf{r}), \quad (4.2\text{-}8)$$

where

$$\nabla = \frac{\partial}{\partial r}\hat{r} + \frac{1}{r}\frac{\partial}{\partial \theta}\hat{\theta} + \frac{1}{r \sin \theta}\frac{\partial}{\partial \psi}\hat{\psi} \qquad (4.2\text{-}9)$$

is the *gradient* expressed in the spherical coordinates (r, θ, ψ), and where $\hat{r}$, $\hat{\theta}$, and $\hat{\psi}$ are unit vectors in the r, θ, and ψ directions, respectively. The radial component $u_r(t, \mathbf{r})$ is given by

$$u_r(t, \mathbf{r}) = \hat{r} \cdot \mathbf{u}(t, \mathbf{r}) = \hat{r} \cdot \nabla\varphi(t, \mathbf{r}), \qquad (4.2\text{-}10)$$

or

$$u_r(t, \mathbf{r}) = \frac{\partial}{\partial r}\varphi(t, \mathbf{r}). \qquad (4.2\text{-}11)$$

Therefore, at $r = a$,

$$u_r(t, a, \theta, \psi) = v_r(t, a, \theta, \psi) = \frac{\partial}{\partial r}\varphi(t, r, \theta, \psi)\bigg|_{r=a}. \qquad (4.2\text{-}12)$$

Substituting Eqs. (4.1-3), (4.2-5), and (4.2-7) into Eq. (4.2-12) yields

$$v_{f,r}(a, \theta, \psi) = k \sum_{m=0}^{\infty} \sum_{n=0}^{m} h_m^{(2)\prime}(ka) P_m^n(\cos \theta)\left[A_{mn}\cos(n\psi) + B_{mn}\sin(n\psi)\right],$$

$$(4.2\text{-}13)$$

where

$$h_m^{(2)\prime}(ka) = \frac{d}{d(kr)}h_m^{(2)}(kr)\bigg|_{r=a}. \qquad (4.2\text{-}14)$$

The constants A_{mn} and B_{mn} are determined from Eq. (4.2-13), which represents the boundary condition at the surface of the sphere. Once the values of these constants are known, they are substituted into Eq. (4.2-5), which is then substituted into Eq. (4.1-3), which yields the velocity potential of the time-harmonic radiated acoustic field due to a vibrating sphere of radius a meters.

4.2.2 Monopole and Dipole Modes of Vibration

In the previous section, the radial component of the velocity vector of the surface of the vibrating sphere was given by Eq. (4.2-7), where $v_{f,r}(a, \theta, \psi)$ was assumed to be a known, real function of θ and ψ. Now, in this section, we shall investigate the simpler problem of

$$v_{f,r}(a, \theta, \psi) = V_0 x(\theta), \tag{4.2-15}$$

where V_0 is a real constant with units of meters per second, and $x(\theta)$ is a real function of θ for $0 \leq \theta \leq \pi$. Both V_0 and $x(\theta)$ are assumed to be known. According to Eq. (4.2-15), the vibration of the surface of the sphere is only a function of θ. Therefore, since there is no dependence on the azimuthal angle ψ, Eq. (4.2-15) describes the *axisymmetric case* of vibration. Substituting Eq. (4.2-15) into Eq. (4.2-13) and rewriting yields

$$V_0 x(\theta) = k \sum_{m=0}^{\infty} A_m h_m^{(2)\prime}(ka) P_m(\cos \theta)$$

$$+ k \sum_{m=0}^{\infty} \sum_{n=1}^{m} h_m^{(2)\prime}(ka) P_m^n(\cos \theta) \left[A_{mn} \cos(n\psi) + B_{mn} \sin(n\psi) \right],$$

$$\tag{4.2-16}$$

where [see Eq. (4.1-55)]

$$P_m^0(\cos \theta) = P_m(\cos \theta), \qquad m = 0, 1, 2, \ldots, \tag{4.2-17}$$

and

$$A_{m0} = A_m, \qquad m = 0, 1, 2, \ldots . \tag{4.2-18}$$

In order to get rid of the ψ dependence on the right-hand side of Eq. (4.2-16), we set

$$A_{mn} = 0, \qquad m = 0, 1, 2, \ldots, \qquad n = 1, 2, \ldots, m, \tag{4.2-19}$$

and

$$B_{mn} = 0, \qquad m = 0, 1, 2, \ldots, \qquad n = 1, 2, \ldots, m. \tag{4.2-20}$$

Therefore, substituting Eqs. (4.2-19) and (4.2-20) into Eq. (4.2-16) yields

$$V_0 x(\theta) = k \sum_{m=0}^{\infty} A_m h_m^{(2)\prime}(ka) P_m(\cos \theta), \tag{4.2-21}$$

which represents the boundary condition for the axisymmetric case of vibration. Now, since the Legendre polynomials $P_m(\cos \theta)$, $m = 0, 1, 2, \ldots$, form an *orthogo-*

nal set of functions with respect to the *weight function* $\sin \theta$ in the closed interval $0 \le \theta \le \pi$, that is, since

$$\int_0^\pi P_m(\cos \theta) P_n(\cos \theta) \sin \theta \, d\theta = \frac{2}{2n+1} \delta_{mn}, \qquad (4.2\text{-}22)$$

where

$$\delta_{mn} = \begin{cases} 1, & m = n, \\ 0, & m \ne n, \end{cases} \qquad (4.2\text{-}23)$$

is the *Kronecker delta*, $x(\theta)$ can be expanded into a *Fourier-Legendre series*, that is,

$$x(\theta) = \sum_{m=0}^\infty a_m P_m(\cos \theta), \qquad (4.2\text{-}24)$$

where

$$a_m = \frac{2m+1}{2} \int_0^\pi x(\theta) P_m(\cos \theta) \sin \theta \, d\theta, \qquad m = 0, 1, 2, \ldots . \quad (4.2\text{-}25)$$

Substituting Eq. (4.2-24) into Eq. (4.2-21) yields

$$V_0 \sum_{m=0}^\infty a_m P_m(\cos \theta) = k \sum_{m=0}^\infty A_m h_m^{(2)\prime}(ka) P_m(\cos \theta), \qquad (4.2\text{-}26)$$

and by matching terms on both sides of Eq. (4.2-26), we finally obtain

$$\boxed{A_m = \frac{a_m V_0}{k h_m^{(2)\prime}(ka)}, \qquad m = 0, 1, 2, \ldots .} \qquad (4.2\text{-}27)$$

It should also be mentioned, for completeness, that Eq. (4.2-27) could have been obtained otherwise than by matching terms. For example, in order to solve for A_m, multiply both sides of Eq. (4.2-26) by $P_n(\cos \theta) \sin \theta$, integrate both sides over θ from 0 to π, and then use Eq. (4.2-22).

Therefore, if we rewrite Eq. (4.2-5) as

$$\varphi_f(r, \theta, \psi) = \sum_{m=0}^\infty A_m h_m^{(2)}(kr) P_m(\cos \theta)$$

$$+ \sum_{m=0}^\infty \sum_{n=1}^m h_m^{(2)}(kr) P_m^n(\cos \theta) [A_{mn} \cos(n\psi) + B_{mn} \sin(n\psi)],$$

$$(4.2\text{-}28)$$

where use has been made of Eqs. (4.2-17) and (4.2-18); then, upon substituting Eqs. (4.2-19), (4.2-20), and (4.2-27) into Eq. (4.2-28), we obtain

$$\varphi_f(r,\theta,\psi) = \frac{V_0}{k} \sum_{m=0}^{\infty} a_m \frac{h_m^{(2)}(kr)}{h_m^{(2)\prime}(ka)} P_m(\cos\theta), \qquad r \geq a, \qquad (4.2\text{-}29)$$

where the coefficients a_m, $m = 0, 1, 2, \ldots$, are given by Eq. (4.2-25). Equation (4.2-29) represents the spatial-dependent part of the velocity potential of the radiated acoustic field corresponding to the axisymmetric case of vibration. The time-harmonic velocity potential of the radiated acoustic field is obtained by substituting Eq. (4.2-29) into Eq. (4.1-3).

Monopole Mode of Vibration
If the real function

$$x(\theta) = 1, \qquad (4.2\text{-}30)$$

then substituting Eqs. (4.2-15) and (4.2-30) into Eq. (4.2-7) yields

$$v_r(t, a, \theta, \psi) = V_0 \exp(+j2\pi ft). \qquad (4.2\text{-}31)$$

Equation (4.2-31) defines the special axisymmetric case of *uniform radiation*, that is, the surface of the sphere is vibrating (expanding and contracting) uniformly in the radial direction with an amplitude of V_0 meters per second at a frequency of f hertz. This special case is also known as the *monopole mode* or *breathing mode* of vibration. In addition, it is sometimes referred to as a *radially pulsating sphere*.

Since [see Eq. (4.1-60)]

$$P_0(\cos\theta) = 1, \qquad (4.2\text{-}32)$$

Eq. (4.2-30) can be rewritten as

$$x(\theta) = P_0(\cos\theta) = 1. \qquad (4.2\text{-}33)$$

Substituting Eq. (4.2-33) into Eq. (4.2-25) yields

$$a_m = \frac{2m+1}{2} \int_0^{\pi} P_m(\cos\theta) P_0(\cos\theta) \sin\theta \, d\theta, \qquad m = 0, 1, 2, \ldots, \qquad (4.2\text{-}34)$$

$$= \frac{2m+1}{2} 2\delta_{m0}, \qquad m = 0, 1, 2, \ldots, \qquad (4.2\text{-}35)$$

or

$$a_m = \begin{cases} 1, & m = 0, \\ 0, & m \neq 0, \end{cases} \qquad (4.2\text{-}36)$$

where use has been made of Eqs. (4.2-22) and (4.2-23). Therefore, upon substituting Eqs. (4.2-36) and (4.2-32) into Eq. (4.2-29), we obtain

$$\boxed{\varphi_f(r, \theta, \psi) = \frac{V_0}{k} \frac{h_0^{(2)}(kr)}{h_0^{(2)\prime}(ka)}, \qquad r \geq a,}$$

$$(4.2\text{-}37)$$

for the monopole mode of vibration. [Compare Eq. (4.2-37) with the velocity potential for the monopole mode of vibration for a vibrating cylinder of radius a meters given by Eq. (3.2-40).] However, since

$$\boxed{h_0^{(2)}(kr) = j \frac{\exp(-jkr)}{kr},}$$

$$(4.2\text{-}38)$$

substituting Eq. (4.2-38) into Eq. (4.2-14) yields

$$h_0^{(2)\prime}(ka) = \left(1 - j\frac{1}{ka}\right)\frac{\exp(-jka)}{ka}, \qquad (4.2\text{-}39)$$

and with the use of Eqs. (4.2-38) and (4.2-39), Eq. (4.2-37) can be rewritten as

$$\boxed{\varphi_f(r, \theta, \psi) = S_0 g_f(\mathbf{r}|\mathbf{0}) \frac{\exp(+jka)}{1 + jka}, \qquad r \geq a,}$$

$$(4.2\text{-}40)$$

where

$$\boxed{S_0 = 4\pi a^2 V_0} \qquad (4.2\text{-}41)$$

is the *source strength* (*volume flow rate*) with units of cubic meters per second, $4\pi a^2$ is the surface area of the sphere with units of square meters, and

$$g_f(\mathbf{r}|\mathbf{0}) = -\frac{\exp(-jkr)}{4\pi r} \qquad (4.2\text{-}42)$$

is the time-independent free-space Green's function for a point source located at the origin in an ideal, homogeneous, fluid medium (see Section 2.6), where

$$r = \sqrt{x^2 + y^2 + z^2}\,. \tag{4.2-43}$$

Note that for the monopole mode of vibration, the radiated acoustic field given by either Eq. (4.2-37) or Eq. (4.2-40) is not a function of θ and ψ, and as a result, the field is said to have *spherical symmetry*. Also note that both Eq. (4.2-38) and Eq. (4.2-39), which were used in the derivation of Eq. (4.2-40), are *exact* expressions. If the circumference of the sphere $2\pi a$ is small compared to a wavelength λ, that is, if $ka \ll 1$, then Eq. (4.2-40) reduces to

$$\boxed{\varphi_f(r,\theta,\psi) \approx S_0 g_f(\mathbf{r}|0), \qquad r \geq a, \quad ka \ll 1.} \tag{4.2-44}$$

Equation (4.2-44) represents the velocity potential of the radiated acoustic field from a *simple omnidirectional point source* (i.e., $ka \ll 1$). Recall that the solution of the inhomogeneous Helmholtz equation with a source distribution equal to the theoretical omnidirectional point source modeled by an impulse function is exactly equal to the free-space Green's function (see Sections 2.6.2 and 2.7). With the use of Eq. (4.2-40), we shall compute the acoustic pressure, the acoustic fluid velocity vector, the time-average intensity vector, and the time-average power produced by the vibrating sphere.

The time-harmonic radiated acoustic pressure can be expressed as

$$p(t,\mathbf{r}) = p_f(\mathbf{r}) \exp(+j2\pi ft), \tag{4.2-45}$$

where (see Example 2.2-1)

$$p_f(\mathbf{r}) = -jk\rho_0 c\varphi_f(\mathbf{r}). \tag{4.2-46}$$

Substituting Eqs. (4.2-40) and (4.2-42) into Eq. (4.2-46) yields

$$\boxed{p_f(r,\theta,\psi) = jk\rho_0 cS_0 \frac{\exp[-jk(r-a)]}{4\pi(1+jka)r}, \qquad r \geq a,} \tag{4.2-47}$$

for the monopole mode of vibration.

Similarly, the time-harmonic radiated acoustic fluid velocity vector can be expressed as

$$\mathbf{u}(t,\mathbf{r}) = \mathbf{u}_f(\mathbf{r}) \exp(+j2\pi ft), \tag{4.2-48}$$

where

$$\mathbf{u}_f(\mathbf{r}) = \nabla\varphi_f(\mathbf{r}). \tag{4.2-49}$$

Substituting Eq. (4.2-40) into Eq. (4.2-49) and using the gradient given by Eq. (4.2-9) yields

$$\mathbf{u}_f(r,\theta,\psi) = -S_0 \frac{\exp(+jka)}{4\pi(1+jka)} \frac{d}{dr} \frac{\exp(-jkr)}{r} \hat{r}, \qquad r \geq a. \quad (4.2\text{-}50)$$

Since

$$\frac{d}{dr} \frac{\exp(-jkr)}{r} = -\left(\frac{1}{r} + jk\right) \frac{\exp(-jkr)}{r}, \qquad (4.2\text{-}51)$$

substituting Eq. (4.2-51) into Eq. (4.2-50) yields

$$\mathbf{u}_f(r,\theta,\psi) = S_0 \frac{\exp[-jk(r-a)]}{4\pi(1+jka)r} \left(\frac{1}{r} + jk\right)\hat{r}, \qquad r \geq a, \quad (4.2\text{-}52)$$

for the monopole mode of vibration. Note that at the surface of the sphere, that is, at $r = a$,

$$\mathbf{u}(t,a,\theta,\psi) = \mathbf{u}_f(a,\theta,\psi)\exp(+j2\pi ft) \qquad (4.2\text{-}53)$$

$$= \frac{S_0}{4\pi a(1+jka)}\left(\frac{1}{a} + jk\right)\exp(+j2\pi ft)\,\hat{r} \qquad (4.2\text{-}54)$$

$$= \frac{S_0}{4\pi a^2}\exp(+j2\pi ft)\,\hat{r} \qquad (4.2\text{-}55)$$

$$= V_0\exp(+j2\pi ft)\,\hat{r}, \qquad (4.2\text{-}56)$$

and as a result, the radial component is given by

$$u_r(t,a,\theta,\psi) = \hat{r}\cdot\mathbf{u}(t,a,\theta,\psi) = V_0\exp(+j2\pi ft), \qquad (4.2\text{-}57)$$

which is identical to the boundary condition given by Eq. (4.2-31).

If Eq. (4.2-52) is rewritten as

$$\mathbf{u}_f(r,\theta,\psi) = u_f(r,\theta,\psi)\hat{r}, \qquad r \geq a, \qquad (4.2\text{-}58)$$

where

$$u_f(r,\theta,\psi) = S_0 \frac{\exp[-jk(r-a)]}{4\pi(1+jka)r}\left(\frac{1}{r} + jk\right), \qquad r \geq a, \quad (4.2\text{-}59)$$

is the *complex speed*, then by comparing Eqs. (4.2-47) and (4.2-59), it can be seen that

$$p_f(r, \theta, \psi) = jk\rho_0 c \left(\frac{1}{r} + jk \right)^{-1} u_f(r, \theta, \psi), \qquad r \geq a, \qquad (4.2\text{-}60)$$

or

$$p_f(r, \theta, \psi) = \rho_0 c \frac{jkr}{1 + jkr} u_f(r, \theta, \psi), \qquad r \geq a. \qquad (4.2\text{-}61)$$

Therefore, the *specific acoustic impedance* of an unbounded, ideal, homogeneous, fluid medium for a time-harmonic wave with spherical symmetry is given by

$$Z = \frac{p_f(r, \theta, \psi)}{u_f(r, \theta, \psi)} = \rho_0 c \frac{jkr}{1 + jkr}, \qquad r \geq a, \qquad (4.2\text{-}62)$$

or

$$Z = \rho_0 c \frac{kr}{\sqrt{1 + (kr)^2}} \exp\left\{ +j\left[\frac{\pi}{2} - \tan^{-1}(kr) \right] \right\}, \qquad r \geq a. \qquad (4.2\text{-}63)$$

Since the specific acoustic impedance is complex, the acoustic pressure and the acoustic fluid speed are out of phase. However, in the far field, that is, when $kr \gg 1$, Eq. (4.2-61) reduces to

$$p_f(r, \theta, \psi) \approx \rho_0 c u_f(r, \theta, \psi), \qquad r \geq a, \quad kr \gg 1, \qquad (4.2\text{-}64)$$

and Eqs. (4.2-62) and (4.2-63) reduce to

$$Z \approx \rho_0 c, \qquad r \geq a, \quad kr \gg 1, \qquad (4.2\text{-}65)$$

which are plane-wave relationships (see Example 2.2-1). Therefore, the radiated acoustic field behaves like a plane wave in the far-field region of the sphere.

Now that we have expressions for both the acoustic pressure and the acoustic fluid velocity vector, we can compute the time-average intensity vector. Recall that the time-average intensity vector for time-harmonic acoustic fields is given by (see Section 1.5)

$$\mathbf{I}_{avg}(\mathbf{r}) = \tfrac{1}{2} \operatorname{Re}\{p_f(\mathbf{r})\mathbf{u}_f^*(\mathbf{r})\}. \qquad (4.2\text{-}66)$$

Substituting Eqs. (4.2-47) and (4.2-52) into Eq. (4.2-66) yields

$$\mathbf{I}_{avg}(r,\theta,\psi) = \tfrac{1}{2} \, \mathrm{Re}\left\{ j \frac{k\rho_0 c S_0^2}{16\pi^2 \left[1 + (ka)^2\right] r^2} \left(\frac{1}{r} - jk\right)\right\}\hat{r}, \qquad r \geq a, \quad (4.2\text{-}67)$$

and upon taking the real part, we obtain

$$\mathbf{I}_{avg}(r,\theta,\psi) = \frac{\rho_0 c k^2 S_0^2}{32\pi^2 \left[1 + (ka)^2\right] r^2}\hat{r}, \qquad r \geq a, \qquad (4.2\text{-}68)$$

for the monopole mode of vibration, where the source strength S_0 is given by Eq. (4.2-41). Note that the magnitude of Eq. (4.2-68) can be expressed as

$$I_{avg}(r,\theta,\psi) = \left|\mathbf{I}_{avg}(r,\theta,\psi)\right| = \frac{\left|p_f(r,\theta,\psi)\right|^2}{2\rho_0 c}, \qquad r \geq a, \qquad (4.2\text{-}69)$$

where $p_f(r,\theta,\psi)$ is given by Eq. (4.2-47). Note that the form of Eq. (4.2-69) is identical to the expression for the time-average intensity of a plane wave (see Example 2.2-1).

The last calculation to be performed is the time-average radiated power. If we enclose the vibrating sphere with another sphere of radius r meters, then the time-average power is given by (see Section 1.5)

$$P_{avg} = \oint_S \mathbf{I}_{avg}(r,\theta,\psi) \cdot d\mathbf{S} \qquad (4.2\text{-}70)$$

where

$$d\mathbf{S} = r^2 \sin\theta \, d\theta \, d\psi \, \hat{r}. \qquad (4.2\text{-}71)$$

Therefore, substituting Eqs. (4.2-68) and (4.2-71) into Eq. (4.2-70) yields

$$P_{avg} = \frac{\rho_0 c k^2 S_0^2}{32\pi^2 \left[1 + (ka)^2\right]} \int_0^{2\pi} \int_0^{\pi} \sin\theta \, d\theta \, d\psi, \qquad (4.2\text{-}72)$$

or

$$P_{avg} = \frac{\rho_0 c k^2 S_0^2}{8\pi \left[1 + (ka)^2\right]}, \qquad (4.2\text{-}73)$$

for the monopole mode of vibration, where the source strength S_0 is given by Eq. (4.2-41).

Dipole Mode of Vibration
If the real function

$$x(\theta) = \cos\theta, \tag{4.2-74}$$

then substituting Eqs. (4.2-15) and (4.2-74) into Eq. (4.2-7) yields

$$v_r(t, a, \theta, \psi) = V_0 \cos\theta \exp(+j2\pi ft). \tag{4.2-75}$$

Equation (4.2-75) defines the special case of the *dipole mode of vibration*. Evaluating Eq. (4.2-75) at $\theta = 0$, $\pi/2$, and π yields

$$v_r(t, a, 0, \psi) = V_0 \exp(+j2\pi ft), \tag{4.2-76}$$

$$v_r(t, a, \pi/2, \psi) = 0, \tag{4.2-77}$$

and

$$v_r(t, a, \pi, \psi) = -V_0 \exp(+j2\pi ft). \tag{4.2-78}$$

By comparing Eqs. (4.2-76) and (4.2-78), it can be seen that while the surface of the sphere at $\theta = 0$ is expanding in the positive $\hat{r}$ direction, the surface of the sphere at $\theta = \pi$ is contracting in the negative $\hat{r}$ direction and vice versa. Hence the term "dipole mode of vibration." Equation (4.2-77) indicates that all of the points on the surface of the sphere at $\theta = \pi/2$ are *nodes*, that is, there is no vibration at these points.

Since [see Eq. (4.1-61)]

$$P_1(\cos\theta) = \cos\theta, \tag{4.2-79}$$

Eq. (4.2-74) can be rewritten as

$$x(\theta) = P_1(\cos\theta) = \cos\theta. \tag{4.2-80}$$

Substituting Eq. (4.2-80) into Eq. (4.2-25) yields

$$a_m = \frac{2m+1}{2} \int_0^\pi P_m(\cos\theta) P_1(\cos\theta) \sin\theta \, d\theta, \qquad m = 0, 1, 2, \ldots, \tag{4.2-81}$$

$$= \frac{2m+1}{2} \frac{2}{3} \delta_{m1}, \tag{4.2-82}$$

$$a_m = \begin{cases} 1, & m = 1, \\ 0, & m \neq 1, \end{cases} \tag{4.2-83}$$

where use has been made of Eqs. (4.2-22) and (4.2-23). Therefore, upon substituting Eqs. (4.2-83) and (4.2-79) into Eq. (4.2-29), we obtain

$$\varphi_f(r,\theta,\psi) = \frac{V_0}{k}\,\frac{h_1^{(2)}(kr)}{h_1^{(2)\prime}(ka)}\cos\theta, \qquad r \geq a, \tag{4.2-84}$$

for the dipole mode of vibration. Compare Eq. (4.2-84) with the velocity potential for the dipole mode of vibration for a vibrating cylinder of radius a meters given by Eq. (3.2-80).

4.3 The Spherical Cavity

In this section we shall derive expressions for the velocity potential and the acoustic pressure inside a *closed* spherical cavity of radius a meters as shown in Fig. 4.3-1. Let us assume that the surface inside the cavity is an *ideal rigid boundary*. Therefore, the normal component of the acoustic fluid velocity vector must be equal to zero at all points on the surface (see Example 2.3-1). This problem is an example of a *Neumann boundary-value problem*. Also recall from Example 2.3-1 that the reflection coefficient of an ideal rigid boundary is equal to one, and as a result, time-average acoustic power cannot escape from inside the

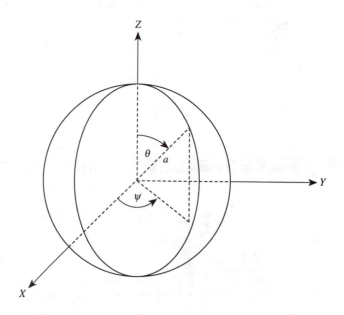

Figure 4.3-1. Spherical cavity of radius a meters.

cavity. The solution of the Helmholtz equation is given by *standing waves* due to reflections.

Since values of the acoustic field inside the cavity at $r = 0$ are required, we shall use the solution of the Helmholtz equation given by Eq. (4.1-34) with $B_m = 0$ as the starting point. Since the value of the velocity potential is also required at $\theta = 0$ and $\theta = \pi$, we set the constant $D_{mn} = 0$ in Eq. (4.1-34) as well. With the use of these two observations, Eq. (4.1-34) reduces to

$$\varphi_{f,mn}(r,\theta,\psi) = j_m(kr)P_m^n(\cos\theta)\left[A_{mn}\cos(n\psi) + B_{mn}\sin(n\psi)\right],$$

$$m,n = 0,1,2,\ldots, \quad (4.3\text{-}1)$$

where

$$A_{mn} = A_m C_{mn} E_n, \quad (4.3\text{-}2)$$

$$B_{mn} = A_m C_{mn} F_n, \quad (4.3\text{-}3)$$

and

$$k = 2\pi f/c = 2\pi/\lambda. \quad (4.3\text{-}4)$$

The values of the unknown constants A_{mn} and B_{mn} are determined by satisfying the boundary conditions.

The time-harmonic acoustic fluid velocity vector is given by

$$\mathbf{u}(t,\mathbf{r}) = \mathbf{u}_f(\mathbf{r})\exp(+j2\pi ft), \quad (4.3\text{-}5)$$

where

$$\mathbf{u}_f(\mathbf{r}) = \nabla\varphi_f(\mathbf{r}) \quad (4.3\text{-}6)$$

and

$$\nabla = \frac{\partial}{\partial r}\hat{r} + \frac{1}{r}\frac{\partial}{\partial\theta}\hat{\theta} + \frac{1}{r\sin\theta}\frac{\partial}{\partial\psi}\hat{\psi} \quad (4.3\text{-}7)$$

is the gradient expressed in the spherical coordinates (r,θ,ψ). The normal component of the acoustic fluid velocity vector can be obtained from Eq. (4.3-5) as follows:

$$u_n(t,\mathbf{r}) = \hat{n}\cdot\mathbf{u}(t,\mathbf{r}) = u_{f,n}(\mathbf{r})\exp(+j2\pi ft), \quad (4.3\text{-}8)$$

where

$$u_{f,n}(\mathbf{r}) = \hat{n}\cdot\nabla\varphi_f(\mathbf{r}), \quad (4.3\text{-}9)$$

and $\hat{n}$ is a unit vector normal (perpendicular) to the boundary.

Since

$$\hat{n} = \hat{r} \tag{4.3-10}$$

is a unit vector that is normal to the boundary at $r = a$, and since the normal component of the acoustic fluid velocity vector must be equal to zero on the rigid boundary at $r = a$, Eq. (4.3-9) becomes

$$u_{f,n}(a,\theta,\psi) = \frac{\partial}{\partial r}\varphi_f(r,\theta,\psi)\bigg|_{r=a} = 0. \tag{4.3-11}$$

Substituting Eq. (4.3-1) into Eq. (4.3-11) yields

$$kj'_m(ka)P_m^n(\cos\theta)\big[A_{mn}\cos(n\psi) + B_{mn}\sin(n\psi)\big] = 0, \tag{4.3-12}$$

where

$$j'_m(ka) = \frac{d}{d(kr)}j_m(kr)\bigg|_{r=a}. \tag{4.3-13}$$

Since Eq. (4.3-12) must hold for all allowed values of θ and ψ at $r = a$,

$$j'_m(ka) = 0. \tag{4.3-14}$$

Note that for a given value of integer m, there are infinitely many values of ka that will satisfy Eq. (4.3-14). If we let $\alpha_{m,l}$ represent the roots of the equation

$$j'_m(\alpha_{m,l}) = 0, \qquad m,l = 0,1,2,\ldots, \tag{4.3-15}$$

where the index l counts the number of roots for a given value of m, then

$$ka = \alpha_{m,l}, \qquad m,l = 0,1,2,\ldots, \tag{4.3-16}$$

or

$$\boxed{k = k_{m,l} = \frac{\alpha_{m,l}}{a}, \qquad m,l = 0,1,2,\ldots,} \tag{4.3-17}$$

and

$$\boxed{f = f_{m,l} = \frac{c}{2\pi a}\alpha_{m,l}, \qquad m,l = 0,1,2,\ldots.} \tag{4.3-18}$$

Therefore, in order to satisfy the boundary condition at $r = a$, the frequency of vibration (in hertz) of the acoustic field inside a spherical cavity with a rigid wall is only allowed certain discrete values rather than a continuum of values. These values are given by Eq. (4.3-18) and are known as the *eigenfrequencies* or natural frequencies of vibration.

Therefore, the time-harmonic velocity potential inside a spherical cavity with a rigid wall corresponding to the (m, n, l) normal mode is given by

$$\varphi_{mn,l}(t, \mathbf{r}) = \varphi_{f,mn,l}(\mathbf{r}) \exp(+j2\pi f_{m,l} t), \qquad m, n, l = 0, 1, 2, \ldots, \quad (4.3\text{-}19)$$

where, upon substituting Eq. (4.3-17) into Eq. (4.3-1), we obtain

$$\varphi_{f,mn,l}(r, \theta, \psi) = j_m(\alpha_{m,l} r/a) P_m^n(\cos \theta) \left[A_{mn,l} \cos(n\psi) + B_{mn,l} \sin(n\psi) \right],$$

$$m, n, l = 0, 1, 2, \ldots. \quad (4.3\text{-}20)$$

The complete time-harmonic normal-mode solution for the velocity potential is obtained by summing the contributions from *all* the normal modes, that is, with the use of Eq. (4.2-4),

$$\varphi(t, \mathbf{r}) = \sum_{m=0}^{\infty} \sum_{n=0}^{m} \sum_{l=0}^{\infty} \varphi_{f,mn,l}(\mathbf{r}) \exp(+j2\pi f_{m,l} t), \qquad (4.3\text{-}21)$$

and upon substituting Eq. (4.3-20) into Eq. (4.3-21), we obtain

$$\varphi(t, r, \theta, \psi) = \sum_{m=0}^{\infty} \sum_{n=0}^{m} \sum_{l=0}^{\infty} j_m(\alpha_{m,l} r/a) P_m^n(\cos \theta)$$

$$\times \left[A_{mn,l} \cos(n\psi) + B_{mn,l} \sin(n\psi) \right] \exp(+j2\pi f_{m,l} t),$$

$$(4.3\text{-}22)$$

where $f_{m,l}$ is given by Eq. (4.3-18).

With the use of Eq. (2.2-28), the time-harmonic acoustic pressure inside a spherical cavity with a rigid wall corresponding to the (m, n, l) normal mode can be expressed as

$$p_{mn,l}(t, \mathbf{r}) = -j2\pi f_{m,l} \rho_0 \varphi_{mn,l}(t, \mathbf{r}), \qquad m, n, l = 0, 1, 2, \ldots. \quad (4.3\text{-}23)$$

Therefore, the complete time-harmonic normal-mode solution for the acoustic pressure is given by

$$p(t, \mathbf{r}) = -j2\pi \rho_0 \sum_{m=0}^{\infty} \sum_{n=0}^{m} \sum_{l=0}^{\infty} f_{m,l} \varphi_{f,mn,l}(\mathbf{r}) \exp(+j2\pi f_{m,l} t), \quad (4.3\text{-}24)$$

and upon substituting Eq. (4.3-20) into Eq. (4.3-24), we obtain

$$
p(t,r,\theta,\psi) = -j2\pi\rho_0 \sum_{m=0}^{\infty} \sum_{n=0}^{m} \sum_{l=0}^{\infty} f_{m,l} j_m(\alpha_{m,l} r/a) P_m^n(\cos\theta)
$$

$$
\times \left[A_{mn,l} \cos(n\psi) + B_{mn,l} \sin(n\psi) \right] \exp(+j2\pi f_{m,l}t),
$$

(4.3-25)

where $f_{m,l}$ is given by Eq. (4.3-18).

Let us conclude this section with the following additional remarks. Although we have satisfied the boundary condition at the surface inside the cavity, the constants $A_{mn,l}$ and $B_{mn,l}$ are still unknown. In order to determine their values, the boundary condition at the location of a sound source inside the cavity must be satisfied.

4.4 Scattering by a Sphere

In this section we shall compute the velocity potential of the *scattered* acoustic field due to a time-harmonic plane wave incident upon a fixed (motionless) sphere of radius a meters. With appropriate changes in boundary conditions, the procedures developed in this section can be adapted in order to solve problems, for example, involving scattering from air or gas bubbles in underwater acoustics, or from blood cells in medical ultrasonics.

The time-harmonic velocity potential of the incident plane wave can be expressed as follows:

$$
\varphi_i(t,\mathbf{R}) = \varphi_{f,i}(\mathbf{R}) \exp(+j2\pi ft),
$$

(4.4-1)

where

$$
\varphi_{f,i}(\mathbf{R}) = A_i \exp(-j\mathbf{k}_i \cdot \mathbf{R})
$$

(4.4-2)

is the spatial-dependent part of the velocity potential; A_i is the amplitude of the incident plane wave;

$$
\mathbf{k}_i = k\hat{n}_i
$$

(4.4-3)

is the incident propagation vector;

$$
k = 2\pi f/c = 2\pi/\lambda
$$

(4.4-4)

is the wave number in radians per meter;

$$\hat{n}_i = u_i \hat{x} + v_i \hat{y} + w_i \hat{z} \qquad (4.4\text{-}5)$$

is the unit vector in the direction of the incident propagation vector, where

$$u_i = \sin \theta_i \cos \psi_i, \qquad (4.4\text{-}6)$$

$$v_i = \sin \theta_i \sin \psi_i, \qquad (4.4\text{-}7)$$

and

$$w_i = \cos \theta_i \qquad (4.4\text{-}8)$$

are the dimensionless direction cosines with respect to the X, Y, and Z axes, respectively; and

$$\mathbf{R} = x\hat{x} + y\hat{y} + z\hat{z}, \qquad (4.4\text{-}9)$$

or $\mathbf{R} = (x, y, z)$, is the position vector measured from the origin of the coordinate system to some arbitrary field point with rectangular coordinates (x, y, z) or, equivalently, with spherical coordinates (r, θ, ψ) (see Fig. 4.4-1).

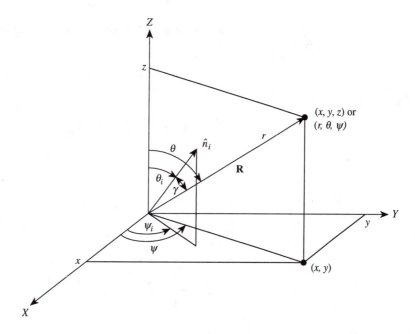

Figure 4.4-1. Unit vector $\hat{n}_i$ and position vector $\mathbf{R}$ with corresponding angles.

The first step in solving this scattering problem is to express $\varphi_{f,i}(\mathbf{R})$ given by Eq. (4.4-2) in terms of the spherical coordinates (r, θ, ψ). Substituting Eq. (4.4-3) and Eqs. (4.4-5) through (4.4-9) into Eq. (4.4-2) yields

$$\varphi_{f,i}(x, y, z) = A_i \exp\{-jk[\sin \theta_i (x \cos \psi_i + y \sin \psi_i) + z \cos \theta_i]\}, \quad (4.4\text{-}10)$$

and since (see Fig. 4.4-1)

$$x = ru = r \sin \theta \cos \psi, \qquad (4.4\text{-}11)$$

$$y = rv = r \sin \theta \sin \psi, \qquad (4.4\text{-}12)$$

and

$$z = rw = r \cos \theta, \qquad (4.4\text{-}13)$$

substituting Eqs. (4.4-11) through (4.4-13) into Eq. (4.4-10) yields

$$\varphi_{f,i}(r, \theta, \psi) = A_i \exp\{-jkr[\sin \theta \sin \theta_i \cos(\psi - \psi_i) + \cos \theta \cos \theta_i]\}, \quad (4.4\text{-}14)$$

where use has been made of the following trigonometric identity:

$$\cos(\alpha - \beta) = \cos \alpha \cos \beta + \sin \alpha \sin \beta. \qquad (4.4\text{-}15)$$

Since

$$\mathbf{k}_i \cdot \mathbf{R} = k\hat{n}_i \cdot \mathbf{R} = kr \cos \gamma, \qquad (4.4\text{-}16)$$

where $|\mathbf{R}| = r$ and γ is the angle between vectors $\hat{n}_i$ and $\mathbf{R}$ (see Fig. 4.4-1), Eq. (4.4-14) can be rewritten as

$$\varphi_{f,i}(r, \theta, \psi) = A_i \exp(-jkr \cos \gamma), \qquad (4.4\text{-}17)$$

where

$$\cos \gamma = \sin \theta \sin \theta_i \cos(\psi - \psi_i) + \cos \theta \cos \theta_i. \qquad (4.4\text{-}18)$$

In order to proceed further, we shall make use of the following identity:

$$\exp(-jkr \cos \gamma) = \sum_{m=0}^{\infty} (-j)^m (2m + 1) j_m(kr) \Bigg[P_m(\cos \theta_i) P_m(\cos \theta) $$
$$+ 2 \sum_{n=1}^{m} \frac{(m - n)!}{(m + n)!} P_m^n(\cos \theta_i) P_m^n(\cos \theta) \cos[n(\psi - \psi_i)] \Bigg].$$

$$(4.4\text{-}19)$$

Therefore, with the use of Eqs. (4.4-17) and (4.4-19), the velocity potential of the incident, time-harmonic plane wave can be expressed in terms of the spherical coordinates (r, θ, ψ) as follows:

$$\varphi_i(t, \mathbf{r}) = \varphi_{f,i}(\mathbf{r}) \exp(+j2\pi ft), \qquad (4.4\text{-}20)$$

where the position vector $\mathbf{r} = (r, \theta, \psi)$, and

$$
\begin{aligned}
\varphi_{f,i}(r, \theta, \psi) = A_i \sum_{m=0}^{\infty} (-j)^m (2m+1) j_m(kr) &\left[P_m(\cos \theta_i) P_m(\cos \theta) \right. \\
&\left. + 2 \sum_{n=1}^{m} \frac{(m-n)!}{(m+n)!} P_m^n(\cos \theta_i) P_m^n(\cos \theta) \cos[n(\psi - \psi_i)] \right].
\end{aligned}
$$

$$(4.4\text{-}21)$$

In addition to the incident plane wave, there is present a *scattered*, outgoing wave traveling away from the sphere. Since the scattered acoustic field is analogous to the acoustic field radiated by a vibrating sphere, by referring to Section 4.2.1, the velocity potential of the time-harmonic scattered acoustic field can be expressed as

$$\varphi_s(t, \mathbf{r}) = \varphi_{f,s}(\mathbf{r}) \exp(+j2\pi ft), \qquad (4.4\text{-}22)$$

where

$$\varphi_{f,s}(r, \theta, \psi) = \sum_{m=0}^{\infty} \sum_{n=0}^{m} h_m^{(2)}(kr) P_m^n(\cos \theta) \left[A_{mn} \cos(n\psi) + B_{mn} \sin(n\psi) \right],$$

$$(4.4\text{-}23)$$

or

$$
\begin{aligned}
\varphi_{f,s}(r, \theta, \psi) = \sum_{m=0}^{\infty} h_m^{(2)}(kr) &\left[A_m P_m(\cos \theta) \right. \\
&\left. + \sum_{n=1}^{m} P_m^n(\cos \theta) \left[A_{mn} \cos(n\psi) + B_{mn} \sin(n\psi) \right] \right],
\end{aligned}
$$

$$(4.4\text{-}24)$$

where

$$k = 2\pi f/c = 2\pi/\lambda. \qquad (4.4\text{-}25)$$

If the sphere is assumed to be *rigid*, then the normal component of the *total* acoustic fluid velocity vector must be equal to zero on the surface of the sphere, that is,

$$u_n(t,r,\theta,\psi)\big|_{r=a} = \hat{n} \cdot \mathbf{u}(t,r,\theta,\psi)\big|_{r=a} = \hat{n} \cdot \nabla\varphi(t,r,\theta,\psi)\big|_{r=a} = 0, \quad (4.4\text{-}26)$$

where $\hat{n}$ is a unit vector normal to the surface of the sphere, and

$$\varphi(t,\mathbf{r}) = \varphi_i(t,\mathbf{r}) + \varphi_s(t,\mathbf{r}). \tag{4.4-27}$$

Since

$$\hat{n} = \hat{r}, \tag{4.4-28}$$

substituting Eqs. (4.2-9), (4.4-27), and (4.4-28) into Eq. (4.4-26) yields the following boundary condition:

$$\frac{\partial}{\partial r}\varphi_i(t,r,\theta,\psi)\bigg|_{r=a} = -\frac{\partial}{\partial r}\varphi_s(t,r,\theta,\psi)\bigg|_{r=a}. \tag{4.4-29}$$

Substituting Eqs. (4.4-20) and (4.4-21) into the left-hand side of Eq. (4.4-29) yields

$$\frac{\partial}{\partial r}\varphi_i(t,r,\theta,\psi)\bigg|_{r=a} = kA_i \sum_{m=0}^{\infty} (-j)^m (2m+1) j'_m(ka)$$

$$\times \left[P_m(\cos\theta_i)P_m(\cos\theta) + 2\sum_{n=1}^{m} \frac{(m-n)!}{(m+n)!} P_m^n(\cos\theta_i)P_m^n(\cos\theta) \right.$$

$$\left. \times [\cos(n\psi)\cos(n\psi_i) + \sin(n\psi)\sin(n\psi_i)] \right] \exp(+j2\pi ft),$$

$$(4.4\text{-}30)$$

where

$$j'_m(ka) = \frac{d}{d(kr)} j_m(kr)\bigg|_{r=a}, \qquad m = 0,1,2,\ldots, \tag{4.4-31}$$

and

$$\cos[n(\psi - \psi_i)] = \cos(n\psi)\cos(n\psi_i) + \sin(n\psi)\sin(n\psi_i). \tag{4.4-32}$$

And upon substituting Eqs. (4.4-22) and (4.4-24) into the right-hand side of Eq. (4.4-29), we obtain

$$-\frac{\partial}{\partial r}\varphi_s(t, r, \theta, \psi)\Bigg|_{r=a} = -k \sum_{m=0}^{\infty} h_m^{(2)'}(ka)\Bigg[A_m P_m(\cos\theta) + \sum_{n=1}^{m} P_m^n(\cos\theta)$$

$$\times\big[A_{mn}\cos(n\psi) + B_{mn}\sin(n\psi)\big]\Bigg]\exp(+j2\pi ft),$$

(4.4-33)

where

$$h_m^{(2)'}(ka) = \frac{d}{d(kr)}h_m^{(2)}(kr)\Bigg|_{r=a}, \qquad m = 0, 1, 2, \dots . \qquad (4.4\text{-}34)$$

Since Eqs. (4.4-30) and (4.4-33) must be equal,

$$A_m = \frac{-(-j)^m(2m+1)A_i j_m'(ka)P_m(\cos\theta_i)}{h_m^{(2)'}(ka)}, \qquad m = 0, 1, 2, \dots, \qquad (4.4\text{-}35)$$

$$A_{mn} = \frac{-(-j)^m(2m+1)(m-n)!\,2A_i j_m'(ka)P_m^n(\cos\theta_i)\cos(n\psi_i)}{(m+n)!\,h_m^{(2)'}(ka)},$$

$$m = 1, 2, 3, \dots, \quad n = 1, 2, \dots, m, \qquad (4.4\text{-}36)$$

$$B_{mn} = \frac{-(-j)^m(2m+1)(m-n)!\,2A_i j_m'(ka)P_m^n(\cos\theta_i)\sin(n\psi_i)}{(m+n)!\,h_m^{(2)'}(ka)},$$

$$m = 1, 2, 3, \dots, \quad n = 1, 2, \dots, m, \qquad (4.4\text{-}37)$$

and

$$A_{0n} = B_{0n} = 0, \qquad n > 0, \qquad (4.4\text{-}38)$$

since [see Eq. (4.1-57)]

$$P_0^n(\cos\theta_i) = 0, \qquad n > 0. \qquad (4.4\text{-}39)$$

Therefore, upon substituting Eq. (4.4-38) into Eq. (4.4-24), we obtain the following expression for the spatial-dependent part of the velocity potential of the scattered acoustic field due to a time-harmonic plane wave incident upon a fixed (motionless), *rigid* sphere:

$$\varphi_{f,s}(r,\theta,\psi) = \sum_{m=0}^{\infty} A_m h_m^{(2)}(kr) P_m(\cos\theta)$$

$$+ \sum_{m=1}^{\infty} \sum_{n=1}^{m} h_m^{(2)}(kr) P_m^n(\cos\theta) \left[A_{mn}\cos(n\psi) + B_{mn}\sin(n\psi) \right],$$

(4.4-40)

where A_m, A_{mn}, and B_{mn} are given by Eqs. (4.4-35) through (4.4-37), respectively, and k is given by Eq. (4.4-4). Note that the first term on the right-hand side of Eq. (4.4-40) is independent of the azimuthal angle ψ.

The corresponding time-harmonic scattered acoustic pressure field and scattered acoustic fluid velocity vector can be obtained from the following equations:

$$p_s(t,\mathbf{r}) = p_{f,s}(\mathbf{r})\exp(+j2\pi ft),$$

(4.4-41)

where

$$p_{f,s}(\mathbf{r}) = -jk\rho_0 c\varphi_{f,s}(\mathbf{r}),$$

(4.4-42)

and

$$\mathbf{u}_s(t,\mathbf{r}) = \mathbf{u}_{f,s}(\mathbf{r})\exp(+j2\pi ft),$$

(4.4-43)

where

$$\mathbf{u}_{f,s}(\mathbf{r}) = \nabla\varphi_{f,s}(\mathbf{r}).$$

(4.4-44)

Recall that the total velocity potential is equal to the sum of the incident and scattered velocity potentials [see Eq. (4.4-27)].

Scattering Cross Section

The last topic to be discussed in this section is the *scattering cross section* of the sphere. The scattering cross section σ_s is defined as that *area* (in square meters) of the incident wavefront that transmits a power equal to the time-average scattered power, that is,

$$I_{\mathrm{avg}_i}\sigma_s = P_{\mathrm{avg}_s},$$

(4.4-45)

or

$$\boxed{\sigma_s = \frac{P_{\text{avg}_s}}{I_{\text{avg}_i}},}$$

(4.4-46)

where

$$P_{\text{avg}_s} = \oint_S \mathbf{I}_{\text{avg}_s}(r, \theta, \psi) \cdot \mathbf{dS}$$

(4.4-47)

is the time-average scattered power,

$$\mathbf{I}_{\text{avg}_s}(r, \theta, \psi) = \tfrac{1}{2} \operatorname{Re}\{p_{f,s}(r, \theta, \psi)\mathbf{u}_{f,s}^*(r, \theta, \psi)\}$$

(4.4-48)

is the time-average intensity vector of the scattered acoustic field,

$$\mathbf{dS} = r^2 \sin \theta \, d\theta \, d\psi \, \hat{r},$$

(4.4-49)

and

$$I_{\text{avg}_i} = \tfrac{1}{2}k^2 \rho_0 c \, |A_i|^2$$

(4.4-50)

is the magnitude of the time-average intensity vector of the time-harmonic incident plane wave (see Example 2.2-1), where A_i is the amplitude of the velocity potential of the incident plane wave [see Eq. (4.4-2)].

Example 4.4-1

In this example we shall obtain the expression for the time-harmonic velocity potential of the scattered acoustic field due to a time-harmonic plane wave traveling in the positive Z direction, perpendicular to the XY plane, incident upon a fixed (motionless), rigid sphere. The incident spherical angles that correspond to this problem are $\theta_i = 0$ and $\psi_i = 0$ (see Fig. 4.4-2). Since $\theta_i = 0$ [see Eqs. (4.1-60) through (4.1-63)],

$$P_m(\cos \theta_i) = P_m(1) = 1, \qquad m = 0, 1, 2, \ldots, \qquad (4.4\text{-}51)$$

and [see Eq. (4.1-58)]

$$P_m^n(\cos \theta_i) = P_m^n(1) = 0, \qquad m = 0, 1, 2, \ldots, \quad n = 1, 2, 3, \ldots .$$

(4.4-52)

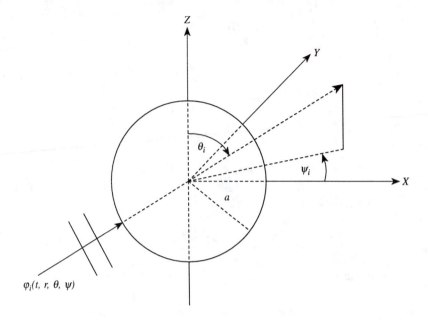

Figure 4.4-2. Time-harmonic plane wave incident upon a sphere of radius a meters.

Therefore, upon substituting Eq. (4.4-51) into Eq. (4.4-35), and substituting Eq. (4.4-52) into Eqs. (4.4-36) and (4.4-37), we obtain

$$
A_m = -(-j)^m (2m + 1) A_i \frac{j'_m(ka)}{h_m^{(2)\prime}(ka)}, \qquad m = 0, 1, 2, \ldots,
$$

$$(4.4\text{-}53)$$

$$
A_{mn} = 0, \qquad m = 1, 2, 3, \ldots, \qquad n = 1, 2, \ldots, m, \qquad (4.4\text{-}54)
$$

and

$$
B_{mn} = 0, \qquad m = 1, 2, 3, \ldots, \qquad n = 1, 2, \ldots, m. \qquad (4.4\text{-}55)
$$

And finally, substituting Eqs. (4.4-53) through (4.4-55) into Eq. (4.4-40) yields

$$\varphi_{f,s}(r,\theta,\psi) = -A_i \frac{j_0'(ka)}{h_0^{(2)'}(ka)} h_0^{(2)}(kr)$$

$$-A_i \sum_{m=1}^{\infty} (-j)^m (2m+1) \frac{j_m'(ka)}{h_m^{(2)'}(ka)} h_m^{(2)}(kr) P_m(\cos\theta),$$

(4.4-56)

where $P_0(\cos\theta) = 1$ [see Eq. (4.1-60)]. The time-harmonic velocity potential of the scattered acoustic field is obtained by substituting Eq. (4.4-56) into Eq. (4.4-22). Note that the first term on the right-hand side of Eq. (4.4-56) is the *omnidirectional term* (i.e., no angular dependence). Also note that for this problem, the acoustic field at $\theta = 0°$ corresponds to *forward scattering*, whereas the acoustic field at $\theta = 180°$ corresponds to *backscatter*.

Problems

4-1 Consider a spherical sound source in air with $\rho_0 = 1.21$ kg/m^3 and $c = 343$ m/sec. If the sphere is in the monopole mode of vibration with $a = 0.1$ m, $V_0 = 1$ m/sec, and $f = 100$ Hz, then compute
(a) the source strength,
(b) the time-average radiated power, and
(c) the source level (see Section 1.6)

$$SL \triangleq 20\log_{10}\left(\frac{\sqrt{2}\,P_0/2}{P_{ref}}\right) \text{ dB re } P_{ref},$$

where

$$P_0 = \left| p_f(r,\theta,\psi) \right|_{r=a+1}$$

is the *peak acoustic pressure amplitude* in pascals measured at a distance of 1 m from the source along its *acoustic axis*, that is, in the direction of the maximum response of the source, and $P_{ref} = 1\ \mu$ Pa (rms) is the *rms reference pressure* to be used in this problem. Note that P_0 can be measured in any direction, since the sound source in this problem is omnidirectional.

4-2 Repeat parts (b) and (c) in Problem 4-1 for a spherical sound source in sea water with $\rho_0 = 1026$ kg/m^3 and $c = 1500$ m/sec.

4-3 For a vibrating sphere of radius a meters in the dipole mode of vibration, compute
(a) the acoustic pressure,
(b) the acoustic fluid velocity vector,
(c) the time-average intensity vector, and
(d) the time-average radiated power.

4-4 Using the velocity potential of the scattered acoustic field derived in Example 4.4-1, compute
(a) the time-average intensity vector in the radial direction and
(b) the time-average power.
Hint: $\mathrm{Re}\{-jh_m^{(2)}(kr)[h_m^{(2)'}(kr)]^*\} = 1/(kr)^2$.

4-5 Find the time-harmonic velocity potential of the scattered acoustic field due to a time-harmonic plane wave traveling in the positive Z direction, perpendicular to the XY plane, incident upon a fixed (motionless), *pressure-release* sphere of radius a meters. The incident spherical angles that correspond to this problem are $\theta_i = 0$ and $\psi_i = 0$ (see Fig. 4.4-2).

Bibliography

M. Abramowitz and I. A. Stegun, editors, *Handbook of Mathematical Functions*, 9th printing, Dover, New York.

L. E. Kinsler, A. R. Frey, A. B. Coppens, and J. V. Sanders, *Fundamentals of Acoustics*, 3rd ed., Wiley, New York, 1982.

P. M. Morse and K. U. Ingard, *Theoretical Acoustics*, Princeton University Press, Princeton, New Jersey, 1987.

T. Myint-U, *Partial Differential Equations of Mathematical Physics*, 2nd ed., North Holland, New York, 1980.

J. A. Stratton, *Electromagnetic Theory*, McGraw-Hill, New York, 1941.

S. Temkin, *Elements of Acoustics*, Wiley, New York, 1981.

Chapter 5

Wave Propagation in Inhomogeneous Media

5.1 The WKB Approximation

In Chapters 2 through 4, various wave propagation problems were solved in the rectangular, cylindrical, and spherical coordinate systems, respectively, when the speed of sound in the fluid medium was constant. Recall that when the speed of sound is constant, the fluid medium is referred to as a homogeneous medium. Now, in this chapter, we shall discuss three different methods of obtaining approximate solutions of the wave equation when the speed of sound is a function of position. When the speed of sound is a function of position, the fluid medium is referred to as an *inhomogeneous* medium. For example, the speed of sound in the ocean is a monotonically increasing function of temperature, salinity, and pressure, that is, it is a function of position in the medium. Its value generally lies between 1450 and 1540 m/sec. In many ocean acoustics problems, the speed of sound is modeled as a function of depth. However, in some cases, it is modeled as a function of both depth and horizontal range, and sometimes it is even modeled as a function of all three spatial coordinates. In some air acoustics problems, the speed of sound is modeled as a function of elevation, or height above the earth's surface.

The first method to be discussed is the *geometrical optics approximation*, or *WKB* (*Wentzel-Kramers-Brillouin*) *approximation*. The remaining two methods, *ray acoustics* and the *parabolic equation approximation*, are discussed in Sections 5.2 and 5.3, respectively.

Let us begin by considering the solution of the following linear, three-dimensional, lossless, homogeneous wave equation that describes the propagation of

307

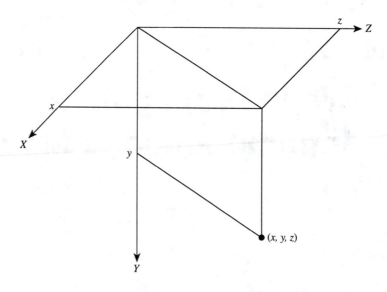

Figure 5.1-1. The rectangular coordinates (x, y, z).

small-amplitude acoustic signals in ideal (nonviscous) fluids (see Section 1.4):

$$\nabla^2 \varphi(t, \mathbf{r}) - \frac{1}{c^2(\mathbf{r})} \frac{\partial^2}{\partial t^2} \varphi(t, \mathbf{r}) = 0, \tag{5.1-1}$$

where

$$\nabla^2 = \frac{\partial^2}{\partial x^2} + \frac{\partial^2}{\partial y^2} + \frac{\partial^2}{\partial z^2} \tag{5.1-2}$$

is the *Laplacian* expressed in the rectangular coordinates (x, y, z), $\varphi(t, \mathbf{r})$ is the *velocity potential* at time t and position $\mathbf{r} = (x, y, z)$ with units of square meters per second, and $c(\mathbf{r})$ is the speed of sound in meters per second, also shown as a function of position $\mathbf{r}$. The rectangular coordinates x (cross-range), y (depth), and z (down-range) are illustrated in Fig. 5.1-1. The wave equation given by Eq. (5.1-1) is referred to as being homogeneous, since there is no source distribution (i.e., no input or forcing function).

We begin the solution of Eq. (5.1-1) by assuming that the velocity potential $\varphi(t, \mathbf{r})$ has a time-harmonic dependence, that is,

$$\varphi(t, \mathbf{r}) = \varphi_f(\mathbf{r}) \exp(+j2\pi ft), \tag{5.1-3}$$

where $\varphi_f(\mathbf{r})$ is the spatial-dependent part of the velocity potential and f is the frequency in hertz. A time-harmonic field is one whose value at any time and point in space depends on a single frequency component. Substituting Eq. (5.1-3) into the time-dependent wave equation given by Eq. (5.1-1) yields the time-independent *lossless Helmholtz equation*

$$\nabla^2\varphi_f(\mathbf{r}) + k^2(\mathbf{r})\varphi_f(\mathbf{r}) = 0, \qquad (5.1\text{-}4)$$

where

$$k(\mathbf{r}) = 2\pi f/c(\mathbf{r}) \qquad (5.1\text{-}5)$$

is the *wave number* in radians per meter. Equation (5.1-4) can also be expressed in terms of the dimensionless *index of refraction*

$$\boxed{n(\mathbf{r}) = c_0/c(\mathbf{r}),} \qquad (5.1\text{-}6)$$

where c_0 is the constant *reference speed of sound* at a source location $\mathbf{r}_0 = (x_0, y_0, z_0)$, that is, $c_0 = c(\mathbf{r}_0)$. Note that $n(\mathbf{r}_0) = 1$. If we let

$$k_0 = 2\pi f/c_0 \qquad (5.1\text{-}7)$$

be the *reference wave number*, then Eq. (5.1-5) can be rewritten as

$$\boxed{k(\mathbf{r}) = k_0 n(\mathbf{r}).} \qquad (5.1\text{-}8)$$

Therefore, substituting Eq. (5.1-8) into Eq. (5.1-4) yields

$$\nabla^2\varphi_f(\mathbf{r}) + k_0^2 n^2(\mathbf{r})\varphi_f(\mathbf{r}) = 0, \qquad (5.1\text{-}9)$$

which is an alternative way of expressing the Helmholtz equation. If we can solve the Helmholtz equation given by either Eq. (5.1-4) or Eq. (5.1-9), then substituting the solution $\varphi_f(\mathbf{r})$ into Eq. (5.1-3) will yield a time-harmonic solution of the wave equation given by Eq. (5.1-1), which is the main objective in this chapter. Time-harmonic solutions are very important, because solutions of the wave equation with arbitrary time dependence can be obtained from time-harmonic solutions by using Fourier transform techniques, as was discussed in Section 2.2.3 and Examples 3.8-1, 3.8-2, and 3.9-1.

The WKB method addresses the problem of trying to solve the Helmholtz equation when the speed of sound is a function of only *one* spatial coordinate. In this section, we shall assume that the speed of sound is an arbitrary function of depth y. Therefore, Eq. (5.1-4) reduces to

$$\nabla^2 \varphi_f(x, y, z) + k^2(y) \varphi_f(x, y, z) = 0, \qquad (5.1\text{-}10)$$

where

$$k(y) = k_0 n(y) = 2\pi f/c(y). \qquad (5.1\text{-}11)$$

The solution of Eq. (5.1-10) will be obtained by using the *method of separation of variables*, that is, we assume a solution of the form (see Section 2.1)

$$\boxed{\varphi_f(x, y, z) = X(x)Y(y)Z(z).} \qquad (5.1\text{-}12)$$

This method can be used because the speed of sound is only a function of *one* spatial coordinate. Substituting Eq. (5.1-12) into Eq. (5.1-10) yields

$$\frac{1}{Y(y)} \frac{d^2}{dy^2} Y(y) + k^2(y) = -\frac{1}{X(x)} \frac{d^2}{dx^2} X(x) - \frac{1}{Z(z)} \frac{d^2}{dz^2} Z(z), \qquad (5.1\text{-}13)$$

where the second-order partial derivatives with respect to x, y, and z have been replaced by ordinary second-order derivatives, since the functions $X(x)$, $Y(y)$, and $Z(z)$ depend on only one of the independent variables x, y, and z, respectively. Since the left-hand side of Eq. (5.1-13) is a function of y and the right-hand side is a function of both x and z, equality is possible only if both sides of Eq. (5.1-13) are equal to a constant (call it κ^2), which is referred to as a *separation constant*. Therefore, since

$$-\frac{1}{X(x)} \frac{d^2}{dx^2} X(x) - \frac{1}{Z(z)} \frac{d^2}{dz^2} Z(z) = \kappa^2, \qquad (5.1\text{-}14)$$

Eq. (5.1-14) can be rewritten as

$$-\frac{1}{X(x)} \frac{d^2}{dx^2} X(x) = \frac{1}{Z(z)} \frac{d^2}{dz^2} Z(z) + \kappa^2 = k_X^2, \qquad (5.1\text{-}15)$$

since equality is possible only if both sides of Eq. (5.1-15) are equal to a constant, which we have called k_X^2, and which is also referred to as a separation constant. From Eq. (5.1-15) we obtain the following second-order, homogeneous (no input or forcing function), ordinary differential equation (ODE):

$$\frac{d^2}{dx^2} X(x) + k_X^2 X(x) = 0, \qquad (5.1\text{-}16)$$

which has the *exact* solution

$$X(x) = A_X \exp[-jk_X(x - x_0)] + B_X \exp[+jk_X(x - x_0)], \quad (5.1\text{-}17)$$

where A_X and B_X are complex constants in general, whose values are determined by satisfying boundary conditions, and x_0 is the x coordinate of a sound source. If k_X is *positive*, then the first term on the right-hand side of Eq. (5.1-17) represents a plane wave traveling in the *positive* X direction, whereas the second term represents a plane wave traveling in the *negative* X direction, since we chose $\exp(+j2\pi ft)$ as our time-harmonic dependence (see Section 2.1). The constant k_X is the propagation-vector component in the X direction with units of radians per meter (see Section 2.2.1).

Also from Eq. (5.1-15), we have the additional second-order homogeneous ODE

$$\frac{d^2}{dz^2} Z(z) + k_Z^2 Z(z) = 0, \quad (5.1\text{-}18)$$

where

$$k_Z^2 = \kappa^2 - k_X^2 \quad (5.1\text{-}19)$$

is a constant, since κ^2 and k_X^2 are constants. Equation (5.1-18) has the *exact* solution

$$Z(z) = A_Z \exp[-jk_Z(z - z_0)] + B_Z \exp[+jk_Z(z - z_0)], \quad (5.1\text{-}20)$$

where A_Z and B_Z are complex constants in general, whose values are determined by satisfying boundary conditions, and z_0 is the z coordinate of a sound source. If k_Z is *positive*, then the first term on the right-hand side of Eq. (5.1-20) represents a plane wave traveling in the *positive* Z direction, whereas the second term represents a plane wave traveling in the *negative* Z direction. The constant k_Z is the propagation-vector component in the Z direction with units of radians per meter (see Section 2.2.1).

Finally, the left-hand side of Eq. (5.1-13) must also be equal to κ^2, that is,

$$\frac{1}{Y(y)} \frac{d^2}{dy^2} Y(y) + k^2(y) = \kappa^2, \quad (5.1\text{-}21)$$

and with the use of Eq. (5.1-19), we obtain the following second-order homogeneous ODE for $Y(y)$:

$$\frac{d^2}{dy^2}Y(y) + k_Y^2(y)Y(y) = 0, \qquad (5.1\text{-}22)$$

where

$$k_Y^2(y) = k^2(y) - k_X^2 - k_Z^2. \qquad (5.1\text{-}23)$$

Note that the *propagation vector* $\mathbf{k}(y)$ can be written as

$$\boxed{\mathbf{k}(y) = k_X \hat{x} + k_Y(y)\hat{y} + k_Z \hat{z},} \qquad (5.1\text{-}24)$$

where k_X and k_Z are the *constant X and Z components*, respectively, and $k_Y(y)$ is the Y component and is a function of depth. Also note that [see Eqs. (5.1-23) and (5.1-24)]

$$|\mathbf{k}(y)| = k(y) = 2\pi f/c(y). \qquad (5.1\text{-}25)$$

In order to obtain a solution of Eq. (5.1-22), we shall follow the WKB method, which begins by assuming that

$$\boxed{Y(y) = a_Y(y)\exp\left[+j\theta_Y(y)\right],} \qquad (5.1\text{-}26)$$

where $a_Y(y)$ and $\theta_Y(y)$ are *real amplitude and phase functions*, respectively. Substituting Eq. (5.1-26) into Eq. (5.1-22) yields

$$a_Y''(y) + \left\{k_Y^2(y) - [\theta_Y'(y)]^2\right\}a_Y(y) + j\left[a_Y(y)\theta_Y''(y) + 2a_Y'(y)\theta_Y'(y)\right] = 0, \qquad (5.1\text{-}27)$$

where the prime and double prime denote first- and second-order derivatives with respect to y, respectively. Since the left-hand side of Eq. (5.1-27) must equal zero, both its real and imaginary parts must be equal to zero, that is,

$$a_Y''(y) + \left\{k_Y^2(y) - [\theta_Y'(y)]^2\right\}a_Y(y) = 0 \qquad (5.1\text{-}28)$$

and

$$a_Y(y)\theta_Y''(y) + 2a_Y'(y)\theta_Y'(y) = 0. \qquad (5.1\text{-}29)$$

Rewriting Eq. (5.1-29) as

$$\frac{1}{a_Y(y)}\frac{d}{dy}a_Y(y) = -\frac{1}{2}\frac{1}{\theta'_Y(y)}\frac{d}{dy}\theta'_Y(y), \qquad (5.1\text{-}30)$$

and multiplying both sides of Eq. (5.1-30) by dy and integrating yields

$$\int_{a_Y(y_0)}^{a_Y(y)}\frac{1}{a_Y(y)}\,da_Y(y) = -\frac{1}{2}\int_{\theta'_Y(y_0)}^{\theta'_Y(y)}\frac{1}{\theta'_Y(y)}\,d\theta'_Y(y), \qquad (5.1\text{-}31)$$

or

$$\ln|a_Y(y)| = -\tfrac{1}{2}\ln|\theta'_Y(y)| + \ln\left[|a_Y(y_0)|\,|\theta'_Y(y_0)|^{1/2}\right], \qquad (5.1\text{-}32)$$

where y_0 is the y coordinate of a sound source. Solving Eq. (5.1-32) for $|a_Y(y)|$ yields

$$|a_Y(y)| = |a_Y(y_0)|\sqrt{\left|\frac{\theta'_Y(y_0)}{\theta'_Y(y)}\right|}, \qquad (5.1\text{-}33)$$

and since $a_Y(y)$ is a real function, if we further assume that $a_Y(y_0)$ is a positive constant, then Eq. (5.1-33) can be rewritten as

$$a_Y(y) = a_Y(y_0)\sqrt{\left|\frac{\theta'_Y(y_0)}{\theta'_Y(y)}\right|}, \qquad (5.1\text{-}34)$$

where it is understood that the positive square root is taken.

The next step is to rewrite Eq. (5.1-28) as follows:

$$1 + \frac{1}{k_Y^2(y)}\left[\frac{a''_Y(y)}{a_Y(y)} - [\theta'_Y(y)]^2\right] = 0. \qquad (5.1\text{-}35)$$

If it is assumed that the phase function changes value much more rapidly than the amplitude function, then

$$\left|\frac{a''_Y(y)}{a_Y(y)}\right| \ll [\theta'_Y(y)]^2, \qquad (5.1\text{-}36)$$

and as a result, Eq. (5.1-35) reduces to

$$[\theta'_Y(y)]^2 \approx k_Y^2(y), \qquad (5.1\text{-}37)$$

or

$$\frac{d}{dy}\theta_Y(y) \approx \pm k_Y(y).$$ (5.1-38)

Multiplying both sides of Eq. (5.1-38) by dy and integrating yields the following expression for the real phase function:

$$\theta_Y(y) \approx \theta_Y(y_0) \pm \int_{y_0}^{y} k_Y(\zeta)\, d\zeta.$$ (5.1-39)

And upon substituting Eq. (5.1-38) into Eq. (5.1-34), we obtain the following expression for the real amplitude function:

$$a_Y(y) \approx \frac{A}{\sqrt{|k_Y(y)|}},$$ (5.1-40)

where

$$A = a_Y(y_0)\sqrt{|k_Y(y_0)|}.$$ (5.1-41)

Therefore, substituting Eqs. (5.1-39) and (5.1-40) into Eq. (5.1-26) and taking into account both signs appearing in Eq. (5.1-39) yields

$$Y(y) \approx \frac{1}{\sqrt{|k_Y(y)|}}\left\{ A_Y \exp\left[-j\int_{y_0}^{y} k_Y(\zeta)\, d\zeta\right] + B_Y \exp\left[+j\int_{y_0}^{y} k_Y(\zeta)\, d\zeta\right]\right\},$$

(5.1-42)

where the real constant A and the complex constant $\exp[+j\theta_Y(y_0)]$ have been absorbed by the complex constants A_Y and B_Y. Equation (5.1-42) is called the *geometrical optics approximation*. It is also known as the *approximate solution of Wentzel, Kramers, and Brillouin (WKB)* or as the *plane-wave ray solution*. It is important to note that the approximate WKB solution given by Eq. (5.1-42) approaches *infinity*, and hence is *invalid*, as the propagation-vector component in the Y direction, $k_Y(y)$, approaches *zero*. Also note that $k_Y(y)$ is equal to zero at the depth of a *turning point*, y_{TP}, that is, $k_Y(y_{TP}) = 0$. More will be said about turning points later in this section. Furthermore, in Section 5.2.5 we shall derive an equation for the magnitude squared of the acoustic pressure that is valid (i.e., finite) at turning points.

If $k_Y(y)$ is *positive*, then the first and second terms on the right-hand side of Eq. (5.1-42) correspond to waves propagating in the *positive* and *negative* Y directions, respectively. The integral

$$\int_{y_0}^{y} k_Y(\zeta)\, d\zeta$$

is referred to as the *WKB phase integral* and represents the *phase* in the Y direction due to a wave traveling from the depth y_0 of a source to the depth y of a field point.

Finally, upon substituting Eqs. (5.1-17), (5.1-20), and (5.1-42) into Eq. (5.1-12), we obtain

$$\varphi_f(x, y, z) \approx \left\{ A_X \exp[-jk_X(x - x_0)] + B_X \exp[+jk_X(x - x_0)] \right\}$$

$$\times \frac{1}{\sqrt{|k_Y(y)|}} \left\{ A_Y \exp\left[-j\int_{y_0}^{y} k_Y(\zeta)\, d\zeta\right] + B_Y \exp\left[+j\int_{y_0}^{y} k_Y(\zeta)\, d\zeta\right] \right\}$$

$$\times \left\{ A_Z \exp[-jk_Z(z - z_0)] + B_Z \exp[+jk_Z(z - z_0)] \right\},$$

$$(5.1\text{-}43)$$

which is the approximate WKB solution of the Helmholtz equation given by Eq. (5.1-10), where [see Eqs. (5.1-11) and (5.1-23)]

$$k_X^2 + k_Y^2(y) + k_Z^2 = k^2(y) = [2\pi f/c(y)]^2. \qquad (5.1\text{-}44)$$

The time-harmonic solution can be obtained by substituting Eq. (5.1-43) into Eq. (5.1-3).

Example 5.1-1 (Homogeneous medium)

In this example we shall consider the case of a homogeneous medium where the speed of sound is constant, that is, $c(y) = c_0$. As a result, the wave number $k(y) = k_0 = 2\pi f/c_0$ is also constant, since the index of . refraction $n(y) = 1$ when $c(y) = c_0$ [see Eqs. (5.1-6) and (5.1-11)]. Upon substituting $k(y) = k_0$ into Eq. (5.1-23), we obtain

$$k_Y^2(y) = k_Y^2 = k_0^2 - k_X^2 - k_Z^2, \qquad (5.1\text{-}45)$$

which indicates that the propagation-vector component in the Y direction is constant, as expected, since k_0, k_X, and k_Z are constants.

Therefore, Eq. (5.1-22) becomes

$$\frac{d^2}{dy^2}Y(y) + k_Y^2 Y(y) = 0. \qquad (5.1\text{-}46)$$

Equation (5.1-46) has the *exact* solution

$$Y(y) = A_Y \exp\left[-jk_Y(y - y_0)\right] + B_Y \exp\left[+jk_Y(y - y_0)\right], \qquad (5.1\text{-}47)$$

where

$$k_Y = \sqrt{k_0^2 - k_X^2 - k_Z^2}. \qquad (5.1\text{-}48)$$

Let us now investigate the approximate WKB solution given by Eq. (5.1-42) for the case of a homogeneous medium. Since

$$k_Y(y) = k_Y \qquad (5.1\text{-}49)$$

for a homogeneous medium, where k_Y is given by Eq. (5.1-48), substituting Eq. (5.1-49) into Eq. (5.1-42) yields [compare with Eq. (5.1-47)]

$$Y(y) \approx \frac{1}{\sqrt{|k_Y|}}\left\{A_Y \exp\left[-jk_Y(y - y_0)\right] + B_Y \exp\left[+jk_Y(y - y_0)\right]\right\}.$$

$$(5.1\text{-}50)$$

By comparing Eq. (5.1-50) with Eq. (5.1-26), we can write that

$$a_Y(y) = \frac{A_Y}{\sqrt{|k_Y|}} \quad \text{or} \quad a_Y(y) = \frac{B_Y}{\sqrt{|k_Y|}} \qquad (5.1\text{-}51)$$

and

$$\theta_Y(y) = \pm k_Y(y - y_0). \qquad (5.1\text{-}52)$$

Therefore,

$$a_Y'(y) = 0, \qquad (5.1\text{-}53)$$

$$a_Y''(y) = 0, \qquad (5.1\text{-}54)$$

and

$$\theta_Y'(y) = \pm k_Y. \qquad (5.1\text{-}55)$$

Substituting Eqs. (5.1-51), (5.1-54), and (5.1-55) into Eq. (5.1-36) yields

$$0 \ll k_Y^2, \tag{5.1-56}$$

which is satisfied as long as k_Y does not equal zero. Note that if $k_Y = 0$ in a homogeneous medium, then there is wave propagation only in the XZ plane, and as a result, a solution for $Y(y)$ is not required.

Example 5.1-2 (Time-average intensity vector)
In this example we shall calculate the time-average intensity vector associated with the time-harmonic WKB solution

$$\varphi(t, x, y, z) = X(x)Y(y)Z(z) \exp(+j2\pi ft), \tag{5.1-57}$$

where Eq. (5.1-12) has been substituted into Eq. (5.1-3) and where $X(x)$, $Y(y)$, and $Z(z)$ are given by Eqs. (5.1-17), (5.1-42), and (5.1-20), respectively.

In order to calculate the time-average intensity vector, we must first calculate the acoustic pressure (see Section 1.4)

$$p(t, x, y, z) = -\rho_0(y) \frac{\partial}{\partial t} \varphi(t, x, y, z) \tag{5.1-58}$$

and the acoustic fluid velocity vector (see Section 1.4)

$$\mathbf{u}(t, x, y, z) = \nabla \varphi(t, x, y, z), \tag{5.1-59}$$

where the equilibrium (ambient) density of the fluid medium $\rho_0(y)$ is shown as a function of depth y. Substituting Eq. (5.1-57) into Eqs. (5.1-58) and (5.1-59) yields

$$p(t, x, y, z) = -j2\pi f\rho_0(y) X(x)Y(y)Z(z) \exp(+j2\pi ft) \tag{5.1-60}$$

and

$$\mathbf{u}(t, x, y, z) = \nabla[X(x)Y(y)Z(z)] \exp(+j2\pi ft), \tag{5.1-61}$$

respectively.

Next, assume that the acoustic field is propagating in the *positive X, Y*, and *Z* directions so that

$$X(x) = A_X \exp[-jk_X(x - x_0)], \tag{5.1-62}$$

$$Y(y) \approx \frac{A_Y}{\sqrt{k_Y(y)}} \exp\left[-j \int_{y_0}^{y} k_Y(\zeta) \, d\zeta\right], \tag{5.1-63}$$

and

$$Z(z) = A_Z \exp[-jk_Z(z - z_0)], \tag{5.1-64}$$

where k_X, $k_Y(y)$, and k_Z are all *positive* in value. Substituting Eqs. (5.1-62) through (5.1-64) into Eq. (5.1-61) yields

$$\mathbf{u}(t, x, y, z) \approx -j X(x) Y(y) Z(z)$$

$$\times \left\{ k_X \hat{x} + \left[k_Y(y) - j \frac{1}{2k_Y(y)} \frac{d}{dy} k_Y(y) \right] \hat{y} + k_Z \hat{z} \right\}$$

$$\times \exp(+j2\pi ft). \tag{5.1-65}$$

Since the pressure and velocity expressions given by Eqs. (5.1-60) and (5.1-65), respectively, are time-harmonic acoustic fields, recall that the time-average intensity vector is given by (see Section 1.5)

$$\mathbf{I}_{avg}(x, y, z) = \tfrac{1}{2} \operatorname{Re}\{ p_f(x, y, z) \mathbf{u}_f^*(x, y, z) \}, \tag{5.1-66}$$

where, by referring to Eqs. (5.1-60) and (5.1-65),

$$p_f(x, y, z) \approx -j2\pi f\rho_0(y) X(x) Y(y) Z(z) \tag{5.1-67}$$

and

$$\mathbf{u}_f(x, y, z) \approx -j X(x) Y(y) Z(z)$$

$$\times \left\{ k_X \hat{x} + \left[k_Y(y) - j \frac{1}{2k_Y(y)} \frac{d}{dy} k_Y(y) \right] \hat{y} + k_Z \hat{z} \right\},$$

$$\tag{5.1-68}$$

respectively. Substituting Eqs. (5.1-67) and (5.1-68) into Eq. (5.1-66) yields the time-average intensity vector

$$\boxed{ \mathbf{I}_{avg}(x, y, z) \approx \frac{|A|^2}{2k_Y(y)} k_0 \rho_0(y) c_0 \mathbf{k}(y) } \tag{5.1-69}$$

with magnitude

$$\boxed{ I_{avg}(x, y, z) \approx \frac{|p_f(x, y, z)|^2}{2\rho_0(y) c(y)}, } \tag{5.1-70}$$

where use has been made of the following expressions:

$$|X(x)Y(y)Z(z)|^2 \approx \frac{|A|^2}{k_Y(y)}, \qquad (5.1\text{-}71)$$

where $|A|^2 = |A_X A_Y A_Z|^2$, and from Eq. (5.1-25), $|\mathbf{k}(y)| = k(y) = 2\pi f/c(y)$. Although Eq. (5.1-70) is in the same form as the intensity expression for a plane wave propagating in a homogeneous medium (see Example 2.2-1), the magnitude of the acoustic pressure $|p_f(x, y, z)|$, the equilibrium (ambient) density $\rho_0(y)$, and the speed of sound $c(y)$ are now functions of position.

Example 5.1-3 (Snell's law)

In this example we shall derive Snell's law by considering three-dimensional wave propagation in an inhomogeneous medium where the speed of sound is an arbitrary function of depth y, as illustrated in Fig. 5.1-2.

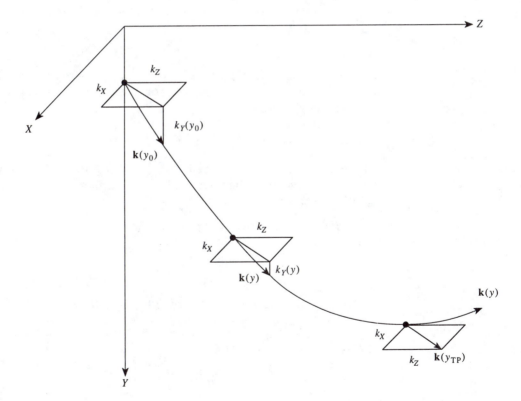

Figure 5.1-2. Wave propagation in three-dimensional space as depicted by the propagation vector $\mathbf{k}(y)$. Note that $\mathbf{k}(y)$ passes through a turning point at $y = y_{TP}$, that is, $k_Y(y_{TP}) = 0$.

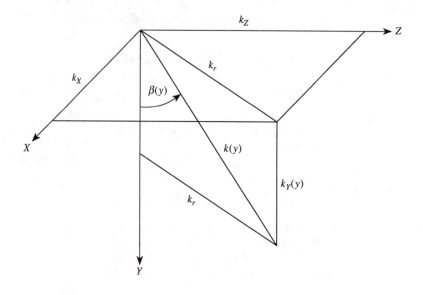

Figure 5.1-3. Geometry defining the propagation-vector component k_r in the horizontal radial direction and the angle $\beta(y)$.

In order to derive Snell's law, refer to Fig. 5.1-3, where it can be seen that the propagation-vector component in the horizontal radial direction is given by

$$k_r = \sqrt{k_X^2 + k_Z^2}. \tag{5.1-72}$$

Note that k_r is a *constant*, since both k_X and k_Z are *constants*, even though the speed of sound is an arbitrary function of depth. From Fig. 5.1-3 it can also be seen that

$$k_r = k(y) \sin \beta(y) \tag{5.1-73}$$

where the angle $\beta(y)$ is a function of depth. Since k_r is a constant, we can write that

$$k(y) \sin \beta(y) = k(y_0) \sin \beta(y_0), \tag{5.1-74}$$

where y_0 is a source depth. Upon substituting Eq. (5.1-11) into Eq. (5.1-74), we obtain *Snell's law*

$$\boxed{\frac{\sin \beta(y)}{c(y)} = \frac{\sin \beta_0}{c_0},} \tag{5.1-75}$$

where $\beta(y)$ is the angle of arrival of the propagation vector or ray path at depth y when the initial angle of propagation or transmission at a source is $\beta_0 = \beta(y_0)$, and where $c(y)$ and $c_0 = c(y_0)$ are the speeds of sound at depths y and y_0, respectively.

Turning Points

As was mentioned earlier, the propagation-vector component in the Y direction is equal to zero at the depth of a turning point y_{TP}, that is, $k_Y(y_{TP}) = 0$. Therefore, at $y = y_{TP}$, the propagation vector is equal to [see Fig. 5.1-2 and Eq. (5.1-24)]

$$\mathbf{k}(y_{TP}) = k_X\hat{x} + k_Z\hat{z}. \tag{5.1-76}$$

In addition, from Fig. 5.1-3,

$$k_Y(y) = k(y)\cos\beta(y). \tag{5.1-77}$$

Therefore, since

$$\boxed{k_Y(y_{TP}) = 0,} \tag{5.1-78}$$

this implies that the angle of arrival at a turning point is

$$\boxed{\beta(y_{TP}) = \pi/2.} \tag{5.1-79}$$

There are two different types of turning points, namely, *upper* and *lower turning points*, as illustrated in Figs. 5.1-4a and b, respectively.

Now that we have derived Snell's law and discussed turning points in more detail, let us discuss the physical significance of Eq. (5.1-36) further. Equation (5.1-36) is a very important assumption, since the approximate WKB solution given by Eq. (5.1-42) depends on Eq. (5.1-36) being satisfied. In Appendix 5A it is shown that Eq. (5.1-36) can be rewritten as

$$\boxed{\left|\frac{c'(y)}{4\pi f \cos^3\beta(y)}\right| \ll \frac{\sqrt{3}}{3},} \tag{5.1-80}$$

where $\beta(y)$ can be determined from Snell's law given by Eq. (5.1-75). Equation (5.1-80) indicates that the approximate WKB solution given by Eq. (5.1-42) is valid (1) for high frequencies f, (2) if the speed of sound $c(y)$ is slowly changing, that is, if $|c'(y)| \ll 1$, and (3) if $\beta(y) \neq \pi/2$, that is, as long as the wave does not pass through a turning point [see Eq. (5.1-79)].

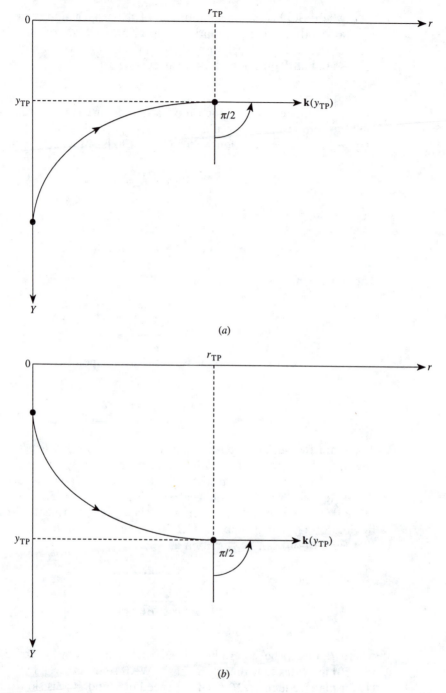

Figure 5.1-4. (*a*) Illustration of a ray passing through an upper turning point. (*b*) Illustration of a ray passing through a lower turning point.

Example 5.1-4 (Depth of a turning point)
When the speed of sound is only a function of depth, $c(y)$ and the plot of $c(y)$ versus depth y are usually referred to as a *sound-speed profile* (SSP). The simplest mathematical model for a SSP, other than a constant, is given by

$$c(y) = c_0 + g(y - y_0), \qquad (5.1\text{-}81)$$

that is, the speed of sound is modeled as a *linear function of depth* with *constant gradient* g, where $c_0 = c(y_0)$ is the speed of sound at a source depth y_0. The parameter g is referred to as the gradient, since $dc(y)/dy = g$. Its value, with units of inverse seconds, can be either positive or negative and indicates the amount by which the speed of sound changes per unit change in depth. In this example we shall determine the depth y_{TP} of a turning point for the SSP given by Eq. (5.1-81).

Evaluating Snell's law given by Eq. (5.1-75) at $y = y_{\mathrm{TP}}$ and using Eq. (5.1-79) yields

$$\boxed{c(y_{\mathrm{TP}}) = c_0/\sin \beta_0.} \qquad (5.1\text{-}82)$$

Note that Eq. (5.1-82) is a general result, valid for any SSP that is a function of depth. Evaluating the SSP given by Eq. (5.1-81) at $y = y_{\mathrm{TP}}$ yields

$$c(y_{\mathrm{TP}}) = g\, y_{\mathrm{TP}} + (c_0 - g\, y_0). \qquad (5.1\text{-}83)$$

Equating the right-hand sides of Eqs. (5.1-82) and (5.1-83) yields the desired result

$$\boxed{y_{\mathrm{TP}} = y_0 + \frac{c_0}{g}\left(\frac{1}{\sin \beta_0} - 1\right).} \qquad (5.1\text{-}84)$$

Example 5.1-5 (WKB solution in the cylindrical coordinate system)
In this example we shall solve the following Helmholtz equation:

$$\nabla^2 \varphi_f(r, \phi, y) + k^2(y)\varphi_f(r, \phi, y) = 0, \qquad (5.1\text{-}85)$$

where

$$\nabla^2 = \frac{\partial^2}{\partial r^2} + \frac{1}{r}\frac{\partial}{\partial r} + \frac{1}{r^2}\frac{\partial^2}{\partial \phi^2} + \frac{\partial^2}{\partial y^2} \qquad (5.1\text{-}86)$$

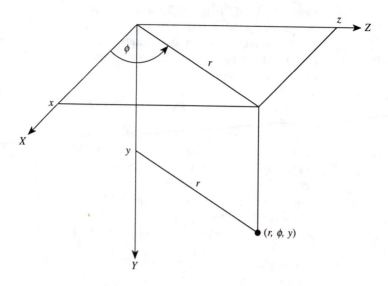

Figure 5.1-5. The cylindrical coordinates (r, ϕ, y).

is the Laplacian expressed in the cylindrical coordinates (r, ϕ, y) (see Fig. 5.1-5) and $k(y)$ is the wave number given by Eq. (5.1-11). If the acoustic field is *axisymmetric*, that is, independent of the azimuthal angle ϕ, then (see Section 3.1)

$$\varphi_f(r, \phi, y) = R(r)Y(y), \qquad (5.1\text{-}87)$$

where

$$R(r) = A_r H_0^{(1)}(k_r r) + B_r H_0^{(2)}(k_r r), \qquad (5.1\text{-}88)$$

$H_0^{(1)}(k_r r)$ and $H_0^{(2)}(k_r r)$ are the zeroth-order Hankel functions of the first and second kind, respectively, k_r is the horizontal radial component of the propagation vector [see Eq. (5.1-72)],

$$r = \sqrt{(x - x_0)^2 + (z - z_0)^2} \qquad (5.1\text{-}89)$$

is the horizontal range, $Y(y)$ is the approximate WKB solution given by Eq. (5.1-42), and

$$k_r^2 + k_Y^2(y) = k^2(y) = [2\pi f / c(y)]^2. \qquad (5.1\text{-}90)$$

5.2 Ray Acoustics

5.2.1 Transport and Eikonal Equations

The WKB method discussed in Section 5.1 was used to solve the Helmholtz equation given by Eq. (5.1-10), where the speed of sound $c(y)$ was assumed to be a function of only one spatial variable, namely, the depth y. As a result, the method of separation of variables could be used to begin the process of obtaining an approximate solution. Now, in this section, we shall investigate the method of *ray acoustics*, which will allow us to obtain an approximate solution of the Helmholtz equation given by Eq. (5.1-4) or, equivalently, Eq. (5.1-9), where the speed of sound is a function of *all three* spatial variables, that is, $c(\mathbf{r}) = c(x, y, z)$. Equation (5.1-9) is repeated below for convenience:

$$\nabla^2 \varphi_f(\mathbf{r}) + k_0^2 n^2(\mathbf{r}) \varphi_f(\mathbf{r}) = 0. \tag{5.1-9}$$

Since the speed of sound is now a function of all three spatial variables, we *cannot* use the method of separation of variables to begin the solution process. Instead, we begin by assuming that the solution of Eq. (5.1-9) is of the form [compare with Eq. (5.1-26)]

$$\boxed{\varphi_f(\mathbf{r}) = a(\mathbf{r}) \exp[+j\theta(\mathbf{r})]} \tag{5.2-1}$$

or

$$\boxed{\varphi_f(\mathbf{r}) = a(\mathbf{r}) \exp[-jk_0 W(\mathbf{r})],} \tag{5.2-2}$$

where $a(\mathbf{r})$ is a *real amplitude function*;

$$\boxed{\theta(\mathbf{r}) = -k_0 W(\mathbf{r})} \tag{5.2-3}$$

is a *real phase function*; k_0 is the reference wave number given by Eq. (5.1-7); $W(\mathbf{r})$, with units of meters, is known as the *eikonal*, a word taken from a Greek word meaning *image*; and $\mathbf{r} = (x, y, z)$. The letter W was chosen for the eikonal because it refers to a wavefront, as shall be discussed later. Substituting Eq. (5.2-2) into Eq. (5.1-9) yields

$$\nabla^2 a(\mathbf{r}) + k_0^2 \left[n^2(\mathbf{r}) - |\nabla W(\mathbf{r})|^2 \right] a(\mathbf{r}) - jk_0 \left[a(\mathbf{r})\nabla^2 W(\mathbf{r}) + 2\nabla a(\mathbf{r}) \cdot \nabla W(\mathbf{r}) \right] = 0. \tag{5.2-4}$$

Since the left-hand side of Eq. (5.2-4) must equal zero, both its real and imaginary parts must be equal to zero, that is,

$$\nabla^2 a(\mathbf{r}) + k_0^2 \left[n^2(\mathbf{r}) - |\nabla W(\mathbf{r})|^2 \right] a(\mathbf{r}) = 0 \qquad (5.2\text{-}5)$$

and

$$\boxed{a(\mathbf{r}) \nabla^2 W(\mathbf{r}) + 2\nabla a(\mathbf{r}) \cdot \nabla W(\mathbf{r}) = 0.} \qquad (5.2\text{-}6)$$

Recalling that $\theta(\mathbf{r})$ and $W(\mathbf{r})$ are related by Eq. (5.2-3), and referring to Eq. (5.1-8), it can be seen that Eqs. (5.2-4) through (5.2-6) are simply three-dimensional generalizations of the one-dimensional results given by Eqs. (5.1-27) through (5.1-29), respectively, obtained by using the WKB method. Equation (5.2-6) is known as the *transport equation*. It is used to determine the amplitude function $a(\mathbf{r})$, which we shall do in Section 5.2-4.

Next, let us rewrite Eq. (5.2-5) as follows:

$$1 + \frac{1}{k_0^2 n^2(\mathbf{r})} \left[\frac{\nabla^2 a(\mathbf{r})}{a(\mathbf{r})} - k_0^2 |\nabla W(\mathbf{r})|^2 \right] = 0. \qquad (5.2\text{-}7)$$

If it is assumed that the phase function changes value much more rapidly than the amplitude function, then [compare with Eq. (5.1-36)]

$$\boxed{\left| \frac{\nabla^2 a(\mathbf{r})}{a(\mathbf{r})} \right| \ll k_0^2 |\nabla W(\mathbf{r})|^2 = |\nabla \theta(\mathbf{r})|^2,} \qquad (5.2\text{-}8)$$

and as a result, Eq. (5.2-7) reduces to

$$\boxed{|\nabla W(\mathbf{r})|^2 = n^2(\mathbf{r})} \qquad (5.2\text{-}9)$$

or, equivalently,

$$\boxed{\left[\frac{\partial}{\partial x} W(\mathbf{r}) \right]^2 + \left[\frac{\partial}{\partial y} W(\mathbf{r}) \right]^2 + \left[\frac{\partial}{\partial z} W(\mathbf{r}) \right]^2 = n^2(\mathbf{r}).} \qquad (5.2\text{-}10)$$

Note that Eqs. (5.2-7) through (5.2-9) are three-dimensional generalizations of the one-dimensional results given by Eqs. (5.1-35) through (5.1-37), respectively, obtained by using the WKB method. Equation (5.2-9) or, equivalently, Eq. (5.2-10) is known as the *eikonal equation*. It is used to determine the eikonal $W(\mathbf{r})$, and hence the phase function $\theta(\mathbf{r})$, and to develop the *ray equations*. The solution of the eikonal equation and the development of the ray equations will be discussed in

Sections 5.2.2 and 5.2.3, respectively. The very important assumption given by Eq. (5.2-8), which was responsible for the eikonal equation, can be rewritten as follows: substituting Eq. (5.2-9) into Eq. (5.2-8) and making use of Eqs. (5.1-5) and (5.1-8) yields

$$\left| \frac{\nabla^2 a(\mathbf{r})}{a(\mathbf{r})} \right| \ll k^2(\mathbf{r}) = \left[\frac{2\pi f}{c(\mathbf{r})} \right]^2. \qquad (5.2\text{-}11)$$

Note that Eq. (5.2-11) is satisfied best at *high frequencies*. In order for Eq. (5.2-2) to be a valid solution of the Helmholtz equation given by Eq. (5.1-9), Eq. (5.2-11) must be satisfied, $a(\mathbf{r})$ and $W(\mathbf{r})$ together must satisfy the transport equation given by Eq. (5.2-6), and $W(\mathbf{r})$ must satisfy the eikonal equation given by Eq. (5.2-9).

Example 5.2-1 (Homogeneous medium)

In this example we shall consider the case of a homogeneous medium where the speed of sound is constant, that is, $c(\mathbf{r}) = c_0$. As a result, $n(\mathbf{r}) = 1$ [see Eq. (5.1-6)] and the Helmholtz equation given by Eq. (5.1-9) reduces to

$$\nabla^2 \varphi_f(\mathbf{r}) + k_0^2 \varphi_f(\mathbf{r}) = 0, \qquad (5.2\text{-}12)$$

where k_0 is given by Eq. (5.1-7). A well-known *exact* solution of Eq. (5.2-12) is given by (see Section 2.2.1)

$$\varphi_f(\mathbf{r}) = A \exp(-j\mathbf{k} \cdot \mathbf{r}) = A \exp(-jk_0 \hat{n} \cdot \mathbf{r}). \qquad (5.2\text{-}13)$$

Assuming a time-harmonic dependence of $\exp(+j2\pi ft)$, Eq. (5.2-13) represents a plane wave propagating in the $\hat{n}$ direction (see Fig. 5.2-1), where A is an arbitrary constant (assumed real in this example),

$$\hat{n} = u\hat{x} + v\hat{y} + w\hat{z} \qquad (5.2\text{-}14)$$

is a unit vector, where u, v, and w are the dimensionless direction cosines with respect to the X, Y, and Z axes, respectively, and

$$\mathbf{r} = x\hat{x} + y\hat{y} + z\hat{z}. \qquad (5.2\text{-}15)$$

By comparing Eqs. (5.2-2) and (5.2-13), it can be seen that the real amplitude function

$$a(\mathbf{r}) = A \qquad (5.2\text{-}16)$$

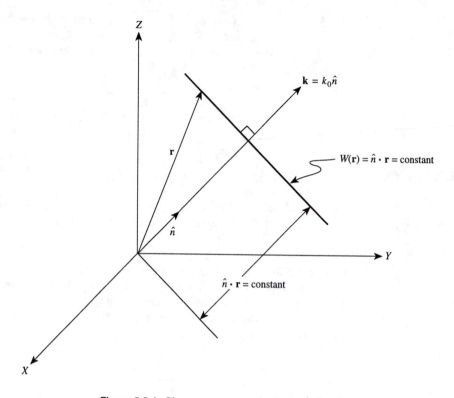

Figure 5.2-1. Plane wave propagating in the $\hat{n}$ direction.

is a constant, and the eikonal

$$W(\mathbf{r}) = \hat{n} \cdot \mathbf{r} = ux + vy + wz. \qquad (5.2\text{-}17)$$

Substituting Eq. (5.2-17) into Eq. (5.2-3) yields the real phase function

$$\theta(\mathbf{r}) = -\mathbf{k} \cdot \mathbf{r} = -k_X x - k_Y y - k_Z z, \qquad (5.2\text{-}18)$$

where

$$\mathbf{k} = k_0 \hat{n} = k_X \hat{x} + k_Y \hat{y} + k_Z \hat{z} \qquad (5.2\text{-}19)$$

is the propagation vector and

$$k_X = k_0 u, \qquad (5.2\text{-}20)$$

$$k_Y = k_0 v, \qquad (5.2\text{-}21)$$

and

$$k_Z = k_0 w \qquad (5.2\text{-}22)$$

are the propagation-vector components in the X, Y, and Z directions, respectively.

From Eq. (5.2-3), it can be seen that for a given value of k_0,

$$W(\mathbf{r}) = \text{constant} \tag{5.2-23}$$

defines a surface of *constant phase* in three-dimensional space, that is, an acoustic field evaluated at any point on this surface will have the same value of phase. Surfaces of constant phase are called *wavefronts*. In the case of a homogeneous medium, wavefronts defined by (see Fig. 5.2-1)

$$W(\mathbf{r}) = \hat{n} \cdot \mathbf{r} = \text{constant} \tag{5.2-24}$$

are *planar* in the rectangular coordinate system. Equation (5.2-24) is, in fact, the equation for a plane in three-dimensional space. Also note from Fig. 5.2-1 that the propagation vector $\mathbf{k}$ is *normal* (perpendicular) to the wavefront.

Next, let us determine whether or not the plane-wave solution given by Eq. (5.2-13) satisfies Eq. (5.2-11) and the transport and eikonal equations. First, let us determine if Eq. (5.2-11) is satisfied. From Eq. (5.2-16) we obtain

$$\nabla^2 a(\mathbf{r}) = \nabla^2 A = 0, \tag{5.2-25}$$

since A is a constant. Substituting Eqs. (5.2-16) and (5.2-25) into Eq. (5.2-11) and noting that $k^2(\mathbf{r}) = k_0^2$ for a homogeneous medium yields $|0/A| = 0 \ll k_0^2$. Therefore, Eq. (5.2-11) is satisfied.

Second, let us determine if the eikonal equation is satisfied. From Eq. (5.2-17) we obtain

$$\nabla W(\mathbf{r}) = u\hat{x} + v\hat{y} + w\hat{z} \tag{5.2-26}$$

or, upon substituting Eq. (5.2-14) into Eq. (5.2-26),

$$\nabla W(\mathbf{r}) = \hat{n}. \tag{5.2-27}$$

Therefore, from Eq. (5.2-26),

$$|\nabla W(\mathbf{r})|^2 = u^2 + v^2 + w^2 = 1, \tag{5.2-28}$$

since u, v, and w are direction cosines, or, from Eq. (5.2-27),

$$|\nabla W(\mathbf{r})|^2 = |\hat{n}|^2 = 1, \tag{5.2-29}$$

since $\hat{n}$ is a unit vector. Therefore, substituting Eq. (5.2-28) into Eq. (5.2-9) and making use of the fact that $n(\mathbf{r}) = 1$ for a homogeneous medium, we see that the eikonal equation is satisfied.

Third, let us determine if the transport equation is satisfied. From Eq. (5.2-16) we obtain

$$\nabla a(\mathbf{r}) = \nabla A = \mathbf{0}, \tag{5.2-30}$$

since A is a constant. From either Eq. (5.2-17) or Eq. (5.2-26) we obtain

$$\nabla^2 W(\mathbf{r}) = \nabla \cdot \nabla W(\mathbf{r}) = 0, \tag{5.2-31}$$

since u, v, and w are not functions of (x, y, z) in a homogeneous medium. Substituting Eqs. (5.2-16), (5.2-26), (5.2-30), and (5.2-31) into Eq. (5.2-6) yields $A(0) + 2(\mathbf{0}) \cdot (u\hat{x} + v\hat{y} + w\hat{z}) = 0$. Therefore, the transport equation is also satisfied.

Let us conclude this example by making one additional observation. Using Eq. (5.2-27), the propagation vector $\mathbf{k}$ corresponding to the plane wave shown in Fig. 5.2-1 and given by Eq. (5.2-19) can be rewritten as

$$\boxed{\mathbf{k} = k_0 \hat{n} = k_0 \nabla W(\mathbf{r}).} \tag{5.2-32}$$

Therefore, $\nabla W(\mathbf{r})$ is normal (perpendicular) to the planar wavefront defined by Eq. (5.2-24). As a result, $\nabla W(\mathbf{r})$ determines the direction of propagation of the wavefront (i.e., the direction of the propagation vector), and therefore, the direction of energy propagation. This property of $\nabla W(\mathbf{r})$ for $W(\mathbf{r}) = $ constant is true in general and not only for plane waves propagating in a homogeneous medium, as we shall discuss next.

5.2.2 Solution of the Eikonal Equation

The eikonal equation given by Eq. (5.2-9) or, equivalently, by Eq. (5.2-10), is a first-order, nonlinear, partial differential equation. As was mentioned previously in Example 5.2-1, for a given value of k_0, Eq. (5.2-23) defines a surface of constant phase in three-dimensional space known as a wavefront. A wavefront in an inhomogeneous medium is an arbitrary surface in general. From the properties of the gradient of a real scalar function, which in this case is the eikonal $W(\mathbf{r})$, $\nabla W(\mathbf{r})$ is normal to the wavefront defined by Eq. (5.2-23) at all points on the wavefront. Each time a different value of the constant in Eq. (5.2-23) is chosen, a different wavefront is defined. An acoustic field evaluated at any point on a particular wavefront will have the same value of phase, although not necessarily the same amplitude value, since the amplitude $a(\mathbf{r})$ is a function of position. The curves

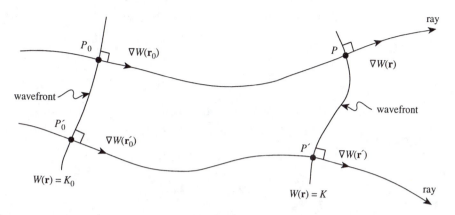

Figure 5.2-2. Illustration of the concepts of wavefronts and rays. The points P_0, P'_0, P, and P' on the ray paths are specified by the position vectors $\mathbf{r}_0$, $\mathbf{r}'_0$, $\mathbf{r}$, and $\mathbf{r}'$, respectively. The parameters K_0 and K are constants.

defined by $\nabla W(\mathbf{r})$, which are normal to the wavefronts, are called *rays* or *ray paths* (see Fig. 5.2-2). Later, in Example 5.2-2, it will be shown that rays define the direction of energy propagation, as might be expected, since they define the direction of wavefront propagation, and hence, the direction of the propagation vector.

Let us begin the solution of the eikonal equation by expressing the vector $\nabla W(\mathbf{r})$ in rectangular coordinates, that is,

$$\nabla W(\mathbf{r}) = \frac{\partial}{\partial x} W(\mathbf{r}) \hat{x} + \frac{\partial}{\partial y} W(\mathbf{r}) \hat{y} + \frac{\partial}{\partial z} W(\mathbf{r}) \hat{z}. \qquad (5.2\text{-}33)$$

Since $\nabla W(\mathbf{r})$ is a vector, it can also be expressed in terms of its magnitude and direction as follows:

$$\nabla W(\mathbf{r}) = |\nabla W(\mathbf{r})| \hat{n}(\mathbf{r}) \qquad (5.2\text{-}34)$$

or, from the eikonal equation given by Eq. (5.2-9),

$$\nabla W(\mathbf{r}) = n(\mathbf{r}) \hat{n}(\mathbf{r}), \qquad (5.2\text{-}35)$$

where $n(\mathbf{r})$ is the *index of refraction* given by Eq. (5.1-6),

$$\boxed{\hat{n}(\mathbf{r}) = u(\mathbf{r}) \hat{x} + v(\mathbf{r}) \hat{y} + w(\mathbf{r}) \hat{z}} \qquad (5.2\text{-}36)$$

is the *unit vector* in the direction of $\nabla W(\mathbf{r})$, and the dimensionless *direction cosines* $u(\mathbf{r})$, $v(\mathbf{r})$, and $w(\mathbf{r})$ are now functions of position along the rays in an inhomogeneous medium and satisfy the relationship

$$u^2(\mathbf{r}) + v^2(\mathbf{r}) + w^2(\mathbf{r}) = 1. \qquad (5.2\text{-}37)$$

Note that if we multiply both sides of Eq. (5.2-35) by k_0 and make use of Eq. (5.1-8), then the *propagation vector* $\mathbf{k}(\mathbf{r})$ can be written as

$$\mathbf{k}(\mathbf{r}) = k(\mathbf{r})\hat{n}(\mathbf{r}) = k_0 \nabla W(\mathbf{r}). \qquad (5.2\text{-}38)$$

Compare Eq. (5.2-38) with the "plane wave in a homogeneous medium" result given by Eq. (5.2-32). Upon substituting Eq. (5.2-36) into Eq. (5.2-35) and equating the result with Eq. (5.2-33), we obtain the following set of equations:

$$\frac{\partial}{\partial x} W(\mathbf{r}) = n(\mathbf{r})u(\mathbf{r}), \qquad (5.2\text{-}39)$$

$$\frac{\partial}{\partial y} W(\mathbf{r}) = n(\mathbf{r})v(\mathbf{r}), \qquad (5.2\text{-}40)$$

$$\frac{\partial}{\partial z} W(\mathbf{r}) = n(\mathbf{r})w(\mathbf{r}). \qquad (5.2\text{-}41)$$

Equations (5.2-39) through (5.2-41) give the x, y, and z components, respectively, of the direction of ray propagation in terms of the index of refraction and the appropriate direction cosines.

Next, consider an *infinitesimal element of arc length ds* at an arbitrary point P with position vector $\mathbf{r}$ along a ray path, as shown in Fig. 5.2-3. The direction cosines with respect to the X, Y, and Z axes are given by the following equations:

$$u(\mathbf{r}) = \cos \alpha(\mathbf{r}) = \sin \theta(\mathbf{r}) \cos \psi(\mathbf{r}) = \frac{dx}{ds}, \qquad (5.2\text{-}42)$$

$$v(\mathbf{r}) = \cos \beta(\mathbf{r}) = \sin \theta(\mathbf{r}) \sin \psi(\mathbf{r}) = \frac{dy}{ds}, \qquad (5.2\text{-}43)$$

$$w(\mathbf{r}) = \cos \gamma(\mathbf{r}) = \cos \theta(\mathbf{r}) = \frac{dz}{ds}, \qquad (5.2\text{-}44)$$

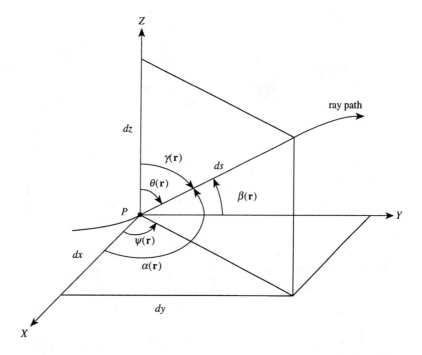

Figure 5.2-3. An infinitesimal element of arc length ds at an arbitrary point P with position vector $\mathbf{r}$ along a ray path, and associated angles.

respectively, where

$$ds = \sqrt{(dx)^2 + (dy)^2 + (dz)^2}.$$
(5.2-45)

Therefore, the unit vector given by Eq. (5.2-36) can be rewritten as

$$\hat{n}(\mathbf{r}) = \frac{dx}{ds}\hat{x} + \frac{dy}{ds}\hat{y} + \frac{dz}{ds}\hat{z},$$
(5.2-46)

where

$$\left(\frac{dx}{ds}\right)^2 + \left(\frac{dy}{ds}\right)^2 + \left(\frac{dz}{ds}\right)^2 = 1.$$
(5.2-47)

In addition, Eqs. (5.2-39) through (5.2-41) can also be rewritten as follows:

$$\frac{\partial}{\partial x} W(\mathbf{r}) = n(\mathbf{r}) \frac{dx}{ds}, \tag{5.2-48}$$

$$\frac{\partial}{\partial y} W(\mathbf{r}) = n(\mathbf{r}) \frac{dy}{ds}, \tag{5.2-49}$$

and

$$\frac{\partial}{\partial z} W(\mathbf{r}) = n(\mathbf{r}) \frac{dz}{ds}, \tag{5.2-50}$$

respectively. Equations (5.2-48) through (5.2-50) will play an important role in the derivation of the ray equations in Section 5.2.3.

Next, compute the *directional derivative dW(**r**)/ds*, which represents the rate of change of the eikonal $W(\mathbf{r})$ with respect to the *arc length s* along a ray path. Expanding $dW(\mathbf{r})/ds$ using the chain rule yields

$$\frac{d}{ds} W(\mathbf{r}) = \frac{\partial}{\partial x} W(\mathbf{r}) \frac{dx}{ds} + \frac{\partial}{\partial y} W(\mathbf{r}) \frac{dy}{ds} + \frac{\partial}{\partial z} W(\mathbf{r}) \frac{dz}{ds}. \tag{5.2-51}$$

By inspecting Eqs. (5.2-33) and (5.2-46), it can be seen that Eq. (5.2-51) can be written as the dot product

$$\frac{d}{ds} W(\mathbf{r}) = \nabla W(\mathbf{r}) \cdot \hat{n}(\mathbf{r}), \tag{5.2-52}$$

and with the use of Eq. (5.2-35),

$$\frac{d}{ds} W(\mathbf{r}) = n(\mathbf{r}) \hat{n}(\mathbf{r}) \cdot \hat{n}(\mathbf{r}), \tag{5.2-53}$$

or

$$\frac{d}{ds} W(\mathbf{r}) = n(\mathbf{r}). \tag{5.2-54}$$

The solution of Eq. (5.2-54) can be expressed as

$$\boxed{W(\mathbf{r}) = W(\mathbf{r}_0) + \int_{\mathbf{r}_0}^{\mathbf{r}} n(x, y, z) \, ds.} \tag{5.2-55}$$

Equation (5.2-55) is also the solution of the eikonal equation given by Eq. (5.2-9).

Besides being referred to as the eikonal, $W(\mathbf{r})$ is also known in optics as the *optical path function*, and is defined as the *optical path length* between two points on a ray path, for example, points P_0 and P with position vectors $\mathbf{r}_0$ and $\mathbf{r}$, respectively, as shown in Fig. 5.2-2. Also, upon substituting Eq. (5.2-55) into Eq. (5.2-3) and using Eq. (5.1-8), we obtain the real phase function

$$\theta(\mathbf{r}) = \theta(\mathbf{r}_0) - \int_{\mathbf{r}_0}^{\mathbf{r}} k(x, y, z)\, ds. \qquad (5.2\text{-}56)$$

Note that it is understood that $x = x(s)$, $y = y(s)$, and $z = z(s)$ in the arguments of the index of refraction and the wave number in Eqs. (5.2-55) and (5.2-56), respectively.

Alternative expressions for both the eikonal and the real phase function can be obtained as follows. Substituting Eqs. (5.2-39) through (5.2-41) into the right-hand side of Eq. (5.2-51) yields

$$\frac{d}{ds} W(\mathbf{r}) = n(\mathbf{r})u(\mathbf{r})\frac{dx}{ds} + n(\mathbf{r})v(\mathbf{r})\frac{dy}{ds} + n(\mathbf{r})w(\mathbf{r})\frac{dz}{ds} \qquad (5.2\text{-}57)$$

with solution

$$W(\mathbf{r}) = W(\mathbf{r}_0) + \int_{x_0}^{x} n(\zeta, y, z)u(\zeta, y, z)\, d\zeta + \int_{y_0}^{y} n(x, \zeta, z)v(x, \zeta, z)\, d\zeta$$
$$+ \int_{z_0}^{z} n(x, y, \zeta)w(x, y, \zeta)\, d\zeta.$$

$$(5.2\text{-}58)$$

And upon substituting Eq. (5.2-58) into Eq. (5.2-3) and using Eq. (5.1-8), we obtain

$$\theta(\mathbf{r}) = \theta(\mathbf{r}_0) - \int_{x_0}^{x} k_X(\zeta, y, z)\, d\zeta - \int_{y_0}^{y} k_Y(x, \zeta, z)\, d\zeta - \int_{z_0}^{z} k_Z(x, y, \zeta)\, d\zeta,$$

$$(5.2\text{-}59)$$

where

$$k_X(\mathbf{r}) = k(\mathbf{r})u(\mathbf{r}), \qquad (5.2\text{-}60)$$

$$k_Y(\mathbf{r}) = k(\mathbf{r})v(\mathbf{r}), \qquad (5.2\text{-}61)$$

$$k_Z(\mathbf{r}) = k(\mathbf{r})w(\mathbf{r}) \qquad (5.2\text{-}62)$$

are the propagation-vector components in the X, Y, and Z directions, respectively. Compare the first and third terms on the right-hand side of Eq. (5.2-59) with the right-hand side of the real phase function given by Eq. (5.1-39) obtained via the WKB method.

Example 5.2-2 (Time-average intensity vector)
In this example we shall calculate the time-average intensity vector associated with the time-harmonic ray-acoustics solution

$$\varphi(t,\mathbf{r}) = a(\mathbf{r}) \exp[-jk_0 W(\mathbf{r})] \exp(+j2\pi ft), \qquad (5.2\text{-}63)$$

which is obtained by substituting Eq. (5.2-2) into Eq. (5.1-3).

In order to calculate the time-average intensity vector, we must first calculate the acoustic pressure (see Section 1.4)

$$p(t,\mathbf{r}) = -\rho_0(\mathbf{r}) \frac{\partial}{\partial t} \varphi(t,\mathbf{r}) \qquad (5.2\text{-}64)$$

and the acoustic fluid velocity vector (see Section 1.4)

$$\mathbf{u}(t,\mathbf{r}) = \nabla\varphi(t,\mathbf{r}), \qquad (5.2\text{-}65)$$

where the equilibrium (ambient) density of the fluid medium $\rho_0(\mathbf{r})$ is shown as a function of position. Substituting Eq. (5.2-63) into Eqs. (5.2-64) and (5.2-65) yields

$$p(t,\mathbf{r}) = p_f(\mathbf{r}) \exp(+j2\pi ft) \qquad (5.2\text{-}66)$$

and

$$\mathbf{u}(t,\mathbf{r}) = \mathbf{u}_f(\mathbf{r}) \exp(+j2\pi ft), \qquad (5.2\text{-}67)$$

where

$$p_f(\mathbf{r}) = -j2\pi f\rho_0(\mathbf{r})a(\mathbf{r}) \exp[-jk_0 W(\mathbf{r})] \qquad (5.2\text{-}68)$$

and

$$\mathbf{u}_f(\mathbf{r}) = -jk_0 a(\mathbf{r}) \exp[-jk_0 W(\mathbf{r})] \nabla W(\mathbf{r}) + \exp[-jk_0 W(\mathbf{r})] \nabla a(\mathbf{r}).$$
$$(5.2\text{-}69)$$

Since the pressure and velocity expressions given by Eqs. (5.2-66) and (5.2-67), respectively, have time-harmonic dependence, the time-average intensity vector can be computed by substituting Eqs. (5.2-68) and

(5.2-69) into Eq. (5.1-66). Doing so yields the desired result:

$$\mathbf{I}_{\text{avg}}(\mathbf{r}) = \tfrac{1}{2}k_0^2\rho_0(\mathbf{r})c_0a^2(\mathbf{r})\,\nabla W(\mathbf{r}).$$

(5.2-70)

Therefore, $\nabla W(\mathbf{r})$ determines the direction of the time-average intensity vector, and since $\nabla W(\mathbf{r})$ also determines the direction of ray propagation, *the ray paths define the direction of energy propagation.*

Finally, taking the magnitude of both sides of Eq. (5.2-70) yields

$$I_{\text{avg}}(\mathbf{r}) = \left|\mathbf{I}_{\text{avg}}(\mathbf{r})\right| = \tfrac{1}{2}k_0^2\rho_0(\mathbf{r})c_0a^2(\mathbf{r})\left|\nabla W(\mathbf{r})\right|, \qquad (5.2\text{-}71)$$

and by making use of Eqs. (5.2-9), (5.1-6), and (5.2-68); Eq. (5.2-71) can be rewritten as

$$I_{\text{avg}}(\mathbf{r}) = \frac{\left|p_f(\mathbf{r})\right|^2}{2\rho_0(\mathbf{r})c(\mathbf{r})}.$$

(5.2-72)

Although Eq. (5.2-72) is in the same form as the intensity expression for a plane wave propagating in a homogeneous medium (see Example 2.2-1), the magnitude of the acoustic pressure $|p_f(\mathbf{r})|$, the equilibrium density $\rho_0(\mathbf{r})$, and the speed of sound $c(\mathbf{r})$ are now functions of position in the case of wave propagation in an inhomogeneous medium.

5.2.3 Ray Equations

In order to develop the ray equations, the solutions of which will enable us to draw the trajectories of ray paths, take the gradient of both sides of Eq. (5.2-54), that is,

$$\nabla\left(\frac{d}{ds}W(\mathbf{r})\right) = \nabla n(\mathbf{r}),$$

(5.2-73)

or

$$\frac{d}{ds}\nabla W(\mathbf{r}) = \nabla n(\mathbf{r}).$$

(5.2-74)

Since $\nabla W(\mathbf{r})$ describes the direction of ray propagation, the expression $d\,\nabla W(\mathbf{r})/ds$ describes *the rate of change of the direction of ray propagation with respect to arc length along a ray path.* This rate of change, as given by Eq. (5.2-74), is equal to the gradient of the index of refraction. Therefore, Eq. (5.2-74) is a representation of the *ray equations in vector form*, since the solution of Eq. (5.2-74) will allow us to draw the trajectories of ray paths by using the information on how the directions of the ray paths change.

With the use of Eqs. (5.2-48) through (5.2-50), the vector form of the ray equations given by Eq. (5.2-74) can be expressed in *scalar form* as follows:

$$\frac{d}{ds}\left(n(\mathbf{r})\frac{dx}{ds}\right) = \frac{\partial}{\partial x}n(\mathbf{r}), \qquad (5.2\text{-}75)$$

$$\frac{d}{ds}\left(n(\mathbf{r})\frac{dy}{ds}\right) = \frac{\partial}{\partial y}n(\mathbf{r}), \qquad (5.2\text{-}76)$$

$$\frac{d}{ds}\left(n(\mathbf{r})\frac{dz}{ds}\right) = \frac{\partial}{\partial z}n(\mathbf{r}). \qquad (5.2\text{-}77)$$

Note that in the case of a homogeneous medium where $c(\mathbf{r}) = c_0$, *all ray paths are straight lines*. This result can be obtained by substituting $n(\mathbf{r}) = 1$ into the ray equations given by Eqs. (5.2-75) through (5.2-77). Doing so yields the result that all three direction cosines are constant, which is the description of a straight line in three-dimensional space.

Example 5.2-3 (Speed of sound as an arbitrary function of depth)
In this example we shall consider the case when the speed of sound is an arbitrary function of depth only, that is, $c(\mathbf{r}) = c(y)$. Therefore, the index of refraction becomes

$$n(\mathbf{r}) = n(y) = c_0/c(y), \qquad (5.2\text{-}78)$$

where $c_0 = c(y_0)$ is the constant reference speed of sound at a source depth y_0. With the use of Eq. (5.2-78), the ray equations reduce to

$$\frac{d}{ds}\left(n(y)\frac{dx}{ds}\right) = 0, \qquad (5.2\text{-}79)$$

$$\frac{d}{ds}\left(n(y)\frac{dy}{ds}\right) = \frac{\partial}{\partial y}n(y) = \frac{d}{dy}n(y), \qquad (5.2\text{-}80)$$

and

$$\frac{d}{ds}\left(n(y)\frac{dz}{ds}\right) = 0. \qquad (5.2\text{-}81)$$

Equations (5.2-79) and (5.2-81) imply that

$$n(y)\frac{dx}{ds} = A \tag{5.2-82}$$

and

$$n(y)\frac{dz}{ds} = B, \tag{5.2-83}$$

where A and B are constants. If we form the ratio of Eqs. (5.2-82) and (5.2-83), then

$$\frac{dx/ds}{dz/ds} = \frac{A}{B} = \text{constant}, \tag{5.2-84}$$

where dx/ds and dz/ds are the direction cosines with respect to the X and Z axes, respectively, at some point P along a ray path. The restriction that the *ratio* of the direction cosines in the X and Z directions must be equal to a constant implies that *when the speed of sound is an arbitrary function of depth y only, a ray path is confined to a plane normal to the XZ plane.* This is illustrated in Fig. 5.2-4, where it can be seen that the orthogonal projection of the ray path onto the XZ plane is a straight line. The ray path implied by Eq. (5.2-84) is confined to a plane that contains this straight line and is normal to the XZ

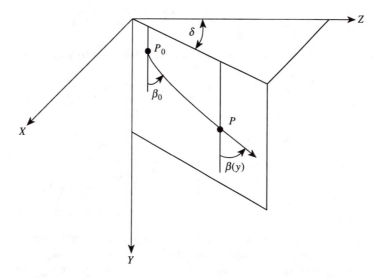

Figure 5.2-4. Ray path confined to a plane normal to the XZ plane.

plane. Note that although the *ratio* of dx/ds and dz/ds is constant along a ray path, dx/ds and dz/ds do change value. For example, in order for the left-hand side of Eq. (5.2-82) to remain constant as the index of refraction changes value as a function of depth, the direction cosine dx/ds must also change value accordingly.

Next, let us analyze the remaining ray equation, Eq. (5.2-80). Since the speed of sound is a function only of depth y in this example,

$$\frac{dy}{ds} = \cos \beta(\mathbf{r}) = \cos \beta(y), \qquad (5.2\text{-}85)$$

and upon substituting Eq. (5.2-85) into Eq. (5.2-80), we obtain

$$\frac{d}{ds}\beta(y) = -\frac{\sin \beta(y)}{n(y)}\frac{d}{dy}n(y), \qquad (5.2\text{-}86)$$

where use has been made of the following application of the chain rule:

$$\frac{d}{ds}n(y) = \left(\frac{d}{dy}n(y)\right)\frac{dy}{ds} = \left(\frac{d}{dy}n(y)\right)\cos \beta(y). \quad (5.2\text{-}87)$$

Since

$$n(y) = c_0/c(y), \qquad (5.2\text{-}88)$$

we have

$$\frac{d}{dy}n(y) = -\frac{c_0}{c^2(y)}\frac{d}{dy}c(y). \qquad (5.2\text{-}89)$$

Substituting Eqs. (5.2-88) and (5.2-89) into Eq. (5.2-86) yields

$$\frac{d}{ds}\beta(y) = \frac{\sin \beta(y)}{c(y)}\frac{d}{dy}c(y), \qquad (5.2\text{-}90)$$

or, using Snell's law [see Eq. (5.1-75)],

$$\boxed{\frac{d}{ds}\beta(y) = \frac{\sin \beta_0}{c_0}\frac{d}{dy}c(y),} \qquad (5.2\text{-}91)$$

where $\sin \beta_0$ and c_0 are constants. Equation (5.2-91) is a very important result. It states that the *curvature* of a ray, which is defined as $d\beta(y)/ds$, is directly proportional to the sound-speed gradient $dc(y)/dy$. When $c(y)$ *increases* with depth, $dc(y)/dy$ will be *positive*,

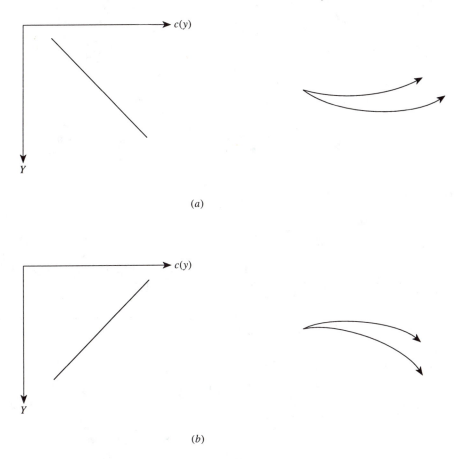

Figure 5.2-5. (*a*) Speed of sound shown increasing with depth; therefore, rays curve upward. (*b*) Speed of sound shown decreasing with depth; therefore, rays curve downward.

and as a result, the curvature will be *positive*, which means that the rays will curve *upward*. Similarly, when $c(y)$ *decreases* with depth, $dc(y)/dy$ will be *negative*, and as a result, the curvature will be *negative*, which means that the rays will curve *downward*. Rays will always curve toward a region of *minimum* sound speed (see Fig. 5.2-5). Note that if $c(y) = c_0$, then according to Eq. (5.2-91), the curvature is zero, which implies that the angle $\beta(y)$ is constant. With $\beta(y)$ constant for all values of depth y, the ray path shown in Fig. 5.2-4 will be a straight line. Therefore, all ray paths are straight lines in a homogeneous fluid medium.

Before proceeding further, let us make several observations that are appropriate at this point in the analysis. If we multiply both sides of Eqs. (5.2-82) and (5.2-83) by the constant reference wave number k_0,

then

$$k(y)\frac{dx}{ds} = k_X(y) = k_X = \text{constant} \qquad (5.2\text{-}92)$$

and

$$k(y)\frac{dz}{ds} = k_Z(y) = k_Z = \text{constant}, \qquad (5.2\text{-}93)$$

since

$$k(y) = k_0 n(y) = 2\pi f/c(y). \qquad (5.2\text{-}94)$$

Equations (5.2-92) and (5.2-93) indicate that the components of the propagation vector in the X and Z directions are constants when the speed of sound is only an arbitrary function of depth y. This is the same result that we obtained when we discussed the WKB method in Section 5.1.

In addition, upon substituting Eqs. (5.2-92) and (5.2-93) into Eq. (5.2-59), and noting that $k_Y(\mathbf{r}) = k_Y(y)$ when $c(\mathbf{r}) = c(y)$, we obtain

$$\boxed{\begin{aligned} \theta(x, y, z) &= \theta(x_0, y_0, z_0) - k_X(x - x_0) \\ &\quad - \int_{y_0}^{y} k_Y(\zeta)\, d\zeta - k_Z(z - z_0), \end{aligned}} \qquad (5.2\text{-}95)$$

which is the real phase function for a fluid medium whose index of refraction (speed of sound) is an arbitrary function of depth y only. Equation (5.2-95) gives the value of phase at a field point (x, y, z) when a source is located at (x_0, y_0, z_0). Compare the first and third terms on the right-hand side of Eq. (5.2-95) with the right-hand side of the real phase function given by Eq. (5.1-39) obtained via the WKB method. In the case of a homogeneous medium, $k_Y(y) = k_Y$, and as a result, Eq. (5.2-95) reduces to

$$\boxed{\begin{aligned} \theta(x, y, z) &= \theta(x_0, y_0, z_0) - k_X(x - x_0) \\ &\quad - k_Y(y - y_0) - k_Z(z - z_0). \end{aligned}} \qquad (5.2\text{-}96)$$

Equation (5.2-96) is the phase function for a plane wave propagating in a homogeneous medium.

We shall conclude this example by deriving equations for several important ray path quantities by using two different approaches. The first approach treats the depth y along a ray path as the independent variable and will enable us to derive well-known expressions for the angle of arrival, travel time, horizontal range, and arc length (path length) as functions of depth along a ray path. However, as we shall soon discover, this approach has problems at turning points.

The second approach treats the horizontal range r along a ray path as the independent variable and is based on solving a system of four first-order ordinary differential equations in order to obtain values for the depth, angle of arrival, travel time, and arc length (path length) as functions of horizontal range along a ray path. The second approach is preferred, since it has no problems at turning points and is therefore well suited for numerical solution and for generating ray-trace plots. Since the ray paths are confined to planes normal to the XZ plane when $c(\mathbf{r}) = c(y)$ (see Fig. 5.2-4), the analysis that follows will make use of Fig. 5.2-6.

Depth as the independent variable. The *angle of arrival* $\beta(y)$ at depth y along a ray path can be obtained from Snell's law [see Eq. (5.1-75)] as follows:

$$\beta(y) = \sin^{-1}[bc(y)], \qquad (5.2\text{-}97)$$

where

$$b = \frac{\sin \beta_0}{c_0}, \qquad (5.2\text{-}98)$$

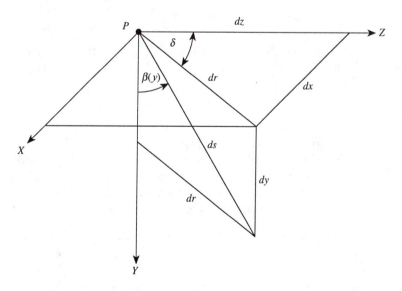

Figure 5.2-6. An infinitesimal element of arc length ds at an arbitrary point P along a ray path.

in seconds per meter, is known as the *ray parameter* and is a *positive constant* for a given ray path. Since each ray path is identified by a different value of β_0, each ray path has its own, different value of b.

The *travel time* τ in seconds along a ray path can be expressed as

$$\tau = \int_0^s \frac{ds}{c(\mathbf{r})} = \int_0^s \frac{ds}{c(y)}, \qquad (5.2\text{-}99)$$

where it is understood that $y = y(s)$. From Fig. 5.2-6, it can be seen that

$$ds = dy/\cos \beta(y). \qquad (5.2\text{-}100)$$

Substituting Eq. (5.2-100) into Eq. (5.2-99) yields

$$\tau = \int_{y_0}^y \frac{d\zeta}{c(\zeta) \cos \beta(\zeta)}. \qquad (5.2\text{-}101)$$

Since

$$\cos \beta(y) = \sqrt{1 - \sin^2 \beta(y)}, \qquad (5.2\text{-}102)$$

substituting Eqs. (5.2-102) and (5.1-75) into Eq. (5.2-101) finally yields

$$\tau = \int_{y_0}^y \frac{d\zeta}{c(\zeta)\sqrt{1 - b^2 c^2(\zeta)}}. \qquad (5.2\text{-}103)$$

It is important to note that the integrand on the right-hand side of Eq. (5.2-103) approaches infinity whenever $y_0 \to y_{TP}$ or $y \to y_{TP}$, since [see Eq. (5.1-82)]

$$c(y_{TP}) = c_0/\sin \beta_0 = 1/b. \qquad (5.2\text{-}104)$$

Therefore, because of the square root, the depth of a turning point y_{TP} is a *branch point* of Eq. (5.2-103). As a result, Eq. (5.2-103) can be used to compute the travel time between two points (r_0, y_0) and (r, y) on a ray path whose initial angle of propagation is β_0 when the horizontal range r along the ray path corresponding to the depth y is less than the horizontal range to the next turning point, so that $y \neq y_{TP}$.

Next, let us derive an expression for the *horizontal range r* in meters along a ray path. Referring back to Fig. 5.2-6, it can be seen that

$$dr = \tan \beta(y) \, dy. \qquad (5.2\text{-}105)$$

Therefore,

$$r = r_0 + \int_{y_0}^{y} \tan \beta(\zeta)\, d\zeta, \tag{5.2-106}$$

where r_0 is the horizontal range of a sound source from the origin. Since

$$\tan \beta(y) = \frac{\sin \beta(y)}{\cos \beta(y)}, \tag{5.2-107}$$

substituting Eqs. (5.2-107), (5.1-75), and (5.2-102) into Eq. (5.2-106) yields

$$\Delta r = r - r_0 = b \int_{y_0}^{y} \frac{c(\zeta)}{\sqrt{1 - b^2 c^2(\zeta)}}\, d\zeta, \tag{5.2-108}$$

where b is given by Eq. (5.2-98). As with Eq. (5.2-103), y_{TP} is a branch point of Eq. (5.2-108). As a result, Eq. (5.2-108) can be used to compute the horizontal range between two points (r_0, y_0) and (r, y) on a ray path whose initial angle of propagation is β_0 when the horizontal range r along the ray path corresponding to the depth y is less than the horizontal range to the next turning point, so that $y \neq y_{\text{TP}}$.

Next, let us calculate the x and z coordinates in meters along a ray path. From Fig. 5.2-6, it can be seen that

$$dx = dr \sin \delta \tag{5.2-109}$$

and

$$dz = dr \cos \delta. \tag{5.2-110}$$

Therefore, upon integrating both sides of Eqs. (5.2-109) and (5.2-110), we obtain

$$\Delta X = x - x_0 = \Delta r \sin \delta \tag{5.2-111}$$

and

$$\Delta Z = z - z_0 = \Delta r \cos \delta, \tag{5.2-112}$$

where

$$\tan \delta = \Delta X / \Delta Z, \tag{5.2-113}$$

x_0 and z_0 are the x and z coordinates of a source, and Δr is given by

Eq. (5.2-108). As a result, the horizontal range can also be expressed as

$$\Delta r = \sqrt{\Delta X^2 + \Delta Z^2}.$$

(5.2-114)

In addition, the *line-of-sight range* R_{LOS} in meters between two points on a ray path is given by

$$R_{LOS} = \sqrt{\Delta r^2 + \Delta Y^2},$$

(5.2-115)

where

$$\Delta Y = y - y_0.$$

(5.2-116)

Note that if a ray path is in the YZ plane, that is, if $\delta = 0°$, then $\Delta X = 0$, $\Delta Z = \Delta r$, and $z = r$.

Finally, let us calculate the *arc length*, or *path length*, s (in meters) along a ray path. Solving Eq. (5.2-100) for s yields

$$s = \int_{y_0}^{y} \frac{d\zeta}{\cos \beta(\zeta)},$$

(5.2-117)

and upon substituting Eqs. (5.2-102) and (5.1-75) into Eq. (5.2-117), we obtain

$$s = \int_{y_0}^{y} \frac{d\zeta}{\sqrt{1 - b^2 c^2(\zeta)}},$$

(5.2-118)

where b is given by Eq. (5.2-98). As with Eqs. (5.2-103) and (5.2-108), y_{TP} is a branch point of Eq. (5.2-118). As a result, Eq. (5.2-118) can be used to compute the arc length between two points (r_0, y_0) and (r, y) on a ray path whose initial angle of propagation is β_0 when the horizontal range r along the ray path corresponding to the depth y is less than the horizontal range to the next turning point, so that $y \neq y_{TP}$. Note that the arc length s is always greater than the line-of-sight range R_{LOS} given by Eq. (5.2-115). Only in the case of free-space propagation in a homogeneous medium where sound rays travel in straight lines will $s = R_{LOS}$.

Horizontal range as the independent variable. If we treat the horizontal range r along a ray path as the independent variable, then we can solve for the depth y, the angle of arrival β, the travel time τ, and the arc length s as functions of r along a ray path by solving a system of four first-order ordinary differential equations, as will be demonstrated next. By using this approach, we shall not have any problems at turning points.

By referring to Eq. (5.2-105), we can write that

$$\frac{dy}{dr} = \frac{1}{\tan \beta(y)} = \cot \beta(y), \qquad (5.2\text{-}119)$$

which is our first desired expression. Therefore, taking the derivative with respect to r of both sides of Eq. (5.2-119) yields

$$\frac{d^2 y}{dr^2} = -\frac{1}{\sin^2 \beta(y)} \frac{d}{dr} \beta(y). \qquad (5.2\text{-}120)$$

Since

$$\frac{d}{dr} \beta(y) = \frac{d}{dy} \beta(y) \frac{dy}{dr}, \qquad (5.2\text{-}121)$$

substituting Eqs. (5.2-121) and (5.2-119) into Eq. (5.2-120) yields

$$\frac{d^2 y}{dr^2} = -\frac{\cos \beta(y)}{\sin^3 \beta(y)} \frac{d}{dy} \beta(y). \qquad (5.2\text{-}122)$$

Next, rewrite Snell's law given by Eq. (5.1-75) as follows:

$$\sin \beta(y) = bc(y), \qquad (5.2\text{-}123)$$

where b is given by Eq. (5.2-98). Taking the derivative with respect to y of both sides of Eq. (5.2-123) yields

$$\frac{d}{dy} \beta(y) = b \frac{c'(y)}{\cos \beta(y)}, \qquad (5.2\text{-}124)$$

where

$$c'(y) = \frac{d}{dy} c(y). \qquad (5.2\text{-}125)$$

Substituting Eqs. (5.2-123) and (5.2-124) into Eq. (5.2-122) finally yields

$$\frac{d^2 y}{dr^2} = -\frac{c'(y)}{b^2 c^3(y)}, \qquad (5.2\text{-}126)$$

which is our second desired expression.

Next, since [see Eq. (5.2-99)]

$$\frac{ds}{d\tau} = c(y),$$

(5.2-127)

we can write that

$$\frac{d\tau}{ds} = \frac{1}{c(y)}.$$

(5.2-128)

However, by referring to Fig. 5.2-6 and using Eq. (5.2-123), it can be shown that

$$\frac{dr}{ds} = \sin \beta(y) = bc(y).$$

(5.2-129)

Therefore, substituting Eq. (5.2-129) into Eq. (5.2-128) yields

$$\frac{d\tau}{dr} = \frac{1}{bc^2(y)},$$

(5.2-130)

which is our third desired expression. Finally, by referring back to Eq. (5.2-129),

$$\frac{ds}{dr} = \frac{1}{bc(y)},$$

(5.2-131)

which is our fourth desired expression.

The next step is to combine Eqs. (5.2-119), (5.2-126), (5.2-130), and (5.2-131) into a system of four first-order ordinary differential equations. If we let

$$x_1 = y,$$

(5.2-132)

$$x_2 = \frac{dy}{dr} = \cot \beta,$$

(5.2-133)

$$x_3 = \tau,$$

(5.2-134)

$$x_4 = s,$$

(5.2-135)

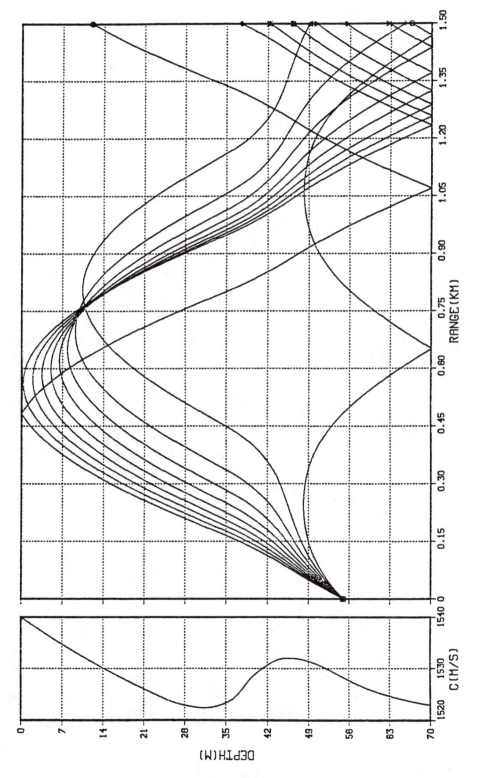

Figure 5.2-7. Ray trace based on solving the system of four first-order ODEs given by Eqs. (5.2-136) through (5.2-139).

then

$$\frac{d}{dr}x_1 = x_2, \qquad\qquad x_1(r_0) = y_0, \qquad\qquad (5.2\text{-}136)$$

$$\frac{d}{dr}x_2 = -\frac{c'(x_1)}{b^2c^3(x_1)}, \qquad x_2(r_0) = \cot \beta_0, \qquad\qquad (5.2\text{-}137)$$

$$\frac{d}{dr}x_3 = \frac{1}{bc^2(x_1)}, \qquad\qquad x_3(r_0) = 0, \qquad\qquad (5.2\text{-}138)$$

$$\frac{d}{dr}x_4 = \frac{1}{bc(x_1)}, \qquad\qquad x_4(r_0) = 0, \qquad\qquad (5.2\text{-}139)$$

where

$$c'(x_1) = \frac{d}{dx_1}c(x_1), \qquad\qquad (5.2\text{-}140)$$

r_0 and y_0 are the horizontal range and depth of a sound source, respectively, β_0 is the initial angle of transmission, and b is the ray parameter given by Eq. (5.2-98). The solution of Eq. (5.2-136) yields the trajectory of a ray path, that is, the depth of a ray as a function of horizontal range. Plotting this information yields a *ray-trace plot* (see Fig. 5.2-7). And finally, the solutions of Eqs. (5.2-137) through (5.2-139) yield the cotangent of the angle of arrival (from which the angle of arrival itself can be obtained), the travel time, and the arc length (path length), respectively, along a ray path as a function of horizontal range.

Example 5.2-4 (**Speed of sound as a linear function of depth**)
The simplest type of SSP, other than a constant speed of sound, was introduced in Example 5.1-4 and is given by

$$c(y) = c_0 + g(y - y_0), \qquad\qquad (5.2\text{-}141)$$

that is, the speed of sound is modeled as a linear function of depth with constant gradient g, where $c_0 = c(y_0)$ is the speed of sound at a source depth y_0. Equation (5.2-141) is a very important mathematical model, since closed-form expressions for the various ray-path quantities can be obtained for a SSP given by Eq. (5.2-141). In addition, any arbitrary depth-dependent SSP can always be approximated by straight-line segments of the form given by Eq. (5.2-141). Figure 5.2-8 illustrates a piecewise linear approximation of a typical SSP. This profile, which can

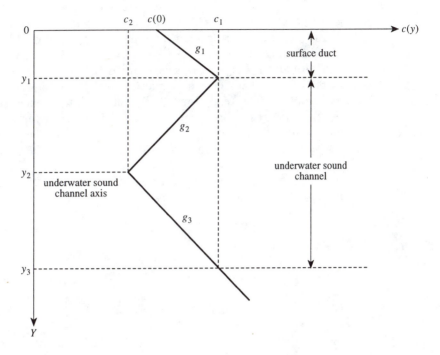

Figure 5.2-8. Piecewise linear approximation of a typical sound-speed profile.

be modeled mathematically as

$$c(y) = \begin{cases} c(0) + g_1 y, & 0 \leq y \leq y_1, \\ c_1 + g_2(y - y_1), & y_1 \leq y \leq y_2, \\ c_2 + g_3(y - y_2), & y \geq y_2, \end{cases} \quad (5.2\text{-}142)$$

is composed of three distinct straight-line segments, each with its own slope (gradient) g_1, g_2, and g_3, where $c(0) = c(y = 0)$, $c_1 = c(y_1)$, and $c_2 = c(y_2)$. Representing an arbitrary depth-dependent SSP by straight-line segments is equivalent to modeling the ocean as a *horizontally stratified medium*. In other words, the ocean medium is broken up into many plane, parallel layers, each layer having its own, different linear SSP with a constant gradient. However, since real-world SSPs are, in general, smooth curves, representing a SSP by many straight-line segments will yield an inaccurate ray-trace plot and inaccurate values for the various ray-path quantities. The accuracy can be improved by increasing the number of layers or, equivalently, the number of straight-line segments.

Typical values of some of the parameters shown in Fig. 5.2-8 are given next. The *surface duct* or *surface sound channel* extends to a depth y_1 that ranges from 10 to 100 m. The depth of the *underwater sound-channel axis* y_2, also known as the *deep sound-channel axis* or *SOFAR* (*sound fixing and ranging*) *axis*, is typically 1000 to 1200 m with extremes of 700 to 2000 m (tropical zone), depending on location. At moderate latitudes (e.g., from 60° S to 60° N) the speed of sound c_2 on the SOFAR axis ranges from 1450 to 1485 m/sec in the Pacific Ocean and from 1450 to 1500 m/sec in the Atlantic Ocean. The gradient g_1 of the surface duct is approximately 0.016 sec^{-1}, and the gradient g_3 is approximately 0.017 sec^{-1}.

Substituting Eq. (5.2-141) into Eq. (5.2-91) yields the following expression for the curvature of a ray:

$$\frac{d}{ds}\beta(y) = bg,$$

(5.2-143)

where the ray parameter b is given by Eq. (5.2-98). Equation (5.2-143) indicates that the curvature along a ray path is *constant*, and as a result, *the ray path is an arc of a circle*. Since the *radius of curvature* R_c in meters is the reciprocal of the magnitude of the curvature, in this example R_c is also constant, that is,

$$R_c = \frac{1}{|d\beta(y)/ds|} = \frac{1}{b|g|},$$

(5.2-144)

where $|g|$ is used to ensure that R_c is positive when g is negative. Note that if $g = 0$, then $c(y) = c_0$ (a constant) and $R_c = \infty$, which means that all ray paths are straight lines in a homogeneous medium.

With the use of the linear SSP given by Eq. (5.2-141), closed-form expressions for the travel time, the horizontal range, and the arc length or path length as functions of depth along a ray path can be obtained either by evaluating the general expressions given by Eqs. (5.2-103), (5.2-108), and (5.2-118), respectively, or by taking an alternative approach, to be discussed next. Substituting Eq. (5.2-124) into Eq. (5.2-101) and using Eq. (5.2-123) yields

$$\tau = \int_{\beta_0}^{\beta(y)} \frac{d\beta(\zeta)}{c'(\zeta)\sin\beta(\zeta)}.$$

(5.2-145)

Since $c'(y) = g$, Eq. (5.2-145) reduces to

$$\tau = \frac{1}{g} \ln \left[\frac{\tan[\beta(y)/2]}{\tan(\beta_0/2)} \right], \tag{5.2-146}$$

where $\beta_0 = \beta(y_0)$ and $\beta(y)$ is given by Eq. (5.2-97).

Next, substituting Eqs. (5.2-107) and (5.2-124) into Eq. (5.2-106) yields

$$r = r_0 + \int_{\beta_0}^{\beta(y)} \frac{\sin \beta(\zeta)}{bc'(\zeta)} \, d\beta(\zeta), \tag{5.2-147}$$

and since $c'(y) = g$, Eq. (5.2-147) reduces to

$$\Delta r = r - r_0 = \frac{1}{bg} [\cos \beta_0 - \cos \beta(y)], \tag{5.2-148}$$

where b is given by Eq. (5.2-98). And upon substituting Eq. (5.2-124) into Eq. (5.2-117), we obtain

$$s = \frac{1}{bg} [\beta(y) - \beta_0], \tag{5.2-149}$$

where $\beta(y)$ and β_0 appearing in the square brackets in Eq. (5.2-149) must be expressed in *radians*.

Two very useful additional formulas will be derived next. Solving for the depth y in Eq. (5.2-141) yields

$$y = y_0 + \frac{c(y) - c_0}{g}, \tag{5.2-150}$$

and upon substituting Eqs. (5.2-123) and (5.2-98) into Eq. (5.2-150), we obtain

$$y = y_0 + \frac{c_0}{g} \left[\frac{\sin \beta(y)}{\sin \beta_0} - 1 \right]. \tag{5.2-151}$$

Therefore, if the angle of arrival is known, the corresponding depth along a ray path can be obtained from Eq. (5.2-151).

Note that Eq. (5.2-148) expresses the horizontal range along a ray path as a function of depth via the angle of arrival $\beta(y)$ given by Eq. (5.2-97). However, solving for $\beta(y)$ from Eq. (5.2-148) yields

$$\beta(y) = \cos^{-1}[\cos \beta_0 - bg(r - r_0)]. \qquad (5.2\text{-}152)$$

Equation (5.2-152) expresses the angle of arrival as a function of the horizontal range along a ray path. The corresponding depth of the ray is obtained by substituting the value of $\beta(y)$ computed from Eq. (5.2-152) into Eq. (5.2-151). Therefore, Eqs. (5.2-152) and (5.2-151) can be used together to generate ray-trace plots with horizontal range r as the independent variable.

Figures 5.2-9 and 5.2-10 illustrate the construction of typical ray paths corresponding to linear SSPs with constant positive and negative gradients, respectively. The ray paths illustrated in Figs. 5.2-9*b* and 5.2-10*b* obey the following equations for a circle:

$$(r - r_{TP})^2 + [y + (R_c - y_{TP})]^2 = R_c^2, \qquad g > 0, \qquad (5.2\text{-}153)$$

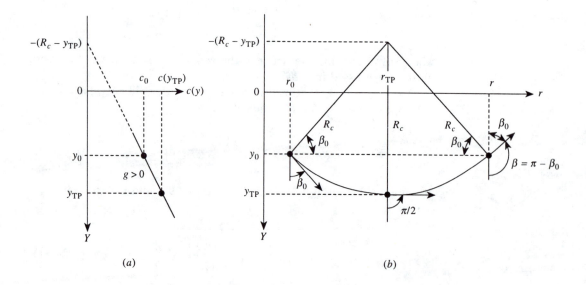

(a)   *(b)*

Figure 5.2-9. (*a*) Linear sound-speed profile with constant positive gradient *g*. (*b*) A typical corresponding ray path.

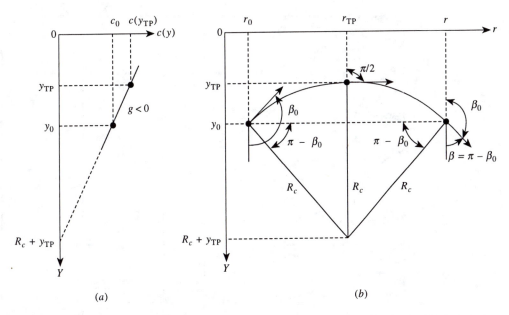

Figure 5.2-10. (*a*) Linear sound-speed profile with constant negative gradient *g*. (*b*) A typical corresponding ray path.

and

$$(r - r_{TP})^2 + [y - (R_c + y_{TP})]^2 = R_c^2, \qquad g < 0, \quad (5.2\text{-}154)$$

respectively, where the radius of curvature R_c is given by Eq. (5.2-144). Also note that

$$c[-(R_c - y_{TP})] = 0, \qquad g > 0, \qquad (5.2\text{-}155)$$

and

$$c(R_c + y_{TP}) = 0, \qquad g < 0, \qquad (5.2\text{-}156)$$

as shown in Figs. 5.2-9*a* and 5.2-10*a*, respectively. The range and depth coordinates of the *first* turning point (r_{TP}, y_{TP}) corresponding to an initial angle of propagation β_0 can be obtained from Eqs. (5.2-148) and (5.2-151), respectively, by setting $\beta(y) = \beta(y_{TP}) = \pi/2$.

And finally, Figs. 5.2-11 through 5.2-13 illustrate ray-trace plots based on the method of piecewise linear approximation. Figure 5.2-11 is an illustration of rays trapped in a surface duct [see Problem 5-3(a)].

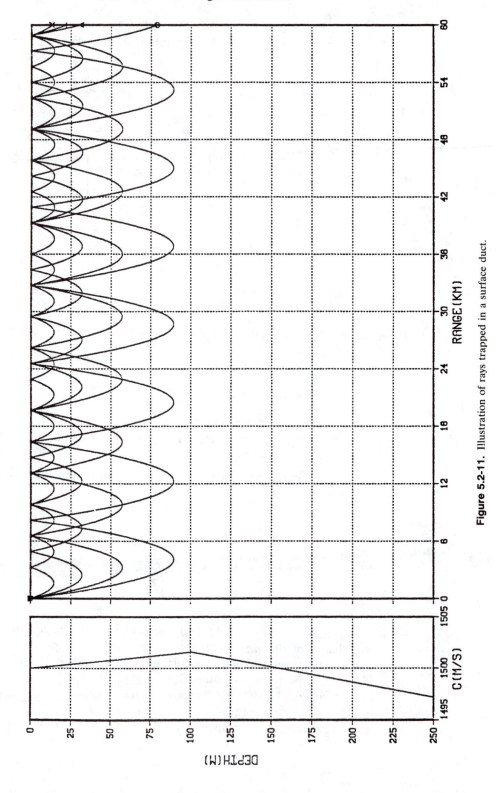

Figure 5.2-11. Illustration of rays trapped in a surface duct.

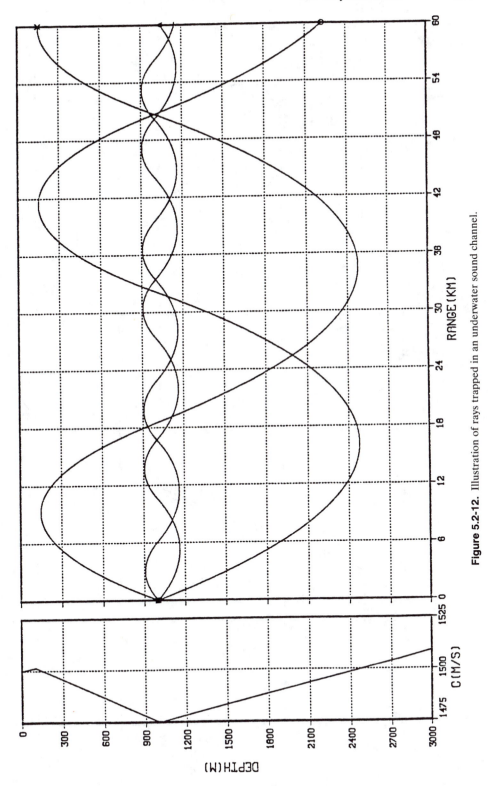

Figure 5.2-12. Illustration of rays trapped in an underwater sound channel.

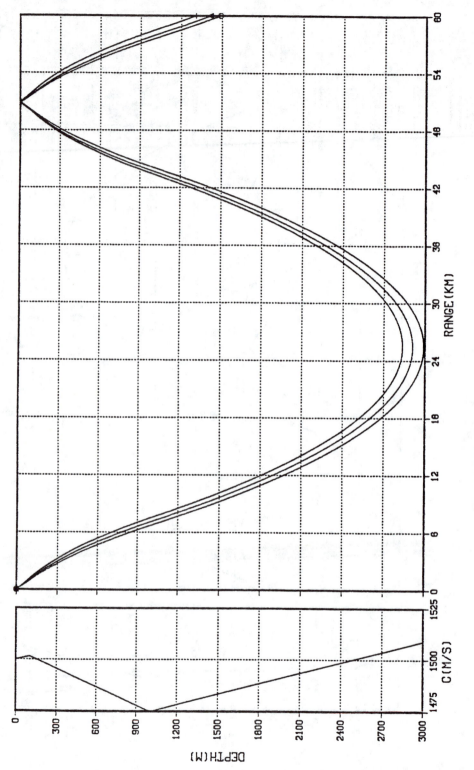

Figure 5.2-13. Illustration of a convergence zone (CZ).

Figure 5.2-12 is an illustration of rays trapped in an underwater sound channel [see Problems 5-3(b) and (c)]. Figure 5.2-13 is an illustration of a *convergence zone* (CZ) [see Problem 5-3(d)]. Note how the three rays shown converge at the ocean surface. By definition, rays associated with a CZ do *not* reflect from the ocean bottom. As might be expected, sound intensity is high within a CZ.

5.2.4 Amplitude Calculations along Rays — Solution of the Transport Equation

In this section we shall obtain an approximate solution for the real amplitude function $a(\mathbf{r})$. We begin by multiplying the transport equation given by Eq. (5.2-6) by $a(\mathbf{r})$. Doing so yields

$$a^2(\mathbf{r}) \, \nabla^2 W(\mathbf{r}) + 2a(\mathbf{r}) \, \nabla a(\mathbf{r}) \cdot \nabla W(\mathbf{r}) = \nabla \cdot \left[a^2(\mathbf{r}) \, \nabla W(\mathbf{r}) \right] = 0. \quad (5.2\text{-}157)$$

Therefore, solving the transport equation is equivalent to solving

$$\nabla \cdot \left[a^2(\mathbf{r}) \, \nabla W(\mathbf{r}) \right] = 0 \qquad (5.2\text{-}158)$$

or, upon substituting Eq. (5.2-70) into Eq. (5.2-158),

$$\nabla \cdot \left[\rho_0^{-1}(\mathbf{r}) \mathbf{I}_{\text{avg}}(\mathbf{r}) \right] = 0, \qquad (5.2\text{-}159)$$

where $\mathbf{I}_{\text{avg}}(\mathbf{r})$ is the time-average intensity vector and $\rho_0(\mathbf{r})$ is the equilibrium (ambient) density of the fluid medium.

The solution of Eq. (5.2-159) can be obtained by using the concept of a *ray tube*, as illustrated in Fig. 5.2-14b. Surfaces S_1 and S_2 shown in Fig. 5.2-14b are assumed to be very small surface areas of the wavefronts that pass through points P_1 and P_2, with position vectors $\mathbf{r}_1 = (x_1, y_1, z_1)$ and $\mathbf{r}_2 = (x_2, y_2, z_2)$, respectively, where $\hat{n}_1$ and $\hat{n}_2$ are unit vectors normal to S_1 and S_2, respectively, pointing in the conventional outward direction. The surface S_3 is formed by the outermost ray paths of the family of ray paths shown in Fig. 5.2-14a, where $\hat{n}_3$ is a unit vector normal to S_3 pointing in the conventional outward direction. Since the divergence of the time-average intensity vector divided by the scalar equilibrium density is equal to zero [see Eq. (5.2-159)], the number of ray paths entering S_1 per unit time is equal to the number of ray paths leaving S_2 per unit time.

Upon integrating both sides of Eq. (5.2-159) over the volume V occupied by the ray-tube segment connecting points P_1 and P_2, and by using the *divergence theorem*, we obtain

$$\int_V \nabla \cdot \rho_0^{-1}(\mathbf{r}) \mathbf{I}_{\text{avg}}(\mathbf{r}) \, dV = \oint_S \rho_0^{-1}(\mathbf{r}) \mathbf{I}_{\text{avg}}(\mathbf{r}) \cdot d\mathbf{S} = 0. \qquad (5.2\text{-}160)$$

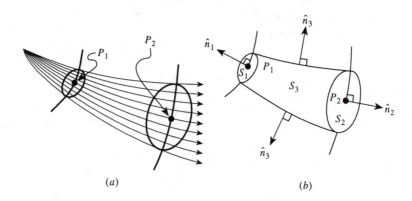

Figure 5.2-14. (*a*) Rays and wavefronts. (*b*) Ray-tube segment formed by the family of ray paths shown in (*a*).

By referring to Fig. 5.2-14*b*, Eq. (5.2-160) can be rewritten as follows:

$$\oint_S \rho_0^{-1}(\mathbf{r})\mathbf{I}_{\text{avg}}(\mathbf{r}) \cdot d\mathbf{S} = \int_{S_1} \rho_0^{-1}(\mathbf{r})\mathbf{I}_{\text{avg}}(\mathbf{r}) \cdot \hat{n}_1 \, dS_1 + \int_{S_2} \rho_0^{-1}(\mathbf{r})\mathbf{I}_{\text{avg}}(\mathbf{r}) \cdot \hat{n}_2 \, dS_2$$

$$+ \int_{S_3} \rho_0^{-1}(\mathbf{r})\mathbf{I}_{\text{avg}}(\mathbf{r}) \cdot \hat{n}_3 \, dS_3 = 0. \qquad (5.2\text{-}161)$$

Since the direction of $\mathbf{I}_{\text{avg}}(\mathbf{r})$ is determined by the direction of the ray paths (see Example 5.2-2), by referring once again to Fig. 5.2-14*b*, it can be seen that $\hat{n}_1$ is in the opposite direction to $\mathbf{I}_{\text{avg}}(\mathbf{r})$, $\hat{n}_2$ is in the same direction as $\mathbf{I}_{\text{avg}}(\mathbf{r})$, and $\hat{n}_3$ is normal to $\mathbf{I}_{\text{avg}}(\mathbf{r})$. As a result,

$$\mathbf{I}_{\text{avg}}(\mathbf{r}) \cdot \hat{n}_1 = -I_{\text{avg}}(\mathbf{r}), \qquad (5.2\text{-}162)$$

$$\mathbf{I}_{\text{avg}}(\mathbf{r}) \cdot \hat{n}_2 = +I_{\text{avg}}(\mathbf{r}), \qquad (5.2\text{-}163)$$

and

$$\mathbf{I}_{\text{avg}}(\mathbf{r}) \cdot \hat{n}_3 = 0, \qquad (5.2\text{-}164)$$

where $I_{\text{avg}}(\mathbf{r})$ is the magnitude of the time-average intensity vector. Therefore, upon substituting Eqs. (5.2-162) through (5.2-164) into Eq. (5.2-161), we obtain

$$\int_{S_2} \rho_0^{-1}(\mathbf{r}) I_{\text{avg}}(\mathbf{r}) \, dS_2 = \int_{S_1} \rho_0^{-1}(\mathbf{r}) I_{\text{avg}}(\mathbf{r}) \, dS_1. \qquad (5.2\text{-}165)$$

Note that if the equilibrium density were equal to a constant, so that it could be canceled from both sides of Eq. (5.2-165), then Eq. (5.2-165) would indicate that the time-average power flowing through surfaces S_1 and S_2 is equal.

Since we have already assumed that S_1 and S_2 are very small surface areas,

$$\rho_0^{-1}(\mathbf{r})\,I_{\mathrm{avg}}(\mathbf{r}) \approx \rho_0^{-1}(\mathbf{r}_1)\,I_{\mathrm{avg}}(\mathbf{r}_1) \qquad \text{on } S_1 \qquad (5.2\text{-}166)$$

and

$$\rho_0^{-1}(\mathbf{r})\,I_{\mathrm{avg}}(\mathbf{r}) \approx \rho_0^{-1}(\mathbf{r}_2)\,I_{\mathrm{avg}}(\mathbf{r}_2) \qquad \text{on } S_2. \qquad (5.2\text{-}167)$$

Therefore,

$$\int_{S_1}\rho_0^{-1}(\mathbf{r})\,I_{\mathrm{avg}}(\mathbf{r})\,dS_1 \approx \rho_0^{-1}(\mathbf{r}_1)\,I_{\mathrm{avg}}(\mathbf{r}_1)\int_{S_1} dS_1 = \rho_0^{-1}(\mathbf{r}_1)\,I_{\mathrm{avg}}(\mathbf{r}_1)S_1 \quad (5.2\text{-}168)$$

and

$$\int_{S_2}\rho_0^{-1}(\mathbf{r})\,I_{\mathrm{avg}}(\mathbf{r})\,dS_2 \approx \rho_0^{-1}(\mathbf{r}_2)\,I_{\mathrm{avg}}(\mathbf{r}_2)\int_{S_2} dS_2 = \rho_0^{-1}(\mathbf{r}_2)\,I_{\mathrm{avg}}(\mathbf{r}_2)S_2, \quad (5.2\text{-}169)$$

where S_1 and S_2 are the wavefront surface areas at the ends of the ray-tube segment. Substituting Eqs. (5.2-168) and (5.2-169) into Eq. (5.2-165) yields

$$\rho_0^{-1}(\mathbf{r}_2)\,I_{\mathrm{avg}}(\mathbf{r}_2)S_2 \approx \rho_0^{-1}(\mathbf{r}_1)\,I_{\mathrm{avg}}(\mathbf{r}_1)S_1, \qquad (5.2\text{-}170)$$

and upon substituting Eq. (5.2-71) into Eq. (5.2-170) and making use of Eqs. (5.2-9) and (5.1-6), we finally obtain

$$a(\mathbf{r}_2) \approx a(\mathbf{r}_1)\left[\frac{c(\mathbf{r}_2)}{c(\mathbf{r}_1)}\frac{S_1}{S_2}\right]^{1/2}. \qquad (5.2\text{-}171)$$

Equation (5.2-171) expresses the amplitude $a(\mathbf{r}_2)$ of the velocity potential at some point P_2 along a ray path in terms of the amplitude $a(\mathbf{r}_1)$ of the velocity potential at some previous point P_1 along the same ray path (see Fig. 5.2-14). When S_2 becomes *smaller* than S_1, the amplitude, and hence the intensity, will *increase*, since the ray tube is *contracting* due to the *focusing* of the rays. In the limit as S_2 approaches zero, the amplitude (intensity) approaches infinity. Points at which S_2 is equal to zero are known as *focal points*. Equation (5.2-171) is *invalid* at focal points. However, when S_2 becomes *larger* than S_1, the amplitude (intensity) will *decrease*, since the ray tube is *spreading* due to the *divergence* of the rays. Note that a *caustic* is a surface composed entirely of focal points. Later, in Section 5.2.5, we shall derive equations for the magnitude squared of the acoustic pressure that are valid (i.e., finite) at both turning points and focal points.

Alternative Amplitude Expressions

Equation (5.2-171) is a very common expression for the amplitude of the velocity potential along a ray path. Alternative expressions for the amplitude can be derived *without the use of a ray tube* as follows. We begin by dividing the transport equation given by Eq. (5.2-6) by $a(\mathbf{r})$. Doing so yields

$$\nabla^2 W(\mathbf{r}) + \frac{2}{a(\mathbf{r})} \nabla a(\mathbf{r}) \cdot \nabla W(\mathbf{r}) = 0. \qquad (5.2\text{-}172)$$

Since

$$\nabla \ln a(\mathbf{r}) = \frac{\nabla a(\mathbf{r})}{a(\mathbf{r})}, \qquad (5.2\text{-}173)$$

substituting Eqs. (5.2-35) and (5.2-173) into Eq. (5.2-172) yields

$$\nabla^2 W(\mathbf{r}) + 2n(\mathbf{r}) \nabla \ln a(\mathbf{r}) \cdot \hat{n}(\mathbf{r}) = 0, \qquad (5.2\text{-}174)$$

since the index of refraction $n(\mathbf{r})$ is a scalar.

Next, compute the directional derivative, $d \ln a(\mathbf{r})/ds$, which represents the rate of change of $\ln a(\mathbf{r})$ with respect to the arc length s along a ray path. Expanding the directional derivative by using the chain rule yields [see Eq. (5.2-52)]

$$\frac{d}{ds} \ln a(\mathbf{r}) = \nabla \ln a(\mathbf{r}) \cdot \hat{n}(\mathbf{r}), \qquad (5.2\text{-}175)$$

and upon substituting Eq. (5.2-175) into Eq. (5.2-174), we obtain the following alternative expression for the transport equation:

$$\boxed{\frac{d}{ds} \ln a(\mathbf{r}) = -\frac{\nabla^2 W(\mathbf{r})}{2n(\mathbf{r})}.} \qquad (5.2\text{-}176)$$

The solution of Eq. (5.2-176) can be expressed as

$$\ln\left[\frac{a(\mathbf{r}_2)}{a(\mathbf{r}_1)}\right] = -\frac{1}{2} \int_{\mathbf{r}_1}^{\mathbf{r}_2} \frac{\nabla^2 W(\mathbf{r})}{n(\mathbf{r})}\, ds, \qquad (5.2\text{-}177)$$

and as a result,

$$a(\mathbf{r}_2) = a(\mathbf{r}_1) \exp\left[-\frac{1}{2} \int_{\mathbf{r}_1}^{\mathbf{r}_2} \frac{\nabla^2 W(\mathbf{r})}{n(\mathbf{r})}\, ds\right]. \qquad (5.2\text{-}178)$$

Equation (5.2-178) can be simplified further.

Since [see Eq. (5.2-35)]

$$\nabla W(\mathbf{r}) = n(\mathbf{r})\hat{n}(\mathbf{r}) \tag{5.2-179}$$

and

$$\nabla^2 W(\mathbf{r}) = \nabla \cdot \nabla W(\mathbf{r}), \tag{5.2-180}$$

substituting Eq. (5.2-179) into Eq. (5.2-180) yields

$$\nabla^2 W(\mathbf{r}) = \nabla \cdot n(\mathbf{r})\hat{n}(\mathbf{r}) = n(\mathbf{r})\nabla \cdot \hat{n}(\mathbf{r}) + \hat{n}(\mathbf{r}) \cdot \nabla n(\mathbf{r}), \tag{5.2-181}$$

where [see Eq. (5.2-36)]

$$\hat{n}(\mathbf{r}) = u(\mathbf{r})\hat{x} + v(\mathbf{r})\hat{y} + w(\mathbf{r})\hat{z} \tag{5.2-182}$$

is the unit vector in the direction of $\nabla W(\mathbf{r})$, and $u(\mathbf{r})$, $v(\mathbf{r})$, and $w(\mathbf{r})$ are the dimensionless direction cosines with respect to the X, Y, and Z axes, respectively. And since [see Eq. (5.2-52)]

$$\frac{d}{ds}n(\mathbf{r}) = \nabla n(\mathbf{r}) \cdot \hat{n}(\mathbf{r}), \tag{5.2-183}$$

substituting Eq. (5.2-183) into Eq. (5.2-181) yields

$$\frac{\nabla^2 W(\mathbf{r})}{n(\mathbf{r})} = \nabla \cdot \hat{n}(\mathbf{r}) + \frac{1}{n(\mathbf{r})}\frac{d}{ds}n(\mathbf{r}). \tag{5.2-184}$$

Substituting Eq. (5.2-184) into Eq. (5.2-178) yields

$$a(\mathbf{r}_2) = a(\mathbf{r}_1) \exp\left[-\frac{1}{2} \int_{n(\mathbf{r}_1)}^{n(\mathbf{r}_2)} \frac{1}{n(\mathbf{r})} \, dn(\mathbf{r}) \right] \exp\left[-\frac{1}{2} \int_{\mathbf{r}_1}^{\mathbf{r}_2} \nabla \cdot \hat{n}(\mathbf{r}) \, ds \right] \tag{5.2-185}$$

$$= a(\mathbf{r}_1) \exp\left[\ln \left| \frac{n(\mathbf{r}_2)}{n(\mathbf{r}_1)} \right|^{-1/2} \right] \exp\left[-\frac{1}{2} \int_{\mathbf{r}_1}^{\mathbf{r}_2} \nabla \cdot \hat{n}(\mathbf{r}) \, ds \right], \tag{5.2-186}$$

and noting that the real index of refraction is always positive, we obtain the following alternative expressions for the amplitude of the velocity potential along a ray path derived without the use of a ray tube:

$$\boxed{a(\mathbf{r}_2) = a(\mathbf{r}_1)\left[\frac{n(\mathbf{r}_1)}{n(\mathbf{r}_2)}\right]^{1/2} \exp\left[-\frac{1}{2} \int_{\mathbf{r}_1}^{\mathbf{r}_2} \nabla \cdot \hat{n}(\mathbf{r}) \, ds \right],} \tag{5.2-187}$$

or, upon substituting Eq. (5.1-6) into Eq. (5.2-187),

$$a(\mathbf{r}_2) = a(\mathbf{r}_1) \left[\frac{c(\mathbf{r}_2)}{c(\mathbf{r}_1)} \right]^{1/2} \exp \left[-\frac{1}{2} \int_{\mathbf{r}_1}^{\mathbf{r}_2} \nabla \cdot \hat{n}(\mathbf{r}) \, ds \right]. \qquad (5.2\text{-}188)$$

By comparing Eqs. (5.2-171) and (5.2-188), it can be seen that the task of computing S_1 and S_2, which are the very small wavefront surface areas that correspond to the ends of the ray-tube segment illustrated in Fig. 5.2-14b, is equivalent to evaluating the integral of the divergence of the unit vector along a ray path, that is,

$$\left[\frac{S_1}{S_2} \right]^{1/2} \equiv \exp \left[-\frac{1}{2} \int_{\mathbf{r}_1}^{\mathbf{r}_2} \nabla \cdot \hat{n}(\mathbf{r}) \, ds \right]. \qquad (5.2\text{-}189)$$

Finally, let us expand the integral

$$\int_{\mathbf{r}_1}^{\mathbf{r}_2} \nabla \cdot \hat{n}(\mathbf{r}) \, ds.$$

Since [see Eq. (5.2-182)]

$$\nabla \cdot \hat{n}(\mathbf{r}) \, ds = \frac{\partial}{\partial x} u(\mathbf{r}) \, ds + \frac{\partial}{\partial y} v(\mathbf{r}) \, ds + \frac{\partial}{\partial z} w(\mathbf{r}) \, ds \qquad (5.2\text{-}190)$$

and [see Eqs. (5.2-42) through (5.2-44)]

$$ds = \frac{dx}{u(\mathbf{r})}, \qquad (5.2\text{-}191)$$

$$ds = \frac{dy}{v(\mathbf{r})}, \qquad (5.2\text{-}192)$$

and

$$ds = \frac{dz}{w(\mathbf{r})}, \qquad (5.2\text{-}193)$$

substituting Eqs. (5.2-191) through (5.2-193) into Eq. (5.2-190) yields

$$\nabla \cdot \hat{n}(\mathbf{r}) \, ds = \frac{1}{u(\mathbf{r})} \frac{\partial}{\partial x} u(\mathbf{r}) \, dx + \frac{1}{v(\mathbf{r})} \frac{\partial}{\partial y} v(\mathbf{r}) \, dy + \frac{1}{w(\mathbf{r})} \frac{\partial}{\partial z} w(\mathbf{r}) \, dz. \qquad (5.2\text{-}194)$$

Therefore, with the use of Eq. (5.2-194),

$$\int_{\mathbf{r}_1}^{\mathbf{r}_2} \nabla \cdot \hat{n}(\mathbf{r})\, ds = \int_{x_1}^{x_2} \frac{1}{u(\zeta, y_2, z_2)} \frac{\partial}{\partial \zeta} u(\zeta, y_2, z_2)\, d\zeta$$

$$+ \int_{y_1}^{y_2} \frac{1}{v(x_2, \zeta, z_2)} \frac{\partial}{\partial \zeta} v(x_2, \zeta, z_2)\, d\zeta \qquad (5.2\text{-}195)$$

$$+ \int_{z_1}^{z_2} \frac{1}{w(x_2, y_2, \zeta)} \frac{\partial}{\partial \zeta} w(x_2, y_2, \zeta)\, d\zeta.$$

Example 5.2-5 (Derivation of the WKB amplitude function)
In this example we shall show that the general expressions for the amplitude of the velocity potential along a ray path based on three-dimensional ray acoustics given by either Eq. (5.2-187) or Eq. (5.2-188), in conjunction with Eq. (5.2-195), reduce to the WKB amplitude function when the speed of sound is only an arbitrary function of depth. Substituting Eqs. (5.2-42) and (5.2-44) into Eqs. (5.2-82) and (5.2-83), respectively, yields

$$u(\mathbf{r}) = u(y) = A/n(y) \qquad (5.2\text{-}196)$$

and

$$w(\mathbf{r}) = w(y) = B/n(y), \qquad (5.2\text{-}197)$$

where A and B are constants, and where it is also true that $v(\mathbf{r}) = v(y)$ when $c(\mathbf{r}) = c(y)$. Therefore,

$$\frac{\partial}{\partial x} u(\mathbf{r}) = \frac{\partial}{\partial x} u(y) = 0, \qquad (5.2\text{-}198)$$

$$\frac{\partial}{\partial y} v(\mathbf{r}) = \frac{\partial}{\partial y} v(y) = \frac{d}{dy} v(y), \qquad (5.2\text{-}199)$$

and

$$\frac{\partial}{\partial z} w(\mathbf{r}) = \frac{\partial}{\partial z} w(y) = 0. \qquad (5.2\text{-}200)$$

Substituting Eqs. (5.2-198) through (5.2-200) into Eq. (5.2-195) yields

$$\int_{\mathbf{r}_1}^{\mathbf{r}_2} \nabla \cdot \hat{n}(\mathbf{r})\, ds = \int_{v(y_1)}^{v(y_2)} \frac{dv(\zeta)}{v(\zeta)} = \ln\left| \frac{v(y_2)}{v(y_1)} \right|, \qquad (5.2\text{-}201)$$

and upon substituting Eqs. (5.2-78) and (5.2-201) into Eq. (5.2-187), we obtain

$$a(\mathbf{r}_2) = a(\mathbf{r}_1) \left[\frac{n(y_1)}{n(y_2)} \right]^{1/2} \exp \left[-\frac{1}{2} \ln \left| \frac{v(y_2)}{v(y_1)} \right| \right]$$

$$= a(\mathbf{r}_1) \left[\frac{n(y_1)}{n(y_2)} \right]^{1/2} \left| \frac{v(y_1)}{v(y_2)} \right|^{1/2}, \tag{5.2-202}$$

where [see Eq. (5.2-43)]

$$v(y) = \cos \beta(y). \tag{5.2-203}$$

If we further substitute Eqs. (5.2-78) and (5.2-203) into Eq. (5.2-202), then we obtain the following expression for the amplitude of the velocity potential along a ray path when the speed of sound is only an arbitrary function of depth:

$$a(\mathbf{r}_2) = a(\mathbf{r}_1) \left[\frac{c(y_2)}{c(y_1)} \right]^{1/2} \left| \frac{\cos \beta(y_1)}{\cos \beta(y_2)} \right|^{1/2}. \tag{5.2-204}$$

Note that the angle $\beta(y)$ can be obtained from Snell's law. Also, by comparing Eqs. (5.2-171) and (5.2-204), it can be seen that

$$\left[\frac{S_1}{S_2} \right]^{1/2} \equiv \left| \frac{\cos \beta(y_1)}{\cos \beta(y_2)} \right|^{1/2}. \tag{5.2-205}$$

Although Eq. (5.2-204) is a valid expression for the amplitude along a ray path, we shall proceed to rewrite it in a different form in order to compare it directly with the WKB amplitude function given by Eq. (5.1-40). If we replace y_1, y_2, $\mathbf{r}_1$, and $\mathbf{r}_2$ with y_0, y, $\mathbf{r}_0$, and $\mathbf{r}$, respectively, in Eq. (5.2-202), and if we multiply and divide the right-hand side of Eq. (5.2-202) by $\sqrt{k_0}$, then

$$a(\mathbf{r}) = a(\mathbf{r}_0) \left| \frac{2\pi v_0 / \lambda_0}{k_0 n(y) v(y)} \right|^{1/2}, \tag{5.2-206}$$

where $n(y_0) = 1$ [see Eq. (5.2-78)], $v_0 = v(y_0)$, and [see Eq. (5.1-7)]

$$k_0 = 2\pi f / c_0 = 2\pi / \lambda_0, \tag{5.2-207}$$

where $c_0 = f\lambda_0$. If we next let

$$f_Y = v_0/\lambda_0, \qquad (5.2\text{-}208)$$

which is the input or transmitted spatial frequency in the Y direction (see Section 2.2.1), and since [see Eqs. (5.2-61) and (5.1-8)]

$$k_Y(y) = k(y)v(y) = k_0 n(y)v(y), \qquad (5.2\text{-}209)$$

Eq. (5.2-206) can be rewritten as

$$a(\mathbf{r}) = a(y) = a(\mathbf{r}_0)\sqrt{\left|\frac{k_Y(y_0)}{k_Y(y)}\right|} \qquad (5.2\text{-}210)$$

or

$$a(\mathbf{r}) = a(y) = \frac{A}{\sqrt{|k_Y(y)|}}, \qquad (5.2\text{-}211)$$

where

$$A = a(\mathbf{r}_0)\sqrt{|k_Y(y_0)|} \qquad (5.2\text{-}212)$$

is a real, positive constant and

$$k_Y(y_0) = 2\pi f_Y. \qquad (5.2\text{-}213)$$

Note that the right-hand side of Eq. (5.2-211) is *identical* with the right-hand side of the amplitude function given by Eq. (5.1-40) obtained by using the WKB method. Recall that the WKB amplitude function is invalid at turning points.

5.2.5 Acoustic Pressure Calculations for Depth-Dependent Speeds of Sound Valid at Turning Points and Focal Points

In this section we shall derive two different sets of equations for the magnitude squared of the acoustic pressure along a ray path using the concept of a ray tube when the speed of sound is an arbitrary function of depth. The first set of equations is valid everywhere except at or near turning points and when the initial angle of propagation is equal to 90°. The second set of equations is valid everywhere (even at or near turning points) except at those depths where the first derivative of the speed of sound is equal to zero and when the initial angle of propagation is equal to 90°. A separate equation valid when the initial angle of propagation is equal to 90° shall also be derived. By using a suitable criterion to be established later, the magnitude squared of the acoustic pressure along a ray path can be computed by switching between the two sets of equations.

We begin the analysis by referring to Fig. 5.2-15, which illustrates a slice through a ray tube, where r is the horizontal range. An omnidirectional point source is located at depth y_0. If the acoustic field is axisymmetric, that is, independent of the azimuthal angle ϕ (see Fig. 5.1-5), and if we invoke conservation of energy, then

$$I_{\text{avg}}(r, y) \, dS = I_{\text{avg}}(r_1, y_1) \, dS_1, \qquad (5.2\text{-}214)$$

where $I_{\text{avg}}(\cdot)$ is the time-average intensity of the acoustic field, and dS_1 and dS are the infinitesimal wavefront surface areas at the beginning and end of the ray

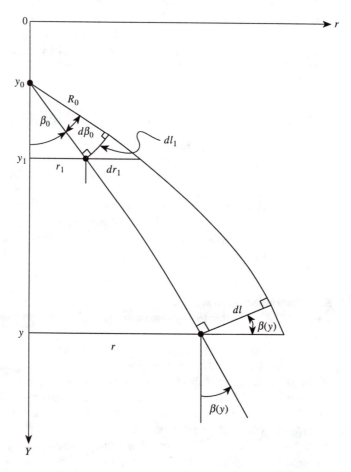

Figure 5.2-15. Illustration of a slice through a ray tube.

tube, respectively, created by rotating the ray-tube slice illustrated in Fig. 5.2-15 by 360° around the Y axis. In addition, if both the speed of sound and the equilibrium (ambient) density are functions of depth y only, then [see Eq. (5.2-72)]

$$I_{\text{avg}}(r, y) = \frac{|p_f(r, y)|^2}{2\rho_0(y)c(y)}.$$ (5.2-215)

In order to proceed further, we need expressions for dS_1 and dS.

By referring to Fig. 5.2-15, we can write that

$$dS_1 = 2\pi r_1 \, dl_1,$$ (5.2-216)

where $2\pi r_1$ is the circumference of a circle with radius

$$r_1 = R_0 \sin \beta_0,$$ (5.2-217)

$\beta_0 = \beta(y_0)$ is the initial angle of propagation at the source depth y_0, and

$$dl_1 = R_0 \, d\beta_0$$ (5.2-218)

is an infinitesimal element of arc length along the wavefront that passes through the point (r_1, y_1). Therefore, substituting Eqs. (5.2-217) and (5.2-218) into Eq. (5.2-216) yields

$$dS_1 = 2\pi R_0^2 \sin \beta_0 \, d\beta_0.$$ (5.2-219)

Also note that

$$y_1 = y_0 + R_0 \cos \beta_0.$$ (5.2-220)

Similarly,

$$dS = 2\pi r \, dl,$$ (5.2-221)

where $2\pi r$ is the circumference of a circle with radius r and

$$dl = \cos \beta(y) \, dr$$ (5.2-222)

is an infinitesimal element of arc length along the wavefront that passes through the point (r, y). Therefore, substituting Eq. (5.2-222) into Eq. (5.2-221) yields

$$dS = 2\pi r \cos \beta(y) \, dr.$$ (5.2-223)

Upon substituting Eqs. (5.2-215), (5.2-219), and (5.2-223) into Eq. (5.2-214), we obtain the following general expression for the magnitude squared of the acoustic pressure along a ray path:

$$\left| p_f(r, y) \right|^2 = \left| p_f(r_1, y_1) \right|^2 \frac{\rho_0(y)c(y)}{\rho_0(y_1)c(y_1)} \frac{R_0^2 \sin \beta_0}{r \left| \cos \beta(y) \, dr/d\beta_0 \right|}, \qquad (5.2\text{-}224)$$

where the magnitude of $\cos \beta(y) \, dr/d\beta_0$ is taken in order to ensure that the magnitude squared of the acoustic pressure is positive. Note that if we let $R_0 = 1$ m, then $\left| p_f(r_1, y_1) \right|$ is related to the source level (SL) of the omnidirectional point source (see Section 1.6). In order to proceed further, we need an expression for $dr/d\beta_0$.

By referring to Eq. (5.2-108), which is rewritten below for convenience,

$$\Delta r = r - r_0 = b \int_{y_0}^{y} \frac{c(\zeta)}{\sqrt{1 - b^2 c^2(\zeta)}} \, d\zeta, \qquad (5.2\text{-}108)$$

it can be seen that the horizontal range r along a ray path can be expressed as a function of β_0 [via the ray parameter b given by Eq. (5.2-98)] and y, that is, $r = r(\beta_0, y)$. Therefore, by treating β_0 and y as *independent variables*, and by using the chain rule,

$$\frac{dr}{d\beta_0} = \frac{d}{d\beta_0} r(\beta_0, y) = \frac{\partial}{\partial \beta_0} r(\beta_0, y) \frac{d\beta_0}{d\beta_0} + \frac{\partial}{\partial y} r(\beta_0, y) \frac{dy}{d\beta_0}, \quad (5.2\text{-}225)$$

and since $dy/d\beta_0 = 0$ when β_0 and y are treated as independent variables,

$$\frac{dr}{d\beta_0} = \frac{\partial}{\partial \beta_0} r(\beta_0, y). \qquad (5.2\text{-}226)$$

Since r is also given by Eq. (5.2-106), which is also rewritten below for convenience,

$$r = r_0 + \int_{y_0}^{y} \tan \beta(\zeta) \, d\zeta, \qquad (5.2\text{-}106)$$

with the use of *Leibnitz's rule*,

$$\frac{\partial r}{\partial \beta_0} = \frac{\partial}{\partial \beta_0} r_0 + \int_{y_0}^{y} \frac{\partial}{\partial \beta_0} \tan \beta(\zeta) \, d\zeta + \tan \beta(y) \frac{\partial}{\partial \beta_0} y - \tan \beta(y_0) \frac{\partial}{\partial \beta_0} y_0,$$

$$(5.2\text{-}227)$$

and since r_0 and y_0 are constants, and $\partial y/\partial \beta_0 = 0$ when β_0 and y are treated as independent variables, we have

$$\frac{\partial r}{\partial \beta_0} = \int_{y_0}^{y} \frac{1}{\cos^2 \beta(\zeta)} \frac{\partial}{\partial \beta_0} \beta(\zeta) \, d\zeta. \qquad (5.2\text{-}228)$$

In order to proceed further, we need an expression for $\partial \beta(y)/\partial \beta_0$.

By referring to Snell's law given by Eq. (5.1-75), we can write that

$$\sin \beta(y) = \frac{c(y)}{c_0} \sin \beta_0, \qquad (5.2\text{-}229)$$

and as a result,

$$\frac{\partial}{\partial \beta_0} \sin \beta(y) = \cos \beta(y) \frac{\partial}{\partial \beta_0} \beta(y) = \frac{c(y)}{c_0} \cos \beta_0 + \frac{\sin \beta_0}{c_0} \frac{\partial}{\partial \beta_0} c(y), \qquad (5.2\text{-}230)$$

and since $\partial c(y)/\partial \beta_0 = 0$ when β_0 and y are treated as independent variables,

$$\frac{\partial}{\partial \beta_0} \beta(y) = \frac{c(y)}{c_0} \frac{\cos \beta_0}{\cos \beta(y)}. \qquad (5.2\text{-}231)$$

Upon substituting Eq. (5.2-229) into Eq. (5.2-231), we obtain

$$\frac{\partial}{\partial \beta_0} \beta(y) = \cot \beta_0 \tan \beta(y). \qquad (5.2\text{-}232)$$

It is interesting to note that the square root of the magnitude of Eq. (5.2-231) is related to the WKB amplitude function given by Eq. (5.2-204).

Substituting Eq. (5.2-232) into Eq. (5.2-228) yields

$$\frac{\partial r}{\partial \beta_0} = \cot \beta_0 \int_{y_0}^{y} \frac{\sin \beta(\zeta)}{\cos^3 \beta(\zeta)} \, d\zeta, \qquad (5.2\text{-}233)$$

and with the use of Eqs. (5.2-102), (5.2-123), and (5.2-98) in Eq. (5.2-233), substituting the resulting expression into Eq. (5.2-226) yields

$$\boxed{\frac{dr}{d\beta_0} = \frac{\partial}{\partial \beta_0} r(\beta_0, y) = \frac{\cos \beta_0}{c_0} \int_{y_0}^{y} \frac{c(\zeta)}{\left[1 - b^2 c^2(\zeta)\right]^{3/2}} \, d\zeta.} \qquad (5.2\text{-}234)$$

Finally, by multiplying both sides of Eq. (5.2-234) by $\cos \beta(y)$, we obtain

$$\cos \beta(y) \frac{dr}{d\beta_0} = \cos \beta(y) \frac{\cos \beta_0}{c_0} \int_{y_0}^{y} \frac{c(\zeta)}{[1 - b^2 c^2(\zeta)]^{3/2}} \, d\zeta. \quad (5.2\text{-}235)$$

Equations (5.2-224) and (5.2-235) represent the first set of equations to be used to compute the magnitude squared of the acoustic pressure along a ray path. Equation (5.2-235) is valid everywhere except at or near a turning point (r_{TP}, y_{TP}), since the depth y_{TP} of a turning point is a branch point (singularity) of the integrand in Eq. (5.2-235) [see Eq. (5.2-104)]. In addition, as $y \to y_{TP}$, we have $\beta(y) \to 90°$ and $\cos \beta(y) \to 0$ as the integrand in Eq. (5.2-235) approaches infinity, resulting in the indeterminate form $0 \times \infty$. Finally, note that Eq. (5.2-235) approaches zero as $\beta_0 \to 90°$, and as a result, Eq. (5.2-224) approaches infinity. In summary, the magnitude squared of the acoustic pressure given by Eqs. (5.2-224) and (5.2-235) is valid everywhere *except at or near a turning point and as long as* $\beta_0 \neq 90°$. A separate equation valid for $\beta_0 = 90°$ will be derived later.

The second set of equations to be derived is required to be finite at or near a turning point. We begin the derivation by substituting Eqs. (5.2-232) and (5.2-105) into Eq. (5.2-228), and with the use of Eq. (5.2-226), we obtain

$$\frac{dr}{d\beta_0} = \frac{\partial}{\partial \beta_0} r(\beta_0, y) = \cot \beta_0 \int_{r_0}^{r} \frac{dr}{\cos^2 \beta(y)}, \quad (5.2\text{-}236)$$

where it is understood that $y = y(r)$ in the integrand on the right-hand side of Eq. (5.2-236). Although Eq. (5.2-236) can be used instead of Eq. (5.2-234), which is preferable if horizontal range r is being used as the independent variable for generating the ray trace, Eq. (5.2-236) has the same problems with turning points and $\beta_0 = 90°$ as Eq. (5.2-234) does.

Continuing with the derivation, note that

$$d \tan \beta(y) = \frac{1}{\cos^2 \beta(y)} \, d\beta(y). \quad (5.2\text{-}237)$$

And since

$$\frac{dr}{d\beta(y)} = \frac{dr}{dy} \frac{dy}{ds} \frac{ds}{d\beta(y)}, \quad (5.2\text{-}238)$$

substituting Eqs. (5.2-90), (5.2-100), and (5.2-105) into Eq. (5.2-238) yields

$$d\beta(y) = \frac{c'(y)}{c(y)} \, dr. \quad (5.2\text{-}239)$$

Therefore, upon substituting Eq. (5.2-239) into Eq. (5.2-237), we obtain

$$\frac{dr}{\cos^2 \beta(y)} = \frac{c(y)}{c'(y)} d \tan \beta(y), \qquad (5.2\text{-}240)$$

and as a result,

$$\int_{r_0}^{r} \frac{dr}{\cos^2 \beta(y)} = \int_{r_0}^{r} \frac{c(y)}{c'(y)} d \tan \beta(y). \qquad (5.2\text{-}241)$$

Evaluating the right-hand side of Eq. (5.2-241) by using the method of integration by parts yields

$$\int_{r_0}^{r} \frac{dr}{\cos^2 \beta(y)} = \frac{c(y)}{c'(y)} \tan \beta(y) - \frac{c_0}{c_0'} \tan \beta_0 - \int_{r_0}^{r} \tan \beta(y) \, d\frac{c(y)}{c'(y)}, \qquad (5.2\text{-}242)$$

where $c_0 = c(y_0)$, $c_0' = c'(y_0)$, $\beta_0 = \beta(y_0)$, and $y_0 = y(r_0)$. Since

$$d\frac{c(y)}{c'(y)} = \frac{[c'(y)]^2 - c(y)c''(y)}{[c'(y)]^2} dy, \qquad (5.2\text{-}243)$$

substituting Eqs. (5.2-243) and (5.2-105) into Eq. (5.2-242) and then substituting the result into Eq. (5.2-236) yields

$$\boxed{\begin{aligned}
\frac{dr}{d\beta_0} &= \frac{\partial}{\partial \beta_0} r(\beta_0, y) = \cot \beta_0 \tan \beta(y) \frac{c(y)}{c'(y)} - \frac{c_0}{c_0'} \\
&\quad + \cot \beta_0 \int_{r_0}^{r} \frac{c(y)c''(y) - [c'(y)]^2}{[c'(y)]^2} dr.
\end{aligned}} \qquad (5.2\text{-}244)$$

Finally, by multiplying both sides of Eq. (5.2-244) by $\cos \beta(y)$, we obtain

$$\boxed{\begin{aligned}
\cos \beta(y) \frac{dr}{d\beta_0} &= \cot \beta_0 \sin \beta(y) \frac{c(y)}{c'(y)} \\
&\quad + \cos \beta(y) \left[\cot \beta_0 \int_{r_0}^{r} \frac{c(y)c''(y) - [c'(y)]^2}{[c'(y)]^2} dr - \frac{c_0}{c_0'} \right].
\end{aligned}}$$

$$(5.2\text{-}245)$$

In Appendix 5B it is shown that Eq. (5.2-245) can also be derived by treating β_0 and $\beta(y)$ as the independent variables, instead of β_0 and y.

Equations (5.2-224) and (5.2-245) represent the second set of equations to be used to compute the magnitude squared of the acoustic pressure along a ray path. Note that at a turning point (r_{TP}, y_{TP}), where $\beta(y_{TP}) = 90°$, Eq. (5.2-245) reduces to

$$\left[\cos \beta(y) \frac{dr}{d\beta_0}\right]_{(r_{TP}, y_{TP})} = \cot \beta_0 \frac{c(y_{TP})}{c'(y_{TP})}, \qquad (5.2\text{-}246)$$

and upon substituting Eq. (5.2-246) into Eq. (5.2-224), we obtain

$$\left|p_f(r_{TP}, y_{TP})\right|^2 = \left|p_f(r_1, y_1)\right|^2 \frac{\rho_0(y_{TP})}{\rho_0(y_1)c(y_1)} \frac{R_0^2 \sin^2 \beta_0}{r_{TP} |\cos \beta_0|} |c'(y_{TP})|, \qquad (5.2\text{-}247)$$

which is finite as long as $\beta_0 \neq 90°$. However, if $\beta_0 = 90°$, then Eq. (5.2-245) reduces to

$$\left[\cos \beta(y) \frac{dr}{d\beta_0}\right]_{\beta_0 = 90°} = -\frac{c_0}{c'_0} \cos \beta(y), \qquad (5.2\text{-}248)$$

which is nonzero as long as $y \neq y_{TP}$. Also note that Eq. (5.2-245) approaches infinity as $c'(y) \to 0$, and as a result, $|p_f(r, y)|^2 \to 0$, which does not make physical sense. In summary, the magnitude squared of the acoustic pressure given by Eqs. (5.2-224) and (5.2-245) is valid everywhere, *even at or near a turning point*, as long as $c'(y) \neq 0$ and $\beta_0 \neq 90°$. When $\beta_0 = 90°$, the magnitude squared of the acoustic pressure is given by Eqs. (5.2-224) and (5.2-248).

In order to determine which set of equations to use, we shall assume that the ray-path calculations are being performed by using horizontal range as the independent variable. Therefore, if the depth y of the ray path at horizontal range r is $y = y(r)$ and if

$$|\cos \beta(y)| > |c'(y)|, \qquad (5.2\text{-}249)$$

then we compute $|p_f(r, y)|^2$ using Eqs. (5.2-224) and (5.2-235). Recall that the expression obtained by multiplying Eq. (5.2-236) by $\cos \beta(y)$ can be used instead of Eq. (5.2-235). However, if

$$|\cos \beta(y)| < |c'(y)|, \qquad (5.2\text{-}250)$$

then we compute $|p_f(r, y)|^2$ using Eqs. (5.2-224) and (5.2-245).

Example 5.2-6 (Acoustic pressure calculations for a depth-dependent, linear SSP and a homogeneous medium)

In this example we shall use Eq. (5.2-244) as the starting point in order to derive the equation for the magnitude squared of the acoustic pressure along a ray path when the speed of sound is a linear function of depth as given by Eq. (5.2-141). Since $c'(y) = g$ and $c''(y) = 0$, Eq. (5.2-244) reduces to

$$\frac{dr}{d\beta_0} = \frac{\partial}{\partial \beta_0} r(\beta_0, y) = -\cot \beta_0 (r - r_0)$$

$$-\frac{c_0}{g}\left[1 - \cot \beta_0 \tan \beta(y)\, \frac{c(y)}{c_0}\right], \quad (5.2\text{-}251)$$

and since [see Eq. (5.1-75)]

$$\frac{c(y)}{c_0} = \frac{\sin \beta(y)}{\sin \beta_0}, \quad (5.2\text{-}252)$$

substituting Eq. (5.2-252) into Eq. (5.2-251) yields

$$\frac{dr}{d\beta_0} = -\cot \beta_0 (r - r_0) - \frac{c_0}{g}\left[\frac{\cos \beta(y) \sin^2 \beta_0 - \cos \beta_0 \sin^2 \beta(y)}{\cos \beta(y) \sin^2 \beta_0}\right].$$

$$(5.2\text{-}253)$$

Next, upon substituting

$$\sin^2 \beta_0 = 1 - \cos^2 \beta_0 \quad (5.2\text{-}254)$$

and

$$\sin^2 \beta(y) = 1 - \cos^2 \beta(y) \quad (5.2\text{-}255)$$

into the numerator of the last term on the right-hand side of Eq. (5.2-253), we obtain

$$\frac{dr}{d\beta_0} = -\cot \beta_0 (r - r_0)$$

$$+ \frac{c_0}{g}\left[\frac{[\cos \beta_0 - \cos \beta(y)][1 + \cos \beta_0 \cos \beta(y)]}{\cos \beta(y) \sin^2 \beta_0}\right]. \quad (5.2\text{-}256)$$

If we further substitute Eqs. (5.2-148) and (5.2-98) into Eq. (5.2-256), then

$$\frac{dr}{d\beta_0} = \frac{\partial}{\partial\beta_0} r(\beta_0, y) = \frac{r - r_0}{\sin\beta_0 \cos\beta(y)} . \qquad (5.2\text{-}257)$$

Assuming that $r_0 = 0$ and substituting Eq. (5.2-257) into Eq. (5.2-224), we finally obtain the following classic result for the magnitude squared of the acoustic pressure along a ray path when the speed of sound is a linear function of depth:

$$\left| p_f(r, y) \right|^2 = \left| p_f(r_1, y_1) \right|^2 \frac{\rho_0(y)c(y)}{\rho_0(y_1)c(y_1)} \frac{r_1^2}{r^2}, \qquad (5.2\text{-}258)$$

where r_1 and y_1 are given by Eqs. (5.2-217) and (5.2-220), respectively. Equation (5.2-258) indicates that the magnitude squared of the acoustic pressure decreases as the reciprocal of the square of the horizontal range r. Note that Eq. (5.2-258) is valid everywhere—there are no problems at turning points or with launch angles equal to 90°.

An alternative derivation of Eq. (5.2-257) can be obtained as follows. By referring to Eq. (5.2-148), and with the use of Eq. (5.2-98), the horizontal range r can be expressed as a function of β_0 and y, that is,

$$r = r(\beta_0, y) = r_0 + \frac{c_0}{g \sin\beta_0} \left[\cos\beta_0 - \cos\beta(y)\right], \quad (5.2\text{-}259)$$

where the angle of arrival $\beta(y)$ is given by Eq. (5.2-97). Therefore, substituting Eq. (5.2-259) into Eq. (5.2-226) yields

$$\frac{dr}{d\beta_0} = \frac{\partial}{\partial\beta_0} r(\beta_0, y)$$

$$= \frac{c_0}{g \sin^2\beta_0} \left[-1 + \cos\beta_0 \cos\beta(y) + \sin\beta_0 \sin\beta(y)\frac{\partial}{\partial\beta_0}\beta(y) \right].$$

$$(5.2\text{-}260)$$

Substituting Eq. (5.2-232) into Eq. (5.2-260) yields

$$\frac{dr}{d\beta_0} = \frac{\partial}{\partial\beta_0} r(\beta_0, y) = \frac{c_0}{g \sin^2\beta_0} \left[\frac{\cos\beta_0 - \cos\beta(y)}{\cos\beta(y)} \right], \quad (5.2\text{-}261)$$

and upon substituting Eqs. (5.2-148) and (5.2-98) into Eq. (5.2-261), we obtain Eq. (5.2-257).

Let us conclude this example by deriving the equation for the magnitude squared of the acoustic pressure along a ray path in a *homogeneous medium*. We begin by multiplying Eq. (5.2-236) by $\cos \beta(y)$, yielding

$$\cos \beta(y) \, \frac{dr}{d\beta_0} = \cos \beta(y) \cot \beta_0 \int_{r_0}^{r} \frac{dr}{\cos^2 \beta(y)}. \qquad (5.2\text{-}262)$$

Since sound rays travel in straight lines in a homogeneous medium, $\beta(y) = \beta_0$, and as a result, $\cos \beta(y) = \cos \beta_0$. Therefore, in the case of a homogeneous medium, Eq. (5.2-262) reduces to [compare with Eq. (5.2-257)]

$$\cos \beta(y) \, \frac{dr}{d\beta_0} = \frac{r - r_0}{\sin \beta_0}. \qquad (5.2\text{-}263)$$

Substituting Eq. (5.2-263) into Eq. (5.2-224), assuming that $r_0 = 0$, and recalling that the equilibrium (ambient) density and speed of sound are constants in a homogeneous medium yields

$$|p_f(r, y)|^2 = |p_f(r_1, y_1)|^2 \frac{R_0^2 \sin^2 \beta_0}{r^2}. \qquad (5.2\text{-}264)$$

Since

$$\sin \beta_0 = \frac{r}{R}, \qquad (5.2\text{-}265)$$

where R is the hypotenuse of the right triangle with base (horizontal range) r and opposite angle β_0, substituting Eq. (5.2-265) into Eq. (5.2-264) yields

$$\boxed{|p_f(r, y)|^2 = |p_f(r_1, y_1)|^2 \frac{R_0^2}{R^2}.} \qquad (5.2\text{-}266)$$

Equation (5.2-266) indicates that in the case of a homogeneous medium, the magnitude squared of the acoustic pressure decreases as the reciprocal of the square of the spherical range R, as expected.

5.3 The Parabolic Equation Approximation

The *parabolic equation approximation* is another method of obtaining an approximate solution of the Helmholtz equation given by Eq. (5.1-9), which is repeated below for convenience:

$$\nabla^2 \varphi_f(\mathbf{r}) + k_0^2 n^2(\mathbf{r}) \varphi_f(\mathbf{r}) = 0, \qquad (5.1\text{-}9)$$

where the speed of sound is a function of all three spatial variables. We begin by assuming that Eq. (5.1-9) has a solution in the form of a "plane wave" propagating in the positive Z direction, that is,

$$\varphi_f(\mathbf{r}) = g(\mathbf{r}) \exp[-jk_0(z - z_0)], \qquad (5.3\text{-}1)$$

where $g(\mathbf{r})$ is a *complex* function in general, k_0 is the constant reference wave number given by Eq. (5.1-7), and z_0 is the Z coordinate of a source [compare Eq. (5.3-1) with Eqs. (5.2-1) or (5.2-2)]. Substituting Eq. (5.3-1) into Eq. (5.1-9) yields

$$\nabla^2 g(\mathbf{r}) - j2k_0 \frac{\partial}{\partial z} g(\mathbf{r}) + k_0^2 [n^2(\mathbf{r}) - 1] g(\mathbf{r}) = 0, \qquad (5.3\text{-}2)$$

where

$$\nabla^2 = \frac{\partial^2}{\partial x^2} + \frac{\partial^2}{\partial y^2} + \frac{\partial^2}{\partial z^2} \qquad (5.3\text{-}3)$$

is the Laplacian in the rectangular coordinates (x, y, z). If it is further assumed that

$$\left| \frac{\partial^2}{\partial z^2} g(\mathbf{r}) \right| \ll 2k_0 \left| \frac{\partial}{\partial z} g(\mathbf{r}) \right|, \qquad (5.3\text{-}4)$$

then Eq. (5.3-2) reduces to

$$\nabla_T^2 g(\mathbf{r}) - j2k_0 \frac{\partial}{\partial z} g(\mathbf{r}) + k_0^2 [n^2(\mathbf{r}) - 1] g(\mathbf{r}) = 0, \qquad (5.3\text{-}5)$$

which is the *parabolic equation approximation* of the Helmholtz equation, where

$$\nabla_T^2 = \frac{\partial^2}{\partial x^2} + \frac{\partial^2}{\partial y^2} \qquad (5.3\text{-}6)$$

is referred to as the *transverse Laplacian* in the rectangular coordinates (x, y). Equation (5.3-5) is usually rewritten as

$$\frac{\partial}{\partial z} g(\mathbf{r}) = -j \frac{1}{2k_0} \nabla_T^2 g(\mathbf{r}) - j \frac{k_0}{2} \left[n^2(\mathbf{r}) - 1 \right] g(\mathbf{r}) \qquad (5.3\text{-}7)$$

and must be solved *numerically* for $g(\mathbf{r})$. When attempting to solve Eq. (5.3-7), one does not have to be concerned about the locations of turning points and focal points as in ray acoustics. The assumption represented by Eq. (5.3-4) is satisfied best at *high frequencies* and is called the *parabolic approximation*, since it leads to the parabolic partial differential equation (PDE) given by Eq. (5.3-7). For numerical computation, a parabolic PDE is easier to solve than an elliptic PDE. The three-dimensional Helmholtz equation is an elliptic PDE.

Equation (5.3-4) is also referred to as the *small-angle approximation*, since the standard solution of Eq. (5.3-7) is a valid representation of $g(\mathbf{r})$, and hence the velocity potential $\varphi_f(\mathbf{r})$ [see Eq. (5.3-1)], only within an angular region of $\theta_0 \le 18°$, where θ_0 is measured from the positive Z axis (see Example 5.3-1). Therefore, standard solutions of the parabolic PDE given by Eq. (5.3-7) are mainly used to describe acoustic fields propagating in an underwater sound channel (see Fig. 5.2-12). Furthermore, the "parabolic approximation" represented by Eq. (5.3-4) is responsible for neglecting $\partial^2 g(\mathbf{r})/\partial z^2$ in favor of $\partial g(\mathbf{r})/\partial z$. The first-order partial derivative with respect to z implies that only one independent solution in the Z direction is allowed, that is, waves propagating in the positive Z direction are accounted for, whereas waves propagating in the negative Z direction are ignored.

Example 5.3-1 (Split-step Fourier transform algorithm for a homogeneous medium)
In the case of a homogeneous medium where $n(\mathbf{r}) = 1$, Eq. (5.3-7) reduces to

$$\frac{\partial}{\partial z} g(\mathbf{r}) = -j \frac{1}{2k_0} \nabla_T^2 g(\mathbf{r}). \qquad (5.3\text{-}8)$$

An *exact* solution of Eq. (5.3-8) is given by

$$g(\mathbf{r}) = \exp\left[-jk_X(x - x_0) \right] \exp\left[-jk_Y(y - y_0) \right]$$

$$\times \exp\left[+j \frac{(k_X^2 + k_Y^2)}{2k_0}(z - z_0) \right], \qquad (5.3\text{-}9)$$

where

$$k_X = k_0 u_0 = k_0 \sin \theta_0 \cos \psi_0 \qquad (5.3\text{-}10)$$

and

$$k_Y = k_0 v_0 = k_0 \sin \theta_0 \sin \psi_0 = k_0 \cos \beta_0 \qquad (5.3\text{-}11)$$

are the constant propagation-vector components in the X and Y directions, respectively, the angles (θ_0, ψ_0) are the constant vertical and azimuthal spherical angles measured with respect to the positive Z and X axes, respectively, the angle β_0 is constant and is measured with respect to the positive Y axis, and (x_0, y_0, z_0) are the rectangular coordinates of a source. Substituting Eqs. (5.3-9) through (5.3-11) into Eq. (5.3-1) yields

$$\varphi_f(\mathbf{r}) = \exp\big[-jk_X(x - x_0)\big] \exp\big[-jk_Y(y - y_0)\big] \exp\big[-j\hat{k}_Z(z - z_0)\big],$$
$$(5.3\text{-}12)$$

where

$$\hat{k}_Z = k_0 \hat{w}_0 \qquad (5.3\text{-}13)$$

and

$$\hat{w}_0 = 1 - \frac{u_0^2 + v_0^2}{2}, \qquad (5.3\text{-}14)$$

or

$$\hat{w}_0 = 1 - \frac{\sin^2 \theta_0}{2}, \qquad (5.3\text{-}15)$$

which is an approximation or estimate of the direction cosine with respect to the Z axis w_0 given by

$$w_0 = \sqrt{1 - (u_0^2 + v_0^2)} = \cos \theta_0. \qquad (5.3\text{-}16)$$

Equations (5.3-14) and (5.3-15) are, in fact, *binomial expansions* of w_0, valid whenever

$$u_0^2 + v_0^2 = \sin^2 \theta_0 \ll 1 \qquad (5.3\text{-}17)$$

or, equivalently,

$$\sin^2 \theta_0 \le 0.1, \qquad (5.3\text{-}18)$$

which implies that

$$\boxed{\theta_0 \leq 18°.}$$ (5.3-19)

Figure 5.3-1 is a plot of $\hat{w}_0$ and w_0 given by Eqs. (5.3-15) and (5.3-16), respectively, versus θ_0. Table 5.3-1 is a tabulation of the *approximation error* $\Delta = \hat{w}_0 - w_0$ versus θ_0 and the *percentage error* $\Delta\%$ with respect to w_0.

In the case of a homogeneous medium where $n(\mathbf{r}) = 1$, it is well known that Eq. (5.3-12) will be an exact solution of the Helmholtz equation given by Eq. (5.1-9) if $\hat{k}_z$ given by Eq. (5.3-13) is replaced by

$$k_z = k_0 w_0 = k_0 \cos \theta_0,$$ (5.3-20)

which is the propagation-vector component in the Z direction. However, as can be seen from Fig. 5.3-1, $\hat{w}_0$ will be a very good approximation of w_0 whenever $\theta_0 \leq 18°$. Therefore, Eq. (5.3-12) will also be a

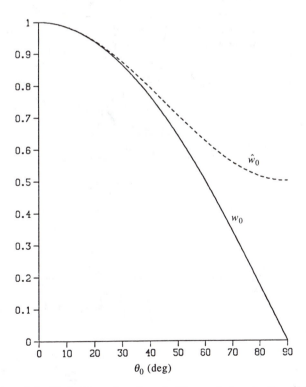

Figure 5.3-1. Plot of $\hat{w}_0$ (dashed curve) and w_0 (solid curve) given by Eqs. (5.3-15) and (5.3-16), respectively, versus θ_0.

Table 5.3-1 Approximation error $\Delta = \widehat{w}_0 - w_0$ versus θ_0 and the percentage error $\Delta\%$ with respect to w_0.

θ_0 (deg)	Δ (10^{-3})	$\Delta\%$
0	0.0	0.0
5	0.007	0.001
10	0.115	0.012
15	0.581	0.06
20	1.818	0.194
25	4.389	0.484
30	8.975	1.036
35	16.353	1.996
40	27.368	3.573
45	42.893	6.066
50	63.8	9.926

very good approximate solution of the Helmholtz equation given by Eq. (5.1-9) when both $n(\mathbf{r}) = 1$ and Eq. (5.3-19) is satisfied. As a result, the assumption represented by Eq. (5.3-4) is also referred to as the *small-angle approximation*.

Besides Eq. (5.3-9), the following three additional *exact* solutions of the parabolic PDE given by Eq. (5.3-8) also exist:

$$g_1(\mathbf{r}) = j\frac{1}{\lambda_0 \Delta Z} \exp\left[-j\frac{k_0}{2\Delta Z}(x^2 + y^2)\right], \qquad (5.3\text{-}21)$$

$$g_2(\mathbf{r}) = j\frac{1}{\lambda_0 \Delta Z} \exp\left\{-j\frac{k_0}{2\Delta Z}\left[(x - x_0)^2 + (y - y_0)^2\right]\right\}, \qquad (5.3\text{-}22)$$

and

$$g_3(\mathbf{r}) = j\frac{1}{\lambda_0 \Delta Z}\int_{-\infty}^{\infty}\int_{-\infty}^{\infty} A(x_0, y_0, z_0)$$

$$\times \exp\left\{-j\frac{k_0}{2\Delta Z}\left[(x - x_0)^2 + (y - y_0)^2\right]\right\} dx_0\, dy_0,$$

$$(5.3\text{-}23)$$

where $A(x_0, y_0, z_0)$ is an arbitrary complex function,

$$k_0 = 2\pi f/c_0 = 2\pi/\lambda_0, \tag{5.3-24}$$

and

$$\Delta Z = z - z_0. \tag{5.3-25}$$

By referring to Eqs. (5.3-21) through (5.3-23), it can be seen that the amplitude decreases as the down-range distance ΔZ increases.

The velocity potential

$$\boxed{\varphi_{f,3}(\mathbf{r}) = g_3(\mathbf{r}) \exp(-jk_0 \Delta Z)} \tag{5.3-26}$$

is not only an approximate solution of the Helmholtz equation given by Eq. (5.1-9) when both $n(\mathbf{r}) = 1$ and Eq. (5.3-19) is satisfied, it is, in fact, the *Fresnel diffraction pattern* of the complex function $A(x_0, y_0, z_0)$. The derivation of the Fresnel diffraction integral in optics is based on the *paraxial assumption*, that is, the angle of propagation θ_0, which is measured with respect to the positive Z axis, is assumed to be small [see Eq. (5.3-19)]. Note that Eq. (5.3-26) can be expressed as

$$\boxed{\varphi_{f,3}(x, y, z) = A(x, y, z) \underset{x,y}{**} \varphi_{f,1}(x, y, z),} \tag{5.3-27}$$

where

$$\boxed{\varphi_{f,1}(\mathbf{r}) = g_1(\mathbf{r}) \exp(-jk_0 \Delta Z)} \tag{5.3-28}$$

and the double asterisk $\underset{x,y}{**}$ denotes a two-dimensional convolution with respect to x and y. Therefore, $\varphi_{f,1}(\mathbf{r})$ given by Eq. (5.3-28) can be interpreted as being *the time-independent free-space spatial impulse response* or *the time-independent free-space Green's function* characterizing Fresnel diffraction in an unbounded, ideal (nonviscous), homogeneous, fluid medium whenever Eq. (5.3-19) is satisfied.

Finally, let us conclude this example by rewriting the two-dimensional convolution expression for the Fresnel diffraction integral given by Eq. (5.3-27) in terms of spatial-domain Fourier transforms. Taking the spatial Fourier transform with respect to x and y of both sides of Eq. (5.3-27) yields

$$\Phi_{f,3}(f_X, f_Y, z) = A(f_X, f_Y, z)\Phi_{f,1}(f_X, f_Y, z), \tag{5.3-29}$$

where

$$\Phi_{f,3}(f_X, f_Y, z) = F_x F_y\{\varphi_{f,3}(x, y, z)\}$$

$$\triangleq \int_{-\infty}^{\infty}\int_{-\infty}^{\infty} \varphi_{f,3}(x, y, z) \exp[+j2\pi(f_X x + f_Y y)] \, dx \, dy,$$

$$(5.3\text{-}30)$$

$$A(f_X, f_Y, z) = F_x F_y\{A(x, y, z)\}$$

$$\triangleq \int_{-\infty}^{\infty}\int_{-\infty}^{\infty} A(x, y, z) \exp[+j2\pi(f_X x + f_Y y)] \, dx \, dy,$$

$$(5.3\text{-}31)$$

and

$$\Phi_{f,1}(f_X, f_Y, z) = F_x F_y\{\varphi_{f,1}(x, y, z)\}$$

$$\triangleq \int_{-\infty}^{\infty}\int_{-\infty}^{\infty} \varphi_{f,1}(x, y, z) \exp[+j2\pi(f_X x + f_Y y)] \, dx \, dy.$$

$$(5.3\text{-}32)$$

If Eqs. (5.3-28) and (5.3-21) are substituted into Eq. (5.3-32), then

$$\Phi_{f,1}(f_X, f_Y, z) = j\frac{\exp(-jk_0 \Delta Z)}{\lambda_0 \Delta Z} F_x F_y\left\{\exp\left[-j\frac{k_0}{2\Delta Z}(x^2 + y^2)\right]\right\},$$

$$(5.3\text{-}33)$$

and since

$$\boxed{\begin{aligned} &F_x F_y\left\{\exp\left[-j\frac{k_0}{2\Delta Z}(x^2 + y^2)\right]\right\} \\ &= -j\lambda_0 \Delta Z \exp\left[+j\frac{2\Delta Z}{k_0}\pi^2(f_X^2 + f_Y^2)\right], \end{aligned}}$$

$$(5.3\text{-}34)$$

substituting Eq. (5.3-34) into Eq. (5.3-33) yields

$$\Phi_{f,1}(f_X, f_Y, z) = \exp(-jk_0 \Delta Z) \exp\left[+j\frac{2\,\Delta Z}{k_0}\pi^2(f_X^2 + f_Y^2)\right].$$

$$(5.3\text{-}35)$$

Since

$$\varphi_{f,3}(x, y, z) = F_{f_X}^{-1}F_{f_Y}^{-1}\{\Phi_{f,3}(f_X, f_Y, z)\}$$

$$\triangleq \int_{-\infty}^{\infty}\int_{-\infty}^{\infty} \Phi_{f,3}(f_X, f_Y, z)$$

$$\times \exp\left[-j2\pi(f_X x + f_Y y)\right] df_X\, df_Y, \quad (5.3\text{-}36)$$

substituting Eqs. (5.3-29), (5.3-31), and (5.3-35) into Eq. (5.3-36) yields

$$\varphi_{f,3}(x, y, z) = \exp(-jk_0 \Delta Z)F_{f_X}^{-1}F_{f_Y}^{-1}\left\{F_{x_0}F_{y_0}\{A(x_0, y_0, z_0)\}\right.$$

$$\left.\times \exp\left[+j\frac{2\,\Delta Z}{k_0}\pi^2(f_X^2 + f_Y^2)\right]\right\},$$

$$(5.3\text{-}37)$$

where k_0 and ΔZ are given by Eqs. (5.3-24) and (5.3-25), respectively. Therefore, if the initial complex acoustic field at $z = z_0$ is $A(x_0, y_0, z_0)$, then the complex acoustic field $\varphi_{f,3}(x, y, z)$ at a distance $\Delta Z = z - z_0$ meters away can be obtained by propagating $A(x_0, y_0, z_0)$ through the homogeneous medium via Eq. (5.3-37), which is the Fresnel diffraction integral. In practical applications, both the forward and inverse two-dimensional spatial Fourier transforms appearing in Eq. (5.3-37) would be evaluated, for example, using two-dimensional forward and inverse FFT (fast-Fourier-transform) computer algorithms. Equation (5.3-37) is, in fact, known as the *split-step Fourier transform algorithm* for a *homogeneous medium*.

Example 5.3-2 (Split-step Fourier transform algorithm for an inhomogeneous medium)

In this example we shall obtain an approximate solution of a parabolic PDE expressed in cylindrical coordinates when the index of refraction (speed of sound) is a function of both horizontal range and depth. We begin by expressing the Helmholtz equation given by Eq. (5.1-9) in terms of the cylindrical coordinates (r, ϕ, y) as follows:

$$\frac{\partial^2}{\partial r^2}\varphi_f(\mathbf{r}) + \frac{1}{r}\frac{\partial}{\partial r}\varphi_f(\mathbf{r}) + \frac{1}{r^2}\frac{\partial^2}{\partial \phi^2}\varphi_f(\mathbf{r})$$

$$+ \frac{\partial^2}{\partial y^2}\varphi_f(\mathbf{r}) + k_0^2 n^2(\mathbf{r})\varphi_f(\mathbf{r}) = 0, \qquad (5.3\text{-}38)$$

where $\mathbf{r} = (r, \phi, y)$, r is the horizontal range, ϕ is the azimuthal angle, and y is the depth (see Fig. 5.1-5). If we assume that the index of refraction is a function of r and y only, that is, if

$$n(\mathbf{r}) = n(r, y) = \frac{c_0}{c(r, y)}, \qquad (5.3\text{-}39)$$

and if we also assume that the acoustic field is *axisymmetric* so that

$$\varphi_f(\mathbf{r}) = \varphi_f(r, y), \qquad (5.3\text{-}40)$$

then Eq. (5.3-38) reduces to

$$\frac{\partial^2}{\partial r^2}\varphi_f(r, y) + \frac{1}{r}\frac{\partial}{\partial r}\varphi_f(r, y) + \frac{\partial^2}{\partial y^2}\varphi_f(r, y)$$

$$+ k_0^2 n^2(r, y)\varphi_f(r, y) = 0. \qquad (5.3\text{-}41)$$

Next, assume that Eq. (5.3-41) has a solution in the form of a "cylindrical wave" propagating in the positive r direction, that is,

$$\varphi_f(r, y) = g(r, y)\frac{\exp[-jk_0(r - r_0)]}{\sqrt{r - r_0}}, \qquad (5.3\text{-}42)$$

where $g(r, y)$ is, in general, a complex function [compare Eq. (5.3-42) with Eq. (5.3-1)], k_0 is the constant reference wave number given by Eq. (5.3-24), and r_0 is the r coordinate of a source. Upon substituting Eq. (5.3-42) into Eq. (5.3-41), we obtain

$$\frac{\partial^2}{\partial r^2}g(r, y) - j2k_0\frac{\partial}{\partial r}g(r, y) + \frac{\partial^2}{\partial y^2}g(r, y)$$

$$+ k_0^2\left[n^2(r, y) - 1 + (2k_0\,\Delta r)^{-2}\right]g(r, y) = 0, \quad (5.3\text{-}43)$$

where

$$\Delta r = r - r_0. \tag{5.3-44}$$

If we now make the parabolic or small-angle assumption

$$\left| \frac{\partial^2}{\partial r^2} g(r, y) \right| \ll 2k_0 \left| \frac{\partial}{\partial r} g(r, y) \right| \tag{5.3-45}$$

and the far-field assumption

$$k_0 \, \Delta r \gg 1, \tag{5.3-46}$$

then Eq. (5.3-43) reduces to

$$\frac{\partial}{\partial r} g(r, y) = -j \frac{1}{2k_0} \frac{\partial^2}{\partial y^2} g(r, y) - j \frac{k_0}{2} \left[n^2(r, y) - 1 \right] g(r, y), \tag{5.3-47}$$

which is a *parabolic PDE* in the cylindrical coordinates r and y.

In order to obtain an approximate solution of Eq. (5.3-47), we begin by taking the spatial Fourier transform with respect to depth y of both sides of Eq. (5.3-47) while treating the index of refraction as a constant. We shall reintroduce the index of refraction as a function of both r and y after obtaining the approximate solution of Eq. (5.3-47). Since (see Appendix 2C)

$$F_y \left\{ \frac{\partial^2}{\partial y^2} g(r, y) \right\} = -(2\pi f_Y)^2 G(r, f_Y), \tag{5.3-48}$$

where $G(r, f_Y) = F_y\{g(r, y)\}$, performing the aforementioned transform yields

$$\frac{\partial}{\partial r} G(r, f_Y) = \left[j \frac{(2\pi f_Y)^2}{2k_0} - j \frac{k_0}{2} (n^2 - 1) \right] G(r, f_Y). \tag{5.3-49}$$

Rewriting Eq. (5.3-49) as

$$\frac{1}{G(r, f_Y)} \frac{\partial}{\partial r} G(r, f_Y) = j \frac{(2\pi f_Y)^2}{2k_0} - j \frac{k_0}{2} (n^2 - 1) \tag{5.3-50}$$

and multiplying both sides of Eq. (5.3-50) by ∂r and integrating yields

$$\int_{G(r_0, f_Y)}^{G(r, f_Y)} \frac{\partial G(r, f_Y)}{G(r, f_Y)} = \int_{r_0}^{r} \left[j\frac{(2\pi f_Y)^2}{2k_0} - j\frac{k_0}{2}(n^2 - 1) \right] \partial r, \quad (5.3\text{-}51)$$

which is equal to

$$\ln\left[\frac{G(r, f_Y)}{G(r_0, f_Y)} \right] = j\frac{(2\pi f_Y)^2}{2k_0}\, \Delta r - j\frac{k_0}{2}(n^2 - 1)\, \Delta r, \quad (5.3\text{-}52)$$

so that

$$G(r, f_Y) = G(r_0, f_Y) \exp\left[+j\frac{(2\pi f_Y)^2}{2k_0}\, \Delta r \right] \exp\left[-j\frac{k_0}{2}(n^2 - 1)\, \Delta r \right].$$

$$(5.3\text{-}53)$$

Since

$$g(r, y) = F_{f_Y}^{-1}\{G(r, f_Y)\} \triangleq \int_{-\infty}^{\infty} G(r, f_Y) \exp(-j2\pi f_Y y)\, df_Y,$$

$$(5.3\text{-}54)$$

substituting Eq. (5.3-53) into Eq. (5.3-54) yields

$$g(r, y) = \exp\left[-j\frac{k_0}{2}(n^2 - 1)\, \Delta r \right]$$

$$\times \int_{-\infty}^{\infty} G(r_0, f_Y) \exp\left[+j\frac{2\,\Delta r}{k_0}(\pi f_Y)^2 \right] \exp(-j2\pi f_Y y)\, df_Y,$$

$$(5.3\text{-}55)$$

or

$$g(r, y) = \exp\left[-j\frac{k_0}{2}[n^2(r, y) - 1]\, \Delta r \right]$$

$$\times F_{f_Y}^{-1}\left\{ F_{y_0}\{g(r_0, y_0)\} \exp\left[+j\frac{2\,\Delta r}{k_0}(\pi f_Y)^2 \right] \right\}, \quad (5.3\text{-}56)$$

where the index of refraction is once again shown explicitly as a function of horizontal range r and depth y. Substituting Eq. (5.3-56) into Eq. (5.3-42) finally yields

$$\varphi_f(r, y) = \frac{\exp\left[-j\dfrac{k_0}{2}[n^2(r, y) + 1] \Delta r\right]}{\sqrt{\Delta r}} \\ \times F_{f_Y}^{-1}\left\{F_{y_0}\{g(r_0, y_0)\} \exp\left[+j\dfrac{2 \Delta r}{k_0}(\pi f_Y)^2\right]\right\}, \qquad (5.3-57)$$

which is one version of the split-step Fourier transform algorithm in the cylindrical coordinates r and y.

Problems

5-1 A sound source is located at a depth of y_0 meters, and the SSP is given by

$$c(y) = c_0 + g(y - y_0),$$

where $y_0 = 1000$ m, $c_0 = 1475$ m/sec, and $g = 0.017$ sec^{-1}.
(a) If the initial angles of transmission at the source are $\beta_0 = 89°$, $87°$, $85°$, $80°$, $75°$, $60°$, and $45°$, then for each value of β_0 given, find the corresponding depth of the first turning point.
(b) What is the angle of arrival $\beta(y)$ in degrees at a depth of $y = 2000$ m if $\beta_0 = 80°$?
(c) If $\beta(y) = 63°$ at a depth of $y = 2500$ m, then what was β_0?
(d) If $\beta_0 = 80°$, then what are the minimum values of source frequency required so that the approximate WKB solution will be a valid representation of the acoustic field at depths of $y = 1500$, 2000, and 2300 m?
(e) If $\beta_0 = 80°$ and the source frequency $f = 1$ kHz, then compute the amplitude of the acoustic field at depths of $y = 1500$, 2000, and 2300 m using the WKB amplitude expression given by Eq. (5.1-40). **Hint:** Use Eq. (5.1-77) to compute $k_Y(y)$.

5-2 A sound source is located at $x_0 = 0$ m, $y_0 = 1000$ m, and $z_0 = 100$ m, and the SSP is given by

$$c(y) = c_0 + g(y - y_0),$$

where $c_0 = 1475$ m/sec and $g = 0.017$ sec^{-1}. Assume ray propagation in a plane normal to the XZ plane with $\delta = 10°$ (see Fig. 5.2-4). If $\beta_0 = 72°$, then find the travel time τ, the horizontal range Δr, the x and z coordinates, and the arc length s between the source and the point on the ray path that returns to the source depth of y_0 meters. **Hint:** See Fig. 5.2-9b.

5-3 Figure 5.2-8 illustrates a piecewise linear approximation of a typical SSP shown as a function of depth y.

(a) If a sound source is located at a depth of y_0 meters, where $0 \leq y_0 \leq y_1$, then derive an expression for β_0 so that the depth of the *lower* turning points, y_{TP}, is y_1. Note that this answer is the *minimum* allowed value for β_0 in order to keep all initially downward propagating rays in the surface duct.

(b) If $y_1 \leq y_0 \leq y_2$, then derive an expression for β_0 so that the depth of the *upper* turning points, y_{TP}, is y_1. **Hint:** $\sin \beta = \sin(\pi - \beta)$. Note that this answer is the *maximum* allowed value for β_0 in order to keep all initially upward propagating rays in the underwater sound channel.

(c) If $y_2 \leq y_0 \leq y_3$, then derive an expression for β_0 so that the depth of the *lower* turning points, y_{TP}, is y_3. Note that this answer is the *minimum* allowed value for β_0 in order to keep all initially downward propagating rays in the underwater sound channel.

(d) If a sound source is located at the ocean surface and the ocean is D meters deep, then using the SSP shown in Fig. 5.2-8, derive an expression for β_0 so that the depth of the *lower* turning points, y_{TP}, is D. Note that launch angles equal to and slightly larger than this value will produce convergence zones. However, if β_0 is too large, then the rays will be trapped in the surface duct.

5-4 Referring to Fig. 5.2-8, let $y_1 = 100$ m, $y_2 = 1000$ m, $c(0) = 1500$ m/sec, $c_2 = 1475$ m/sec, $g_1 = 0.016$ sec^{-1}, and $g_3 = 0.017$ sec^{-1}.

(a) Find c_1.

(b) Find g_2.

(c) Find y_3.

(d) Compute the width of the underwater sound channel.

5-5 Referring to Fig. 5.2-8, let $y_1 = 100$ m, $c(0) = 1500$ m/sec, and $g_1 = 0.016$ sec^{-1}.

(a) If the source depth $y_0 = y_1$ and $\beta_0 = 90°$, then what is the angle of arrival, $\beta(y)$, at the surface of the ocean? **Hint:** $\sin \beta = \sin(\pi - \beta)$.

(b) Assuming ray propagation in the YZ plane and using your answer from part (a), if the z coordinate of the source is $z_0 = 100$ m, then find the travel time τ, the horizontal range Δr, the z coordinate, and the arc length s between the source and the *first* surface bounce.

(c) What is the horizontal range Δr between the first and second surface bounces?

5-6 Referring to Fig. 5.2-8, let $y_1 = 100$ m, $y_2 = 1000$ m, $y_3 = 2564.7$ m, $c_2 = 1475$ m/sec, $g_2 = -0.0296$ sec^{-1}, and $g_3 = 0.017$ sec^{-1}.

(a) If the source depth $y_0 = y_2$, then using the parameter values given in this problem, evaluate β_0 from Problem 5-3(b).

(b) Assuming ray propagation in the YZ plane and using your answer from part (a), if the z coordinate of the source is $z_0 = 100$ m, then find the travel time τ, the horizontal range Δr, the z coordinate, and the arc length s between the source and the first upper turning point at $y_{TP} = y_1$.

(c) If the source depth $y_0 = y_2$, then using the parameter values given in this problem, evaluate β_0 from Problem 5-3(c).

(d) Assuming ray propagation in the YZ plane and using your answer from part (c), if the z coordinate of the source is $z_0 = 100$ m, then find the travel time τ, the horizontal range Δr, the z coordinate, and the arc length s between the source and the first lower turning point at $y_{TP} = y_3$.

5-7 For a SSP given by $c(y) = c_0 + g(y - y_0)$, verify that:
(a) $c[-(R_c - y_{TP})] = 0$ when $g > 0$.
(b) $c(R_c + y_{TP}) = 0$ when $g < 0$. **Hint:** For $g < 0$, replace g with $-|g|$.

5-8 Verify that Eqs. (5.3-9) and (5.3-23) are exact solutions of Eq. (5.3-8) by direct substitution.

5-9 Verify by direct substitution that Eq. (5.3-53) is a solution of Eq. (5.3-49).

Appendix 5A

In this appendix we shall rewrite Eq. (5.1-36) in terms of the source frequency f, the speed of sound $c(y)$, and the angle of arrival $\beta(y)$. If it is assumed that $k_Y(y)$ is positive, then it can be shown using Eq. (5.1-40) that

$$a''_Y(y) \approx \frac{A}{2}\left[\tfrac{3}{2}[k_Y(y)]^{-5/2}[k'_Y(y)]^2 - [k_Y(y)]^{-3/2}k''_Y(y)\right]. \quad (5A\text{-}1)$$

If we rewrite Eq. (5.1-36) as

$$\frac{1}{[\theta'_Y(y)]^2}\left|\frac{a''_Y(y)}{a_Y(y)}\right| \ll 1, \quad (5A\text{-}2)$$

and then substitute Eqs. (5.1-37), (5.1-40), and (5A-1) into Eq. (5A-2), we obtain the following inequality:

$$3\left|\left[\frac{1}{2k_Y^2(y)}k'_Y(y)\right]^2 - \frac{1}{6}\frac{k''_Y(y)}{k_Y^3(y)}\right| \ll 1. \quad (5A\text{-}3)$$

Assuming that

$$|k''_Y(y)| \ll |k'_Y(y)|, \quad (5A\text{-}4)$$

Eq. (5A-3) reduces to

$$\left| \frac{1}{2k_Y^2(y)} k_Y'(y) \right| \ll \frac{\sqrt{3}}{3}. \tag{5A-5}$$

The next step is to rewrite the left-hand side of Eq. (5A-5).

Taking the derivative with respect to y of Eq. (5.1-23) yields

$$\frac{d}{dy} k_Y^2(y) = 2k(y)k'(y), \tag{5A-6}$$

and since

$$\frac{d}{dy} k_Y^2(y) = 2k_Y(y)k_Y'(y), \tag{5A-7}$$

equating the right-hand sides of Eqs. (5A-6) and (5A-7), solving for $k_Y'(y)$, and multiplying and dividing the right-hand side of the resulting expression by $k(y)$ yields

$$k_Y'(y) = \frac{k^2(y)k'(y)}{k_Y(y)k(y)}. \tag{5A-8}$$

Substituting Eq. (5.1-11) into Eq. (5A-8) further yields

$$k_Y'(y) = -\frac{k^2(y)c'(y)}{k_Y(y)c(y)}, \tag{5A-9}$$

or

$$\frac{k_Y'(y)}{2k_Y^2(y)} = -\frac{k^2(y)c'(y)}{2k_Y^3(y)c(y)}. \tag{5A-10}$$

And upon substituting Eqs. (5A-10), (5.1-77), and (5.1-11) into Eq. (5A-5), we finally obtain the desired result

$$\boxed{\left| \frac{c'(y)}{4\pi f \cos^3 \beta(y)} \right| \ll \frac{\sqrt{3}}{3},} \tag{5A-11}$$

where $\beta(y)$ can be determined from Snell's law given by Eq. (5.1-75).

Appendix 5B

In this appendix we shall derive Eq. (5.2-245) by working with the rotated coordinate system $r'Y'$ shown in Fig. 5B-1. The new coordinates (r', y') are related to the original coordinates (r, y) by the following transformation equations (see Fig. 5B-1):

$$r' = r \cos \beta(y) - y \sin \beta(y) \qquad (5B\text{-}1)$$

and

$$y' = r \sin \beta(y) + y \cos \beta(y), \qquad (5B\text{-}2)$$

where $\beta(y)$ is the angle of rotation of the axes. By comparing Fig. 5.2-15 with Fig. 5B-1, it can be seen that

$$dl = dr' = dr \cos \beta(y), \qquad (5B\text{-}3)$$

and upon substituting Eq. (5B-3) into Eq. (5.2-224), we obtain the following alternative general expression for the magnitude squared of the acoustic pressure

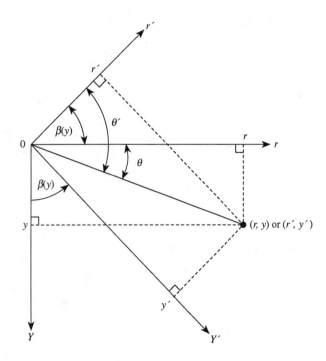

Figure 5B-1. Coordinate axes r' and Y' obtained by rotating the coordinate axes r and Y by an angle $\beta(y)$.

along a ray path:

$$\boxed{\left|p_f(r,y)\right|^2 = \left|p_f(r_1,y_1)\right|^2 \frac{\rho_0(y)c(y)}{\rho_0(y_1)c(y_1)} \frac{R_0^2 \sin \beta_0}{r|dr'/d\beta_0|},} \qquad \text{(5B-4)}$$

where the magnitude of $dr'/d\beta_0$ is taken in order to ensure that the magnitude squared of the acoustic pressure is positive. In order to proceed further, we need an expression for $dr'/d\beta_0$.

We shall obtain it by treating β_0 and $\beta(y)$ as the independent variables, instead of β_0 and y. Then the chain rule yields

$$\frac{dr'}{d\beta_0} = \frac{d}{d\beta_0}r'[\beta_0, \beta(y)]$$

$$= \frac{\partial}{\partial \beta_0}r'[\beta_0, \beta(y)]\frac{d\beta_0}{d\beta_0} + \frac{\partial}{\partial \beta(y)}r'[\beta_0, \beta(y)]\frac{d}{d\beta_0}\beta(y), \quad \text{(5B-5)}$$

and since $d\beta(y)/d\beta_0 = 0$ when β_0 and $\beta(y)$ are treated as independent variables,

$$\frac{dr'}{d\beta_0} = \frac{\partial}{\partial \beta_0}r'[\beta_0, \beta(y)]. \qquad \text{(5B-6)}$$

From Eq. (5B-1)

$$\frac{\partial r'}{\partial \beta_0} = r\frac{\partial}{\partial \beta_0}\cos \beta(y) + \cos \beta(y)\frac{\partial r}{\partial \beta_0} - y\frac{\partial}{\partial \beta_0}\sin \beta(y) - \sin \beta(y)\frac{\partial y}{\partial \beta_0},$$
$$\text{(5B-7)}$$

and since $\partial \cos \beta(y)/\partial \beta_0 = 0$ and $\partial \sin \beta(y)/\partial \beta_0 = 0$ when β_0 and $\beta(y)$ are treated as independent variables,

$$\frac{\partial r'}{\partial \beta_0} = \cos \beta(y)\frac{\partial r}{\partial \beta_0} - \sin \beta(y)\frac{\partial y}{\partial \beta_0}. \qquad \text{(5B-8)}$$

We shall compute $\partial y/\partial \beta_0$ next.

If we rewrite Snell's law as

$$c(y) = \frac{c_0}{\sin \beta_0}\sin \beta(y), \qquad \text{(5B-9)}$$

then

$$\frac{\partial}{\partial \beta_0}c(y) = \frac{c_0}{\sin \beta_0}\frac{\partial}{\partial \beta_0}\sin \beta(y) - \cos \beta_0 \frac{c_0}{\sin^2 \beta_0}\sin \beta(y), \quad \text{(5B-10)}$$

and since $\partial \sin \beta(y)/\partial \beta_0 = 0$ when β_0 and $\beta(y)$ are treated as independent variables, upon using Snell's law,

$$\frac{\partial}{\partial \beta_0} c(y) = -\cot \beta_0 \, c(y). \tag{5B-11}$$

Since

$$\frac{\partial}{\partial \beta_0} c(y) = \frac{\partial}{\partial y} c(y) \frac{\partial y}{\partial \beta_0} = \frac{d}{dy} c(y) \frac{\partial y}{\partial \beta_0}, \tag{5B-12}$$

equating the right-hand sides of Eqs. (5B-11) and (5B-12) yields

$$\frac{\partial y}{\partial \beta_0} = -\cot \beta_0 \frac{c(y)}{c'(y)}, \tag{5B-13}$$

where

$$c'(y) = \frac{d}{dy} c(y). \tag{5B-14}$$

And upon substituting Eq. (5B-13) into Eq. (5B-8), we obtain

$$\frac{\partial r'}{\partial \beta_0} = \cos \beta(y) \frac{\partial r}{\partial \beta_0} + \cot \beta_0 \sin \beta(y) \frac{c(y)}{c'(y)}. \tag{5B-15}$$

We shall compute $\partial r/\partial \beta_0$ next.
  Since [see Eq. (5.2-239)]

$$dr = \frac{c(y)}{c'(y)} d\beta(y), \tag{5B-16}$$

solving for the horizontal range r yields

$$r = r_0 + \int_{\beta_0}^{\beta(y)} \frac{c(\zeta)}{c'(\zeta)} d\beta(\zeta), \tag{5B-17}$$

which implies that r is a function of β_0 and $\beta(y)$, that is, $r = r[\beta_0, \beta(y)]$. Therefore, with the use of *Leibnitz's rule*,

$$\frac{\partial r}{\partial \beta_0} = \frac{\partial}{\partial \beta_0} r_0 + \int_{\beta_0}^{\beta(y)} \frac{\partial}{\partial \beta_0} \frac{c(\zeta)}{c'(\zeta)} d\beta(\zeta) + \frac{c(y)}{c'(y)} \frac{\partial}{\partial \beta_0} \beta(y) - \frac{c(y_0)}{c'(y_0)} \frac{\partial}{\partial \beta_0} \beta_0, \tag{5B-18}$$

and since r_0 is a constant and $\partial\beta(y)/\partial\beta_0 = 0$ when β_0 and $\beta(y)$ are treated as independent variables,

$$\frac{\partial r}{\partial\beta_0} = \int_{\beta_0}^{\beta(y)} \frac{\partial}{\partial\beta_0} \frac{c(\zeta)}{c'(\zeta)} d\beta(\zeta) - \frac{c_0}{c_0'}, \tag{5B-19}$$

where $c_0 = c(y_0)$ and $c_0' = c'(y_0)$. Since

$$\frac{\partial}{\partial\beta_0} \frac{c(y)}{c'(y)} = \frac{\partial}{\partial y} \frac{c(y)}{c'(y)} \frac{\partial y}{\partial\beta_0} = \left[-\frac{c(y)c''(y)}{[c'(y)]^2} + 1 \right] \frac{\partial y}{\partial\beta_0}, \tag{5B-20}$$

where

$$c''(y) = \frac{d^2}{dy^2} c(y), \tag{5B-21}$$

substituting Eq. (5B-13) into Eq. (5B-20) yields

$$\frac{\partial}{\partial\beta_0} \frac{c(y)}{c'(y)} = \cot\beta_0 \left[\frac{c^2(y)c''(y) - c(y)[c'(y)]^2}{[c'(y)]^3} \right]. \tag{5B-22}$$

Therefore, substituting Eqs. (5B-16) and (5B-22) into Eq. (5B-19) yields

$$\frac{\partial r}{\partial\beta_0} = \cot\beta_0 \int_{r_0}^{r} \frac{c(y)c''(y) - [c'(y)]^2}{[c'(y)]^2} dr - \frac{c_0}{c_0'}, \tag{5B-23}$$

where it is understood that $y = y(r)$ in the integrand on the right-hand side. And upon substituting Eq. (5B-23) into Eq. (5B-15), with the use of Eq. (5B-6), we finally obtain the desired result:

$$\frac{dr'}{d\beta_0} = \frac{\partial}{\partial\beta_0} r'[\beta_0, \beta(y)] = \cos\beta(y) \left[\cot\beta_0 \int_{r_0}^{r} \frac{c(y)c''(y) - [c'(y)]^2}{[c'(y)]^2} dr - \frac{c_0}{c_0'} \right]$$

$$+ \cot\beta_0 \sin\beta(y) \frac{c(y)}{c'(y)}. \tag{5B-24}$$

Note that the right-hand sides of Eqs. (5B-24) and (5.2-245) are identical.

Bibliography

M. Born and E. Wolf, *Principles of Optics*, 6th ed., Pergamon, Oxford, 1980.

C. A. Boyles, *Acoustic Waveguides*, Wiley, New York, 1984.

L. Brekhovskikh, *Waves in Layered Media*, 2nd ed., Academic Press, New York, 1980.

L. Brekhovskikh and Yu. Lysanov, *Fundamentals of Ocean Acoustics*, 2nd ed., Springer-Verlag, Berlin, 1991.

C. S. Clay and H. Medwin, *Acoustical Oceanography*, Wiley, New York, 1977.

S. Cornbleet, "Geometrical optics reviewed: A new light on an old subject," *Proc. IEEE*, **71**, 471–502 (1983).

J. A. DeSanto, editor, *Ocean Acoustics*, Springer-Verlag, Berlin, 1979.

F. B. Jensen, W. A. Kuperman, M. B. Porter, and H. Schmidt, *Computational Ocean Acoustics*, AIP Press, Woodbury, New York, 1994.

J. B. Keller and J. S. Papadakis, editors, *Wave Propagation and Underwater Acoustics*, Springer-Verlag, Berlin, 1977.

L. E. Kinsler, A. R. Frey, A. B. Coppens, and J. V. Sanders, *Fundamentals of Acoustics*, 3rd ed., Wiley, New York, 1982.

M. V. Klein, *Optics*, Wiley, New York, 1970.

H. R. Krol, "Intensity calculations along a single ray," *J. Acoust. Soc. Am.*, **53**, 864–868 (1973).

S. T. McDaniel, "Application of the parabolic approximation to predict acoustical propagation in the ocean," *Am. J. Phys.*, **47**(1), 63–68 (1979).

C. B. Moler and L. P. Solomon, "Use of splines and numerical integration in geometrical acoustics," *J. Acoust. Soc. Am.*, **48**, 739–744 (1970).

C. B. Officer, *Introduction to the Theory of Sound Transmission*, McGraw-Hill, New York, 1958.

F. D. Tappert, "The parabolic approximation method," in *Wave Propagation and Underwater Acoustics* (edited by J. B. Keller and J. S. Papadakis), pp. 224–287, Springer-Verlag, Berlin, 1977.

I. Tolstoy and C. S. Clay, *Ocean Acoustics*, McGraw-Hill, New York, 1966.

B. J. Uscinski, *The Elements of Wave Propagation in Random Media*, McGraw-Hill, New York, 1977.

L. J. Ziomek, *Underwater Acoustics—A Linear Systems Theory Approach*, Academic Press, Orlando, Florida, 1985.

L. J. Ziomek, "Three-dimensional ray acoustics: New expressions for the amplitude, eikonal, and phase functions," *IEEE J. Oceanic Eng.*, **14**, 396–399 (1989).

L. J. Ziomek, "The RRA algorithm: Recursive ray acoustics for three-dimensional speeds of sound," *IEEE J. Oceanic Eng.*, **18**, 25–30 (1993).

L. J. Ziomek, "Sound-pressure level calculations using the RRA algorithm for depth-dependent speeds of sound valid at turning points and focal points," *IEEE J. Oceanic Eng.*, **19**, 242–248 (1994).

Part II

Space-Time Signal Processing

Chapter 6

Complex Aperture Theory

6.1 Coupling Transmitted and Received Electrical Signals to the Fluid Medium

In the field of optics, a rectangular or circular hole in an opaque screen is referred to as a rectangular or circular aperture. The complex aperture function, that is, the magnitude and phase of the light distribution within the aperture, determines the Fresnel and Fraunhofer diffraction patterns of the aperture. The complex aperture function can describe the light distribution within, for example, a single rectangular hole or an array of rectangular holes in an opaque screen. In electromagnetics, the meaning of the word "aperture" has been extended to refer to either a single electromagnetic antenna or an array of electromagnetic antennas.

Similarly, in acoustics, the word "aperture" is used to refer to either a single electroacoustic transducer or an array of electroacoustic transducers. When used in the *active mode*, as a *transmitter*, an electroacoustic transducer converts electrical signals (voltages and currents) into acoustic signals (sound waves). An everyday example of an electroacoustic transducer being used in the active mode is a speaker in a radio. When used in the *passive mode*, as a *receiver*, an electroacoustic transducer converts acoustic signals into electrical signals. In this chapter we shall use the principles of complex aperture theory in order to derive many basic equations that can be used to describe the performance of either a single electroacoustic transducer or an array of electroacoustic transducers. Array theory will be discussed later, in Chapter 7.

Consider the transmit aperture shown in Fig. 6.1-1. The aperture is arbitrary in shape and occupies a volume V in general. This physical situation is known as a

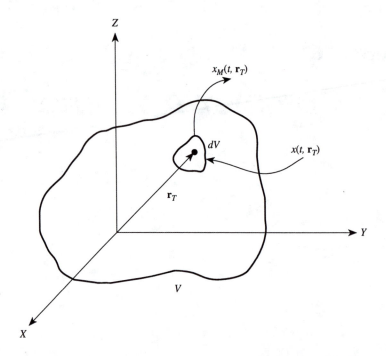

Figure 6.1-1. Arbitrarily shaped transmit aperture shown occupying a volume V.

volume aperture. The subscripts T and M refer to the transmit aperture and fluid medium, respectively. Imagine applying an input electrical signal $x(t, \mathbf{r}_T)$ to the transmit aperture at time t and spatial location $\mathbf{r}_T$, where $\mathbf{r}_T$ is a position vector. If we treat the infinitesimal volume element dV of the aperture located at $\mathbf{r}_T$ as a *linear filter* with *impulse response* $\alpha_T(t, \mathbf{r}_T)$, then the output acoustic signal $x_M(t, \mathbf{r}_T)$ from the transmit aperture, which is also the input acoustic signal to the fluid medium, can be expressed as a *time-domain convolution integral* as follows:

$$x_M(t, \mathbf{r}_T) = \int_{-\infty}^{\infty} x(\tau, \mathbf{r}_T)\alpha_T(t - \tau, \mathbf{r}_T)\, d\tau \qquad (6.1\text{-}1)$$

or

$$x_M(t, \mathbf{r}_T) = x(t, \mathbf{r}_T) \underset{t}{*} \alpha_T(t, \mathbf{r}_T) \qquad (6.1\text{-}2)$$

where $*$ denotes convolution with respect to time t. The input acoustic signal to the fluid medium, $x_M(t, \mathbf{r}_T)$, is also known as the *source distribution*. Taking the Fourier transform of both sides of Eq. (6.1-2) with respect to t yields

$$X_M(f, \mathbf{r}_T) = X(f, \mathbf{r}_T)A_T(f, \mathbf{r}_T), \qquad (6.1\text{-}3)$$

where

$$X_M(f, \mathbf{r}_T) = F_t\{x_M(t, \mathbf{r}_T)\} = \int_{-\infty}^{\infty} x_M(t, \mathbf{r}_T) \exp(-j2\pi ft) \, dt \quad (6.1\text{-}4)$$

is the complex frequency spectrum of the input acoustic signal or source distribution,

$$X(f, \mathbf{r}_T) = F_t\{x(t, \mathbf{r}_T)\} = \int_{-\infty}^{\infty} x(t, \mathbf{r}_T) \exp(-j2\pi ft) \, dt \quad (6.1\text{-}5)$$

is the complex frequency spectrum of the input electrical signal, and

$$A_T(f, \mathbf{r}_T) = F_t\{\alpha_T(t, \mathbf{r}_T)\} = \int_{-\infty}^{\infty} \alpha_T(t, \mathbf{r}_T) \exp(-j2\pi ft) \, dt \quad (6.1\text{-}6)$$

is the *complex frequency response* at spatial location $\mathbf{r}_T$ of the transmit aperture, also known as the *complex transmit aperture function*, where f represents *input* or *transmitted* frequencies in hertz. Since

$$x_M(t, \mathbf{r}_T) = F_f^{-1}\{X_M(f, \mathbf{r}_T)\} = \int_{-\infty}^{\infty} X_M(f, \mathbf{r}_T) \exp(+j2\pi ft) \, df, \quad (6.1\text{-}7)$$

substituting Eq. (6.1-3) into Eq. (6.1-7) yields

$$\boxed{x_M(t, \mathbf{r}_T) = \int_{-\infty}^{\infty} X(f, \mathbf{r}_T) A_T(f, \mathbf{r}_T) \exp(+j2\pi ft) \, df.} \quad (6.1\text{-}8)$$

Compared with Eq. (6.1-1), Eq. (6.1-8) is a more useful representation of the input acoustic signal or source distribution for our purposes. Equation (6.1-8) represents the *coupling* of the input electrical signal to the fluid medium, that is, the production of the input acoustic signal or source distribution due to the input electrical signal via the transmit aperture, whose performance is described, in part, by the complex transmit aperture function.

A similar situation exists at the receive aperture shown in Fig. 6.1-2. As before, the aperture is arbitrary in shape and occupies a volume V in general. The subscripts R and M refer to the receive aperture and fluid medium, respectively. By following the same reasoning used in the development of Eqs. (6.1-1) through (6.1-8), the output electrical signal $y(t, \mathbf{r}_R)$ from the receive aperture at time t and spatial location $\mathbf{r}_R$ can be expressed as a time-domain convolution integral as follows:

$$y(t, \mathbf{r}_R) = \int_{-\infty}^{\infty} y_M(\tau, \mathbf{r}_R) \alpha_R(t - \tau, \mathbf{r}_R) \, d\tau \quad (6.1\text{-}9)$$

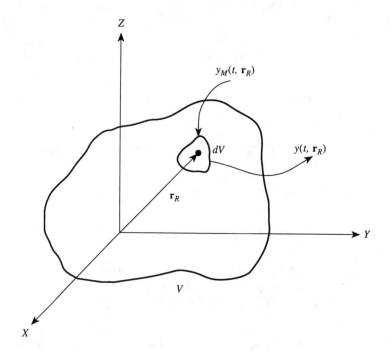

Figure 6.1-2. Arbitrarily shaped receive aperture shown occupying a volume V.

or

$$y(t, \mathbf{r}_R) = y_M(t, \mathbf{r}_R) \underset{t}{*} \alpha_R(t, \mathbf{r}_R), \tag{6.1-10}$$

where $y_M(t, \mathbf{r}_R)$ is the output acoustic signal from the fluid medium, which is also the input acoustic signal to the receive aperture at time t and spatial location $\mathbf{r}_R$, and $\alpha_R(t, \mathbf{r}_R)$ is the *impulse response* of the infinitesimal volume element dV of the receive aperture located at $\mathbf{r}_R$. Taking the Fourier transform of both sides of Eq. (6.1-10) with respect to t yields

$$Y(\eta, \mathbf{r}_R) = Y_M(\eta, \mathbf{r}_R) A_R(\eta, \mathbf{r}_R) \tag{6.1-11}$$

where

$$Y(\eta, \mathbf{r}_R) = F_t\{y(t, \mathbf{r}_R)\} = \int_{-\infty}^{\infty} y(t, \mathbf{r}_R) \exp(-j2\pi\eta t) \, dt \tag{6.1-12}$$

is the complex frequency spectrum of the output electrical signal,

$$Y_M(\eta, \mathbf{r}_R) = F_t\{y_M(t, \mathbf{r}_R)\} = \int_{-\infty}^{\infty} y_M(t, \mathbf{r}_R) \exp(-j2\pi\eta t) \, dt \tag{6.1-13}$$

is the complex frequency spectrum of the output acoustic signal from the fluid medium, which is also the input acoustic signal to the receive aperture, and

$$A_R(\eta, \mathbf{r}_R) = F_t\{\alpha_R(t, \mathbf{r}_R)\} = \int_{-\infty}^{\infty} \alpha_R(t, \mathbf{r}_R) \exp(-j2\pi\eta t) \, dt \quad (6.1\text{-}14)$$

is the *complex frequency response* at spatial location $\mathbf{r}_R$ of the receive aperture, also known as the *complex receive aperture function*, where η represents *output* or *received* frequencies in hertz and, in general, $\eta \neq f$. For example, if there is any motion, the received frequencies will not be equal to the transmitted frequencies because of Doppler shift. Taking the inverse Fourier transform of both sides of Eq. (6.1-11) with respect to η yields

$$y(t, \mathbf{r}_R) = \int_{-\infty}^{\infty} Y_M(\eta, \mathbf{r}_R) A_R(\eta, \mathbf{r}_R) \exp(+j2\pi\eta t) \, d\eta. \quad (6.1\text{-}15)$$

Note that Eq. (6.1-15) is analogous to Eq. (6.1-8). Equation (6.1-15) represents the *coupling* of the fluid medium to the output electrical signal, that is, the production of the output electrical signal due to the input acoustic signal via the receive aperture, whose performance is described, in part, by the complex receive aperture function.

6.2 Near-Field and Far-Field Directivity Functions of Volume Apertures

6.2.1 Near-Field Directivity Functions

Recall from Section 2.7.1 that the *exact* solution of the linear, three-dimensional, lossless, inhomogeneous wave equation

$$\nabla^2 \varphi(t, \mathbf{r}) - \frac{1}{c^2} \frac{\partial^2}{\partial t^2} \varphi(t, \mathbf{r}) = x_M(t, \mathbf{r}) \quad (6.2\text{-}1)$$

is given by

$$\varphi(t, \mathbf{r}) = -\frac{1}{4\pi} \int_{V_0} \frac{x_M[t - (|\mathbf{r} - \mathbf{r}_0|/c), \mathbf{r}_0]}{|\mathbf{r} - \mathbf{r}_0|} \, dV_0, \quad (6.2\text{-}2)$$

where $\varphi(t, \mathbf{r})$ is the scalar velocity potential in square meters per second, c is the constant speed of sound in meters per second, and $x_M(t, \mathbf{r})$ is the source distribution in inverse seconds. If the coupling equation for the source distribution given by Eq. (6.1-8) is substituted into Eq. (6.2-2), then

$$\varphi(t,\mathbf{r}) = \int_{-\infty}^{\infty} \int_{V_0} X(f,\mathbf{r}_0) A_T(f,\mathbf{r}_0) g_f(\mathbf{r}|\mathbf{r}_0) \, dV_0 \exp(+j2\pi ft) \, df, \quad (6.2\text{-}3)$$

where

$$g_f(\mathbf{r}|\mathbf{r}_0) \triangleq -\frac{\exp(-jk|\mathbf{r}-\mathbf{r}_0|)}{4\pi|\mathbf{r}-\mathbf{r}_0|} \qquad (6.2\text{-}4)$$

is the time-independent free-space Green's function, or free-space spatial impulse response, of an unbounded, ideal (nonviscous), homogeneous, fluid medium (see Section 2.6),

$$k = 2\pi f/c = 2\pi/\lambda \qquad (6.2\text{-}5)$$

is the wave number in radians per meter, λ is the wavelength in meters,

$$\mathbf{r} = x\hat{x} + y\hat{y} + z\hat{z} \qquad (6.2\text{-}6)$$

is the position vector to a field point, and

$$\mathbf{r}_0 = x_0\hat{x} + y_0\hat{y} + z_0\hat{z} \qquad (6.2\text{-}7)$$

is the position vector to a source point (see Fig. 6.2-1). Also recall from Section 1.4 that the acoustic pressure $p(t,\mathbf{r})$ in pascals (Pa) and the acoustic fluid velocity vector $\mathbf{u}(t,\mathbf{r})$ in meters per second can be obtained from the velocity potential as follows:

$$p(t,\mathbf{r}) = -\rho_0 \frac{\partial}{\partial t} \varphi(t,\mathbf{r}), \qquad (6.2\text{-}8)$$

where ρ_0 is the ambient (equilibrium) density of the fluid medium in kilograms per cubic meter, and

$$\mathbf{u}(t,\mathbf{r}) = \nabla\varphi(t,\mathbf{r}), \qquad (6.2\text{-}9)$$

where

$$\nabla = \frac{\partial}{\partial x}\hat{x} + \frac{\partial}{\partial y}\hat{y} + \frac{\partial}{\partial z}\hat{z} \qquad (6.2\text{-}10)$$

is the *gradient* expressed in the rectangular coordinates (x, y, z). Note that ρ_0 is constant, since we are dealing with a homogeneous medium.

The solution of the wave equation given by Eq. (6.2-3) is an exact expression. However, by approximating the time-independent free-space Green's function via

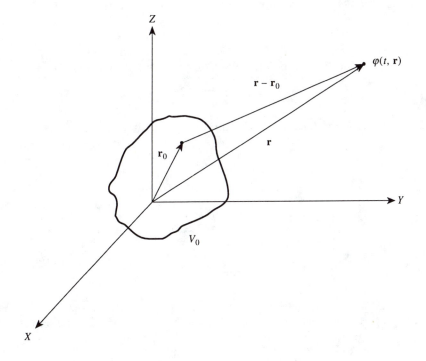

Figure 6.2-1. Source distribution shown occupying a volume V_0.

Fresnel and *Fraunhofer expansions*, both the near-field and far-field directivity functions or beam patterns of a transmit volume aperture can be derived, respectively, along with corresponding near-field and far-field expressions for the velocity potential. We begin the analysis by noting that the range term $|\mathbf{r} - \mathbf{r}_0|$ appears both as an amplitude term and as a phase term in the Green's function given by Eq. (6.2-4). Since

$$|\mathbf{r} - \mathbf{r}_0| = \sqrt{(\mathbf{r} - \mathbf{r}_0) \cdot (\mathbf{r} - \mathbf{r}_0)} \,, \qquad (6.2\text{-}11)$$

it can be shown that

$$|\mathbf{r} - \mathbf{r}_0| = r\sqrt{1 + b} \,, \qquad (6.2\text{-}12)$$

where

$$b = \left(\frac{r_0}{r}\right)^2 - 2\frac{\hat{r} \cdot \mathbf{r}_0}{r} \,, \qquad (6.2\text{-}13)$$

$$r_0 = |\mathbf{r}_0| = \sqrt{x_0^2 + y_0^2 + z_0^2} \qquad (6.2\text{-}14)$$

is the magnitude of the position vector to a source point,

$$r = |\mathbf{r}| = \sqrt{x^2 + y^2 + z^2}$$ (6.2-15)

is the magnitude of the position vector to a field point, and

$$\hat{r} = u\hat{x} + v\hat{y} + w\hat{z}$$ (6.2-16)

is the *unit vector* in the direction of $\mathbf{r}$, that is,

$$\mathbf{r} = r\hat{r},$$ (6.2-17)

where

$$u = \sin\theta\cos\psi,$$ (6.2-18)

$$v = \sin\theta\sin\psi,$$ (6.2-19)

and

$$w = \cos\theta$$ (6.2-20)

are the dimensionless direction cosines with respect to the X, Y, and Z axes, respectively (see Fig. 6.2-2). Note that u, v, and w take on values between -1 and 1, since $0 \le \theta \le \pi$ and $0 \le \psi \le 2\pi$, and that

$$u^2 + v^2 + w^2 = 1.$$ (6.2-21)

Also note that the parameter b given by Eq. (6.2-13) is dimensionless. We are now in a position to approximate the Green's function given by Eq. (6.2-4).

Consider the spherical spreading amplitude term $1/|\mathbf{r} - \mathbf{r}_0|$. In order to approximate the range term $|\mathbf{r} - \mathbf{r}_0|$, we shall use a *binomial expansion* of the square root in Eq. (6.2-12), that is,

$$|\mathbf{r} - \mathbf{r}_0| = r\sqrt{1 + b} \approx r\left(1 + \frac{b}{2} - \frac{b^2}{8} + \cdots\right), \qquad |b| < 1. \qquad (6.2\text{-}22)$$

Note that Eq. (6.2-22) is valid only if $|b| < 1$, where b is given by Eq. (6.2-13). In Appendix 6A it is shown that $|b| < 1$ whenever

$$r > \begin{cases} r_0\left(\sqrt{1 + \cos^2\phi} + \cos\phi\right), & 0 \le \phi \le \pi/2, \\ r_0\left(\sqrt{1 + \cos^2\phi} - \cos\phi\right), & \pi/2 \le \phi \le \pi, \end{cases} \qquad (6.2\text{-}23)$$

where ϕ is the angle between $\hat{r}$ and $\mathbf{r}_0$. Therefore, assuming that $|b| < 1$ and using only the *first term* of the binomial expansion in Eq. (6.2-22), we have $|\mathbf{r} - \mathbf{r}_0| \approx r$;

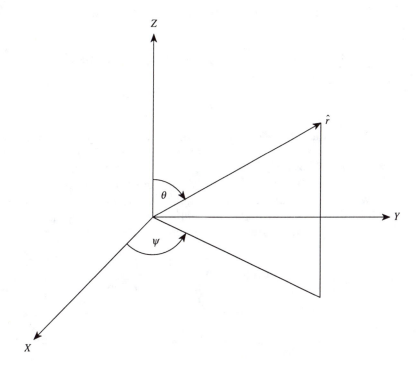

Figure 6.2-2. The unit vector $\hat{r}$ and the spherical angles θ and ψ. Note that ψ is measured in a counterclockwise direction.

and as a result,

$$\frac{1}{|\mathbf{r} - \mathbf{r}_0|} \approx \frac{1}{r}.\tag{6.2-24}$$

Although using only the first term of the binomial expansion is satisfactory for approximating the amplitude term, it is *not* satisfactory for approximating the phase term $\exp(-jk|\mathbf{r} - \mathbf{r}_0|)$, since small changes in the range $|\mathbf{r} - \mathbf{r}_0|$ can lead to large changes in phase.

   In order to approximate $\exp(-jk|\mathbf{r} - \mathbf{r}_0|)$, we shall use the *first three terms* of the binomial expansion as shown in Eq. (6.2-22). If terms involving r_0 raised to powers greater than two are neglected, then Eq. (6.2-22) reduces to

$$|\mathbf{r} - \mathbf{r}_0| \approx r - \hat{r} \cdot \mathbf{r}_0 + \frac{r_0^2 - (\hat{r} \cdot \mathbf{r}_0)^2}{2r}, \qquad |b| < 1.\tag{6.2-25}$$

Therefore, upon substituting Eqs. (6.2-24) and (6.2-25) into Eq. (6.2-4), we obtain

$$g_f(\mathbf{r}|\mathbf{r}_0) \approx -\frac{\exp(-jkr)}{4\pi r}\exp(+jk\hat{r}\cdot\mathbf{r}_0)\exp\left[-jk\frac{r_0^2-(\hat{r}\cdot\mathbf{r}_0)^2}{2r}\right], \qquad |b|<1,$$

(6.2-26)

which is a *near-field expansion* of the time-independent free-space Green's function involving all terms up to the second power in r_0.

The *Fresnel expansion* of $g_f(\mathbf{r}|\mathbf{r}_0)$ can also be obtained from Eq. (6.2-22) by using only the *first two terms* of the binomial expansion instead of the first three, which is equivalent to neglecting the dot-product term $(\hat{r}\cdot\mathbf{r}_0)^2$ in Eqs. (6.2-25) and (6.2-26). Therefore, the Fresnel expansion, or Fresnel approximation, of the time-independent free-space Green's function is given by

$$\boxed{g_f(\mathbf{r}|\mathbf{r}_0) \approx -\frac{\exp(-jkr)}{4\pi r}\exp(+jk\hat{r}\cdot\mathbf{r}_0)\exp\left(-jk\frac{r_0^2}{2r}\right), \qquad |b|<1.}$$
(6.2-27)

Neglecting $(\hat{r}\cdot\mathbf{r}_0)^2$ in Eq. (6.2-26) in order to obtain Eq. (6.2-27) requires that

$$r_0^2 \gg (\hat{r}\cdot\mathbf{r}_0)^2$$

(6.2-28)

or, equivalently, that

$$r_0^2 \geq K(\hat{r}\cdot\mathbf{r}_0)^2,$$

(6.2-29)

where the constant $K > 1$ controls the region of validity of the Fresnel approximation. And upon taking the positive square root of both sides of Eq. (6.2-29), we obtain

$$r_0 \geq \sqrt{K}\,|\hat{r}\cdot\mathbf{r}_0|,$$

(6.2-30)

where the magnitude of the dot product is used because r_0 must be positive. Since we already introduced ϕ as the angle between $\hat{r}$ and $\mathbf{r}_0$, where $\hat{r}$ is the unit vector in the direction of the position vector $\mathbf{r}$ to a field point, and $\mathbf{r}_0$ is the position vector to a source point, Eq. (6.2-30) can be expressed as

$$1 \geq \sqrt{K}\,|\cos\phi|, \qquad 0 \leq \phi \leq \pi,$$

(6.2-31)

or

$$\phi_{\min} \leq \phi \leq \phi_{\max}, \qquad 0 \leq \phi \leq \pi,$$

(6.2-32)

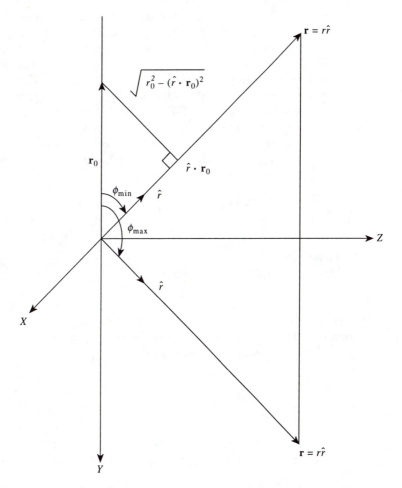

Figure 6.2-3. Geometrical representation of the Fresnel approximation.

where

$$\phi_{min} = \cos^{-1}\left(1/\sqrt{K}\right) \tag{6.2-33}$$

and

$$\phi_{max} = \pi - \phi_{min} \tag{6.2-34}$$

since $|\cos(\pi - \phi)| = |\cos\phi|$ (see Fig. 6.2-3). Since a factor of 10 is a typical choice to represent "much greater than," if we let $K = 10$, then Eq. (6.2-32) becomes

$$\boxed{72° \le \phi \le 108°.} \tag{6.2-35}$$

For example, for the geometry shown in Fig. 6.2-3, if the source distribution lies in the *XY* plane, then the Fresnel approximation is only valid inside an angular region no greater than 18° from the *Z* axis. This is analogous to the *paraxial assumption* that is made in the derivation of the Fresnel diffraction integral in optics. It is also interesting to note that this 18° criterion is the same result that we obtained in our discussion of the solution of the parabolic partial differential equation for a homogeneous fluid medium in Example 5.3-1.

Since we have just shown that in order for the Fresnel approximation to be valid Eq. (6.2-35) must be satisfied, we next set the radial distance to a source point r_0 equal to its *maximum* value (call it R) and substitute either $\phi = 72°$ or $\phi = 108°$ into Eq. (6.2-23) to obtain

$$\boxed{r > 1.356R.}\qquad(6.2\text{-}36)$$

Equation (6.2-36) stipulates the *minimum* range that a field point has to be from an aperture in order to guarantee that $|b| < 1$ so that the Fresnel approximation is valid.

Next, by using the Fresnel expansion of the time-independent free-space Green's function given by Eq. (6.2-27), we shall derive a criterion that establishes the boundary between the near-field and far-field regions of an aperture. We begin by examining the quadratic phase factor

$$\exp\left(-jk\frac{r_0^2}{2r}\right)$$

that appears on the right-hand side of Eq. (6.2-27). This phase factor accounts for the effects of *wavefront curvature*, and since $k = 2\pi/\lambda$, it can be rewritten as

$$\exp\left(-j\pi\frac{r_0^2}{\lambda r}\right).$$

We shall consider this quadratic phase factor to be *insignificant* if

$$\pi\frac{r_0^2}{\lambda r} < 1,\qquad(6.2\text{-}37)$$

or

$$r > \pi r_0^2/\lambda.\qquad(6.2\text{-}38)$$

In order to obtain a conservative criterion based on Eq. (6.2-38), we consider the worst case, which is when the numerator becomes the largest. This corresponds to setting r_0 equal to its maximum value R. Therefore, if

$$r > \pi R^2/\lambda > 1.356R, \qquad (6.2\text{-}39)$$

then we shall consider the quadratic phase factor to be *insignificant*, since it will be approximately equal to $\exp(j0) = 1$, a constant, which indicates no phase variation as a function of r_0 and r. Equation (6.2-39) is the *far-field criterion*. However, if

$$1.356R < r < \pi R^2/\lambda, \qquad (6.2\text{-}40)$$

then we shall consider the quadratic phase factor to be *significant*, that is, a nonnegligible function of r_0 and r. Equation (6.2-40) is the *near-field criterion*. Note that Eq. (6.2-36) has been incorporated into Eqs. (6.2-39) and (6.2-40). Also note that both the far-field and near-field criteria depend on the ratio between an aperture's size and the wavelength.

Since we have a Fresnel expansion of $g_f(\mathbf{r}|\mathbf{r}_0)$ given by Eq. (6.2-27), we can proceed with the derivation of the near-field directivity function. Substituting Eq. (6.2-27) into the exact solution of the wave equation given by Eq. (6.2-3) yields the following approximate solution of the wave equation:

$$\varphi(t,\mathbf{r}) \approx -\frac{1}{4\pi r}\int_{-\infty}^{\infty}\int_{V_0} X(f,\mathbf{r}_0)\,A_T(f,\mathbf{r}_0)\,\exp\left(-jk\frac{r_0^2}{2r}\right)\exp(+jk\hat{r}\cdot\mathbf{r}_0)\,dV_0$$

$$\times \exp\left[+j2\pi f\left(t-\frac{r}{c}\right)\right]df, \qquad (6.2\text{-}41)$$

where the expression $t-(r/c)$ is known as the *retarded time*. Retarded time is a measure of the amount of time that has elapsed since the acoustic field transmitted by a source first appears at a receiver located r meters from the source. With the use of Eqs. (6.2-5), (6.2-7), and (6.2-16), we can write that

$$\exp(+jk\hat{r}\cdot\mathbf{r}_0) = \exp[+j2\pi(f_X x_0 + f_Y y_0 + f_Z z_0)], \qquad (6.2\text{-}42)$$

where

$$f_X = u/\lambda, \qquad (6.2\text{-}43)$$

$$f_Y = v/\lambda, \qquad (6.2\text{-}44)$$

and

$$f_Z = w/\lambda \qquad (6.2\text{-}45)$$

are the *spatial frequencies* in the X, Y, and Z directions, respectively, with units of cycles per meter (see Section 2.2.1) and u, v, and w are the dimensionless direction cosines with respect to the X, Y, and Z axes, respectively, given by

Eqs. (6.2-18) through (6.2-20). If we substitute Eq. (6.2-42) into Eq. (6.2-41) and if we designate the volume integral by I, then

$$I = \int_{-\infty}^{\infty}\int_{-\infty}^{\infty}\int_{-\infty}^{\infty} X(f,\mathbf{r}_0) A_T(f,\mathbf{r}_0) \exp\left(-jk\frac{r_0^2}{2r}\right)$$

$$\times \exp\left[+j2\pi(f_X x_0 + f_Y y_0 + f_Z z_0)\right] dx_0\, dy_0\, dz_0. \qquad (6.2\text{-}46)$$

Equation (6.2-46) can be interpreted as being a *three-dimensional spatial Fourier transform*, that is,

$$I = F_{\mathbf{r}_0}\left\{ X(f,\mathbf{r}_0)\left[A_T(f,\mathbf{r}_0) \exp\left(-jk\frac{r_0^2}{2r}\right)\right]\right\}, \qquad (6.2\text{-}47)$$

where $F_{\mathbf{r}_0}\{\cdot\}$ is shorthand notation for a three-dimensional spatial Fourier transform with respect to x_0, y_0, and z_0; that is, $F_{\mathbf{r}_0}\{\cdot\} = F_{x_0}F_{y_0}F_{z_0}\{\cdot\}$. In Eq. (6.2-47) it can be seen that the complex transmit aperture function and quadratic phase factor have been grouped together, separate from the transmitted electrical signal.

Recall that the time-domain Fourier transform of the product of two time functions is equal to a convolution integral in the frequency domain. Similarly, the spatial Fourier transform of the product of two spatial functions is equal to a convolution integral in the spatial-frequency domain. Therefore, Eq. (6.2-47) can be expressed as follows:

$$I = X_M(f,r,\mathbf{v}) = X(f,\mathbf{v}) \underset{\mathbf{v}}{*} D_T(f,r,\mathbf{v}), \qquad (6.2\text{-}48)$$

where

$$X_M(f,r,\mathbf{v}) = \int_{-\infty}^{\infty} X(f,\boldsymbol{\alpha}) D_T(f,r,\mathbf{v}-\boldsymbol{\alpha})\, d\boldsymbol{\alpha} \qquad (6.2\text{-}49)$$

is the near-field complex frequency and angular spectrum of the input acoustic signal to the fluid medium; $\mathbf{v} = (\nu_X, \nu_Y, \nu_Z)$ is a three-dimensional vector whose components are spatial frequencies in the X, Y, and Z directions, respectively;

$$X(f,\boldsymbol{\alpha}) = F_{\mathbf{r}_0}\{X(f,\mathbf{r}_0)\} = \int_{-\infty}^{\infty} X(f,\mathbf{r}_0) \exp(+j2\pi\boldsymbol{\alpha}\cdot\mathbf{r}_0)\, d\mathbf{r}_0 \qquad (6.2\text{-}50)$$

is the complex frequency and angular spectrum of the transmitted electrical signal; and

$$\boxed{\begin{aligned} D_T(f,r,\boldsymbol{\alpha}) &= F_{\mathbf{r}_0}\left\{ A_T(f,\mathbf{r}_0) \exp\left(-jk\frac{r_0^2}{2r}\right)\right\} \\[2mm] &= \int_{-\infty}^{\infty} A_T(f,\mathbf{r}_0) \exp\left(-jk\frac{r_0^2}{2r}\right) \exp(+j2\pi\boldsymbol{\alpha}\cdot\mathbf{r}_0)\, d\mathbf{r}_0 \end{aligned}} \qquad (6.2\text{-}51)$$

is the *near-field directivity function* or *beam pattern* of complex transmit aperture function $A_T(f, \mathbf{r}_0)$ based on the Fresnel approximation [see Eqs. (6.2-35) and (6.2-40)] where

$$\boldsymbol{\alpha} = (f_X, f_Y, f_Z) \tag{6.2-52}$$

and

$$\mathbf{r}_0 = (x_0, y_0, z_0). \tag{6.2-53}$$

Recall that the complex transmit aperture function $A_T(f, \mathbf{r}_0)$ is the complex frequency response at spatial location $\mathbf{r}_0$ of the transmit aperture. Note that the right-hand side of Eq. (6.2-49) is shorthand notation for a three-dimensional convolution integral, since $d\boldsymbol{\alpha} = df_X \, df_Y \, df_Z$, and that the right-hand sides of Eqs. (6.2-50) and (6.2-51) are also shorthand notation for triple integrals, since $d\mathbf{r}_0 = dx_0 \, dy_0 \, dz_0$. From Eq. (6.2-51) it can be seen that the near-field directivity function is a function of frequency f, the range r to a field point, and the spherical angles θ and ψ, since the spatial frequencies f_X, f_Y, and f_Z are related to the direction cosines u, v, and w, respectively, which are related to the spherical angles θ and ψ. The phrase "angular spectrum" is used because the spatial frequencies can ultimately be expressed in terms of the spherical angles. The form of Eq. (6.2-51) is also known as a *Fresnel diffraction integral* in optics or as a *three-dimensional spatial Fresnel transform*.

If Eq. (6.2-48) is substituted into Eq. (6.2-41) in place of the volume integral I given by Eq. (6.2-46), then the near-field expression for the velocity potential becomes

$$\varphi(t, \mathbf{r}) = \varphi(t, r, \theta, \psi) \approx -\frac{1}{4\pi r} \int_{-\infty}^{\infty} X_M(f, r, \boldsymbol{v}) \exp\left[+j2\pi f\left(t - \frac{r}{c}\right)\right] df,$$

$$\tag{6.2-54}$$

where $X_M(f, r, \boldsymbol{v})$ is given by Eq. (6.2-49) and $\boldsymbol{v} = (v_X, v_Y, v_Z)$.

Before we begin our discussion of far-field directivity functions in Section 6.2.2, it is important to note that a near-field expression or near-field beam pattern is valid in the far field as well. However, a far-field expression or far-field beam pattern is only valid in the far field.

6.2.2   Far-Field Directivity Functions

If a field point at range r is in the far-field region of an aperture, that is, if Eq. (6.2-39) is satisfied, then the *Fraunhofer expansion* or *Fraunhofer approximation* of the time-independent free-space Green's function can be obtained from the Fresnel expansion given by Eq. (6.2-27) by neglecting the quadratic phase factor.

Doing so yields the following Fraunhofer, or far-field, expansion:

$$g_f(\mathbf{r}|\mathbf{r}_0) \approx -\frac{\exp(-jkr)}{4\pi r}\exp(+jk\hat{r}\cdot\mathbf{r}_0), \qquad |b| < 1. \qquad (6.2\text{-}55)$$

Note that if Eq. (6.2-39) is satisfied and if, in fact, $r > 2.414R$ [set $\phi = 0$ or $\phi = \pi$ in Eq. (6.2-23)], then $|b| < 1$ for all values of ϕ, that is, $0 \le \phi \le \pi$.

Similarly, the far-field directivity function can be obtained from the near-field directivity function given by Eq. (6.2-51) by neglecting the quadratic phase factor. Therefore, the *far-field directivity function*, or *beam pattern*, of the complex transmit aperture function $A_T(f,\mathbf{r}_0)$ is given by

$$D_T(f,\boldsymbol{\alpha}) = F_{\mathbf{r}_0}\{A_T(f,\mathbf{r}_0)\} = \int_{-\infty}^{\infty} A_T(f,\mathbf{r}_0)\exp(+j2\pi\boldsymbol{\alpha}\cdot\mathbf{r}_0)\,d\mathbf{r}_0, \qquad (6.2\text{-}56)$$

where $d\mathbf{r}_0 = dx_0\,dy_0\,dz_0$. In addition,

$$A_T(f,\mathbf{r}_0) = F_{\boldsymbol{\alpha}}^{-1}\{D_T(f,\boldsymbol{\alpha})\} = \int_{-\infty}^{\infty} D_T(f,\boldsymbol{\alpha})\exp(-j2\pi\boldsymbol{\alpha}\cdot\mathbf{r}_0)\,d\boldsymbol{\alpha}, \qquad (6.2\text{-}57)$$

where $F_{\boldsymbol{\alpha}}^{-1}\{\cdot\} = F_{f_X}^{-1}F_{f_Y}^{-1}F_{f_Z}^{-1}\{\cdot\}$ and $d\boldsymbol{\alpha} = df_X\,df_Y\,df_Z$. Therefore, *an aperture function and its far-field beam pattern form a spatial Fourier transform pair*. Note that the far-field beam pattern is *not* a function of the range r to a field point as the near-field beam pattern is. The form of Eq. (6.2-56) is also known as a *Fraunhofer diffraction integral* in optics or as a *three-dimensional spatial Fourier transform*.

In the far field, the near-field expression for the velocity potential given by Eq. (6.2-54) reduces to the following far-field expression:

$$\varphi(t,\mathbf{r}) = \varphi(t,r,\theta,\psi) \approx -\frac{1}{4\pi r}\int_{-\infty}^{\infty} X_M(f,\boldsymbol{\nu})\exp\left[+j2\pi f\left(t - \frac{r}{c}\right)\right]df,$$

$$(6.2\text{-}58)$$

where

$$X_M(f,\boldsymbol{\nu}) = \int_{-\infty}^{\infty} X(f,\boldsymbol{\alpha})D_T(f,\boldsymbol{\nu}-\boldsymbol{\alpha})\,d\boldsymbol{\alpha} \qquad (6.2\text{-}59)$$

is the complex frequency and angular spectrum of the input acoustic signal to the fluid medium obtained from Eq. (6.2-49) by replacing the near-field directivity function with the far-field directivity function.

Next, let us derive an expression for the complex frequency and angular spectrum of the output electrical signal from a receive aperture. Recall that the complex frequency spectrum of the output electrical signal from a receive aperture is given by Eq. (6.1-11), which is rewritten below for convenience:

$$Y(\eta, \mathbf{r}_R) = Y_M(\eta, \mathbf{r}_R) A_R(\eta, \mathbf{r}_R). \qquad (6.1\text{-}11)$$

Taking the spatial Fourier transform of both sides of Eq. (6.1-11) with respect to $\mathbf{r}_R$ yields

$$Y(\eta, \boldsymbol{\gamma}) = Y_M(\eta, \boldsymbol{\gamma}) \underset{\gamma}{*} D_R(\eta, \boldsymbol{\gamma}), \qquad (6.2\text{-}60)$$

or

$$\boxed{Y(\eta, \boldsymbol{\gamma}) = \int_{-\infty}^{\infty} Y_M(\eta, \boldsymbol{\beta}) D_R(\eta, \boldsymbol{\gamma} - \boldsymbol{\beta}) \, d\boldsymbol{\beta},} \qquad (6.2\text{-}61)$$

where

$$Y(\eta, \boldsymbol{\gamma}) = F_{\mathbf{r}_R}\{Y(\eta, \mathbf{r}_R)\} = \int_{-\infty}^{\infty} Y(\eta, \mathbf{r}_R) \exp(+j2\pi\boldsymbol{\gamma} \cdot \mathbf{r}_R) \, d\mathbf{r}_R \quad (6.2\text{-}62)$$

is the complex frequency and angular spectrum of the output electrical signal from the receive aperture; $\boldsymbol{\gamma} = (\gamma_X, \gamma_Y, \gamma_Z)$ is a three-dimensional vector whose components are spatial frequencies in the X, Y, and Z directions, respectively;

$$Y_M(\eta, \boldsymbol{\beta}) = F_{\mathbf{r}_R}\{Y_M(\eta, \mathbf{r}_R)\} = \int_{-\infty}^{\infty} Y_M(\eta, \mathbf{r}_R) \exp(+j2\pi\boldsymbol{\beta} \cdot \mathbf{r}_R) \, d\mathbf{r}_R \quad (6.2\text{-}63)$$

is the complex frequency and angular spectrum of the output acoustic signal from the fluid medium, which is also the input acoustic signal to the receive aperture; $\boldsymbol{\beta} = (\beta_X, \beta_Y, \beta_Z)$ is a three-dimensional vector whose components are spatial frequencies in the X, Y, and Z directions, respectively; and

$$D_R(\eta, \boldsymbol{\beta}) = F_{\mathbf{r}_R}\{A_R(\eta, \mathbf{r}_R)\} = \int_{-\infty}^{\infty} A_R(\eta, \mathbf{r}_R) \exp(+j2\pi\boldsymbol{\beta} \cdot \mathbf{r}_R) \, d\mathbf{r}_R \quad (6.2\text{-}64)$$

is the far-field directivity function or beam pattern of the complex receive aperture function $A_R(\eta, \mathbf{r}_R)$, where $d\boldsymbol{\beta} = d\beta_X \, d\beta_Y \, d\beta_Z$ and $d\mathbf{r}_R = dx_R \, dy_R \, dz_R$. Note that the right-hand side of Eq. (6.2-61) is shorthand notation for a three-dimensional convolution integral. Equation (6.2-61) will be used in conjunction with our discussion of FFT beamforming in Section 8.1.

Example 6.2-1

Assume that an *identical* input electrical signal is applied at all spatial locations $\mathbf{r}_0$ of a transmit aperture, that is,

$$x(t, \mathbf{r}_0) = x(t), \qquad (6.2\text{-}65)$$

so that

$$X(f, \mathbf{r}_0) = X(f). \qquad (6.2\text{-}66)$$

Substituting Eq. (6.2-66) into Eq. (6.2-50) yields

$$X(f, \boldsymbol{\alpha}) = X(f) \int_{-\infty}^{\infty} \exp(+j2\pi\boldsymbol{\alpha} \cdot \mathbf{r}_0) \, d\mathbf{r}_0 = X(f) F_{\mathbf{r}_0}\{1\}, \qquad (6.2\text{-}67)$$

or

$$\boxed{X(f, \boldsymbol{\alpha}) = X(f)\delta(\boldsymbol{\alpha}),} \qquad (6.2\text{-}68)$$

since

$$\boxed{F_{\mathbf{r}_0}\{1\} = \delta(\boldsymbol{\alpha}).} \qquad (6.2\text{-}69)$$

Substituting Eq. (6.2-68) into Eq. (6.2-49) and making use of the sifting property of impulse functions yields

$$X_M(f, r, \boldsymbol{v}) = X(f) D_T(f, r, \boldsymbol{v}), \qquad (6.2\text{-}70)$$

and upon substituting Eq. (6.2-70) into Eq. (6.2-54), the near-field velocity-potential expression reduces to

$$\varphi(t, r, \theta, \psi) \approx -\frac{1}{4\pi r} \int_{-\infty}^{\infty} X(f) D_T(f, r, \boldsymbol{v}) \exp\left[+j2\pi f\left(t - \frac{r}{c}\right)\right] df.$$

$$(6.2\text{-}71)$$

Similarly, substituting Eq. (6.2-68) into Eq. (6.2-59) yields

$$\boxed{X_M(f, \boldsymbol{v}) = X(f) D_T(f, \boldsymbol{v}),} \qquad (6.2\text{-}72)$$

and upon substituting Eq. (6.2-72) into Eq. (6.2-58), the far-field velocity-potential expression reduces to

$$\varphi(t,r,\theta,\psi) \approx -\frac{1}{4\pi r} \int_{-\infty}^{\infty} X(f)D_T(f,\boldsymbol{\nu}) \exp\left[+j2\pi f\left(t - \frac{r}{c}\right)\right] df.$$

(6.2-73)

If it is further assumed that the input electrical signal is time harmonic, that is, if

$$x(t,\mathbf{r}_0) = x(t) = \exp(+j2\pi f_0 t),$$

(6.2-74)

then

$$X(f,\mathbf{r}_0) = X(f) = \delta(f - f_0).$$

(6.2-75)

Substituting Eq. (6.2-75) into Eqs. (6.2-71) and (6.2-73) yields

$$\varphi(t,r,\theta,\psi) \approx -D_T(f_0,r,\boldsymbol{\nu}_0)\frac{\exp\left[+j2\pi f_0\left(t - \frac{r}{c}\right)\right]}{4\pi r}$$

(6.2-76)

in the near field, and

$$\varphi(t,r,\theta,\psi) \approx -D_T(f_0,\boldsymbol{\nu}_0)\frac{\exp\left[+j2\pi f_0\left(t - \frac{r}{c}\right)\right]}{4\pi r}$$

(6.2-77)

in the far field, respectively, where $\boldsymbol{\nu}_0 = (\nu_{X_0}, \nu_{Y_0}, \nu_{Z_0})$, $\nu_{X_0} = u/\lambda_0$, $\nu_{Y_0} = v/\lambda_0$, and $\nu_{Z_0} = w/\lambda_0$. The wavelength $\lambda_0 = c/f_0$ is the wavelength in the fluid medium, and u, v, and w are the direction cosines given by Eqs. (6.2-18) through (6.2-20), respectively. Equation (6.2-77) will be used to derive the equation for the directivity index of an aperture in Section 6.7.

Example 6.2-1 raises an interesting point. If we decide that any electronics used for purposes of amplitude shading and/or beam steering are to be considered as part of the aperture, then the assumption that an identical input electrical signal is applied at all spatial locations of a transmit aperture is true in most practical cases. Because of this assumption, the convolution-integral expression for $X_M(f,\boldsymbol{\nu})$ given by Eq. (6.2-59) reduced to the product given by Eq. (6.2-72). However, the situation is different at the receive aperture. Since the acoustic field $Y_M(\boldsymbol{\eta},\boldsymbol{\beta})$ incident upon a receive aperture is not, in general, identical at all spatial locations of the aperture, the convolution integral given by Eq. (6.2-61) must be used.

6.3 Linear Apertures and Far-Field Directivity Functions

6.3.1 Amplitude Windows

Before we begin our discussion, a few comments concerning notation are in order. For the remainder of Chapter 6, and in Chapter 7, we shall drop the subscript notation T and R which has been used to distinguish between transmit and receive apertures, respectively, since we shall now be concerned with apertures in general. Also, the position vectors $\mathbf{r}_T$ and $\mathbf{r}_R$, which were used to identify the spatial locations of the transmit and receive apertures, respectively, will be replaced by the position vector $\mathbf{r}_a$.

Recall that the far-field directivity function of a general volume aperture is given by [see Eq. (6.2-56)]

$$D(f, \boldsymbol{\alpha}) = F_{\mathbf{r}_a}\{A(f, \mathbf{r}_a)\} = \int_{-\infty}^{\infty} A(f, \mathbf{r}_a) \exp(+j2\pi\boldsymbol{\alpha}\cdot\mathbf{r}_a)\, d\mathbf{r}_a, \quad (6.3\text{-}1)$$

where

$$\mathbf{r}_a = (x_a, y_a, z_a), \qquad (6.3\text{-}2)$$

$$\boldsymbol{\alpha} = (f_X, f_Y, f_Z), \qquad (6.3\text{-}3)$$

and $d\mathbf{r}_a = dx_a\, dy_a\, dz_a$. Note that the complex aperture function can be expressed as

$$A(f, \mathbf{r}_a) = a(f, \mathbf{r}_a) \exp[+j\theta(f, \mathbf{r}_a)], \qquad (6.3\text{-}4)$$

where $a(f, \mathbf{r}_a)$ is the *amplitude* and $\theta(f, \mathbf{r}_a)$ is the *phase* of the complex frequency response at spatial location $\mathbf{r}_a$ of the aperture. Both $a(f, \mathbf{r}_a)$ and $\theta(f, \mathbf{r}_a)$ are *real functions*. The function $a(f, \mathbf{r}_a)$ is also known as the *amplitude window*.

Now consider the case of a *linear aperture* of length L meters lying along the X axis as shown in Fig. 6.3-1. A linear aperture can represent either a single electroacoustic transducer or a linear array of many individual electroacoustic transducers. For example, if a linear aperture represents a single electroacoustic transducer that is being used in the active mode, that is, as a transmitter, then this physical situation corresponds to a *continuous line source*. Also shown in Fig. 6.3-1 is a field point with spherical coordinates (r, θ, ψ). The field point is assumed to be in the far-field region of the aperture, at a range $r > \pi R^2/\lambda > 1.356R$ meters, where $R = L/2$ meters is the maximum radial extent of the aperture [see Eq. (6.2-39)].

Since the linear aperture illustrated in Fig. 6.3-1 is shown lying along the X axis, the position vector that describes the spatial location of the aperture is given by

$$\mathbf{r}_a = (x_a, 0, 0), \qquad (6.3\text{-}5)$$

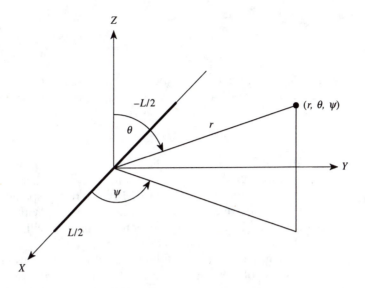

Figure 6.3-1. Linear aperture of length L meters lying along the X axis. Also shown is a field point with spherical coordinates (r, θ, ψ).

and as a result,

$$A(f, \mathbf{r}_a) = A(f, x_a), \qquad (6.3\text{-}6)$$

$$\boldsymbol{\alpha} \cdot \mathbf{r}_a = (f_X, f_Y, f_Z) \cdot (x_a, 0, 0) = f_X x_a, \qquad (6.3\text{-}7)$$

and

$$d\mathbf{r}_a = dx_a. \qquad (6.3\text{-}8)$$

Therefore, upon substituting Eqs. (6.3-6) through (6.3-8) into Eq. (6.3-1), we obtain the following one-dimensional spatial Fourier transform expression for the far-field directivity function (beam pattern) of a linear aperture lying along the X axis:

$$\boxed{D(f, f_X) = F_{x_a}\{A(f, x_a)\} = \int_{-L/2}^{L/2} A(f, x_a) \exp(+j2\pi f_X x_a)\, dx_a,} \qquad (6.3\text{-}9)$$

where

$$A(f, x_a) = a(f, x_a) \exp\left[+j\theta(f, x_a)\right] \qquad (6.3\text{-}10)$$

and [see Eq. (6.2-43)]

$$f_X = \frac{u}{\lambda} = \frac{\sin\theta\cos\psi}{\lambda}. \tag{6.3-11}$$

Since the spatial frequency f_X can be expressed in terms of θ and ψ, the far-field directivity function can ultimately be expressed as a function of frequency f and the spherical angles θ and ψ, that is,

$$D(f, f_X) \rightarrow D(f, \theta, \psi). \tag{6.3-12}$$

Note that if a linear aperture lies along either the Y or Z axis instead of the X axis, then we can simply replace x_a and f_X with either y_a and f_Y, or z_a and f_Z, respectively, in Eqs. (6.3-9) and (6.3-10), where the spatial frequencies f_Y and f_Z are given by Eqs. (6.2-44) and (6.2-45), respectively. With the use of Eq. (6.3-9), we shall compute closed-form expressions for the normalized far-field beam patterns of several hypothetical amplitude windows.

Although the various amplitude windows to be discussed next are hypothetical in the sense of not representing actual frequency responses of electroacoustic transducers, they are very common functions and will serve well our purposes of demonstrating far-field beam-pattern calculations, beamwidth calculations, and the relationship between beamwidth and sidelobe levels of far-field beam patterns. Besides, sampled versions of these continuous amplitude windows are, in fact, used to amplitude weight arrays of electroacoustic transducers, as will be discussed in Chapter 7. In addition, time-domain versions of these continuous amplitude windows can also be used to reduce the sidelobe levels of the normalized auto-ambiguity functions of common transmitted electrical signals, as will be discussed in Chapter 8.

In the examples that follow, it is assumed that the linear aperture represents a single electroacoustic transducer of length L meters and that the phase response of the transducer is zero, that is,

$$A(f, x_a) = a(f, x_a), \qquad \theta(f, x_a) = 0. \tag{6.3-13}$$

Therefore, according to Eq. (6.3-13), the complex frequency response of the transducer is equal to the amplitude response (amplitude window) $a(f, x_a)$. We shall examine the effects of a nonzero phase response later, in Section 6.3.3.

Example 6.3-1 (The rectangular amplitude window)
In this example we shall derive the normalized far-field beam pattern of the rectangular amplitude window illustrated in Fig. 6.3-2. The rectangular amplitude window is defined as follows:

$$\text{rect}(x/L) \triangleq \begin{cases} 1, & |x| \le L/2, \\ 0, & |x| > L/2. \end{cases} \tag{6.3-14}$$

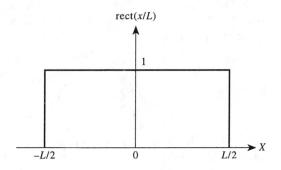

Figure 6.3-2. Rectangular amplitude window.

Imagine applying an input electrical signal in the form of a cosine wave at frequency f to a single electroacoustic transducer (i.e., a continuous line source) lying along the X axis. If a pressure probe is then placed very near to the transducer without touching it, the radiated acoustic pressure measured along the length L of the transducer is the response of the transducer at frequency f as a function of spatial coordinate x along the transducer. Therefore, if

$$A(f, x_a) = a(f, x_a) = \mathrm{rect}(x_a/L), \qquad (6.3\text{-}15)$$

then the amplitude of the complex frequency response of the transducer is constant along the entire length L of the transducer, regardless of the value of frequency f. In addition, the phase of the complex frequency response is zero. Equation (6.3-15) represents the simplest mathematical model for the complex aperture function $A(f, x_a)$.

Substituting Eqs. (6.3-15) and (6.3-14) into Eq. (6.3-9) yields

$$D(f, f_X) = F_{x_a}\{\mathrm{rect}(x_a/L)\} = \int_{-L/2}^{L/2} \exp(+j2\pi f_X x_a)\, dx_a, \quad (6.3\text{-}16)$$

or

$$\boxed{D(f, f_X) = F_{x_a}\{\mathrm{rect}(x_a/L)\} = L\,\mathrm{sinc}(f_X L)} \qquad (6.3\text{-}17)$$

where

$$\boxed{\mathrm{sinc}(x) \triangleq \frac{\sin(\pi x)}{\pi x}.} \qquad (6.3\text{-}18)$$

Equation (6.3-17) is the *unnormalized* far-field directivity function of the rectangular amplitude window. The *normalized* far-field directivity function is defined as follows:

$$D_N(f, f_X) \triangleq \frac{D(f, f_X)}{D_{max}},$$ (6.3-19)

where D_{max} is the *normalization factor*, that is, it is the maximum value of the unnormalized far-field directivity function. Therefore, by referring to Eq. (6.3-17),

$$D_{max} = D(f, 0) = L,$$ (6.3-20)

since sinc(0) = 1 is the maximum value of the sinc function. Substituting Eqs. (6.3-17) and (6.3-20) into Eq. (6.3-19) yields the following expression for the *normalized* far-field beam pattern of the rectangular amplitude window:

$$D_N(f, f_X) = \text{sinc}(f_X L).$$ (6.3-21)

Note that $D_N(f, 0) = 1$. Since $f_X = u/\lambda$, the normalized far-field beam pattern of the rectangular amplitude window given by Eq. (6.3-21) can also be expressed as

$$D_N(f, u) = \text{sinc}\left(\frac{L}{\lambda} u\right),$$ (6.3-22)

where the wavelength λ has been suppressed in the argument of D_N because frequency f and wavelength λ are related by $c = f\lambda$, that is, $D_N(f, f_X) = D_N(f, u/\lambda) \rightarrow D_N(f, u)$. The magnitude of the normalized far-field beam pattern given by Eq. (6.3-22) is plotted as a function of direction cosine u in Fig. 6.3-3. The plot is shown only for positive values of u, since the magnitude of the beam pattern is symmetric about $u = 0$. The level of the first sidelobe is approximately -13 dB.

From Fig. 6.3-3 it can be seen that the width of the main lobe of the normalized (and unnormalized) far-field beam pattern depends on the ratio λ/L. Therefore, the beamwidth is directly proportional to the wavelength λ (inversely proportional to the frequency f) and inversely proportional to the length L of the aperture (transducer). As a result, the beamwidth can be decreased by keeping the length L of the aperture constant while increasing the frequency f, or by keeping the

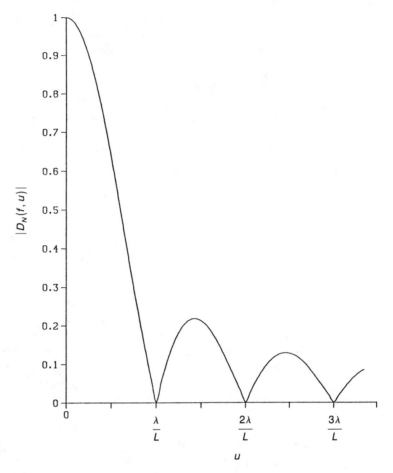

Figure 6.3-3. Magnitude of the normalized far-field beam pattern of the rectangular amplitude window plotted as a function of direction cosine u.

frequency f constant while increasing the length L of the aperture. Beamwidth will be discussed in detail in Sections 6.3.2 and 6.3.3.

From a theoretical point of view, the normalized far-field beam pattern given by Eq. (6.3-22) can be evaluated at any value of u. However, it is only the interval $-1 \leq u \leq 1$ that pertains to the real aperture problem, since u is a direction cosine given by

$$u = \sin \theta \cos \psi, \qquad (6.3-23)$$

and as a result, $-1 \leq u \leq 1$, since $0 \leq \theta \leq \pi$ and $0 \leq \psi \leq 2\pi$. The interval $-1 \leq u \leq 1$ is called the *visible region* of an aperture (transducer) lying along the X axis. The value $u = 0$, which implies that

$\psi = \pi/2$ or $\psi = 3\pi/2$, corresponds to a field point being at *broadside*, or *on-axis*, relative to an aperture (transducer) lying along the X axis (see Fig. 6.3-1). And $u = \pm 1$, which implies that $\theta = \pi/2$ and $\psi = 0$ (for $u = +1$) and $\theta = \pi/2$ and $\psi = \pi$ (for $u = -1$), corresponds to a field point being at the *horizon*, or at *end-fire* geometry, relative to an aperture (transducer) lying along the X axis (see Fig. 6.3-1).

With the use of Eq. (6.3-23), the normalized far-field beam pattern of the rectangular amplitude window given by Eq. (6.3-22) can finally be expressed as

$$D_N(f,\theta,\psi) = \mathrm{sinc}\!\left(\frac{L}{\lambda}\sin\theta\cos\psi\right). \qquad (6.3\text{-}24)$$

Note that by setting $\theta = \pi/2$ in Eq. (6.3-24), we obtain

$$D_N(f,\theta,\psi) = \mathrm{sinc}\!\left(\frac{L}{\lambda}\cos\psi\right), \qquad \theta = \frac{\pi}{2}, \qquad (6.3\text{-}25)$$

which is the normalized *horizontal* far-field beam pattern of the rectangular amplitude window (see Fig. 6.3-4). As can be seen from Fig. 6.3-4, as the ratio L/λ *decreases*, the beamwidth *increases*. Horizontal beam patterns are discussed further in Example 6.3-7.

Successive Differentiation

The far-field directivity function of the rectangular amplitude window could also have been computed by using the method of *successive differentiation* instead of by direct integration. The method of successive differentiation is most advantageous for computing far-field beam patterns when an amplitude window is composed of straight line segments, although the method can be applied to arbitrarily shaped windows as well. This method is based on the spatial Fourier transform of the nth-order partial derivative of a complex aperture function.

Consider the following analysis: Since

$$D(f,f_X) = F_{x_a}\{A(f,x_a)\}, \qquad (6.3\text{-}26)$$

we have

$$A(f,x_a) = F_{f_X}^{-1}\{D(f,f_X)\} = \int_{-\infty}^{\infty} D(f,f_X)\exp(-j2\pi f_X x_a)\,df_X. \qquad (6.3\text{-}27)$$

If we take the partial derivative with respect to x_a of both sides of Eq. (6.3-27),

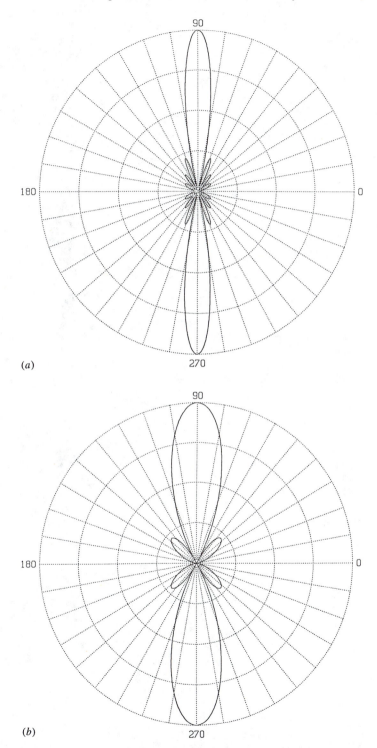

Figure 6.3-4. Polar plot as a function of the azimuthal (bearing) angle ψ of the magnitude of the normalized horizontal far-field beam pattern of the rectangular amplitude window given by Eq. (6.3-25) for (a) $L/\lambda = 4$, (b) $L/\lambda = 2$, (c) $L/\lambda = 1$, and (d) $L/\lambda = 0.5$.

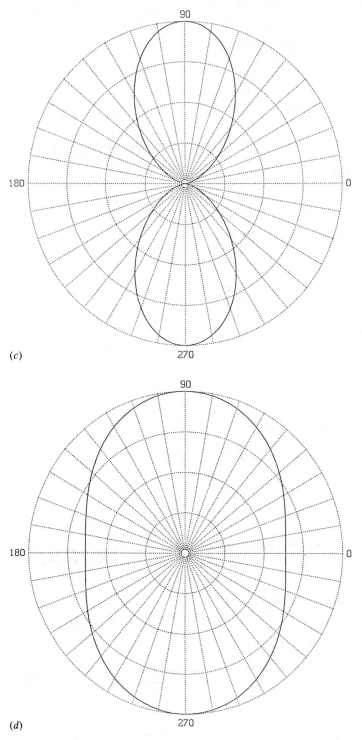

(c)

(d)

Figure 6.3-4. *Continued*

then we obtain

$$\frac{\partial}{\partial x_a} A(f, x_a) = \int_{-\infty}^{\infty} (-j2\pi f_X) D(f, f_X) \exp(-j2\pi f_X x_a) \, df_X, \quad (6.3\text{-}28)$$

which can be expressed as

$$\frac{\partial}{\partial x_a} A(f, x_a) = F_{f_X}^{-1}\{-j2\pi f_X D(f, f_X)\}. \quad (6.3\text{-}29)$$

From Eq. (6.3-29), we can write that

$$D(f, f_X) = \frac{1}{-j2\pi f_X} F_{x_a}\left\{\frac{\partial}{\partial x_a} A(f, x_a)\right\} \quad (6.3\text{-}30)$$

or, in general,

$$\boxed{D(f, f_X) = \frac{1}{(-j2\pi f_X)^n} F_{x_a}\left\{\frac{\partial^n}{\partial x_a^n} A(f, x_a)\right\},} \quad (6.3\text{-}31)$$

where $\partial^n/\partial x_a^n$ represents the nth-order partial derivative with respect to x_a. The result given by Eq. (6.3-31) is the basis for the method of successive differentiation. We shall demonstrate this method by using it to compute the normalized far-field beam pattern of the triangular amplitude window (also known as the Bartlett or Fejer window).

Example 6.3-2 (**The triangular amplitude window**)
In this example we shall derive the normalized far-field beam pattern of the triangular amplitude window illustrated in Fig. 6.3-5. The triangular amplitude window is defined as follows:

$$\text{tri}(x/L) \triangleq \begin{cases} 1 - \dfrac{|x|}{L/2}, & |x| \leq \dfrac{L}{2}, \\ 0, & |x| > \dfrac{L}{2}. \end{cases} \quad (6.3\text{-}32)$$

Therefore, if

$$A(f, x_a) = a(f, x_a) = \text{tri}(x_a/L), \quad (6.3\text{-}33)$$

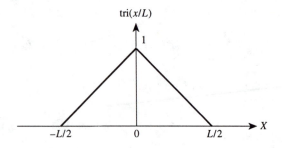

Figure 6.3-5. Triangular amplitude window.

then the amplitude of the complex frequency response of the transducer is triangular in shape along the length L of the transducer, regardless of the value of frequency f. In addition, the phase of the complex frequency response is zero.

As was mentioned previously, the method of successive differentiation is most advantageous when an amplitude window is piecewise linear, since single or multiple differentiations of such a window will ultimately yield impulse functions that have simple spatial Fourier transforms. For example, if we differentiate the triangular amplitude window twice, then we obtain the impulse functions shown in Fig. 6.3-6b. As a result, we can express the second-order partial derivative of Eq. (6.3-33) as follows:

$$\frac{\partial^2}{\partial x_a^2} A(f, x_a) = \frac{2}{L}\left[\delta\left(x_a + \frac{L}{2}\right) - 2\delta(x_a) + \delta\left(x_a - \frac{L}{2}\right)\right]. \quad (6.3\text{-}34)$$

Since

$$\boxed{F_{x_a}\{\delta(x_a \pm d)\} = \exp(\mp j2\pi f_X d),} \quad (6.3\text{-}35)$$

taking the spatial Fourier transform of both sides of Eq. (6.3-34) yields

$$F_{x_a}\left\{\frac{\partial^2}{\partial x_a^2} A(f, x_a)\right\} = \frac{2}{L}\left[\exp(-j\pi f_X L) - 2 + \exp(+j\pi f_X L)\right]$$

$$(6.3\text{-}36)$$

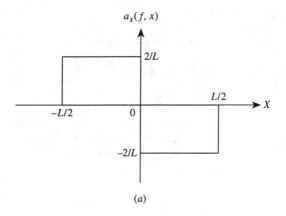

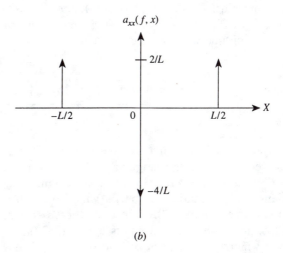

Figure 6.3-6. Partial derivatives of the triangular amplitude window. (a) First-order partial derivative $a_x(f, x) = \partial a(f, x)/\partial x$. ($b$) Second-order partial derivative $a_{xx}(f, x) = \partial^2 a(f, x)/\partial x^2$.

or, using Euler's formula,

$$F_{x_a}\left\{\frac{\partial^2}{\partial x_a^2}A(f, x_a)\right\} = -\frac{4}{L}[1 - \cos(\pi f_X L)]. \qquad (6.3\text{-}37)$$

If we now substitute Eq. (6.3-37) into Eq. (6.3-31) with $n = 2$, and if we make use of the identity

$$2\sin^2\alpha = 1 - \cos(2\alpha), \qquad (6.3\text{-}38)$$

then we obtain

$$\boxed{D(f, f_X) = F_{x_a}\!\left\{\operatorname{tri}\!\left(\frac{x_a}{L}\right)\right\} = \frac{L}{2}\operatorname{sinc}^2\!\left(\frac{f_X L}{2}\right),} \qquad (6.3\text{-}39)$$

which is the *unnormalized* far-field directivity function of the triangular amplitude window (see Example 6.3-5 for an alternative way to obtain the same result). By referring to Eq. (6.3-39), the normalization factor is given by

$$D_{\max} = D(f, 0) = L/2, \qquad (6.3\text{-}40)$$

since sinc(0) = 1 is the maximum value of the sinc function. Substituting Eqs. (6.3-39) and (6.3-40) into Eq. (6.3-19) yields the following expression for the *normalized* far-field beam pattern of the triangular amplitude window (see Fig. 6.3-7):

$$\boxed{D_N(f, f_X) = \operatorname{sinc}^2(f_X L/2).} \qquad (6.3\text{-}41)$$

The levels of the first sidelobes of the normalized far-field beam patterns o the rectangular and triangular amplitude windows are approximately -13 an -27 dB, respectively. However, the main lobe of the beam pattern of the triangular amplitude window is wider than that of the rectangular amplitude window (see Fig. 6.3-7). Note that whereas the rectangular amplitude window is discontinuous at the end points $x = \pm L/2$ (see Fig. 6.3-2), the triangular amplitude window approaches zero at $x = \pm L/2$ in a comparatively smooth fashion (see Fig. 6.3-5). As a general rule, the more smoothly an amplitude window approaches zero at the end points $x = \pm L/2$, the lower will be the sidelobe levels of its normalized far-field beam pattern. By "more smoothly" we mean "with a smaller slope." However, a reduction in sidelobe levels generally results in an increase in the width of the beam pattern's main lobe. This trend will become more apparent as we consider additional amplitude windows.

Multiplication and Convolution

Another useful method for the computation of far-field beam patterns is the method of *multiplication and convolution*. This method is applicable when a complex aperture function is equal to the product of two functions. For example, if

$$A(f, x_a) = A_1(f, x_a) A_2(f, x_a), \qquad (6.3\text{-}42)$$

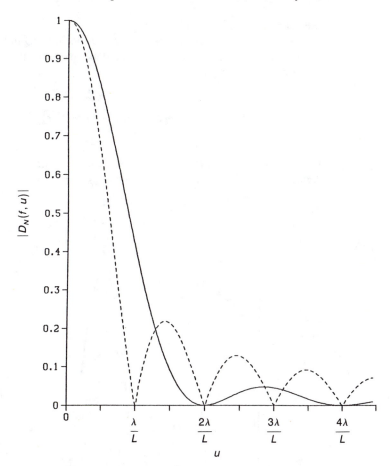

Figure 6.3-7. Magnitude of the normalized far-field beam patterns of the rectangular amplitude window (dashed curve) and triangular amplitude window (solid curve) plotted as a function of direction cosine u.

then the far-field directivity function is given by

$$D(f, f_X) = F_{x_a}\{A(f, x_a)\} = F_{x_a}\{A_1(f, x_a)\} \underset{f_X}{*} F_{x_a}\{A_2(f, x_a)\} \quad (6.3\text{-}43)$$

or

$$D(f, f_X) = D_1(f, f_X) \underset{f_X}{*} D_2(f, f_X), \quad (6.3\text{-}44)$$

where $\underset{f_X}{*}$ denotes convolution with respect to the spatial frequency f_X, and D_1 and D_2 are the far-field beam patterns of the complex aperture functions A_1 and A_2, respectively.

Example 6.3-3

In this example we shall demonstrate the method of multiplication and convolution by computing the far-field beam pattern of the following aperture function:

$$A(f, x_a) = \cos(2\pi x_a/d)\,\mathrm{rect}(x_a/L). \qquad (6.3\text{-}45)$$

This function is shown in Fig. 6.3-8 for $L = 5d/2$, where d is the *spatial period* in meters. Note that this function is analogous to a

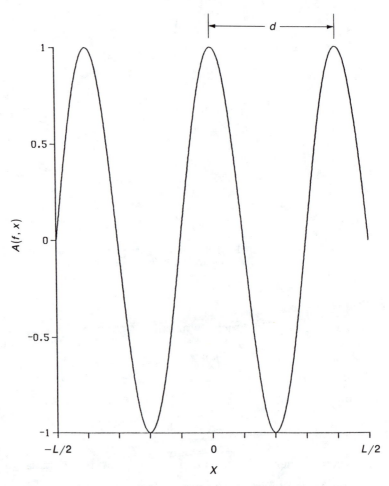

Figure 6.3-8. Plot of the aperture function given by Eq. (6.3-45) for $L = 5d/2$, where d is the spatial period in meters.

time-domain, rectangular-envelope, continuous-wave (CW) pulse. Comparing Eqs. (6.3-42) and (6.3-45), we let

$$A_1(f, x_a) = \cos(2\pi x_a/d) \qquad (6.3\text{-}46)$$

and

$$A_2(f, x_a) = \text{rect}(x_a/L). \qquad (6.3\text{-}47)$$

Since

$$\cos(2\pi x_a/d) = 0.5[\exp(+j2\pi x_a/d) + \exp(-j2\pi x_a/d)] \qquad (6.3\text{-}48)$$

and

$$\boxed{F_{x_a}\{\exp(\pm j2\pi f_X' x_a)\} = \delta(f_X \pm f_X'),} \qquad (6.3\text{-}49)$$

we have

$$\boxed{F_{x_a}\{\cos(2\pi x_a/d)\} = 0.5\left[\delta\left(f_X + \frac{1}{d}\right) + \delta\left(f_X - \frac{1}{d}\right)\right].} \qquad (6.3\text{-}50)$$

Therefore, by referring to Eqs. (6.3-46) and (6.3-50),

$$D_1(f, f_X) = F_{x_a}\{A_1(f, x_a)\} = 0.5\left[\delta\left(f_X + \frac{1}{d}\right) + \delta\left(f_X - \frac{1}{d}\right)\right],$$

$$(6.3\text{-}51)$$

and by referring to Eqs. (6.3-47) and (6.3-17),

$$D_2(f, f_X) = F_{x_a}\{A_2(f, x_a)\} = L \, \text{sinc}(f_X L). \qquad (6.3\text{-}52)$$

Substituting Eqs. (6.3-45), (6.3-51), and (6.3-52) into Eq. (6.3-43) yields

$$D(f, f_X) = F_{x_a}\{\cos(2\pi x_a/d)\, \text{rect}(x_a/L)\} \qquad (6.3\text{-}53)$$

$$= \frac{L}{2}\left[\delta\left(f_X + \frac{1}{d}\right) + \delta\left(f_X - \frac{1}{d}\right)\right]\underset{f_X}{*}\text{sinc}(f_X L) \qquad (6.3\text{-}54)$$

$$= \frac{L}{2}\left[\delta\left(f_X + \frac{1}{d}\right)\underset{f_X}{*}\text{sinc}(f_X L) + \delta\left(f_X - \frac{1}{d}\right)\underset{f_X}{*}\text{sinc}(f_X L)\right],$$

$$(6.3\text{-}55)$$

or

$$D(f, f_X) = F_{x_a}\left\{\cos\left(\frac{2\pi x_a}{d}\right)\text{rect}\left(\frac{x_a}{L}\right)\right\}$$

$$= \frac{L}{2}\left\{\text{sinc}\left[\left(f_X + \frac{1}{d}\right)L\right] + \text{sinc}\left[\left(f_X - \frac{1}{d}\right)L\right]\right\}.$$

(6.3-56)

The magnitude of the unnormalized far-field directivity function given by Eq. (6.3-56) is shown in Fig. 6.3-9 for $L = 5d/2$. Note that as

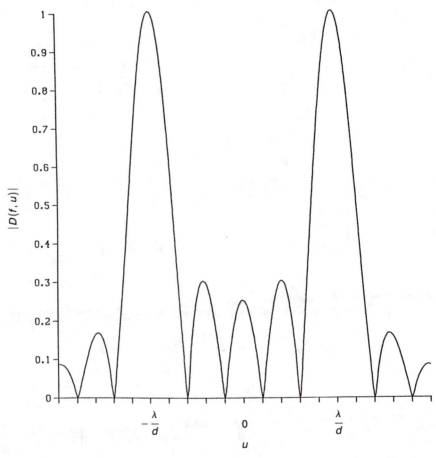

Figure 6.3-9. Magnitude of the unnormalized far-field directivity function given by Eq. (6.3-56) plotted as a function of direction cosine u for $L = 5d/2$.

$d \to \infty$, the aperture function given by Eq. (6.3-45) approaches $\mathrm{rect}(x_a/L)$ and the beam pattern given by Eq. (6.3-56) approaches $L\,\mathrm{sinc}(f_X L)$, which is the unnormalized far-field directivity function of the rectangular amplitude window.

Equation (6.3-56) is a very important result, because it can be used to obtain the far-field directivity functions of the cosine, Hanning, Hamming, and Blackman amplitude windows, as we shall demonstrate next.

Example 6.3-4 (**The cosine amplitude window**)
In this example we shall derive the normalized far-field beam pattern of the cosine amplitude window illustrated in Fig. 6.3-10. The cosine

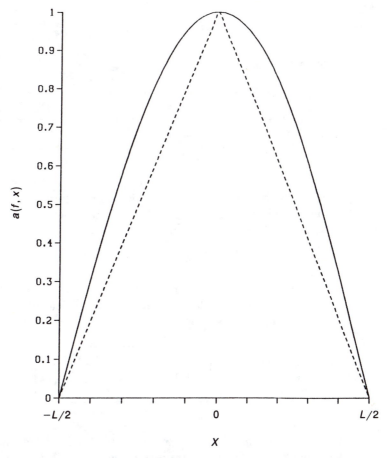

Figure 6.3-10. Cosine amplitude window (solid curve) and triangular amplitude window (dashed curve).

amplitude window is given by

$$\boxed{a(f, x_a) = \cos(\pi x_a/L)\,\text{rect}(x_a/L).}\qquad(6.3\text{-}57)$$

Therefore, if

$$A(f, x_a) = a(f, x_a) = \cos(\pi x_a/L)\,\text{rect}(x_a/L),\quad(6.3\text{-}58)$$

then the amplitude of the complex frequency response of the transducer is in the shape of a half-period cosine function along the length L of the transducer, regardless of the value of frequency f. The phase of the complex frequency response is zero. Since Eq. (6.3-58) can be obtained by substituting $d = 2L$ into Eq. (6.3-45), substituting $d = 2L$ into Eq. (6.3-56) and making use of the identity

$$\sin\left(\pi f_X L \pm \frac{\pi}{2}\right) = \pm\cos(\pi f_X L)\qquad(6.3\text{-}59)$$

yields

$$\boxed{D(f, f_X) = F_{x_a}\{\cos(\pi x_a/L)\,\text{rect}(x_a/L)\} = \frac{2L}{\pi}\,\frac{\cos(\pi f_X L)}{1 - (2f_X L)^2},}$$

$$(6.3\text{-}60)$$

which is the *unnormalized* far-field beam pattern of the cosine amplitude window. By referring to Eq. (6.3-60), the normalization factor is given by

$$D_{\max} = D(f, 0) = 2L/\pi.\qquad(6.3\text{-}61)$$

Substituting Eqs. (6.3-60) and (6.3-61) into Eq. (6.3-19) yields the following expression for the *normalized* far-field beam pattern of the cosine amplitude window (see Fig. 6.3-11):

$$\boxed{D_N(f, f_X) = \frac{\cos(\pi f_X L)}{1 - (2f_X L)^2}.}\qquad(6.3\text{-}62)$$

The level of the first sidelobe of the normalized far-field beam pattern of the cosine amplitude window is approximately -23 dB, whereas the levels for the rectangular and triangular amplitude windows are approximately -13 and

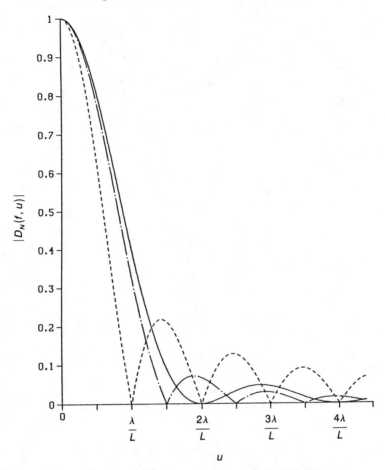

Figure 6.3-11. Magnitude of the normalized far-field beam patterns of the rectangular amplitude window (dashed curve), triangular amplitude window (solid curve), and cosine amplitude window (dot-dash curve) plotted as a function of direction cosine u.

-27 dB, respectively. This is not surprising, since the cosine amplitude window approaches zero at the end points $x = \pm L/2$ in a less abrupt manner (with a smaller slope) than the rectangular amplitude window, but more abruptly (with a larger slope) than the triangular amplitude window (see Fig. 6.3-10).

Hanning, Hamming, and Blackman Amplitude Windows
The next amplitude window to be discussed is the *Hanning* amplitude window, which is also known as the *cosine squared* or *raised cosine* window. Figure 6.3-12 illustrates the Hanning amplitude window, which is given by

$$\boxed{a(f, x_a) = \cos^2(\pi x_a/L)\,\text{rect}(x_a/L)} \qquad (6.3\text{-}63)$$

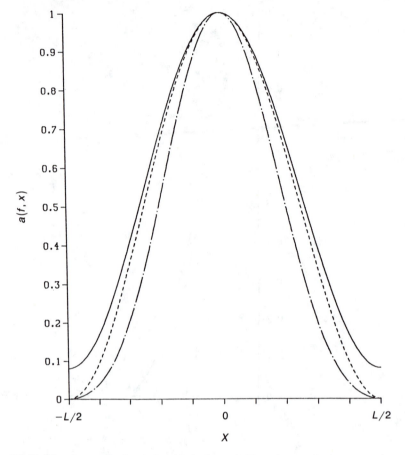

Figure 6.3-12. Hanning amplitude window (dashed curve), Hamming amplitude window (solid curve), and Blackman amplitude window (dot-dash curve).

or, since

$$\cos^2 \alpha = 0.5[1 + \cos(2\alpha)], \tag{6.3-64}$$

$$a(f, x_a) = 0.5[1 + \cos(2\pi x_a/L)]\,\text{rect}(x_a/L). \tag{6.3-65}$$

Therefore, if

$$A(f, x_a) = a(f, x_a) = 0.5\,\text{rect}(x_a/L) + 0.5\cos(2\pi x_a/L)\,\text{rect}(x_a/L),$$

$$\tag{6.3-66}$$

then the unnormalized far-field beam pattern of the Hanning amplitude window is given by

$$D(f, f_X) = 0.5L \, \text{sinc}(f_X L) + 0.5 F_{x_a}\{\cos(2\pi x_a/L) \, \text{rect}(x_a/L)\}, \quad (6.3\text{-}67)$$

where use has been made of Eq. (6.3-17). The spatial Fourier transform of the second term on the right-hand side of Eq. (6.3-67) can be obtained by substituting $d = L$ into Eq. (6.3-56) (see Problem 6-14).

The last two amplitude windows to be discussed are the *Hamming* and *Blackman* amplitude windows (see Fig. 6.3-12). The Hamming amplitude window is given by

$$\boxed{a(f, x_a) = [0.54 + 0.46 \cos(2\pi x_a/L)] \, \text{rect}(x_a/L),} \quad (6.3\text{-}68)$$

and the Blackman amplitude window is given by

$$\boxed{a(f, x_a) = [0.42 + 0.5 \cos(2\pi x_a/L) + 0.08 \cos(4\pi x_a/L)] \, \text{rect}(x_a/L).}$$

$$(6.3\text{-}69)$$

Note that the Hamming amplitude window is discontinuous at the end points $x = \pm L/2$. The unnormalized far-field beam patterns of the Hamming and Blackman amplitude windows can be obtained by using Eq. (6.3-17) and substituting $d = L$ and $d = L/2$ into Eq. (6.3-56) (see Problem 6-14). The levels of the first sidelobes of the normalized far-field beam patterns of the Hanning, Hamming, and Blackman amplitude windows are approximately -31, -44, and -58 dB, respectively (see Fig. 6.3-13). From Fig. 6.3-13, it can be seen that although the level of the first sidelobe of the Hamming beam pattern is less than the level of the first sidelobe of the Hanning beam pattern, the width of the main lobe of the Hamming beam pattern is also less than the width of the main lobe of the Hanning beam pattern. This is an exception to the general rule that beamwidth increases as sidelobe level decreases.

The method of multiplication and convolution is also useful when a complex aperture function can be expressed as the convolution of two functions. For example, if

$$A(f, x_a) = A_1(f, x_a) \underset{x_a}{*} A_2(f, x_a), \quad (6.3\text{-}70)$$

where $\underset{x_a}{*}$ denotes convolution with respect to x_a, then the far-field directivity function is given by

$$D(f, f_X) = F_{x_a}\{A(f, x_a)\} = F_{x_a}\{A_1(f, x_a)\} F_{x_a}\{A_2(f, x_a)\} \quad (6.3\text{-}71)$$

or

$$D(f, f_X) = D_1(f, f_X) D_2(f, f_X), \quad (6.3\text{-}72)$$

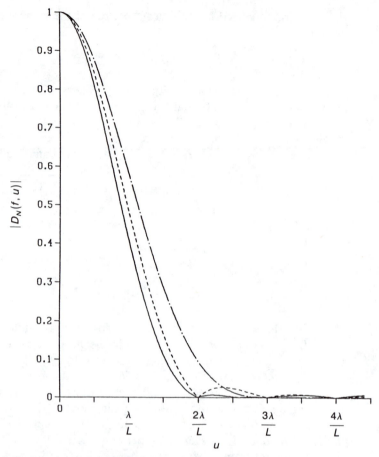

Figure 6.3-13. Magnitude of the normalized far-field beam patterns of the Hanning amplitude window (dashed curve), Hamming amplitude window (solid curve), and Blackman amplitude window (dot-dash curve) plotted as a function of direction cosine u.

where D_1 and D_2 are the far-field beam patterns of the complex aperture functions A_1 and A_2, respectively.

Example 6.3-5 (**Trapezoidal amplitude window**)

In this example we shall derive the unnormalized far-field directivity function of the trapezoidal amplitude window illustrated in Fig. 6.3-14. Since a trapezoid is equal to the convolution of two rectangles, the trapezoidal amplitude window shown in Fig. 6.3-14 can be expressed as

$$a(f, x_a) = h_1 \operatorname{rect}(x_a/L_1) \underset{x_a}{*} h_2 \operatorname{rect}(x_a/L_2), \qquad L_1 > L_2, \quad (6.3\text{-}73)$$

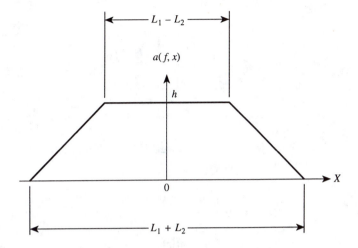

Figure 6.3-14. Trapezoidal amplitude window.

where the height h is given by

$$h = h_1 h_2 L_2, \qquad L_1 > L_2. \tag{6.3-74}$$

Therefore, if

$$A(f, x_a) = a(f, x_a) = h_1 \, \mathrm{rect}(x_a/L_1) \underset{x_a}{*} h_2 \, \mathrm{rect}(x_a/L_2), \; L_1 > L_2,$$
$$\tag{6.3-75}$$

then with the use of Eq. (6.3-17), the unnormalized far-field directivity function of the trapezoidal amplitude window shown in Fig. 6.3-14 is given by

$$D(f, f_X) = h_1 h_2 L_1 L_2 \, \mathrm{sinc}(f_X L_1) \, \mathrm{sinc}(f_X L_2), \; L_1 > L_2. \tag{6.3-76}$$

Note that if $h_2 = h_1 = 1$ and $L_2 = L_1$, then $h = L_1$, and as a result, the trapezoid shown in Fig. 6.3-14 reduces to a triangle of height $h = L_1$ and base $2L_1$, that is,

$$L_1 \, \mathrm{tri}\!\left(\frac{x_a}{2L_1}\right) = \mathrm{rect}(x_a/L_1) \underset{x_a}{*} \mathrm{rect}(x_a/L_1). \tag{6.3-77}$$

Therefore,

$$F_{x_a}\!\left\{ L_1 \, \mathrm{tri}\!\left(\frac{x_a}{2L_1}\right) \right\} = L_1^2 \, \mathrm{sinc}^2(f_X L_1), \tag{6.3-78}$$

or

$$F_{x_a}\left\{\text{tri}\left(\frac{x_a}{2L_1}\right)\right\} = L_1 \, \text{sinc}^2(f_X L_1). \qquad (6.3\text{-}79)$$

If $L_1 = L/2$, then Eq. (6.3-79) reduces to

$$F_{x_a}\left\{\text{tri}\left(\frac{x_a}{L}\right)\right\} = \frac{L}{2} \, \text{sinc}^2\left(\frac{f_X L}{2}\right), \qquad (6.3\text{-}80)$$

which is the unnormalized far-field directivity function of the triangular amplitude window [see Eq. (6.3-39)].

All of the amplitude windows discussed in this section were *even* functions of the spatial coordinate x_a, and as a result, their far-field beam patterns were *real* and *even* functions of the direction cosine u. In general, if an amplitude window is an *even* function of a spatial coordinate, then the corresponding far-field beam pattern will be a *real* and *even* function of the appropriate direction cosine. Similarly, if an amplitude window is an *odd* function of a spatial coordinate, then the corresponding far-field beam pattern will be an *imaginary* and *odd* function of the appropriate direction cosine (see Problem 6-10).

6.3.2 Beamwidth

The term "beamwidth" refers to some measure of the width of the main lobe of a far-field directivity function (beam pattern). *The beamwidth of a far-field beam pattern is the same whether the beam pattern is normalized or unnormalized.* However, it is customary to work with normalized far-field beam patterns. The most common measure of beamwidth is the 3-dB beamwidth. Since the maximum value of the magnitude of a *normalized* far-field beam pattern is equal to 1, or 0 dB, the 3-dB beamwidth is defined as the width of the main lobe between those two points that correspond to magnitude values equal to -3 dB. Note that

$$20 \log_{10}(\sqrt{2}/2) = 10 \log_{10} \tfrac{1}{2} = -3.01 \text{ dB}. \qquad (6.3\text{-}81)$$

The 3-dB beamwidth is also referred to as the *half-power beamwidth* [note the argument $\tfrac{1}{2}$ in Eq. (6.3-81)]. Figure 6.3-15 shows the magnitude of a typical normalized far-field beam pattern of a linear aperture lying along the X axis, plotted as a function of direction cosine u. The 3-dB beamwidth, designated Δu, is given by

$$\Delta u = u_+ - u_- > 0 \qquad (6.3\text{-}82)$$

or, since the magnitude of the beam pattern is symmetric in u space,

$$\Delta u = 2u_+, \qquad (6.3\text{-}83)$$

where $u_+ > 0$ and $u_- < 0$ as shown.

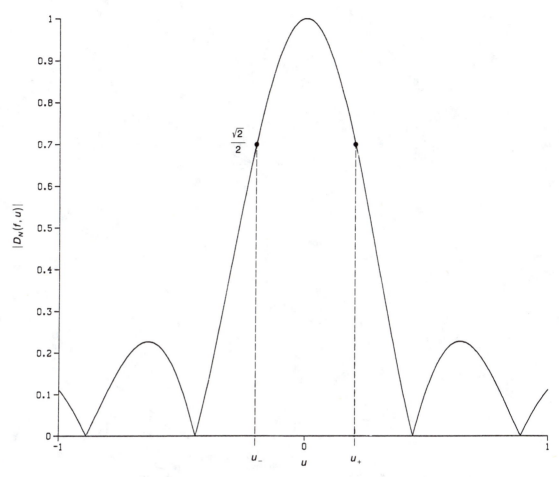

Figure 6.3-15. Magnitude of a typical normalized far-field beam pattern of a linear aperture lying along the X axis, plotted as a function of direction cosine u.

The problem now is to express the 3-dB beamwidth Δu in terms of the 3-dB beamwidths $\Delta \theta$ and $\Delta \psi$ associated with the spherical angles θ and ψ. Note that Δu is dimensionless, whereas $\Delta \theta$ and $\Delta \psi$ have units of degrees. Since $u = \sin \theta \cos \psi$ in general, let (see Fig. 6.3-16)

$$u_+ = \sin \theta_+ \cos \psi_+ > 0, \qquad (6.3\text{-}84)$$

where

$$\theta_+ = \frac{\Delta \theta}{2} \qquad (6.3\text{-}85)$$

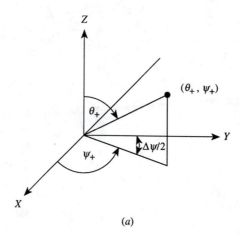

(a)

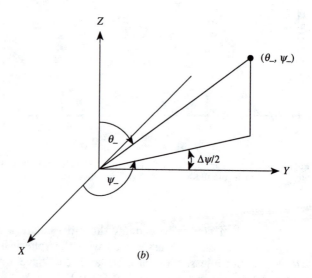

(b)

Figure 6.3-16. Representation of the spherical angles (a) (θ_+, ψ_+) and (b) (θ_-, ψ_-).

and

$$\psi_+ = \frac{\pi}{2} - \frac{\Delta\psi}{2};$$ (6.3-86)

and let

$$u_- = \sin\theta_- \cos\psi_- < 0,$$ (6.3-87)

where

$$\theta_- = \theta_+ = \frac{\Delta\theta}{2}$$ (6.3-88)

and

$$\psi_- = \frac{\pi}{2} + \frac{\Delta\psi}{2}.$$ (6.3-89)

With ψ_+ and ψ_- given by Eqs. (6.3-86) and (6.3-89), respectively, it is assured that $u_+ > 0$ and $u_- < 0$, since

$$\cos\psi_+ = \cos\left(\frac{\pi}{2} - \frac{\Delta\psi}{2}\right) = \sin\left(\frac{\Delta\psi}{2}\right) > 0$$ (6.3-90)

and

$$\cos\psi_- = \cos\left(\frac{\pi}{2} + \frac{\Delta\psi}{2}\right) = -\sin\left(\frac{\Delta\psi}{2}\right) < 0.$$ (6.3-91)

Substituting Eqs. (6.3-85) and (6.3-90) into Eq. (6.3-84) yields

$$u_+ = \sin\left(\frac{\Delta\theta}{2}\right) \sin\left(\frac{\Delta\psi}{2}\right),$$ (6.3-92)

and substituting Eqs. (6.3-88) and (6.3-91) into Eq. (6.3-87) yields

$$u_- = -\sin\left(\frac{\Delta\theta}{2}\right) \sin\left(\frac{\Delta\psi}{2}\right).$$ (6.3-93)

Finally, upon substituting Eqs. (6.3-92) and (6.3-93) into Eq. (6.3-82), we obtain

$$\Delta u = 2\sin\left(\frac{\Delta\theta}{2}\right) \sin\left(\frac{\Delta\psi}{2}\right).$$ (6.3-94)

However, we now have only *one equation* but *two unknowns*, $\Delta\theta$ and $\Delta\psi$. Therefore, when dealing with a linear aperture, we must restrict ourselves to either the *vertical* or the *horizontal* far-field beam pattern when making a beamwidth calculation, as is demonstrated in the next two examples.

Example 6.3-6 (Vertical beam pattern)

Consider a linear aperture lying along the X axis. Therefore, by definition, the vertical beam pattern lies in the XZ plane (see Fig. 6.3-17). In the XZ plane, $u = \sin\theta$ when $\psi = 0$, and $u = -\sin\theta$ when $\psi = \pi$. As a result,

$$u_+ = \sin\theta_+ > 0, \qquad \psi_+ = 0, \tag{6.3-95}$$

and

$$u_- = -\sin\theta_- < 0, \qquad \psi_- = \pi, \tag{6.3-96}$$

where

$$\theta_+ = \theta_- = \frac{\Delta\theta}{2}. \tag{6.3-97}$$

Therefore,

$$\Delta u = u_+ - u_- = 2\sin\left(\frac{\Delta\theta}{2}\right) > 0 \tag{6.3-98}$$

or,

$$\boxed{\Delta\theta = 2\sin^{-1}\left(\frac{\Delta u}{2}\right), \qquad \frac{\Delta u}{2} \le 1,} \tag{6.3-99}$$

where $\Delta\theta$ is the 3-dB beamwidth in degrees of a normalized vertical far-field beam pattern in the XZ plane.

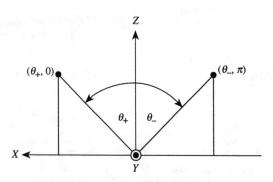

Figure 6.3-17. Representation of the spherical angles $(\theta_+, \psi_+ = 0)$ and $(\theta_-, \psi_- = \pi)$.

Example 6.3-7 (Horizontal beam pattern)

Consider a linear aperture lying along the X axis. Therefore, by definition, the horizontal beam pattern lies in the XY plane (see Fig. 6.3-18). In the XY plane, $u = \cos \psi$, since $\theta = \pi/2$. As a result,

$$u_+ = \cos \psi_+ > 0, \qquad \theta_+ = \pi/2, \qquad (6.3\text{-}100)$$

and

$$u_- = \cos \psi_- < 0, \qquad \theta_- = \pi/2, \qquad (6.3\text{-}101)$$

where

$$\psi_+ = \frac{\pi}{2} - \frac{\Delta\psi}{2} \qquad (6.3\text{-}102)$$

and

$$\psi_- = \frac{\pi}{2} + \frac{\Delta\psi}{2}. \qquad (6.3\text{-}103)$$

Therefore, with the use of Eqs. (6.3-90) and (6.3-91),

$$\Delta u = u_+ - u_- = 2\sin\left(\frac{\Delta\psi}{2}\right) > 0, \qquad (6.3\text{-}104)$$

or

$$\boxed{\Delta\psi = 2\sin^{-1}\left(\frac{\Delta u}{2}\right), \qquad \frac{\Delta u}{2} \leq 1,} \qquad (6.3\text{-}105)$$

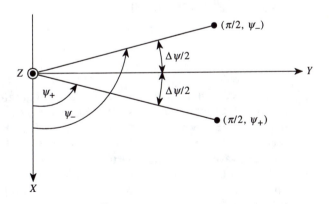

Figure 6.3-18. Representation of the spherical angles $(\theta_+ = \pi/2, \psi_+)$ and $(\theta_- = \pi/2, \psi_-)$.

where $\Delta\psi$ is the 3-dB beamwidth in degrees of a normalized horizontal far-field beam pattern in the XY plane. Note that the right-hand sides of Eqs. (6.3-99) and (6.3-105) are identical.

By examining Eqs. (6.3-99) and (6.3-105), it can be seen that in order to obtain numerical values for $\Delta\theta$ and $\Delta\psi$, we need to know Δu. Consider the next example.

Example 6.3-8

In this example we shall compute the 3-dB beamwidth Δu in direction-cosine space of the normalized far-field directivity function of the rectangular amplitude window given by Eq. (6.3-22). With the use of Eq. (6.3-18), evaluating Eq. (6.3-22) at $u = u_+$ yields

$$D_N(f, u_+) = \mathrm{sinc}\left(\frac{L}{\lambda}u_+\right) = \frac{\sin(\pi L u_+/\lambda)}{\pi L u_+/\lambda} = \frac{\sqrt{2}}{2}, \quad (6.3\text{-}106)$$

since u_+ is the coordinate of one of the 3-dB-down points in direction-cosine space. Equation (6.3-106) can be rewritten as

$$\sin(\pi x) - \frac{\sqrt{2}}{2}\pi x = 0, \quad (6.3\text{-}107)$$

where

$$x = L u_+/\lambda, \quad (6.3\text{-}108)$$

and upon substituting Eq. (6.3-83) into Eq. (6.3-108), we obtain

$$\Delta u = 2\frac{\lambda}{L}x. \quad (6.3\text{-}109)$$

Since the solution (root) of Eq. (6.3-107) is $x \approx 0.443$, substituting this result into Eq. (6.3-109) yields

$$\boxed{\Delta u \approx 0.886\frac{\lambda}{L}} \quad (6.3\text{-}110)$$

for the rectangular amplitude window. Therefore, by substituting Eq. (6.3-110) into Eqs. (6.3-99) and (6.3-105), the 3-dB beamwidths of the normalized vertical and horizontal far-field beam patterns of the rectangular amplitude window can be computed as functions of frequency (wavelength) and aperture length. It is important to note that this procedure can be used to calculate Δu for *any* normalized far-field directivity function of a linear aperture (see Problem 6-16), including the normalized far-field directivity functions of linear arrays. Arrays are discussed in Chapter 7.

6.3.3 Beam Steering

In Section 6.3.1 we assumed that the phase response of a linear aperture was equal to zero and concentrated our efforts on calculating the far-field beam patterns of various amplitude windows. Now, in this section, we shall study what happens to a far-field beam pattern when the phase response $\theta(f, x_a)$ is nonzero, in particular, when $\theta(f, x_a)$ is a linear function of x_a.

Let us begin the analysis by expressing $\theta(f, x_a)$ in terms of the following polynomial:

$$\theta(f, x_a) = \theta_0(f) + \theta_1(f)x_a + \theta_2(f)x_a^2 + \cdots + \theta_N(f)x_a^N, \quad (6.3\text{-}111)$$

where $\theta_0(f)$ represents a *constant* value of phase (independent of x_a) across the aperture, $\theta_1(f)x_a$ represents a *linear* phase variation across the aperture and is responsible for *beam steering* or *beam tilting*, and $\theta_2(f)x_a^2$ represents a *quadratic* phase variation across the aperture and is responsible for *focusing* in the Fresnel zone. The remaining, higher-order terms in Eq. (6.3-111) are not considered. Note that Eq. (6.3-111) could be used as a mathematical model if we were trying to fit a polynomial to actual phase response data using, for example, the method of nonlinear least-squares estimation.

Next, let $D(f, f_X)$ be the far-field beam pattern of the amplitude window $a(f, x_a)$, that is,

$$D(f, f_X) = F_{x_a}\{a(f, x_a)\} = \int_{-\infty}^{\infty} a(f, x_a) \exp(+j2\pi f_X x_a) \, dx_a, \quad (6.3\text{-}112)$$

and let $D'(f, f_X)$ be the far-field beam pattern of the aperture function $A(f, x_a)$ given by Eq. (6.3-10), that is,

$$D'(f, f_X) = F_{x_a}\{a(f, x_a) \exp[+j\theta(f, x_a)]\}$$

$$= \int_{-\infty}^{\infty} a(f, x_a) \exp[+j\theta(f, x_a)] \exp(+j2\pi f_X x_a) \, dx_a. \quad (6.3\text{-}113)$$

Now, assume that the phase response across the aperture is a linear function of x_a, that is,

$$\theta(f, x_a) = \theta_1(f)x_a, \quad (6.3\text{-}114)$$

where

$$\theta_1(f) = -2\pi f_X', \quad (6.3\text{-}115)$$

$$f_X' = u'/\lambda = fu'/c, \quad (6.3\text{-}116)$$

and

$$u' = \sin \theta' \cos \psi'. \tag{6.3-117}$$

Note that the linear phase response given by Eq. (6.3-114) is an *odd* function of x_a and is a straight line that passes through the origin with slope given by Eq. (6.3-115). If Eqs. (6.3-114) and (6.3-115) are substituted into Eq. (6.3-113), then

$$D'(f, f_X) = F_{x_a}\{a(f, x_a) \exp(-j2\pi f_X' x_a)\}$$

$$= \int_{-\infty}^{\infty} a(f, x_a) \exp\left[+j2\pi(f_X - f_X') x_a\right] dx_a, \tag{6.3-118}$$

and by comparing Eq. (6.3-118) with Eq. (6.3-112),

$$D'(f, f_X) = D(f, f_X - f_X'), \tag{6.3-119}$$

or

$$F_{x_a}\{a(f, x_a) \exp(-j2\pi f_X' x_a)\} = D(f, f_X - f_X'), \tag{6.3-120}$$

where $D(f, f_X)$ is given by Eq. (6.3-112). Equation (6.3-119) can also be expressed as

$$D'(f, u) = D(f, u - u'), \tag{6.3-121}$$

since $D(f, f_X) = D(f, u/\lambda) \rightarrow D(f, u)$. Therefore, a linear phase response (linear phase variation) across the length of the aperture will cause the beam pattern $D(f, u)$ to be steered in the direction $u = u'$ in direction-cosine space, which is equivalent to steering or tilting the beam pattern to $\theta = \theta'$ and $\psi = \psi'$ [see Eq. (6.3-117)]. Equation (6.3-121) indicates that the far-field directivity function $D'(f, u)$ is simply a translated or shifted version of $D(f, u)$ (see Fig. 6.3-19). When the beam pattern $D(f, u)$ is steered, its shape and beamwidth remain unchanged when plotted as a function of direction cosine u. However, they do change when the beam pattern is plotted as a function of the spherical angles θ and ψ, as shall be demonstrated next.

Beamwidth at an Arbitrary Beam Tilt Angle

When a far-field beam pattern is steered or tilted from broadside or on axis toward the horizon or end fire, its *beamwidth will increase from a minimum value at broadside to a maximum value at end fire*.

In order to compute the beamwidth at an arbitrary beam tilt angle, let us restrict ourselves to the normalized horizontal far-field beam pattern of a linear aperture lying along the X axis (see Example 6.3-7), where ψ' is the beam tilt angle in the XY plane (see Fig. 6.3-20). In the XY plane, $u = \cos \psi$, since

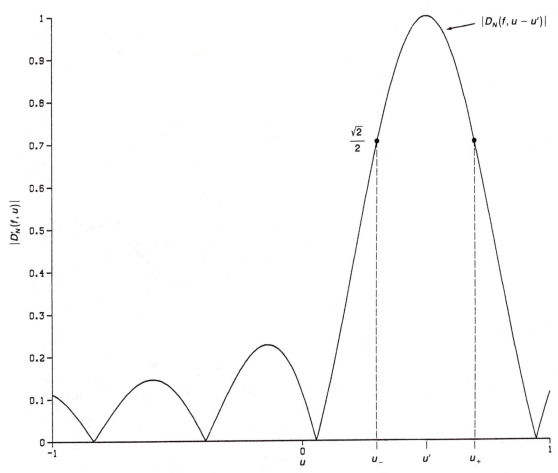

Figure 6.3-19. Magnitude of a typical normalized far-field beam pattern of a linear aperture lying along the X axis, steered in the direction $u = u'$ in direction-cosine space.

$\theta = \pi/2$. Therefore, by referring to Figs. 6.3-19 through 6.3-21,

$$u_+ = \cos \psi_+ > 0, \qquad \theta_+ = \pi/2, \qquad (6.3\text{-}122)$$

$$u_- = \cos \psi_- > 0, \qquad \theta_- = \pi/2, \qquad (6.3\text{-}123)$$

and

$$u' = \cos \psi', \qquad (6.3\text{-}124)$$

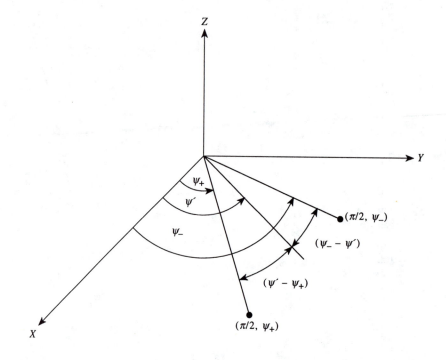

Figure 6.3-20. Representation of the spherical angles $(\theta_+ = \pi/2, \psi_+)$, $(\theta_- = \pi/2, \psi_-)$ and the beam tilt angle ψ'. Note that the *half-beamwidth angles* $\psi' - \psi_+$ and $\psi_- - \psi'$ are not equal in general.

where

$$u_+ > u_-,\tag{6.3-125}$$

$$u_+ = u' + \frac{\Delta u}{2},\tag{6.3-126}$$

and

$$u_- = u' - \frac{\Delta u}{2}.\tag{6.3-127}$$

In addition, by referring to Figs. 6.3-20 and 6.3-21, it can be seen that the 3-dB beamwidth in degrees of a normalized horizontal far-field beam pattern in the *XY* plane is given by

$$\Delta\psi = \psi_- - \psi_+,\tag{6.3-128}$$

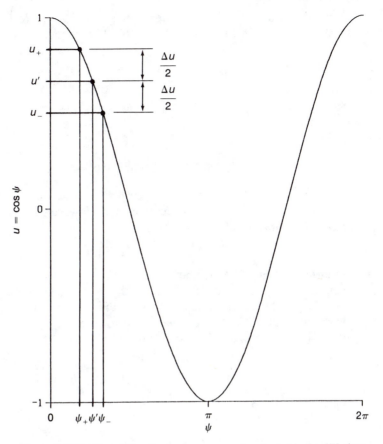

Figure 6.3-21. Plot of direction cosine $u = \cos\psi$ evaluated in the XY plane.

and with the use of Eqs. (6.3-122) through (6.3-124), (6.3-126), and (6.3-127),

$$\Delta\psi = \cos^{-1}\!\left(\cos\psi' - \frac{\Delta u}{2}\right) - \cos^{-1}\!\left(\cos\psi' + \frac{\Delta u}{2}\right), \qquad \left|\cos\psi' \pm \frac{\Delta u}{2}\right| \leq 1,$$

$$(6.3\text{-}129)$$

where Δu can be obtained by using the procedure outlined in Example 6.3-8. In order to use Eq. (6.3-129), the inequality on the right-hand side must hold for both the plus and minus signs. Although the *half-beamwidth angles* $\psi' - \psi_+$ and $\psi_- - \psi'$ are not equal in general for arbitrary beam tilt angles ψ', they are equal to one another and to $\Delta\psi/2$ for $\psi' = 0$, $\pi/2$, π, $3\pi/2$, and 2π, since $\cos\psi$ is symmetric about those values (see Fig. 6.3-21). Note that $\psi' = \pi/2$ corresponds to an *untilted* horizontal beam pattern for a linear aperture lying along the X axis (see Fig. 6.3-20). With $\psi' = \pi/2$, Eq. (6.3-129) reduces to Eq. (6.3-105), since

$$\cos^{-1} x = \frac{\pi}{2} - \sin^{-1} x, \tag{6.3-130}$$

$$\cos^{-1}(-x) = \frac{\pi}{2} + \sin^{-1} x, \tag{6.3-131}$$

and as a result,

$$\cos^{-1}(-x) - \cos^{-1} x = 2\sin^{-1} x. \tag{6.3-132}$$

Also note that $\psi' = 0$ and $\psi' = \pi$ corresponds to steering the horizontal beam pattern to the horizon or end-fire geometry. However, for $\psi' = 0$ and $\psi' = \pi$, Eq. (6.3-129) is invalid. Therefore, a different approach must be used in order to compute the beamwidth at end-fire geometry.

Beamwidth at End-Fire Geometry
Restricting ourselves once again to the normalized horizontal far-field beam pattern of a linear aperture lying along the X axis, consider the case where the beam tilt angle ψ' in the XY plane is zero (see Fig. 6.3-20). With $\psi' = 0$ we have $u' = 1$ [see Eq. (6.3-124)], and as a result, Eq. (6.3-127) becomes (see Fig. 6.3-22)

$$u_- = 1 - \frac{\Delta u}{2}. \tag{6.3-133}$$

Also, with $\psi' = 0$,

$$\psi_- = \frac{\Delta\psi}{2}, \tag{6.3-134}$$

since $\cos\psi$ is symmetric about $\psi = \psi' = 0$ (see Fig. 6.3-21). Therefore, substituting Eq. (6.3-134) into Eq. (6.3-123) yields

$$u_- = \cos\left(\frac{\Delta\psi}{2}\right), \tag{6.3-135}$$

and upon equating the right-hand sides of Eqs. (6.3-133) and (6.3-135), we finally obtain

$$\boxed{\Delta\psi = 2\cos^{-1}\left(1 - \frac{\Delta u}{2}\right),} \tag{6.3-136}$$

where $\Delta\psi$ is the 3-dB beamwidth in degrees of a normalized horizontal far-field beam pattern in the XY plane at end-fire geometry.

Let us end our discussion in this section by illustrating the distortion a beam pattern undergoes due to beam steering. By referring to Eq. (6.3-22), the normalized far-field beam pattern of the rectangular amplitude window steered to $u = u'$

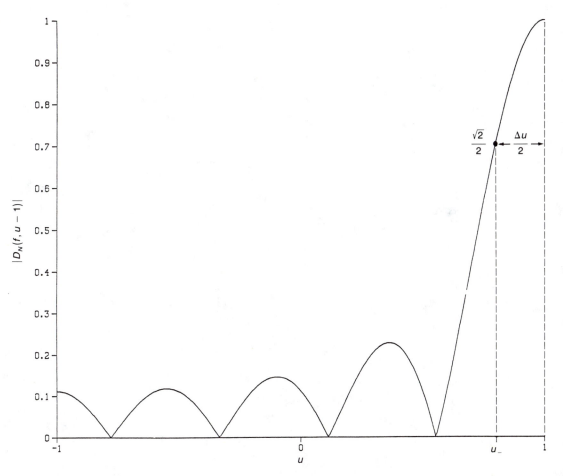

Figure 6.3-22. Magnitude of a typical normalized far-field beam pattern of a linear aperture lying along the X axis, steered in the direction $u' = 1$ (end fire).

is given by

$$D_N(f, u - u') = \text{sinc}\left(\frac{L}{\lambda}(u - u')\right). \qquad (6.3\text{-}137)$$

The corresponding horizontal beam pattern is given by

$$D_N(f, \theta, \psi) = \text{sinc}\left(\frac{L}{\lambda}(\cos\psi - \cos\psi')\right), \qquad \theta = \frac{\pi}{2}. \qquad (6.3\text{-}138)$$

Plots of Eq. (6.3-138) for different values of ψ' are shown in Fig. 6.3-23. Note how

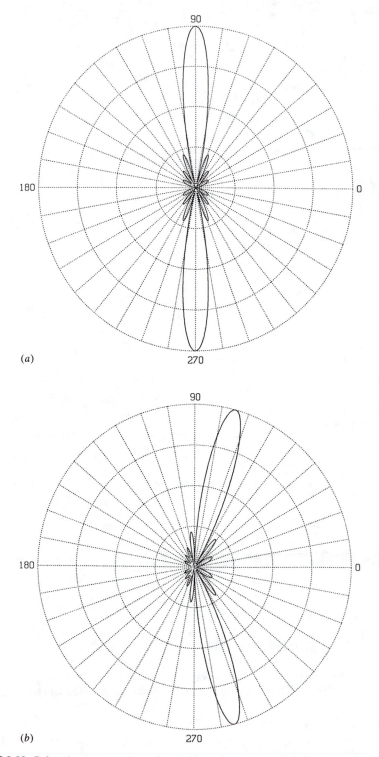

Figure 6.3-23. Polar plot, as a function of the azimuthal (bearing) angle ψ, of the magnitude of the normalized horizontal far-field beam pattern of the rectangular amplitude window given by Eq. (6.3-138) for $L/\lambda = 4$ and for beam tilt angle (a) $\psi' = 90°$ (broadside), (b) $\psi' = 75°$, (c) $\psi' = 60°$, (d) $\psi' = 45°$, (e) $\psi' = 30°$, and (f) $\psi' = 0°$ (end fire).

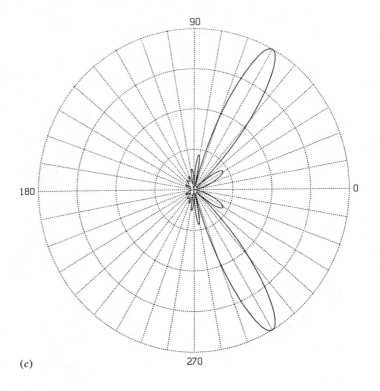

(c)

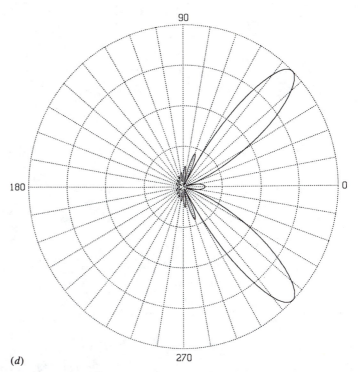

(d)

Figure 6.3-23. *Continued*

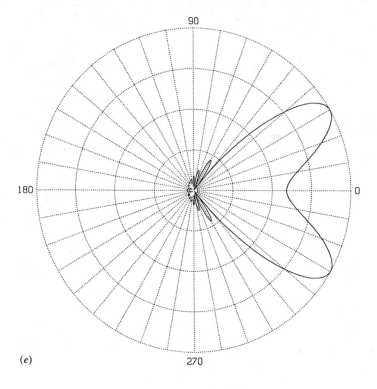

(e)

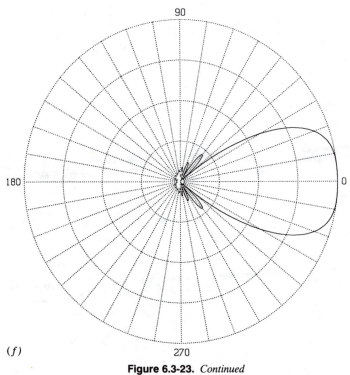

(f)

Figure 6.3-23. *Continued*

the beamwidth increases and how the main lobe becomes more asymmetrical about the beam tilt axis as the beam pattern is steered from broadside ($\psi' = \pi/2$) to end fire ($\psi' = 0$).

6.4 Linear Apertures and Near-Field Directivity Functions

6.4.1 Aperture Focusing

Recall that the near-field directivity function of a general volume aperture, based on the Fresnel approximation [see Eqs. (6.2-35) and (6.2-40)], is given by [see Eq. (6.2-51)]

$$D(f, r, \boldsymbol{\alpha}) = F_{\mathbf{r}_a}\left\{ A(f, \mathbf{r}_a) \exp\left(-jk\frac{r_a^2}{2r}\right) \right\}$$

$$= \int_{-\infty}^{\infty} A(f, \mathbf{r}_a) \exp\left(-jk\frac{r_a^2}{2r}\right) \exp(+j2\pi\boldsymbol{\alpha}\cdot\mathbf{r}_a)\, d\mathbf{r}_a, \quad (6.4\text{-}1)$$

where

$$A(f, \mathbf{r}_a) = a(f, \mathbf{r}_a) \exp\left[+j\theta(f, \mathbf{r}_a)\right] \tag{6.4-2}$$

is the complex aperture function,

$$\mathbf{r}_a = (x_a, y_a, z_a), \tag{6.4-3}$$

$$r_a = |\mathbf{r}_a| = \sqrt{x_a^2 + y_a^2 + z_a^2} \tag{6.4-4}$$

is the magnitude of the position vector to an aperture point,

$$r = |\mathbf{r}| = \sqrt{x^2 + y^2 + z^2} \tag{6.4-5}$$

is the magnitude of the position vector to a field point,

$$\boldsymbol{\alpha} = (f_X, f_Y, f_Z), \tag{6.4-6}$$

and $d\mathbf{r}_a = dx_a\, dy_a\, dz_a$.

Now consider the case of a *linear aperture* of length L meters lying along the X axis as shown in Fig. 6.3-1. Recall that a linear aperture can represent either a single electroacoustic transducer or a linear array of many individual electroacoustic transducers. Also shown in Fig. 6.3-1 is a field point with spherical coordinates (r, θ, ψ). The field point is assumed to be in the near-field (Fresnel) region of the aperture, at a range $1.356R < r < \pi R^2/\lambda$ meters, where $R = L/2$ meters is the maximum radial extent of the aperture [see Eq. (6.2-40)].

Since the linear aperture illustrated in Fig. 6.3-1 is shown lying along the X axis, the position vector that describes the spatial location of the aperture is given by

$$\mathbf{r}_a = (x_a, 0, 0), \tag{6.4-7}$$

and as a result,

$$A(f, \mathbf{r}_a) = A(f, x_a), \tag{6.4-8}$$

$$r_a^2 = |\mathbf{r}_a|^2 = x_a^2, \tag{6.4-9}$$

$$\boldsymbol{\alpha} \cdot \mathbf{r}_a = (f_X, f_Y, f_Z) \cdot (x_a, 0, 0) = f_X x_a, \tag{6.4-10}$$

and

$$d\mathbf{r}_a = dx_a. \tag{6.4-11}$$

Therefore, upon substituting Eqs. (6.4-7) through (6.4-11) into Eq. (6.4-1), we obtain the following one-dimensional spatial Fresnel transform expression for the near-field directivity function (beam pattern) of a linear aperture lying along the X axis:

$$\boxed{\begin{aligned} D(f, r, f_X) &= F_{x_a}\left\{ A(f, x_a) \exp\left(-jk\frac{x_a^2}{2r} \right) \right\} \\ &= \int_{-L/2}^{L/2} A(f, x_a) \exp\left(-jk\frac{x_a^2}{2r} \right) \exp(+j2\pi f_X x_a)\, dx_a, \end{aligned}} \tag{6.4-12}$$

where

$$A(f, x_a) = a(f, x_a) \exp\left[+j\theta(f, x_a) \right] \tag{6.4-13}$$

and [see Eq. (6.2-43)]

$$f_X = \frac{u}{\lambda} = \frac{\sin\theta\cos\psi}{\lambda}. \tag{6.4-14}$$

Since the spatial frequency f_X can be expressed in terms of θ and ψ, the near-field directivity function can ultimately be expressed as a function of frequency f, the range r to a field point, and the spherical angles θ and ψ, that is,

$$D(f, r, f_X) \rightarrow D(f, r, \theta, \psi). \tag{6.4-15}$$

Note that if a linear aperture lies along either the Y or Z axis instead of the X axis, then one can simply replace x_a and f_X either with y_a and f_Y or with z_a and

f_Z, respectively, in Eqs. (6.4-12) and (6.4-13), where f_Y and f_Z are given by Eqs. (6.2-44) and (6.2-45), respectively.

In order to discuss the concept of *aperture focusing*, let us investigate what happens to the near-field beam pattern given by Eq. (6.4-12) when the phase response $\theta(f, x_a)$ across the length of the aperture is a quadratic function of x_a, that is [see Eq. (6.3-111)],

$$\theta(f, x_a) = \theta_2(f)x_a^2, \tag{6.4-16}$$

where

$$\theta_2(f) = \frac{k}{2r'} \tag{6.4-17}$$

and

$$k = 2\pi f/c = 2\pi/\lambda \tag{6.4-18}$$

is the wave number. Note that the quadratic phase response given by Eq. (6.4-16) is an *even* function of x_a and is a parabola that passes through the origin. If Eqs. (6.4-13), (6.4-16), and (6.4-17) are substituted into Eq. (6.4-12), then

$$D(f, r, f_X) = \int_{-L/2}^{L/2} a(f, x_a) \exp\left[-jk\frac{x_a^2}{2}\left(\frac{1}{r} - \frac{1}{r'}\right)\right] \exp(+j2\pi f_X x_a)\, dx_a,$$
$$\tag{6.4-19}$$

and if Eq. (6.4-19) is evaluated at the near-field range $r = r'$, then it reduces to

$$D(f, r', f_X) = \int_{-L/2}^{L/2} a(f, x_a) \exp(+j2\pi f_X x_a)\, dx_a, \tag{6.4-20}$$

or

$$D(f, r', f_X) = F_{x_a}\{a(f, x_a)\} = D(f, f_X), \tag{6.4-21}$$

where $D(f, f_X)$ is the far-field beam pattern of the amplitude window $a(f, x_a)$. Therefore, Eq. (6.4-21) indicates that a far-field beam pattern can be *focused* to the near-field range $r = r'$ from the aperture when the phase response (phase variation) across the aperture is given by Eqs. (6.4-16) through (6.4-18). The quadratic phase variation compensates for wavefront curvature at near-field distances from the aperture. The physical situation represented by Eq. (6.4-21) is known as *aperture focusing*.

A good way to think about aperture focusing is to compare it with taking a photograph. If the subject is far enough away (in the far field), we do not concern ourselves with the actual range to the subject—we simply set the focus on the camera to infinity and take the picture. However, if the subject is nearby (in the near field), we do concern ourselves with the actual range to the subject, since we must refocus the camera whenever the subject moves.

6.4.2 Beam Steering and Aperture Focusing

Let us investigate what happens to the near-field beam pattern given by Eq. (6.4-12) when the phase response (phase variation) $\theta(f, x_a)$ across the length of the aperture is equal to the sum of a linear and a quadratic function of x_a, that is [see Eq. (6.3-111)],

$$\theta(f, x_a) = \theta_1(f) x_a + \theta_2(f) x_a^2, \tag{6.4-22}$$

where

$$\theta_1(f) = -2\pi f'_X, \tag{6.4-23}$$

$$\theta_2(f) = \frac{k}{2r'}, \tag{6.4-24}$$

$$f'_X = \frac{u'}{\lambda} = \frac{\sin \theta' \cos \psi'}{\lambda}, \tag{6.4-25}$$

and

$$k = 2\pi f/c = 2\pi/\lambda. \tag{6.4-26}$$

If Eq. (6.4-13) and Eqs. (6.4-22) through (6.4-24) are substituted into Eq. (6.4-12), then

$$D(f, r, f_X) = \int_{-L/2}^{L/2} a(f, x_a) \exp\left[-jk\frac{x_a^2}{2}\left(\frac{1}{r} - \frac{1}{r'}\right)\right]$$

$$\times \exp\left[+j2\pi(f_X - f'_X)x_a\right] dx_a, \tag{6.4-27}$$

and if Eq. (6.4-27) is evaluated at the near-field range $r = r'$, then it reduces to

$$D(f, r', f_X) = \int_{-L/2}^{L/2} a(f, x_a) \exp\left[+j2\pi(f_X - f'_X)x_a\right] dx_a, \tag{6.4-28}$$

or

$$D(f, r', f_X) = F_{x_a}\{a(f, x_a) \exp(-j2\pi f'_X x_a)\} = D(f, f_X - f'_X). \tag{6.4-29}$$

Equation (6.4-29) indicates that the far-field beam pattern of the amplitude window $a(f, x_a)$ has been *focused* to the near-field range $r = r'$ from the aperture, and *steered* in the direction $u = u'$ in direction-cosine space, which is equivalent to steering (tilting) the beam pattern to $\theta = \theta'$ and $\psi = \psi'$ [see Eq. (6.4-25)].

6.5 Planar Apertures and Far-Field Directivity Functions

Consider the case of a *planar aperture* of arbitrary shape lying in the *XY* plane as shown in Fig. 6.5-1. A planar aperture can represent either a single electroacoustic transducer or a planar array of many individual electroacoustic transducers. Also shown in Fig. 6.5-1 is a field point with spherical coordinates (r, θ, ψ). The field point is assumed to be in the far-field region of the aperture, at a range $r > \pi R^2/\lambda > 1.356R$ meters, where R is the maximum radial extent of the aperture [see Eq. (6.2-39)].

Since the planar aperture illustrated in Fig. 6.5-1 is shown lying in the *XY* plane, the position vector that describes the spatial location of the aperture is given by

$$\mathbf{r}_a = (x_a, y_a, 0), \tag{6.5-1}$$

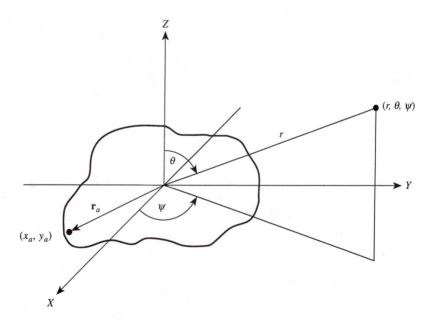

Figure 6.5-1. Planar aperture of arbitrary shape lying in the *XY* plane. Also shown is a field point with spherical coordinates (r, θ, ψ).

and as a result,

$$A(f, \mathbf{r}_a) = A(f, x_a, y_a), \tag{6.5-2}$$

$$\boldsymbol{\alpha} \cdot \mathbf{r}_a = (f_X, f_Y, f_Z) \cdot (x_a, y_a, 0) = f_X x_a + f_Y y_a, \tag{6.5-3}$$

and

$$d\mathbf{r}_a = dx_a \, dy_a. \tag{6.5-4}$$

Therefore, upon substituting Eqs. (6.5-1) through (6.5-4) into Eq. (6.3-1), we obtain the following two-dimensional spatial Fourier transform expression for the far-field directivity function (beam pattern) of a planar aperture lying in the XY plane:

$$
\begin{aligned}
D(f, f_X, f_Y) &= F_{x_a} F_{y_a}\{A(f, x_a, y_a)\} \\
&= \int_{-\infty}^{\infty} \int_{-\infty}^{\infty} A(f, x_a, y_a) \exp[+j2\pi(f_X x_a + f_Y y_a)] \, dx_a \, dy_a,
\end{aligned}
$$

$$\tag{6.5-5}$$

where

$$A(f, x_a, y_a) = a(f, x_a, y_a) \exp[+j\theta(f, x_a, y_a)], \tag{6.5-6}$$

$$f_X = \frac{u}{\lambda} = \frac{\sin\theta\cos\psi}{\lambda}, \tag{6.5-7}$$

and

$$f_Y = \frac{v}{\lambda} = \frac{\sin\theta\sin\psi}{\lambda}. \tag{6.5-8}$$

Since the spatial frequencies f_X and f_Y can be expressed in terms of θ and ψ, the far-field directivity function can ultimately be expressed as a function of frequency f and the spherical angles θ and ψ, that is,

$$D(f, f_X, f_Y) \rightarrow D(f, \theta, \psi). \tag{6.5-9}$$

Figure 6.5-2 shows the magnitude of a typical normalized far-field beam pattern of a planar aperture lying in the XY plane, plotted as a function of direction cosines u and v, since

$$D(f, f_X, f_Y) \rightarrow D(f, u, v). \tag{6.5-10}$$

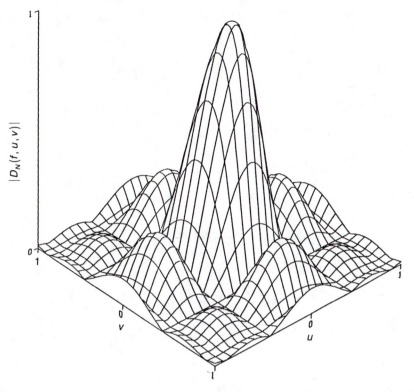

Figure 6.5-2. Magnitude of a typical normalized far-field beam pattern of a planar aperture lying in the XY plane, plotted as a function of direction cosines u and v.

6.5.1 Beam Steering

Let $D(f, f_X, f_Y)$ be the far-field beam pattern of the amplitude window $a(f, x_a, y_a)$, that is,

$$D(f, f_X, f_Y) = F_{x_a} F_{y_a} \{a(f, x_a, y_a)\}$$

$$= \int_{-\infty}^{\infty} \int_{-\infty}^{\infty} a(f, x_a, y_a) \exp\left[+j2\pi(f_X x_a + f_Y y_a)\right] dx_a\, dy_a, \quad (6.5\text{-}11)$$

and let $D'(f, f_X, f_Y)$ be the far-field beam pattern of the aperture function $A(f, x_a, y_a)$ given by Eq. (6.5-6), that is,

$$D'(f, f_X, f_Y) = F_{x_a} F_{y_a} \{a(f, x_a, y_a) \exp\left[+j\theta(f, x_a, y_a)\right]\}$$

$$= \int_{-\infty}^{\infty} \int_{-\infty}^{\infty} a(f, x_a, y_a) \exp\left[+j\theta(f, x_a, y_a)\right]$$

$$\times \exp\left[+j2\pi(f_X x_a + f_Y y_a)\right] dx_a\, dy_a. \quad (6.5\text{-}12)$$

Now, assume that the phase response across the aperture is a linear function of both x_a and y_a, that is,

$$\theta(f, x_a, y_a) = -2\pi f_X' x_a - 2\pi f_Y' y_a, \qquad (6.5\text{-}13)$$

where

$$f_X' = \frac{u'}{\lambda} = \frac{\sin \theta' \cos \psi'}{\lambda} \qquad (6.5\text{-}14)$$

and

$$f_Y' = \frac{v'}{\lambda} = \frac{\sin \theta' \sin \psi'}{\lambda}. \qquad (6.5\text{-}15)$$

If Eq. (6.5-13) is substituted into Eq. (6.5-12), then

$$
\begin{aligned}
D'(f, f_X, f_Y) \\
= F_{x_a} F_{y_a}\{a(f, x_a, y_a) \exp[-j2\pi(f_X' x_a + f_Y' y_a)]\} \\
= \int_{-\infty}^{\infty} \int_{-\infty}^{\infty} a(f, x_a, y_a) \exp\{+j2\pi[(f_X - f_X') x_a + (f_Y - f_Y') y_a]\} \, dx_a \, dy_a,
\end{aligned}
$$

$$(6.5\text{-}16)$$

and by comparing Eq. (6.5-16) with Eq. (6.5-11),

$$D'(f, f_X, f_Y) = D(f, f_X - f_X', f_Y - f_Y'), \qquad (6.5\text{-}17)$$

or

$$F_{x_a} F_{y_a}\{a(f, x_a, y_a) \exp[-j2\pi(f_X' x_a + f_Y' y_a)]\} = D(f, f_X - f_X', f_Y - f_Y'), \qquad (6.5\text{-}18)$$

where $D(f, f_X, f_Y)$ is given by Eq. (6.5-11). Equation (6.5-17) can also be expressed as

$$D'(f, u, v) = D(f, u - u', v - v'), \qquad (6.5\text{-}19)$$

since $D(f, f_X, f_Y) = D(f, u/\lambda, v/\lambda) \rightarrow D(f, u, v)$. Therefore, a linear phase response (linear phase variation) across the aperture will cause the beam pattern $D(f, u, v)$ to be steered in the direction $u = u'$ and $v = v'$ in direction-cosine space, which is equivalent to steering or tilting the beam pattern to $\theta = \theta'$ and $\psi = \psi'$. Equation (6.5-19) indicates that the far-field directivity function $D'(f, u, v)$ is simply a translated or shifted version of $D(f, u, v)$ (see Fig. 6.5-3). When the beam pattern $D(f, u, v)$ is steered, its shape and beamwidth remain unchanged

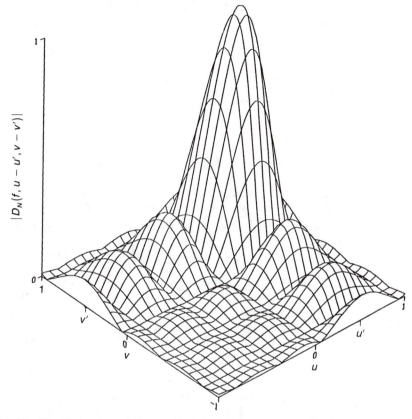

Figure 6.5-3. Magnitude of a typical normalized far-field beam pattern of a planar aperture lying in the XY plane, steered in the direction $u = u'$ and $v = v'$ in direction-cosine space.

when plotted as a function of direction cosines u and v. However, they do change when the beam pattern is plotted as a function of the spherical angles θ and ψ (see Section 6.3.3).

6.5.2 Separable Functions — Rectangular Coordinates

A function of two independent variables is called *separable* if it can be written as the product of two functions, each of which depends on only one of the independent variables. For example, the function $g(x, y)$ is separable in the rectangular coordinates x and y if

$$g(x, y) = g_X(x) g_Y(y). \qquad (6.5\text{-}20)$$

If the complex aperture function $A(f, x_a, y_a)$ is separable in the rectangular coordinates x_a and y_a, then

$$A(f, x_a, y_a) = A_X(f, x_a) A_Y(f, y_a), \qquad (6.5\text{-}21)$$

where

$$A_X(f, x_a) = a_X(f, x_a) \exp[+j\theta_X(f, x_a)] \qquad (6.5\text{-}22)$$

and

$$A_Y(f, y_a) = a_Y(f, y_a) \exp[+j\theta_Y(f, y_a)]. \qquad (6.5\text{-}23)$$

Equation (6.5-21) indicates that the complex frequency response of the aperture in the X direction is independent of the complex frequency response of the aperture in the Y direction. If Eq. (6.5-21) is substituted into Eq. (6.5-5), then

$$D(f, f_X, f_Y) = F_{x_a}\{A_X(f, x_a)\} F_{y_a}\{A_Y(f, y_a)\}$$

$$= \int_{-\infty}^{\infty} A_X(f, x_a) \exp(+j2\pi f_X x_a)\, dx_a$$

$$\times \int_{-\infty}^{\infty} A_Y(f, y_a) \exp(+j2\pi f_Y y_a)\, dy_a, \qquad (6.5\text{-}24)$$

or

$$\boxed{D(f, f_X, f_Y) = D_X(f, f_X) D_Y(f, f_Y),} \qquad (6.5\text{-}25)$$

where

$$D_X(f, f_X) = F_{x_a}\{A_X(f, x_a)\} \qquad (6.5\text{-}26)$$

and

$$D_Y(f, f_Y) = F_{y_a}\{A_Y(f, y_a)\}. \qquad (6.5\text{-}27)$$

Therefore, according to Eq. (6.5-25), if the complex aperture function $A(f, x_a, y_a)$ is separable in the rectangular coordinates x_a and y_a, then the corresponding far-field beam pattern $D(f, f_X, f_Y)$ is equal to the product of the individual beam patterns $D_X(f, f_X)$ and $D_Y(f, f_Y)$.

Example 6.5-1 (The rectangular piston)
One of the most common examples of a planar aperture lying in the XY plane is a single electroacoustic transducer, rectangular in shape, with sides equal to L_X and L_Y meters, as shown in Fig. 6.5-4. The simplest

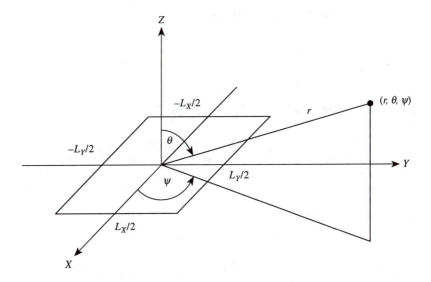

Figure 6.5-4. A single electroacoustic transducer, rectangular in shape, with sides equal to L_X and L_Y meters, lying in the XY plane. Also shown is a field point with spherical coordinates (r, θ, ψ).

mathematical model for the complex frequency response (complex aperture function) for this rectangular shaped transducer is given by

$$A(f, x_a, y_a) = \text{rect}(x_a/L_X)\,\text{rect}(y_a/L_Y), \qquad (6.5\text{-}28)$$

which is separable in the rectangular coordinates x_a and y_a. A single electroacoustic transducer, rectangular in shape, with complex aperture function given by Eq. (6.5-28), is referred to as a *rectangular piston*. The unnormalized and normalized far-field directivity functions of a rectangular piston lying in the XY plane are given by

$$\boxed{D(f, f_X, f_Y) = L_X L_Y \,\text{sinc}(f_X L_X)\,\text{sinc}(f_Y L_Y)} \qquad (6.5\text{-}29)$$

and

$$\boxed{D_N(f, f_X, f_Y) = \text{sinc}(f_X L_X)\,\text{sinc}(f_Y L_Y),} \qquad (6.5\text{-}30)$$

respectively, where $L_X L_Y$ is the area of the piston in square meters. Note that Eqs. (6.5-28) through (6.5-30) are two-dimensional generalizations of the one-dimensional results obtained for the rectangular amplitude window discussed in Example 6.3-1. Also note that the maximum radial extent of a rectangular-shaped transducer lying in the XY plane is given by

$$R = \sqrt{(L_X/2)^2 + (L_Y/2)^2} \,. \qquad (6.5\text{-}31)$$

6.5.3 Separable Functions — Polar Coordinates

Using Eq. (6.5-5) as the starting point, consider the problem of computing the far-field beam pattern of a planar aperture that is circular in shape, with radius a meters and complex aperture function $A(f, r_a, \phi_a)$ expressed in the polar coordinates r_a and ϕ_a (see Fig. 6.5-5). By referring to Fig. 6.5-5, we can write that

$$x_a = r_a \cos \phi_a, \qquad (6.5\text{-}32)$$

$$y_a = r_a \sin \phi_a, \qquad (6.5\text{-}33)$$

$$dx_a \, dy_a \rightarrow r_a \, dr_a \, d\phi_a, \qquad (6.5\text{-}34)$$

and

$$A(f, x_a, y_a) = A(f, r_a \cos \phi_a, r_a \sin \phi_a) \rightarrow A(f, r_a, \phi_a). \qquad (6.5\text{-}35)$$

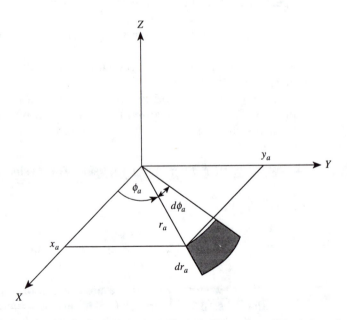

Figure 6.5-5. Relationship between the rectangular coordinates (x_a, y_a) and the polar coordinates (r_a, ϕ_a). Also shown is the infinitesimal area $r_a \, dr_a \, d\phi_a$.

Upon substituting Eqs. (6.5-7), (6.5-8), and (6.5-32) through (6.5-35) into Eq. (6.5-5), and by using the trigonometric identity

$$\cos(\alpha - \beta) = \cos \alpha \cos \beta + \sin \alpha \sin \beta, \tag{6.5-36}$$

we obtain

$$D(f,\theta,\psi) = \int_0^{2\pi} \int_0^a A(f,r_a,\phi_a) \exp\left[+j\frac{2\pi r_a}{\lambda} \sin\theta \cos(\psi - \phi_a)\right] r_a \, dr_a \, d\phi_a, \tag{6.5-37}$$

where

$$A(f,r_a,\phi_a) = a(f,r_a,\phi_a) \exp\left[+j\theta(f,r_a,\phi_a)\right]. \tag{6.5-38}$$

Note that $2\pi r_a$ is the circumference of a circle with radius r_a meters. Equation (6.5-37) will be used in Section 7.4.2 in order to derive an expression for the far-field directivity function of a planar array of concentric circular arrays.

Equation (6.5-37) is the far-field beam pattern of the general complex aperture function given by Eq. (6.5-38). Now, let us assume that $A(f,r_a,\phi_a)$ is separable in the polar coordinates r_a and ϕ_a, that is,

$$A(f,r_a,\phi_a) = A_r(f,r_a) A_\phi(f,\phi_a), \tag{6.5-39}$$

where

$$A_r(f,r_a) = a_r(f,r_a) \exp\left[+j\theta_r(f,r_a)\right] \tag{6.5-40}$$

and

$$A_\phi(f,\phi_a) = a_\phi(f,\phi_a) \exp\left[+j\theta_\phi(f,\phi_a)\right]. \tag{6.5-41}$$

Equation (6.5-39) indicates that the complex frequency response of the aperture in the r direction is independent of that in the ϕ direction. If Eq. (6.5-39) is substituted into Eq. (6.5-37), then

$$D(f,\theta,\psi) = \int_0^a A_r(f,r_a) r_a \int_0^{2\pi} A_\phi(f,\phi_a) \exp(+jb \sin \alpha) \, d\phi_a \, dr_a, \tag{6.5-42}$$

where

$$b = \frac{2\pi r_a}{\lambda} \sin\theta, \tag{6.5-43}$$

$$\alpha = \psi - \phi_a + \frac{\pi}{2}, \tag{6.5-44}$$

and

$$\sin \alpha = \sin\left(\psi - \phi_a + \frac{\pi}{2}\right) = \cos(\psi - \phi_a). \tag{6.5-45}$$

Note that

$$\exp(+jb \sin \alpha) = \sum_{n=-\infty}^{\infty} J_n(b) \exp(+jn\alpha), \tag{6.5-46}$$

where

$$J_n(b) = \frac{1}{2\pi} \int_{-\pi}^{\pi} \exp[\pm j(b \sin \beta - n\beta)] \, d\beta, \tag{6.5-47}$$

or

$$J_n(b) = \frac{1}{\pi} \int_0^{\pi} \cos(b \sin \beta - n\beta) \, d\beta, \tag{6.5-48}$$

is the *nth-order Bessel function of the first kind*. If Eqs. (6.5-46), (6.5-43), and (6.5-44) are substituted into Eq. (6.5-42), and we note that

$$\exp(+jn\pi/2) = j^n, \tag{6.5-49}$$

then

$$D(f,\theta,\psi) = \sum_{n=-\infty}^{\infty} j^n c_n(f) A_r(f,\theta;n) \exp(+jn\psi), \tag{6.5-50}$$

where

$$c_n(f) = \frac{1}{2\pi} \int_0^{2\pi} A_\phi(f,\phi_a) \exp(-jn\phi_a) \, d\phi_a, \tag{6.5-51}$$

$$A_r(f,\theta;n) = H_n\{A_r(f,r_a)\} = 2\pi \int_0^a A_r(f,r_a) J_n\left(\frac{2\pi r_a}{\lambda} \sin \theta\right) r_a \, dr_a \tag{6.5-52}$$

is the *nth-order Hankel transform* of $A_r(f,r_a)$, and

$$J_{-n}(x) = (-1)^n J_n(x), \tag{6.5-53}$$

where n is an integer and x is an arbitrary, real argument. Recall that since the planar aperture was assumed to be circular with radius a meters, the aperture function is equal to zero for $r_a > a$, that is,

$$A_r(f, r_a) = 0, \qquad r_a > a. \qquad (6.5\text{-}54)$$

Therefore, when the complex aperture function is separable in polar coordinates [see Eq. (6.5-39)], its far-field directivity function can be computed using Eqs. (6.5-50) through (6.5-52). Note that since $J_n(x) \to 0$ as $n \to \infty$, the infinite summation indicated by Eq. (6.5-50) never need be evaluated in practice. Instead, only a finite summation need be evaluated from $n = -N$ to $n = N$, where N is chosen so that $J_n(x)$ for $|n| > N$ is negligible according to one's own standards. Also note that if we define the spatial frequency in the radial direction as

$$f_r \triangleq \sqrt{f_X^2 + f_Y^2} = \frac{\sin \theta}{\lambda}, \qquad (6.5\text{-}55)$$

then the nth-order Hankel transform of $A_r(f, r_a)$ can be rewritten as

$$A_r(f, f_r; n) = H_n\{A_r(f, r_a)\} = 2\pi \int_0^\infty A_r(f, r_a) J_n(2\pi f_r r_a) r_a \, dr_a, \qquad (6.5\text{-}56)$$

and similarly, the *nth-order inverse Hankel transform* is given by

$$A_r(f, r_a) = H_n^{-1}\{A_r(f, f_r; n)\} = 2\pi \int_0^\infty A_r(f, f_r; n) J_n(2\pi f_r r_a) f_r \, df_r,$$

$$(6.5\text{-}57)$$

where

$$A_r(f, f_r; n) \to A_r(f, \theta; n). \qquad (6.5\text{-}58)$$

Example 6.5-2 (Circular symmetry)

Consider an aperture function that is separable in polar coordinates, with ϕ component given by

$$A_\phi(f, \phi_a) = \exp(+jm\phi_a), \qquad (6.5\text{-}59)$$

where m is an integer. Equation (6.5-59) is equivalent to a cosine variation in ϕ_a and is analogous to a time-harmonic dependence. If Eq. (6.5-59) is substituted into Eq. (6.5-51), then

$$c_n(f) = \frac{1}{2\pi} \int_0^{2\pi} \exp[+j(m - n)\phi_a] \, d\phi_a = \delta_{mn}, \qquad (6.5\text{-}60)$$

where

$$\delta_{mn} = \begin{cases} 1, & m = n, \\ 0, & m \neq n, \end{cases} \qquad (6.5\text{-}61)$$

is the *Kronecker delta*. Substituting Eqs. (6.5-60) and (6.5-61) into Eq. (6.5-50) yields the far-field directivity function

$$D(f,\theta,\psi) = j^m A_r(f,\theta;m)\exp(+jm\psi), \qquad (6.5\text{-}62)$$

where [see Eq. (6.5-52)]

$$A_r(f,\theta;m) = H_m\{A_r(f,r_a)\}. \qquad (6.5\text{-}63)$$

Now, if $m = 0$, then $A_\phi(f,\phi_a) = 1$, and as a result,

$$A(f,r_a,\phi_a) = A_r(f,r_a). \qquad (6.5\text{-}64)$$

The aperture function given by Eq. (6.5-64) is said to be *circularly symmetric*, since it is not a function of the polar angle ϕ_a. With $m = 0$, the far-field beam pattern given by Eq. (6.5-62) reduces to

$$D(f,\theta,\psi) = A_r(f,\theta;0) = H_0\{A_r(f,r_a)\}, \qquad (6.5\text{-}65)$$

where

$$H_0\{A_r(f,r_a)\} = 2\pi \int_0^a A_r(f,r_a) J_0\left(\frac{2\pi r_a}{\lambda}\sin\theta\right) r_a\, dr_a \quad (6.5\text{-}66)$$

is the zeroth-order Hankel transform of $A_r(f,r_a)$, also known as the *Fourier-Bessel transform* of $A_r(f,r_a)$. Note that in the circularly symmetric case, the far-field beam pattern is *axisymmetric*, that is, it is not a function of the azimuthal angle ψ to a field point.

Example 6.5-3 (The circular piston)
Analogous to the rectangular piston discussed in Example 6.5-1, another very common example of a planar aperture lying in the XY plane is a single electroacoustic transducer, circular in shape, with radius a meters, as shown in Fig. 6.5-6. The simplest mathematical model for the complex frequency response (complex aperture function) for this circular shaped transducer is given by

$$A(f,r_a,\phi_a) = A_r(f,r_a) = \text{circ}(r_a/a) = \begin{cases} 1, & r_a \leq a, \\ 0, & r_a > a, \end{cases} \qquad (6.5\text{-}67)$$

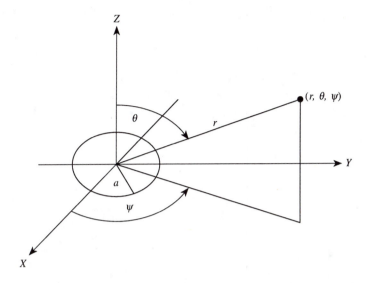

Figure 6.5-6. A single electroacoustic transducer, circular in shape, with radius a meters, lying in the XY plane. Also shown is a field point with spherical coordinates (r, θ, ψ).

which is circularly symmetric. A single electroacoustic transducer, circular in shape, with complex aperture function given by Eq. (6.5-67), is referred to as a *circular piston*. With the use of Eqs. (6.5-65) and (6.5-66), and the identity

$$\int_0^x J_0(\alpha)\alpha \, d\alpha = x J_1(x), \qquad (6.5\text{-}68)$$

the unnormalized far-field beam pattern of a circular piston lying in the XY plane is given by

$$D(f, \theta, \psi) = a \frac{J_1\left(\dfrac{2\pi a}{\lambda} \sin \theta\right)}{\dfrac{\sin \theta}{\lambda}}, \qquad (6.5\text{-}69)$$

where $2\pi a$ is the circumference of the piston in meters. Since the normalization factor is

$$D_{\max} = D(f, \theta, \psi)|_{\theta=0°} = \pi a^2, \qquad (6.5\text{-}70)$$

which is the area of the piston in square meters, the normalized

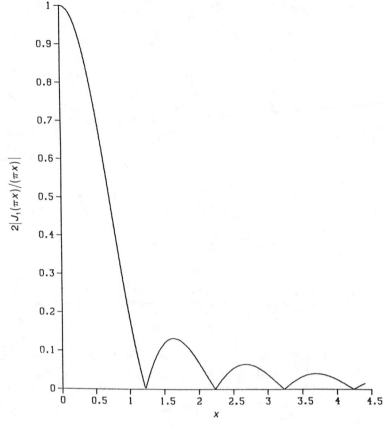

Figure 6.5-7. Magnitude of the normalized far-field beam pattern of a circular piston lying in the *XY* plane, where $x = (2a/\lambda)\sin\theta$.

far-field beam pattern of a circular piston lying in the *XY* plane is given by

$$D_N(f,\theta,\psi) = 2\frac{J_1\left(\dfrac{2\pi a}{\lambda}\sin\theta\right)}{\dfrac{2\pi a}{\lambda}\sin\theta}. \qquad (6.5\text{-}71)$$

Note that the far-field beam pattern is *axisymmetric*, that is, it is not a function of the azimuthal angle ψ to a field point. Figure 6.5-7 is a plot of the magnitude of Eq. (6.5-71), where the level of the first sidelobe is approximately -18 dB.

Example 6.5-4

In this example we shall compute the 3-dB beamwidth $\Delta\theta$ of the normalized vertical far-field directivity function of a circular piston lying in the XY plane using the procedure discussed in Example 6.3-8. From Eq. (6.5-71) we can write that

$$D_N(f, \theta_+) = 2 \frac{J_1\left(\dfrac{2\pi a}{\lambda} \sin \theta_+\right)}{\dfrac{2\pi a}{\lambda} \sin \theta_+} = \frac{\sqrt{2}}{2}, \qquad (6.5\text{-}72)$$

or

$$J_1(\pi x) - \frac{\sqrt{2}}{4}\pi x = 0, \qquad (6.5\text{-}73)$$

where [see Eq. (6.3-85)]

$$x = \frac{2a}{\lambda} \sin \theta_+ = \frac{2a}{\lambda} \sin\left(\frac{\Delta\theta}{2}\right), \qquad (6.5\text{-}74)$$

so that

$$\Delta\theta = 2 \sin^{-1}\left(x \frac{\lambda}{2a}\right). \qquad (6.5\text{-}75)$$

The solution (root) of Eq. (6.5-73) is $x \approx 0.514$, and upon substituting this result into Eq. (6.5-75), we obtain

$$\boxed{\Delta\theta = 2 \sin^{-1}\left(0.257 \frac{\lambda}{a}\right)} \qquad (6.5\text{-}76)$$

where a is the radius of the circular piston in meters.

Let us conclude this example by comparing Eq. (6.5-76) for a circular piston with the 3-dB beamwidth $\Delta\theta$ of the normalized vertical far-field directivity function in the XZ plane of a rectangular piston lying in the XY plane. Since direction cosine $v = 0$ in the XZ plane, the normalized far-field directivity function of a rectangular piston lying in the XY plane reduces to the normalized far-field directivity function of a linear aperture lying along the X axis with a rectangular amplitude window for a complex frequency response (see Examples 6.5-1 and 6.3-8). Therefore, substituting Eq. (6.3-110) from Example 6.3-8 into

Eq. (6.3-99) yields

$$\Delta\theta = 2\sin^{-1}\left(0.443\frac{\lambda}{L_X}\right),$$

(6.5-77)

where L_X is the length (in meters) of the rectangular piston in the X direction. In order to make a fair comparison, consider rectangular and circular pistons of the same area. If we let $L_Y = L_X$, then the area of the rectangular piston is L_X^2. If we then let

$$L_X^2 = \pi a^2,$$

(6.5-78)

where πa^2 is the area of the circular piston, then

$$L_X = \sqrt{\pi}\,a,$$

(6.5-79)

and upon substituting Eq. (6.5-79) into Eq. (6.5-77), we finally obtain

$$\Delta\theta = 2\sin^{-1}\left(0.250\frac{\lambda}{a}\right)$$

(6.5-80)

for the rectangular piston. By comparing Eq. (6.5-76) for a circular piston with Eq. (6.5-80) for a rectangular piston, it can be seen that a circular piston will have a slightly larger 3-dB beamwidth. However, recall that the levels of the first sidelobes of the normalized far-field directivity functions of a circular piston and a rectangular amplitude window are approximately -18 and -13 dB, respectively. The slight increase in beamwidth is offset by an additional 5 dB sidelobe suppression provided by a circular piston.

6.6 Planar Apertures and Near-Field Directivity Functions

6.6.1 Beam Steering and Aperture Focusing

Consider the case of a *planar aperture* of arbitrary shape lying in the XY plane as shown in Fig. 6.5-1. Recall that a planar aperture can represent either a single electroacoustic transducer or a planar array of many individual electroacoustic transducers. Also shown in Fig. 6.5-1 is a field point with spherical coordinates (r, θ, ψ). The field point is assumed to be in the near-field (Fresnel) region of the aperture, at a range $1.356R < r < \pi R^2/\lambda$ meters, where R meters is the maximum radial extent of the aperture [see Eq. (6.2-40)].

Since the planar aperture illustrated in Fig. 6.5-1 is shown lying in the XY plane, the position vector that describes the spatial location of the aperture is

given by

$$\mathbf{r}_a = (x_a, y_a, 0), \tag{6.6-1}$$

and as a result,

$$A(f, \mathbf{r}_a) = A(f, x_a, y_a), \tag{6.6-2}$$

$$r_a^2 = |\mathbf{r}_a|^2 = x_a^2 + y_a^2, \tag{6.6-3}$$

$$\boldsymbol{\alpha} \cdot \mathbf{r}_a = (f_X, f_Y, f_Z) \cdot (x_a, y_a, 0) = f_X x_a + f_Y y_a, \tag{6.6-4}$$

and

$$d\mathbf{r}_a = dx_a \, dy_a. \tag{6.6-5}$$

Therefore, upon substituting Eqs. (6.6-1) through (6.6-5) into Eq. (6.4-1), we obtain the following two-dimensional spatial Fresnel transform expression for the near-field directivity function (beam pattern) of a planar aperture lying in the XY plane based on the Fresnel approximation [see Eqs. (6.2-35) and (6.2-40)]:

$$
\begin{aligned}
D(f, r, f_X, f_Y) &= F_{x_a} F_{y_a} \left\{ A(f, x_a, y_a) \exp\left[-jk \frac{(x_a^2 + y_a^2)}{2r} \right] \right\} \\
&= \int_{-\infty}^{\infty} \int_{-\infty}^{\infty} A(f, x_a, y_a) \exp\left[-jk \frac{(x_a^2 + y_a^2)}{2r} \right] \\
&\quad \times \exp\left[+j2\pi (f_X x_a + f_Y y_a) \right] dx_a \, dy_a,
\end{aligned}
\tag{6.6-6}
$$

where

$$A(f, x_a, y_a) = a(f, x_a, y_a) \exp\left[+j\theta(f, x_a, y_a) \right], \tag{6.6-7}$$

$$f_X = \frac{u}{\lambda} = \frac{\sin\theta \cos\psi}{\lambda}, \tag{6.6-8}$$

and

$$f_Y = \frac{v}{\lambda} = \frac{\sin\theta \sin\psi}{\lambda}. \tag{6.6-9}$$

Since the spatial frequencies f_X and f_Y can be expressed in terms of θ and ψ, the near-field directivity function can ultimately be expressed as a function of frequency f, the range r to a field point, and the spherical angles θ and ψ, that is,

$$D(f, r, f_X, f_Y) \rightarrow D(f, r, \theta, \psi). \tag{6.6-10}$$

Next, assume that the phase response (phase variation) across the aperture is equal to the sum of linear and quadratic functions of x_a and y_a, that is,

$$\theta(f, x_a, y_a) = -2\pi f_X' x_a - 2\pi f_Y' y_a + k\frac{(x_a^2 + y_a^2)}{2r'}, \qquad (6.6\text{-}11)$$

where

$$f_X' = \frac{u'}{\lambda} = \frac{\sin\theta'\cos\psi'}{\lambda}, \qquad (6.6\text{-}12)$$

$$f_Y' = \frac{v'}{\lambda} = \frac{\sin\theta'\sin\psi'}{\lambda}, \qquad (6.6\text{-}13)$$

and

$$k = 2\pi f/c = 2\pi/\lambda \qquad (6.6\text{-}14)$$

is the wave number. If Eqs. (6.6-7) and (6.6-11) are substituted into Eq. (6.6-6), then

$$D(f, r, f_X, f_Y) = \int_{-\infty}^{\infty}\int_{-\infty}^{\infty} a(f, x_a, y_a)\exp\left[-jk\frac{(x_a^2 + y_a^2)}{2}\left(\frac{1}{r} - \frac{1}{r'}\right)\right]$$

$$\times \exp\{+j2\pi[(f_X - f_X')x_a + (f_Y - f_Y')y_a]\}\,dx_a\,dy_a, \qquad (6.6\text{-}15)$$

and if Eq. (6.6-15) is evaluated at the near-field range $r = r'$, then Eq. (6.6-15) reduces to

$$D(f, r', f_X, f_Y) = \int_{-\infty}^{\infty}\int_{-\infty}^{\infty} a(f, x_a, y_a)$$

$$\times \exp\{+j2\pi[(f_X - f_X')x_a + (f_Y - f_Y')y_a]\}\,dx_a\,dy_a, \qquad (6.6\text{-}16)$$

or

$$D(f, r', f_X, f_Y) = F_{x_a}F_{y_a}\{a(f, x_a, y_a)\exp[-j2\pi(f_X' x_a + f_Y' y_a)]\}$$

$$= D(f, f_X - f_X', f_Y - f_Y'). \qquad (6.6\text{-}17)$$

Equation (6.6-17) indicates that the far-field beam pattern of the amplitude window $a(f, x_a, y_a)$ has been *focused* to the near-field range $r = r'$ from the aperture, and *steered* in the direction $u = u'$ and $v = v'$ in direction cosine space, which is equivalent to steering (tilting) the beam pattern to $\theta = \theta'$ and $\psi = \psi'$.

6.7 Directivity Index

In this section we shall derive an equation for the *directivity* of an aperture, where the aperture is characterized by its far-field beam pattern. Recall that the term "aperture" can refer to either a single electroacoustic transducer or an array of electroacoustic transducers. The *directivity index* is simply the decibel equivalent of the directivity.

The directivity of an aperture, when used in the active mode as a transmitter, is a measure of its ability to concentrate the available acoustic power into a preferred direction. When the aperture is used in the passive mode as a receiver, its directivity is a measure of its ability to distinguish between several sound sources located at different spatial locations within the medium. Therefore, directivity is basically a measure of the beamwidth and sidelobe levels of a far-field beam pattern. In the analysis that follows, it is assumed that the aperture is being used as a transmitter. Note that if the transducer or transducers that make up the aperture are *reversible*, that is, if they can be used as either transmitters or receivers, then the transmit and receive far-field beam patterns of the aperture are *identical*.

In order to compute the directivity of an aperture when used as a transmitter (sound source), we must calculate the time-average intensity vector of the radiated acoustic field, and then the time-average radiated power. To do so, we must first calculate the acoustic pressure and the acoustic fluid velocity vector due to the sound source. We begin the analysis by using the following result obtained in Example 6.2-1: when an identical time-harmonic input electrical signal is applied at all spatial locations of an arbitrarily shaped volume transmit aperture, the scalar velocity potential (in square meters per second) of the radiated acoustic field in the far field is given by

$$\varphi(t,r,\theta,\psi) \approx -D(f,\theta,\psi)\frac{\exp\left[+j2\pi f\left(t-\dfrac{r}{c}\right)\right]}{4\pi r}, \tag{6.7-1}$$

where $D(f,\theta,\psi)$ is the unnormalized far-field beam pattern of the aperture. Also recall from Section 1.4 that the acoustic pressure $p(t,r,\theta,\psi)$ in pascals (Pa) and the acoustic fluid velocity vector $\mathbf{u}(t,r,\theta,\psi)$ in meters per second can be obtained from the velocity potential as follows:

$$p(t,r,\theta,\psi) = -\rho_0\frac{\partial}{\partial t}\varphi(t,r,\theta,\psi), \tag{6.7-2}$$

where ρ_0 is the ambient (equilibrium) density of the fluid medium in kilograms per cubic meter, and

$$\mathbf{u}(t,r,\theta,\psi) = \nabla\varphi(t,r,\theta,\psi), \tag{6.7-3}$$

where

$$\nabla = \frac{\partial}{\partial r}\hat{r} + \frac{1}{r}\frac{\partial}{\partial\theta}\hat{\theta} + \frac{1}{r\sin\theta}\frac{\partial}{\partial\psi}\hat{\psi} \qquad (6.7\text{-}4)$$

is the gradient expressed in the spherical coordinates (r, θ, ψ), and $\hat{r}$, $\hat{\theta}$, and $\hat{\psi}$ are unit vectors in the r, θ, and ψ directions, respectively. Note that ρ_0 is constant, since we are dealing with a homogeneous medium. Therefore, substituting Eq. (6.7-1) into Eq. (6.7-2) yields

$$p(t, r, \theta, \psi) \approx jk\rho_0 cD(f, \theta, \psi)\frac{\exp\left[+j2\pi f\left(t - \frac{r}{c}\right)\right]}{4\pi r}, \qquad (6.7\text{-}5)$$

where $\rho_0 c$ is the characteristic impedance of the medium. Equation (6.7-5) is an approximate far-field expression for the radiated acoustic pressure due to an arbitrarily shaped volume transmit aperture. Note, however, that if the size of the aperture is small compared to a wavelength, then the beam pattern reduces to a constant, that is, the source is omnidirectional, and Eq. (6.7-5) reduces to the form of the radiated acoustic pressure due to an omnidirectional, spherical point source (see Section 4.2.2).

Next, let us compute the *radial* component of the acoustic fluid velocity vector. As we shall show later, only the radial component contributes to the time-average radiated power. By referring to Eqs. (6.7-3) and (6.7-4), the radial component of the acoustic fluid velocity vector is given by

$$\mathbf{u}_r(t, r, \theta, \psi) = \frac{\partial}{\partial r}\varphi(t, r, \theta, \psi)\hat{r}, \qquad (6.7\text{-}6)$$

and upon substituting Eq. (6.7-1) into Eq. (6.7-6), we obtain

$$\mathbf{u}_r(t, r, \theta, \psi) \approx -D(f, \theta, \psi)\frac{\partial}{\partial r}\left[\frac{\exp(-jkr)}{r}\right]\frac{\exp(+j2\pi ft)}{4\pi}\hat{r}, \qquad (6.7\text{-}7)$$

where

$$k = 2\pi f/c. \qquad (6.7\text{-}8)$$

Since

$$\frac{\partial}{\partial r}\left[\frac{\exp(-jkr)}{r}\right] = -k^2\left[\frac{1}{(kr)^2} + j\frac{1}{kr}\right]\exp(-jkr), \qquad (6.7\text{-}9)$$

and since it was assumed that the field point is in the far-field region of the aperture, that is, $kr \gg 1$, Eq. (6.7-9) reduces to

$$\frac{\partial}{\partial r}\left[\frac{\exp(-jkr)}{r}\right] \approx -jk\frac{\exp(-jkr)}{r}, \qquad kr \gg 1. \qquad (6.7\text{-}10)$$

Therefore, substituting Eq. (6.7-10) into Eq. (6.7-7) yields

$$\mathbf{u}_r(t,r,\theta,\psi) = u_r(t,r,\theta,\psi)\hat{r} \approx jkD(f,\theta,\psi)\frac{\exp\left[+j2\pi f\left(t-\dfrac{r}{c}\right)\right]}{4\pi r}\hat{r},\ kr \gg 1,$$

$$(6.7\text{-}11)$$

where $u_r(t,r,\theta,\psi)$ is the complex acoustic fluid *speed* in meters per second in the radial direction. Note that the acoustic pressure given by Eq. (6.7-5) can be expressed as

$$p(t,r,\theta,\psi) = \rho_0 c u_r(t,r,\theta,\psi), \qquad kr \gg 1. \qquad (6.7\text{-}12)$$

Equation (6.7-12) is in the form of a plane-wave relationship between the acoustic pressure and the acoustic fluid speed (see Section 2.2.1). Therefore, the radiated acoustic field behaves like a plane wave in the radial direction in the far-field region of the aperture.

Now that we have expressions for both the acoustic pressure and the acoustic fluid velocity vector in the radial direction, we can calculate the time-average intensity vector in the radial direction. Since the pressure and velocity-vector expressions given by Eqs. (6.7-5) and (6.7-11) are time-harmonic acoustic fields, that is, since

$$p(t,r,\theta,\psi) = p_f(r,\theta,\psi)\exp(+j2\pi ft) \qquad (6.7\text{-}13)$$

and

$$\mathbf{u}_r(t,r,\theta,\psi) = \mathbf{u}_{f,r}(r,\theta,\psi)\exp(+j2\pi ft), \qquad (6.7\text{-}14)$$

where

$$p_f(r,\theta,\psi) \approx jk\rho_0 cD(f,\theta,\psi)\frac{\exp(-jkr)}{4\pi r} \qquad (6.7\text{-}15)$$

and

$$\mathbf{u}_{f,r}(r,\theta,\psi) \approx jkD(f,\theta,\psi)\frac{\exp(-jkr)}{4\pi r}\hat{r}, \qquad kr \gg 1, \qquad (6.7\text{-}16)$$

the time-average intensity vector in the radial direction in watts per square meter can be computed by using the following equation (see Section 1.5):

$$\mathbf{I}_{r_{avg}}(r,\theta,\psi) = \tfrac{1}{2}\,\mathrm{Re}\{p_f(r,\theta,\psi)\mathbf{u}^*_{f,r}(r,\theta,\psi)\}. \qquad (6.7\text{-}17)$$

Therefore, upon substituting Eqs. (6.7-15) and (6.7-16) into Eq. (6.7-17), we obtain [compare with Eq. (4.2-68)]

$$\boxed{\mathbf{I}_{r_{avg}}(r,\theta,\psi) \approx \frac{\rho_0 c k^2}{32\pi^2 r^2}\,|D(f,\theta,\psi)|^2\hat{r}, \qquad kr \gg 1.} \qquad (6.7\text{-}18)$$

The next quantity to be calculated is the time-average radiated power. If we enclose the arbitrarily shaped volume transmit aperture with a sphere of radius r meters, then the time-average power in watts is given by (see Section 1.5)

$$P_{avg} = \oint_S \mathbf{I}_{r_{avg}}(r,\theta,\psi)\cdot d\mathbf{S}, \qquad (6.7\text{-}19)$$

where

$$d\mathbf{S} = r^2\sin\theta\,d\theta\,d\psi\,\hat{r}. \qquad (6.7\text{-}20)$$

Therefore, substituting Eqs. (6.7-18) and (6.7-20) into Eq. (6.7-19) yields [compare with Eqs. (4.2-72) and (4.2-73)]

$$\boxed{P_{avg} \approx \frac{\rho_0 c k^2}{32\pi^2}\int_0^{2\pi}\int_0^{\pi}|D(f,\theta,\psi)|^2\sin\theta\,d\theta\,d\psi, \qquad kr \gg 1.} \qquad (6.7\text{-}21)$$

It is important to note that even if we computed all three components of the acoustic fluid velocity vector instead of just the radial component, and even if the time-average intensity vector had three components, the time-average power would still be given by Eq. (6.7-21) because of the dependence of $d\mathbf{S}$ on the unit vector $\hat{r}$ in the radial direction [see Eq. (6.7-20)]. Only the radial component of the time-average intensity vector would remain after the dot product was taken between the time-average intensity vector and $d\mathbf{S}$.

The *directivity* D of an aperture is defined as follows:

$$\boxed{D \triangleq \frac{I_{max}}{I_{ref}},} \qquad (6.7\text{-}22)$$

where I_{max} is the maximum value of the magnitude of the far-field ($kr \gg 1$), time-average intensity vector in the radial direction of the acoustic field produced by the aperture, and I_{ref} is a reference intensity. By referring to Eq. (6.7-18),

$$I_{\max} = \frac{\rho_0 c k^2}{32\pi^2 r^2} |D_{\max}|^2, \qquad (6.7\text{-}23)$$

where $D_{\max}$ is the maximum value of the unnormalized far-field directivity function of the aperture, also known as the normalization factor (see Section 6.3.1).

The reference intensity is defined as the time-average intensity of the acoustic field produced by an omnidirectional, spherical point source that radiates the same time-average power as the arbitrarily shaped source, that is,

$$I_{\text{ref}} \triangleq \frac{P_{\text{avg}}}{4\pi r^2}, \qquad (6.7\text{-}24)$$

and upon substituting Eq. (6.7-21) into Eq. (6.7-24), we obtain

$$I_{\text{ref}} = \frac{\rho_0 c k^2}{32\pi^2 r^2} \frac{|D_{\max}|^2}{4\pi} \int_0^{2\pi} \int_0^{\pi} |D_N(f,\theta,\psi)|^2 \sin\theta \, d\theta \, d\psi, \qquad (6.7\text{-}25)$$

where

$$D_N(f,\theta,\psi) = \frac{D(f,\theta,\psi)}{D_{\max}} \qquad (6.7\text{-}26)$$

is the normalized far-field directivity function of the aperture. Therefore, substituting Eqs. (6.7-23) and (6.7-25) into Eq. (6.7-22) yields the following general expression for the directivity of an arbitrarily shaped volume transmit aperture (source):

$$D = \frac{4\pi}{\int_0^{2\pi} \int_0^{\pi} |D_N(f,\theta,\psi)|^2 \sin\theta \, d\theta \, d\psi}. \qquad (6.7\text{-}27)$$

Note that the directivity D is dimensionless. The *directivity index* DI is defined as follows:

$$\boxed{\text{DI} \triangleq 10\log_{10} D \text{ dB},} \qquad (6.7\text{-}28)$$

which is the decibel equivalent of the directivity.

Values of the directivity range from a minimum of $D = 1$ (DI = 0 dB) for an omnidirectional source ($|D_N(f,\theta,\psi)| = 1$) such as a radially pulsating sphere (see Section 4.2.2), to large numbers for highly directional sources. For example, for an omnidirectional ($|D_N(f,\theta,\psi)| = 1$) *hemispherical* source mounted on an infinite baffle in the *XY* plane, the directivity $D = 2$ (DI = 3 dB), since the limits of integration in Eq. (6.7-27) are from 0 to $\pi/2$ in the θ direction and from 0 to 2π in the ψ direction. In general, the directivity of an aperture increases with increasing frequency or physical size, since the beamwidth of the corresponding far-field beam pattern decreases with increasing frequency or size.

Finally, it should be emphasized again that both the directivity and the directivity index of either a single electroacoustic transducer or an array of electroacoustic transducers can be computed from Eqs. (6.7-27) and (6.7-28), respectively, as long as the normalized far-field beam pattern is known. Arrays are discussed next, in Chapter 7.

Problems

6-1   Verify Eqs. (6.2-12) and (6.2-13).

6-2   Verify Eq. (6.2-25).

6-3   Show that Eq. (6.3-16) is equal to Eq. (6.3-17).

6-4   Verify Eq. (6.3-20).

6-5   Compute the normalized far-field directivity functions of the following aperture functions:
(a) $A(f, y_a) = \text{rect}(y_a/L)$,
(b) $A(f, z_a) = \text{rect}(z_a/L)$.
Express your answers in terms of the spherical angles θ and ψ.

6-6   Using Eq. (6.3-28) as the starting point, show that the third-order partial derivative of the aperture function $A(f, x_a)$ with respect to x_a is related to its far-field beam pattern $D(f, f_X)$ by the following expression:

$$D(f, f_X) = \frac{1}{(-j2\pi f_X)^3} F_{x_a} \left\{ \frac{\partial^3}{\partial x_a^3} A(f, x_a) \right\}.$$

6-7   Verify Eq. (6.3-35).

6-8   Derive Eq. (6.3-39) by direct integration instead of using the method of successive differentiation.

6-9   **The sum beam pattern.** Using the method of successive differentiation, compute the *unnormalized* far-field directivity function of the rectangular amplitude window shown in Fig. 6.3-2. Compare your result with Eq. (6.3-17). When an aperture function is modeled by a rectangular amplitude window,

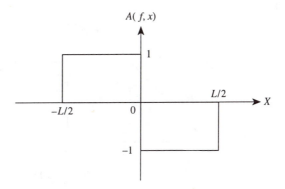

Figure P6-10

the resulting far-field beam pattern is referred to as a *sum beam pattern*, since the aperture function can be thought of as being equal to the *sum* of the left half (from $-L/2$ to 0) and the right half (from 0 to $L/2$) of a rectangular amplitude window.

6-10 **The difference beam pattern.** Using the method of successive differentiation, compute the *unnormalized* far-field directivity function of the aperture function shown in Fig. P6-10. Note that since this aperture function is an odd function of x, its far-field beam pattern will be imaginary and odd. When an aperture function is modeled by the function shown in Fig. P6-10, the resulting far-field beam pattern is referred to as a *difference beam pattern*, since the aperture function shown can be thought of as being equal to the *difference* between the left half (from $-L/2$ to 0) and the right half (from 0 to $L/2$) of a rectangular amplitude window. The two halves of the aperture function shown in Fig. P6-10 have the same magnitude but are 180° out of phase.

6-11 (a) Using the method of successive differentiation, compute the *unnormalized* far-field directivity function of the trapezoidal amplitude window shown in Fig. P6-11. Use the following trigonometric identity:

$$\cos \alpha - \cos \beta = -2 \sin\left(\frac{\alpha + \beta}{2}\right) \sin\left(\frac{\alpha - \beta}{2}\right).$$

(b) Using your answer for part (a), what is the equation for the beam pattern when $d = 0$ and $h = 1$?

(c) Using your answer for part (a), what is the equation for the beam pattern when $d = L/2$ and $h = 1$?

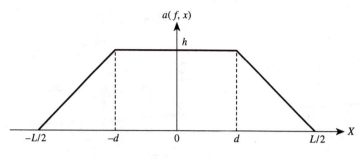

Figure P6-11

6-12 Show that

$$F_{f_X}^{-1}\{\delta(f_X \pm f_X')\} = \exp(\pm j2\pi f_X' x_a).$$

From this result we can write that [see Eq. (6.3-49)]

$$F_{x_a}\{\exp(\pm j2\pi f_X' x_a)\} = \delta(f_X \pm f_X'),$$

and if $f_X' = 0$, then

$$F_{x_a}\{1\} = \delta(f_X)$$

and

$$F_{f_X}^{-1}\{\delta(f_X)\} = 1.$$

6-13 Show that

$$g(x) \underset{x}{*} \delta(x \pm x_0) = g(x \pm x_0),$$

where $\underset{x}{*}$ denotes convolution with respect to x, and x_0 is an arbitrary constant.

6-14 Derive expressions for the *normalized* far-field directivity functions of the Hanning, Hamming, and Blackman amplitude windows. Use the following trigonometric identity: $\sin(\alpha \pm \pi) = -\sin\alpha$.

6-15 Compute the 3-dB beamwidth of the normalized horizontal far-field beam pattern of a rectangular amplitude window for the following values of the ratio L/λ: $L/\lambda = 0.5, 1, 2,$ and 4. What is your conclusion about the relationship between beamwidth, length of the linear aperture, and frequency?

6-16 Using your answers for Problem 6-14 and the procedure outlined in Example 6.3-8, compute the 3-dB beamwidths Δu in direction-cosine space of the normalized far-field beam patterns of the triangular, cosine, Hanning, Hamming, and Blackman amplitude windows.

6-17 Using your answers for Problem 6-16, compute the 3-dB beamwidths of the normalized horizontal far-field beam patterns of the triangular, cosine, Hanning, Hamming, and Blackman amplitude windows for $L/\lambda = 1$, 2, and 4.

6-18 Compute the 3-dB beamwidths of the normalized horizontal far-field beam patterns of the rectangular, triangular, cosine, Hanning, Hamming, and Blackman amplitude windows for the following beam tilt angles: $\psi' = 90°$ (broadside), $75°$, $60°$, $45°$, $30°$, and $0°$ (end fire). Use $L/\lambda = 4$.

6-19 Consider a linear aperture lying along the X axis with a linear phase variation across its length.
(a) Show that the 3-dB beamwidth of the normalized vertical far-field beam pattern in the XZ plane is given by

$$\Delta\theta = \sin^{-1}\left(\sin\theta' + \frac{\Delta u}{2}\right) - \sin^{-1}\left(\sin\theta' - \frac{\Delta u}{2}\right), \qquad \left|\sin\theta' \pm \frac{\Delta u}{2}\right| \le 1,$$

where θ' is the beam tilt angle in the XZ plane. Note that $\theta' = 0$ corresponds to an untilted vertical beam pattern, and in this case $\Delta\theta$ reduces to Eq. (6.3-99), since

$$\sin^{-1}(-x) = -\sin^{-1}(x).$$

Hint: Follow a procedure analogous to the one used to derive Eq.(6.3-129).
(b) Show that the 3-dB beamwidth of the normalized vertical far-field beam pattern in the XZ plane steered to end-fire geometry is given by

$$\Delta\theta = 2\cos^{-1}\left(1 - \frac{\Delta u}{2}\right).$$

Hint: Follow a procedure analogous to the one used to derive Eq. (6.3-136).

6-20 The spherical coordinates of a sound source with respect to a linear aperture lying along the X axis have been determined to be $r_S = 50$ m, $\theta_S = 15°$, and $\psi_S = 63°$. It is also given that $L = 2$ m, $c = 1500$ m/sec, and $f = 25$ kHz.

(a) Is the sound source in the aperture's near field or far field?

(b) What must the phase variation across the length of the aperture be if the aperture's far-field beam pattern is to be focused (if required) and steered to coordinates (r_S, θ_S, ψ_S)?

(c) The sound source changes its coordinates to $r_S = 70$ m, $\theta_S = 33°$, and $\psi_S = 47°$. Repeat parts (a) and (b).

(d) Repeat parts (a) through (c) for a linear aperture lying along the Y axis.

6-21 Derive an expression for the normalized far-field beam pattern of the complex aperture function

$$A(f, x_a, y_a) = A_X(f, x_a) A_Y(f, y_a)$$

when

(a) $A_X(f, x_a) = \mathrm{rect}(x_a/L_X)$ and $A_Y(f, y_a) = \mathrm{tri}(y_a/L_Y)$.

(b) $A_X(f, x_a) = \mathrm{tri}(x_a/L_X)$ and $A_Y(f, y_a) = \mathrm{tri}(y_a/L_Y)$.

(c) $A_X(f, x_a) = \mathrm{rect}(x_a/L_X)\exp(-j2\pi f'_X x_a)$ and $A_Y(f, y_a) = \mathrm{tri}(y_a/L_Y)\exp(-j2\pi f'_Y y_a)$.

(d) $A_X(f, x_a) = \mathrm{tri}(x_a/L_X)$ and $A_Y(f, y_a) = \cos(\pi y_a/L_Y)\mathrm{rect}(y_a/L_Y)$.

(e) $A_X(f, x_a) = \cos(\pi x_a/L_X)\mathrm{rect}(x_a/L_X)\exp(-j2\pi f'_X x_a)$ and $A_Y(f, y_a) = \cos(\pi y_a/L_Y)\mathrm{rect}(y_a/L_Y)\exp(-j2\pi f'_Y y_a)$.

6-22 Verify

(a) Eq. (6.5-60),

(b) Eq. (6.5-69), and

(c) Eq. (6.5-70). Note that for this part,

$$J_0(0) = 1,$$

$$J_n(0) = 0, \quad n \neq 0,$$

and

$$2J'_n(x) = J_{n-1}(x) - J_{n+1}(x),$$

where

$$J'_n(x) = \frac{d}{dx}J_n(x).$$

6-23 Consider a thin, planar electroacoustic transducer, circular in shape with radius b meters, lying in the XY plane. The complex frequency response or complex aperture function of the transducer is circularly symmetric and is

given by

$$A_r(f, r_a) = \begin{cases} 1, & 0 \le r_a < a, \\ -1, & a \le r_a \le b, \end{cases}$$

where $b > a$. Find the unnormalized far-field directivity function of this transducer.

6-24 Consider a rectangular and a circular piston of equal areas lying in the XY plane. With $L_Y = L_X$ and for $a/\lambda = 0.5$, 1, 2, and 4, compute the 3-dB beamwidth in degrees of the normalized vertical far-field beam pattern in the XZ plane of the:
(a) circular piston.
(b) rectangular piston.

6-25 The spherical coordinates of a sound source with respect to a planar aperture lying in the XY plane have been determined to be $r_S = 72$ m, $\theta_S = 18°$, and $\psi_S = 103°$. It is also given that $L_X = 2$ m, $L_Y = 3$ m, $c = 1500$ m/sec, and $f = 15$ kHz.
(a) Is the sound source in the aperture's near field or far field?
(b) What must the phase variation across the aperture be if the aperture's far-field beam pattern is to be focused (if required) and steered to coordinates (r_S, θ_S, ψ_S)?
(c) The sound source changes its coordinates to $r_S = 120$ m, $\theta_S = 38°$, and $\psi_S = 53°$. Repeat parts (a) and (b).

6-26 Verify Eq. (6.7-9).

6-27 Consider a rectangular piston lying in the XY plane. Compute the far-field radiated acoustic pressure at the field point with spherical coordinates $r = 120$ m, $\theta = 49°$, and $\psi = 88°$ at time $t = 1.884 \, r/c$ sec. Use $L_Y = L_X = 0.3$ m, $\rho_0 = 1026$ kg/m^3, $c = 1500$ m/sec, and $f = 15$ kHz. **Hint:** Compute the real part of the complex acoustic pressure.

Appendix 6A

In this appendix we shall derive a criterion that will guarantee that $|b| < 1$, where [see Eq. (6.2-13)]

$$b = \left(\frac{r_0}{r}\right)^2 - 2\frac{\hat{r} \cdot \mathbf{r}_0}{r}, \qquad \text{(6A-1)}$$

or, expanding the dot product,

$$b = \frac{r_0^2}{r^2} - 2\frac{r_0}{r} \cos \phi, \tag{6A-2}$$

where ϕ is the angle between $\hat{r}$ and $\mathbf{r}_0$. With the use of Eq. (6A-2), $|b| < 1$ can be expressed as

$$\left| \frac{r_0^2}{r^2} - 2\frac{r_0}{r} \cos \phi \right| < 1. \tag{6A-3}$$

If we restrict ourselves for the moment to $\pi/2 \le \phi \le \pi$, then the absolute-value sign on the left-hand side of Eq. (6A-3) can be removed, that is,

$$\frac{r_0^2}{r^2} - 2\frac{r_0}{r} \cos \phi < 1, \qquad \pi/2 \le \phi \le \pi, \tag{6A-4}$$

or

$$r^2 + 2r_0 r \cos \phi > r_0^2, \qquad \pi/2 \le \phi \le \pi. \tag{6A-5}$$

Completing the square on the left-hand side of Eq. (6A-5) yields

$$r^2 + 2r_0 r \cos \phi + r_0^2 \cos^2 \phi > r_0^2 + r_0^2 \cos^2 \phi, \qquad \pi/2 \le \phi \le \pi, \tag{6A-6}$$

$$(r + r_0 \cos \phi)^2 > r_0^2(1 + \cos^2 \phi), \qquad \pi/2 \le \phi \le \pi, \tag{6A-7}$$

and, finally,

$$r > r_0\left(\sqrt{1 + \cos^2 \phi} - \cos \phi\right), \qquad \pi/2 \le \phi \le \pi. \tag{6A-8}$$

For $0 \le \phi \le \pi/2$, we substitute the identity

$$\cos(\pi - \phi) = -\cos \phi \tag{6A-9}$$

into Eq. (6A-8), and as a result,

$$r > \begin{cases} r_0\left(\sqrt{1 + \cos^2 \phi} + \cos \phi\right), & 0 \le \phi \le \pi/2, \\ r_0\left(\sqrt{1 + \cos^2 \phi} - \cos \phi\right), & \pi/2 \le \phi \le \pi. \end{cases} \tag{6A-10}$$

Therefore, if Eq. (6A-10) is satisfied, then $|b| < 1$.

Bibliography

J. D. Gaskill, *Linear Systems, Fourier Transforms, and Optics*, Wiley, New York, 1978.

J. W. Goodman, *Introduction to Fourier Optics*, McGraw-Hill, New York, 1968.

D. Middleton, "A statistical theory of reverberation and similar first-order scattered fields. Part I: Waveforms and the general process," *IEEE Trans. Inf. Theory*, **IT-13**, 372–392 (1967).

A. Papoulis, *Systems and Transforms with Applications in Optics*, McGraw-Hill, New York, 1968.

A. Papoulis, *Signal Analysis*, McGraw-Hill, New York, 1977.

B. D. Steinberg, *Principles of Aperture and Array System Design*, Wiley, New York, 1976.

L. J. Ziomek, "Three necessary conditions for the validity of the Fresnel phase approximation for the near-field beam pattern of an aperture," *IEEE J. Oceanic Engr.*, **18**, 73–75 (1993).

Chapter 7

Array Theory

7.1 Linear Arrays and Far-Field Directivity Functions

7.1.1 Product Theorem

An array can be thought of as a *sampled aperture*, that is, an aperture that is excited only at points or in localized areas. An array consists of individual electroacoustic transducers called *elements*. When an array is used in the *active mode*, the electroacoustic transducers are used as *transmitters*, converting electrical signals (voltages and currents) into acoustic signals (sound waves). When an array is used in the *passive mode*, the electroacoustic transducers are used as *receivers*, converting acoustic signals into electrical signals for further signal processing. In underwater acoustic applications, when an electroacoustic transducer is used as a receiver, it is referred to as a *hydrophone*. In seismic applications, it is referred to as a *geophone*.

From complex aperture theory we know that an aperture function and its corresponding far-field directivity function (beam pattern) form a spatial Fourier transform pair. In the case of a *linear array* lying along the X axis, the far-field directivity function is given by [see Eqs. (6.3-9) through (6.3-11)]

$$D(f, f_X) = F_{x_a}\{A(f, x_a)\} = \int_{-\infty}^{\infty} A(f, x_a) \exp(+j2\pi f_X x_a) \, dx_a, \quad (7.1\text{-}1)$$

where $A(f, x_a)$ is the complex frequency response (complex aperture function) of the array and

$$f_X = \frac{u}{\lambda} = \frac{\sin\theta \cos\psi}{\lambda}. \tag{7.1-2}$$

Since the spatial frequency f_X can be expressed in terms of θ and ψ, the far-field directivity function can ultimately be expressed as a function of frequency f and the spherical angles θ and ψ, that is, $D(f, f_X) \to D(f, \theta, \psi)$. We shall consider a field point to be in the far-field region of an array if it is at a range

$$r > \pi R^2/\lambda > 1.356R \tag{7.1-3}$$

from the origin of the array, where R is the maximum radial extent of the array [see Eq. (6.2-39)]. The spherical coordinates (r, θ, ψ) are shown in Fig. 6.3-1. Therefore, in order to compute the far-field beam pattern of an array, we must first specify a functional form for its complex aperture function.

Even Number of Elements

Consider a linear array composed of an even number of elements lying along the X axis. If the elements are not identical, that is, if each element has a different complex frequency response, and if the elements are not equally spaced, as shown in Fig. 7.1-1, then the complex frequency response (complex aperture function) of this array can be expressed as

$$A(f, x_a) = \sum_{n=-N/2}^{-1} c_n(f)e_n(f, x_a - x_n) + \sum_{n=1}^{N/2} c_n(f)e_n(f, x_a - x_n) \tag{7.1-4}$$

or

$$A(f, x_a) = \sum_{n=1}^{N/2} [c_{-n}(f)e_{-n}(f, x_a - x_{-n}) + c_n(f)e_n(f, x_a - x_n)], \tag{7.1-5}$$

where N is the total even number of elements, x_n is the x coordinate of the center of element n, $c_n(f)$ is the frequency-dependent *complex weight* associated with element n, and $e_n(f, x_a)$ is the complex frequency response (complex aperture function) of element n, also known as the *element function*. Note that, in general, $x_{-n} \neq x_n$. As is illustrated in Fig. 7.1-1, each element is a single electroacoustic transducer L meters in length. Linear arrays of planar elements, such as rectangular and circular pistons, will be discussed later, in Section 7.4.1.

Equation (7.1-4) indicates that the complex frequency response of the array is equal to the *linear superposition* of the complex-weighted frequency responses of all the individual elements in the array. The complex weights are used to control the complex frequency response of the array and, thus, the array's far-field directivity function via amplitude and phase weighting, as will be discussed later. Substituting Eq. (7.1-5) into Eq. (7.1-1) yields the following expression for the

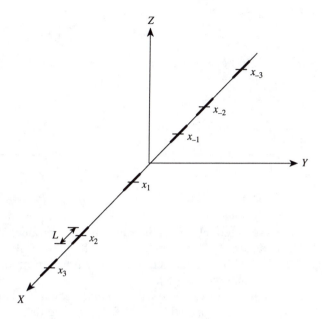

Figure 7.1-1. Linear array composed of an even number of unevenly spaced elements lying along the X axis. Each element is L meters in length.

far-field beam pattern:

$$D(f, f_X) = \sum_{n=1}^{N/2} \left[c_{-n}(f) E_{-n}(f, f_X) \exp(+j2\pi f_X x_{-n}) \right.$$

$$\left. + c_n(f) E_n(f, f_X) \exp(+j2\pi f_X x_n) \right], \qquad (7.1\text{-}6)$$

where

$$E_n(f, f_X) = F_{x_a}\{e_n(f, x_a)\} \qquad (7.1\text{-}7)$$

is the far-field beam pattern of element (electroacoustic transducer) n.

If all the elements in the array are *identical* (i.e., all the elements have the same complex frequency response), but still not equally spaced, then Eq. (7.1-6) reduces to

$$\boxed{D(f, f_X) = E(f, f_X) S(f, f_X),} \qquad (7.1\text{-}8)$$

where

$$E(f, f_X) = F_{x_a}\{e(f, x_a)\} \tag{7.1-9}$$

is the far-field beam pattern of one of the identical elements in the array, and for N even,

$$S(f, f_X) = \sum_{n=1}^{N/2} \left[c_{-n}(f) \exp(+j2\pi f_X x_{-n}) + c_n(f) \exp(+j2\pi f_X x_n) \right].$$

$$\tag{7.1-10}$$

Equation (7.1-8) is referred to as the *product theorem* for linear arrays. It states that the far-field directivity function $D(f, f_X)$ of a linear array of *identical*, complex-weighted elements is equal to the product of $E(f, f_X)$, which is the far-field directivity function of *one* of the identical elements in the array, and $S(f, f_X)$, which is the far-field directivity function of an equivalent linear array of identical, complex-weighted, omnidirectional point elements. Although the product theorem requires identical elements, equal spacing of the elements is not required. The interpretation of $S(f, f_X)$ will be discussed next.

Recall from Chapters 2 and 3 that an impulse function was used as a mathematical model for an omnidirectional point source. Similarly, the mathematical model that we shall use for the complex frequency response (complex aperture function) of an omnidirectional point element is given by

$$e(f, x_a) = \delta(x_a). \tag{7.1-11}$$

Substituting Eq. (7.1-11) into Eq. (7.1-9) yields

$$E(f, f_X) = F_{x_a}\{\delta(x_a)\} = 1, \tag{7.1-12}$$

which indicates that the far-field beam pattern of an omnidirectional point element is equal to a *constant*, independent of the spherical angles θ and ψ, as expected. The term "omnidirectional" means that an electroacoustic transducer transmits and/or receives equally well in all directions. Upon substituting Eq. (7.1-12) into Eq. (7.1-8), we obtain the following expression for the far-field directivity function of a linear array of an even number of identical, complex-weighted, omnidirectional point elements:

$$D(f, f_X) = S(f, f_X), \tag{7.1-13}$$

where $S(f, f_X)$ is given by Eq. (7.1-10).

Next, let us take the inverse spatial Fourier transform of the product theorem given by Eq. (7.1-8). Doing so yields

$$A(f, x_a) = e(f, x_a) \underset{x_a}{*} s(f, x_a), \tag{7.1-14}$$

where $\underset{x_a}{*}$ denotes convolution with respect to x_a, and since

$$F_{f_X}^{-1}\{\exp(+j2\pi f_X x_n)\} = \delta(x_a - x_n), \qquad (7.1\text{-}15)$$

we have

$$s(f, x_a) = F_{f_X}^{-1}\{S(f, f_X)\} = \sum_{n=1}^{N/2} [c_{-n}(f)\delta(x_a - x_{-n}) + c_n(f)\delta(x_a - x_n)]. \qquad (7.1\text{-}16)$$

Equation (7.1-16) is the complex frequency response (complex aperture function) of a linear array of N (even) identical, complex-weighted, omnidirectional point elements. It is interesting to note that Eq. (7.1-16) is in the form of the impulse response of a FIR (finite impulse response) filter. Therefore, a linear array of N (even) identical, complex-weighted elements can be thought of as a spatial FIR filter, and $s(f, x_a)$ can be thought of as the spatial impulse response of the array. For example, if the element function (input signal) is the impulse function given by Eq. (7.1-11), then the aperture function (output signal) given by Eq. (7.1-14) reduces to

$$A(f, x_a) = \delta(x_a) \underset{x_a}{*} s(f, x_a) = s(f, x_a), \qquad (7.1\text{-}17)$$

which is the spatial impulse response of the array.

Finally, if all the elements in the array are *equally spaced*, then

$$x_n = (n - 0.5)d, \qquad n = 1, 2, \ldots, N/2, \qquad (7.1\text{-}18a)$$

and

$$x_{-n} = -x_n, \qquad n = 1, 2, \ldots, N/2, \qquad (7.1\text{-}18b)$$

where d is the *interelement spacing* in meters.

Odd Number of Elements

Now consider a linear array composed of an odd number of elements lying along the X axis. If the elements are not identical, that is, if each element has a different complex frequency response, and if the elements are not equally spaced, as shown in Fig. 7.1-2, then by using the principle of linear superposition, the complex frequency response (complex aperture function) of this array can be expressed as

$$A(f, x_a) = \sum_{n=-N'}^{N'} c_n(f)e_n(f, x_a - x_n), \qquad (7.1\text{-}19)$$

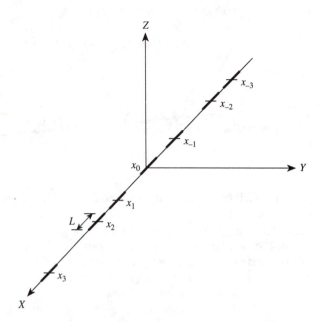

Figure 7.1-2. Linear array composed of an odd number of unevenly spaced elements lying along the X axis. Each element is L meters in length.

where

$$N' = \frac{N - 1}{2},\qquad(7.1\text{-}20)$$

N is the total odd number of elements, x_n is the x coordinate of the center of element n, and $x_0 = 0$, since one element is centered at the origin of the array. Note that, in general, $x_{-n} \neq x_n$. Substituting Eq. (7.1-19) into Eq. (7.1-1) yields the following expression for the far-field beam pattern:

$$D(f, f_X) = \sum_{n=-N'}^{N'} c_n(f) E_n(f, f_X) \exp(+j2\pi f_X x_n),\qquad(7.1\text{-}21)$$

where $E_n(f, f_X)$ is the far-field beam pattern of element (electroacoustic transducer) n [see Eq. (7.1-7)].

If all the elements in the array are *identical* (i.e., all the elements have the same complex frequency response), but still not equally spaced, then Eq. (7.1-21) reduces to the *product theorem* for linear arrays [see Eq. (7.1-8)]:

$$D(f, f_X) = E(f, f_X) S(f, f_X),\qquad(7.1\text{-}22)$$

where $E(f, f_X)$ is the far-field beam pattern of *one* of the identical elements in the array [see Eq. (7.1-9)], and for N odd, the far-field beam pattern of an equivalent linear array of identical, complex-weighted, omnidirectional point elements is given by

$$S(f, f_X) = \sum_{n=-N'}^{N'} c_n(f) \exp(+j2\pi f_X x_n), \qquad (7.1\text{-}23)$$

where N' is given by Eq. (7.1-20). As we stressed before, although the product theorem requires identical elements, equal spacing of the elements is not required.

Taking the inverse spatial Fourier transform of the product theorem given by Eq. (7.1-22) yields [see Eq. (7.1-14)]

$$A(f, x_a) = e(f, x_a) \underset{x_a}{*} s(f, x_a), \qquad (7.1\text{-}24)$$

where

$$s(f, x_a) = F_{f_X}^{-1}\{S(f, f_X)\} = \sum_{n=-N'}^{N'} c_n(f)\delta(x_a - x_n) \qquad (7.1\text{-}25)$$

is the complex frequency response (complex aperture function) of a linear array of N (odd) identical, complex-weighted, omnidirectional point elements. And since Eq. (7.1-25) is in the form of the impulse response of a FIR filter, such an array can be thought of as a spatial FIR filter, and $s(f, x_a)$ can be thought of as the spatial impulse response of the array.

Finally, if all the elements in the array are *equally spaced*, then

$$x_n = nd, \qquad n = -N',\ldots,0,\ldots, N', \qquad (7.1\text{-}26)$$

where d is the *interelement spacing* in meters.

At this point in our discussion, let us make the following two observations. First, one can simulate the case when one or more elements (electroacoustic transducers) in the array are broken, that is, not radiating or receiving, by setting $c_n(f) = 0$ for those elements. One can then investigate the effect this situation has on the far-field beam pattern of the array. Second, if the length L of the individual elements is small compared to a wavelength, then the far-field beam pattern of the array $D(f, f_X)$ is, in effect, dependent on $S(f, f_X)$ alone, since $E(f, f_X)$ is approximately equal to a constant, that is, the elements are omnidirectional for all practical purposes (see Example 7.1-1).

Example 7.1-1 (Omnidirectional elements)

Consider a linear array of identical elements where the complex frequency response of the elements $e(f, x_a)$ is modeled by the rectangular amplitude window (see Example 6.3-1), that is,

$$e(f, x_a) = \text{rect}(x_a/L), \tag{7.1-27}$$

with corresponding unnormalized far-field beam pattern

$$E(f, f_X) = L \, \text{sinc}(f_X L), \tag{7.1-28}$$

where L is the length of an individual element and

$$\text{sinc}(x) \triangleq \frac{\sin(\pi x)}{\pi x}. \tag{7.1-29}$$

Since

$$\sin \alpha \approx \alpha, \qquad \alpha \le 0.0873 \text{ rad}, \tag{7.1-30}$$

we have

$$\sin(\pi f_X L) \approx \pi f_X L \tag{7.1-31}$$

whenever

$$\pi f_X L = \pi u L / \lambda \le 0.0873, \tag{7.1-32}$$

or, setting direction cosine u equal to its maximum positive value of unity,

$$L/\lambda \le 0.0278. \tag{7.1-33}$$

Therefore, if Eq. (7.1-33) is satisfied, then Eq. (7.1-28) reduces to

$$E(f, f_X) = L \, \text{sinc}(f_X L) \approx L, \qquad L/\lambda \le 0.0278. \tag{7.1-34}$$

Since the far-field beam pattern given by Eq. (7.1-34) is equal to a constant, the element is omnidirectional, that is, its far-field beam pattern is independent of the spherical angles θ and ψ. In other words, the element radiates and/or receives equally well in all directions.

Let us next examine the complex weight $c_n(f)$. In general,

$$\boxed{c_n(f) = a_n(f) \exp[+j\theta_n(f)],} \tag{7.1-35}$$

where $a_n(f)$ and $\theta_n(f)$ are *real*, frequency-dependent amplitude and phase weights, respectively. Now, let us assume that the complex weight obeys the following property:

$$c_{-n}(f) = c_n^*(f) \tag{7.1-36}$$

where the asterisk denotes complex conjugate. Since

$$c_{-n}(f) = a_{-n}(f) \exp[+j\theta_{-n}(f)] \tag{7.1-37}$$

and

$$c_n^*(f) = a_n(f) \exp[-j\theta_n(f)], \tag{7.1-38}$$

if Eq. (7.1-36) is true, then

$$a_{-n}(f) = a_n(f) \tag{7.1-39}$$

and

$$\theta_{-n}(f) = -\theta_n(f), \tag{7.1-40}$$

that is, the amplitude weight is an *even* function of the index n, and the phase weight is an *odd* function of the index n. Since an odd function passes through the origin ($n = 0$ in this case),

$$\theta_0(f) = 0, \tag{7.1-41}$$

and as a result,

$$c_0(f) = a_0(f). \tag{7.1-42}$$

Recall from linear aperture theory that all of the continuous amplitude windows we discussed in Section 6.3.1 were even functions of x_a, and that the linear phase variation responsible for beam steering was an odd function of x_a (see Section 6.3.3). Amplitude and phase weighting will be discussed further in Sections 7.1.2 and 7.1.3, respectively.

The equations for the far-field beam pattern $S(f, f_X)$ for N even and N odd can be simplified if the elements are equally spaced and the complex weights obey Eq. (7.1-36). For example, if Eqs. (7.1-18a), (7.1-18b), and (7.1-36) are substituted into Eq. (7.1-10), then

$$S(f, f_X) = \sum_{n=1}^{N/2} \{[c_n(f) \exp[+j2\pi f_X(n - 0.5)d]]^*$$

$$+ [c_n(f) \exp[+j2\pi f_X(n - 0.5)d]]\}, \tag{7.1-43}$$

and since

$$Z^* + Z = 2\,\text{Re}\{Z\},\tag{7.1-44}$$

Eq. (7.1-43) reduces to

$$S(f, f_X) = 2\,\text{Re}\left\{\sum_{n=1}^{N/2} c_n(f)\exp\left[+j2\pi f_X(n-0.5)d\right]\right\},\tag{7.1-45}$$

or

$$\boxed{S(f, f_X) = 2\sum_{n=1}^{N/2} a_n(f)\cos\left[2\pi f_X(n-0.5)d + \theta_n(f)\right].}\tag{7.1-46}$$

Equation (7.1-46) is the far-field directivity function of a linear array of N (even) identical, equally spaced, complex-weighted, omnidirectional point elements lying along the X axis, where the complex weights obey the symmetry property given by Eq. (7.1-36).

Similarly, if Eqs. (7.1-26) and (7.1-36) are substituted into Eq. (7.1-23), then

$$S(f, f_X) = c_0(f) + \sum_{n=1}^{N'} \left\{\left[c_n(f)\exp(+j2\pi f_X nd)\right]^* + \left[c_n(f)\exp(+j2\pi f_X nd)\right]\right\}.$$

$$\tag{7.1-47}$$

With the use of Eqs. (7.1-42) and (7.1-44), Eq. (7.1-47) reduces to

$$S(f, f_X) = a_0(f) + 2\,\text{Re}\left\{\sum_{n=1}^{N'} c_n(f)\exp(+j2\pi f_X nd)\right\},\tag{7.1-48}$$

or

$$\boxed{S(f, f_X) = a_0(f) + 2\sum_{n=1}^{N'} a_n(f)\cos\left[2\pi f_X nd + \theta_n(f)\right],}\tag{7.1-49}$$

where N' is given by Eq. (7.1-20). Equation (7.1-49) is the far-field directivity function of a linear array of N (odd) identical, equally spaced, complex-weighted, omnidirectional point elements lying along the X axis, where the complex weights obey the symmetry property given by Eq. (7.1-36).

Finally, if a linear array lies along either the Y or Z axis instead of the X axis, then simply replace x_a and f_X either with y_a and f_Y or with z_a and f_Z, respectively, in the appropriate equations, where the spatial frequencies f_Y and f_Z are given by Eqs. (6.2-44) and (6.2-45), respectively.

7.1.2 Amplitude Weighting

Consider a linear array of N (odd) identical, equally spaced, *amplitude-weighted* elements lying along the X axis. From the product theorem, the far-field directivity function of this array is given by

$$D(f, f_X) = E(f, f_X)S(f, f_X),\qquad (7.1\text{-}50)$$

where

$$S(f, f_X) = \sum_{n=-N'}^{N'} a_n(f)\exp(+j2\pi f_X nd),\qquad (7.1\text{-}51)$$

since no phase weighting is being done, that is, since $\theta_n(f) = 0$, and N' is given by Eq. (7.1-20).

The most common set of amplitude weights are simply sampled values of the continuous amplitude windows already discussed in Section 6.3.1. For example, consider the following set of equations, where $n = -(N-1)/2,\ldots,0,\ldots,$ $(N-1)/2$ and N is odd:

(1) rectangular amplitude weight:

$$a_n(f) = 1,\qquad (7.1\text{-}52)$$

(2) triangular amplitude weight:

$$a_n(f) = 1 - \frac{|n|}{(N-1)/2},\qquad (7.1\text{-}53)$$

(3) cosine amplitude weight:

$$a_n(f) = \cos\!\left(\frac{\pi n}{N-1}\right),\qquad (7.1\text{-}54)$$

(4) Hanning amplitude weight:

$$a_n(f) = 0.5 + 0.5\cos\!\left(\frac{2\pi n}{N-1}\right),\qquad (7.1\text{-}55)$$

(5) Hamming amplitude weight:

$$a_n(f) = 0.54 + 0.46\cos\!\left(\frac{2\pi n}{N-1}\right),\qquad (7.1\text{-}56)$$

(6) Blackman amplitude weight:

$$a_n(f) = 0.42 + 0.5\cos\!\left(\frac{2\pi n}{N-1}\right) + 0.08\cos\!\left(\frac{4\pi n}{N-1}\right).\qquad (7.1\text{-}57)$$

Although the amplitude weight $a_n(f)$ is, in general, frequency dependent, the amplitude weights given by Eqs. (7.1-52) through (7.1-57) are not functions of frequency.

The process of amplitude weighting an array of elements is also known as amplitude shading or as applying an amplitude window to the array. Amplitude weighting can be implemented by connecting each element in the array in series with an electronic amplifier, whose gain is determined by the amplitude weight. However, the use of electronic amplifiers is really not necessary, since amplitude weights, as well as phase weights, can be implemented by using digital signal processing (digital beamforming), as will be discussed in Section 8.1. Note that the amplitude weights given by Eqs. (7.1-52) through (7.1-57) are also used in digital filter design and spectrum analysis problems as well.

Since the amplitude weights given by Eqs. (7.1-52) through (7.1-57) are even functions of the index n, Eq. (7.1-49) can be used to evaluate $S(f, f_X)$ instead of Eq. (7.1-51) by setting $\theta_n(f) = 0$ in Eq. (7.1-49). Note that Eq. (7.1-49) only requires $(N - 1)/2 + 1 = (N + 1)/2$ values of $a_n(f)$ in order to evaluate $S(f, f_X)$, whereas Eq. (7.1-51) requires all N values. However, as is discussed in Example 7.1-2, Eq. (7.1-51) can be used to obtain closed-form expressions for $S(f, f_X)$ for the amplitude weights given by Eq. (7.1-52) and Eqs. (7.1-54) through (7.1-57).

Example 7.1-2

In this example we shall evaluate Eq. (7.1-51) for the following set of amplitude weights:

$$a_n(f) = \cos\left(\frac{b\pi n}{N - 1}\right), \qquad n = -\frac{N - 1}{2}, \dots, 0, \dots, \frac{N - 1}{2},$$

$$(7.1\text{-}58)$$

where b is an arbitrary constant and N is odd. Note that the amplitude weights given by Eq. (7.1-52) and Eqs. (7.1-54) through (7.1-57) all involve one or more terms of the form of Eq. (7.1-58), where b takes on the values 0, 1, 2, and 4. Since Eq. (7.1-58) can be rewritten as

$$a_n(f) = 0.5 \exp\left(+j\frac{b\pi n}{N - 1}\right) + 0.5 \exp\left(-j\frac{b\pi n}{N - 1}\right), \quad (7.1\text{-}59)$$

and since

$$\sum_{k=-K}^{K} \exp(+jk\alpha) = \frac{\sin[(2K + 1)\alpha/2]}{\sin(\alpha/2)}, \qquad (7.1\text{-}60)$$

upon substituting Eq. (7.1-59) into Eq. (7.1-51), and with the use of

Eq. (7.1-60), we obtain

$$S(f, f_X) = \frac{\sin\left(\pi f_X Nd + \frac{b\pi N}{2(N-1)}\right)}{2\sin\left(\pi f_X d + \frac{b\pi}{2(N-1)}\right)} + \frac{\sin\left(\pi f_X Nd - \frac{b\pi N}{2(N-1)}\right)}{2\sin\left(\pi f_X d - \frac{b\pi}{2(N-1)}\right)}.$$

(7.1-61)

Since Eq. (7.1-58) is analogous to the continuous amplitude window discussed in Example 6.3-3, it is not surprising that Eq. (7.1-61) is analogous to Eq. (6.3-56).

Equation (7.1-61) is a very important result because it can be used to obtain the far-field directivity functions $S(f, f_X)$ of the rectangular, cosine, Hanning, Hamming, and Blackman amplitude weights. For example, if $b = 0$, then Eq. (7.1-58) reduces to

$$a_n(f) = \cos 0 = 1, \qquad n = -\frac{N-1}{2}, \ldots, 0, \ldots, \frac{N-1}{2}, \qquad (7.1\text{-}62)$$

which is the set of rectangular amplitude weights. And with $b = 0$, Eq. (7.1-61) reduces to

$$S(f, f_X) = \frac{\sin(\pi f_X Nd)}{\sin(\pi f_X d)}, \qquad (7.1\text{-}63\text{a})$$

which is the far-field directivity function of a linear array of N (odd) identical, equally spaced, omnidirectional point elements with *rectangular amplitude weighting*, lying along the X axis. Since the spatial frequency f_X is given by Eq. (7.1-2), Eq. (7.1-63a) can also be expressed as follows:

$$S(f, u) = \frac{\sin(\pi u Nd/\lambda)}{\sin(\pi u d/\lambda)} \qquad (7.1\text{-}63\text{b})$$

and

$$S(f, \theta, \psi) = \frac{\sin\left(\pi N \frac{d}{\lambda} \sin\theta \cos\psi\right)}{\sin\left(\pi \frac{d}{\lambda} \sin\theta \cos\psi\right)}. \qquad (7.1\text{-}63\text{c})$$

Finally, for completeness, note that

$$S(f, f_X) = \frac{2}{N-1} \frac{\sin^2[\pi f_X (N-1)d/2]}{\sin^2(\pi f_X d)} \qquad (7.1\text{-}64)$$

is the far-field directivity function of a linear array of N (odd) identical, equally spaced, omnidirectional point elements with *triangular amplitude weighting*, lying along the X axis. Equation (7.1-64) *cannot* be obtained directly from Eq. (7.1-61).

Example 7.1-3
Consider a linear array of N (odd) identical, equally spaced elements with *rectangular amplitude weighting*, lying along the X axis. If the complex frequency response (complex aperture function) of the elements is modeled by the *continuous rectangular amplitude window*, then from the product theorem, the unnormalized far-field directivity function of the array is given by

$$D(f, f_X) = L \operatorname{sinc}(f_X L) \frac{\sin(\pi f_X N d)}{\sin(\pi f_X d)}. \qquad (7.1\text{-}65)$$

If the array is composed of omnidirectional point elements with rectangular amplitude weighting, then

$$D(f, f_X) = \frac{\sin(\pi f_X N d)}{\sin(\pi f_X d)} \qquad (7.1\text{-}66)$$

since $E(f, f_X) = 1$.

Example 7.1-4
Consider a linear array of N (odd) identical, equally spaced elements with *triangular amplitude weighting*, lying along the X axis. If the complex frequency response (complex aperture function) of the elements is modeled by the *continuous rectangular amplitude window*, then from the product theorem, the unnormalized far-field directivity function of the array is given by

$$D(f, f_X) = L \operatorname{sinc}(f_X L) \frac{2}{N-1} \frac{\sin^2[\pi f_X (N-1)d/2]}{\sin^2(\pi f_X d)}. \qquad (7.1\text{-}67)$$

If the array is composed of omnidirectional point elements with triangular amplitude weighting, then

$$D(f, f_X) = \frac{2}{N-1} \frac{\sin^2[\pi f_X(N-1)d/2]}{\sin^2(\pi f_X d)} \qquad (7.1\text{-}68)$$

since $E(f, f_X) = 1$.

If closed-form expressions are available for $E(f, f_X)$ and $S(f, f_X)$, then a closed-form expression can be obtained for the unnormalized far-field directivity function $D(f, f_X) = E(f, f_X)S(f, f_X)$ of a linear array. As a result, the dimensionless 3-dB beamwidth Δu in direction-cosine space can be obtained for the normalized far-field beam patterns $E_N(f, f_X)$ and $S_N(f, f_X)$ separately, or for the normalized far-field beam pattern of the array $D_N(f, f_X) = E_N(f, f_X)S_N(f, f_X)$ using the procedure outlined in Example 6.3-8. Once Δu has been determined, the 3-dB beamwidths $\Delta \theta$ and $\Delta \psi$, in degrees, of the normalized vertical and horizontal far-field beam patterns, respectively, can be obtained using the results in Examples 6.3-6 and 6.3-7, respectively (see Problems 7-1 and 7-2).

Dolph-Chebyshev Method of Amplitude Weighting

We know from our discussion of linear apertures in Section 6.3.1 that different amplitude windows yield far-field beam patterns with different sidelobe levels and beamwidths, and as sidelobe levels decrease, beamwidth generally increases. This principle also applies to the far-field directivity functions of amplitude-weighted arrays. However, the Dolph-Chebyshev method of amplitude weighting makes it possible to *optimize* the far-field beam patterns of arrays so that for any specified sidelobe level relative to the level of the main lobe, the *narrowest* possible main-lobe beamwidth is achieved, or for any specified main-lobe beamwidth, the *lowest* possible sidelobe level is achieved. It is important to note that sidelobe level and main-lobe beamwidth are subject to the same physical limitations that apply to all amplitude weights, such as the number and spacing of the elements, the operating frequency, and the directional properties of the elements themselves as characterized by the far-field beam pattern $E(f, f_X)$. One disadvantage of the Dolph-Chebyshev method is that all of the sidelobes are at the same level—they do not fall off in value. The Dolph-Chebyshev method is demonstrated in Example 7.1-5.

Example 7.1-5 (The Dolph-Chebyshev method)
Suppose we are given a linear array of $N = 7$ identical, equally spaced elements lying along the X axis and it is desired to amplitude weight this array using the Dolph-Chebyshev method. Although we are using an odd number of elements in this example, it is important to note that this method applies to even N as well. The far-field beam pattern of

this array is given by

$$D(f, f_X) = E(f, f_X)S(f, f_X), \qquad (7.1\text{-}69)$$

and since the amplitude weights produced by the Dolph-Chebyshev method are even functions of the index n [see Eq. (7.1-49)],

$$S(f, f_X) = a_0(f) + 2 \sum_{n=1}^{3} a_n(f) \cos(2\pi f_X n d), \qquad (7.1\text{-}70)$$

since $N = 7$ and $\theta_n(f) = 0$. Note that if N were even, then we would use Eq. (7.1-46) to start the analysis instead of Eq. (7.1-49). If we let

$$\alpha = \pi f_X d, \qquad (7.1\text{-}71)$$

then Eq. (7.1-70) can be rewritten as

$$S(f, f_X) = a_0(f) + 2a_1(f) \cos(2\alpha) + 2a_2(f) \cos(4\alpha)$$
$$+ 2a_3(f) \cos(6\alpha). \qquad (7.1\text{-}72)$$

Since the Dolph-Chebyshev method depends on certain properties of *Chebyshev polynomials*, we must digress for a moment to discuss these properties before continuing further.

The nth-degree Chebyshev polynomial is defined as follows:

$$T_n(x) = \begin{cases} \cos(n \cos^{-1} x), & |x| \leq 1, \\ \cosh(n \cosh^{-1} x), & |x| > 1. \end{cases} \qquad (7.1\text{-}73)$$

If we also let [see Eq. (7.1-71)]

$$\alpha = \cos^{-1} x \qquad (7.1\text{-}74)$$

so that

$$x = \cos \alpha, \qquad (7.1\text{-}75)$$

then

$$T_n(x) = \cos(n\alpha), \qquad |x| \leq 1. \qquad (7.1\text{-}76)$$

By comparing Eq. (7.1-76) with Eq. (7.1-72), it can be seen that the far-field beam pattern $S(f, f_X)$ can be expressed as a sum of Chebyshev polynomials and that the sum of these polynomials is itself a polynomial. The degree of this polynomial, which is called the *array polyno-*

mial, is $n = N - 1$, or one less than the total number of elements in the array.

All Chebyshev polynomials possess the following important properties:

(1) For $n \geq 2$,

$$T_n(x) = 2xT_{n-1}(x) - T_{n-2}(x),\qquad(7.1\text{-}77)$$

where

$$T_0(x) = 1 \qquad(7.1\text{-}78)$$

and

$$T_1(x) = x. \qquad(7.1\text{-}79)$$

(2) $T_n(x)$ has n real roots in the interval $-1 < x < 1$.
(3) $T_n(x)$ has unit-magnitude maxima and minima that occur alternately at

$$x_k = \cos(k\pi/n),\qquad k = 1, 2, \ldots, n - 1, \qquad(7.1\text{-}80)$$

where $-1 < x_k < 1$ and $|T_n(x_k)| = 1$. Therefore, Chebyshev polynomials have an "equal ripple" characteristic in the interval $-1 < x < 1$.
(4) At $x = \pm 1$, $|T_n(\pm 1)| = 1$. However, $|T_n(x)| > 1$ for $|x| > 1$, and

$$\left| \frac{d}{dx} T_n(x) \right|_{x = \pm 1} = n^2. \qquad(7.1\text{-}81)$$

Chebyshev polynomials possess additional properties, but they are not pertinent to the present problem.

In the Dolph-Chebyshev method, the magnitude of the level of the main lobe corresponds to the value of $T_n(x_0)$ where $x_0 > 1$, and the magnitude of the level of the sidelobes is numerically equal to the absolute value of the interior maxima and minima, that is, unity. Note that it is the "equal ripple" characteristic of Chebyshev polynomials, as discussed in property (3), that is responsible for generating sidelobes all at the same level in the Dolph-Chebyshev method of amplitude weighting. Therefore, the ratio

$$\frac{\text{main-lobe level}}{\text{sidelobe level}} = T_n(x_0) = r > 1, \qquad(7.1\text{-}82)$$

which can be made as large as desired, is used to determine the value of x_0, that is, we solve for x_0 in $T_n(x_0) = r$. By referring back to Eq. (7.1-73), we can write that

$$T_n(x_0) = \cosh(n \cosh^{-1} x_0) = r, \qquad |x_0| > 1, \qquad(7.1\text{-}83)$$

and as a result,

$$\boxed{x_0 = \cosh\left(\frac{1}{n} \cosh^{-1} r\right), \qquad |x_0| > 1.} \qquad (7.1\text{-}84)$$

Returning now to our example, since we have $N = 7$ elements, the degree n of the array polynomial is

$$n = N - 1 = 6. \qquad (7.1\text{-}85)$$

Also, suppose we want the main-lobe level to be 30 dB above the sidelobe level, that is,

$$20 \log_{10} r = 30 \text{ dB}, \qquad (7.1\text{-}86)$$

or

$$r = 31.623. \qquad (7.1\text{-}87)$$

Therefore, upon substituting Eqs. (7.1-85) and (7.1-87) into Eq. (7.1-84), we obtain

$$x_0 = 1.248. \qquad (7.1\text{-}88)$$

Since the array polynomial given by Eq. (7.1-72) is a sixth-degree polynomial, we also need an expression for the sixth-degree Chebyshev polynomial $T_6(x)$. From the recurrence relationship given by Eq. (7.1-77), we obtain the following equation for $T_6(x)$ (see Fig. 7.1-3):

$$T_6(x) = 32x^6 - 48x^4 + 18x^2 - 1. \qquad (7.1\text{-}89)$$

However, since $|x| = |\cos \alpha| \leq 1$ [see Eq. (7.1-75)], and since we need to evaluate $T_6(x)$ at $x = x_0 = 1.248 > 1$, we need to *scale* x by x_0, that is, replace x with $x_0 x$ in Eq. (7.1-89), so that

$$T_6(x_0 x) = 32x_0^6 x^6 - 48x_0^4 x^4 + 18x_0^2 x^2 - 1, \qquad |x| \leq 1. \quad (7.1\text{-}90)$$

Therefore, when x is equal to its maximum positive value of unity, $T_6(x_0 x)$ will be equal to $T_6(x_0)$.

The next step is to express the array polynomial given by Eq. (7.1-72) in terms of x. Since

$$\cos(2\alpha) = 2\cos^2 \alpha - 1, \qquad (7.1\text{-}91)$$

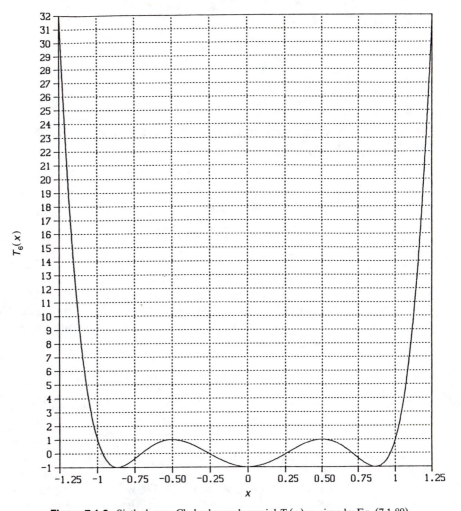

Figure 7.1-3. Sixth-degree Chebyshev polynomial $T_6(x)$ as given by Eq. (7.1-89).

substituting Eq. (7.1-75) into Eq. (7.1-91) yields

$$\cos(2\alpha) = 2x^2 - 1. \qquad (7.1\text{-}92)$$

Similarly,

$$\cos(4\alpha) = 8\cos^4 \alpha - 8\cos^2 \alpha + 1, \qquad (7.1\text{-}93)$$

or

$$\cos(4\alpha) = 8x^4 - 8x^2 + 1; \qquad (7.1\text{-}94)$$

and

$$\cos(6\alpha) = 32\cos^6\alpha - 48\cos^4\alpha + 18\cos^2\alpha - 1, \quad (7.1\text{-}95)$$

or

$$\cos(6\alpha) = 32x^6 - 48x^4 + 18x^2 - 1. \quad (7.1\text{-}96)$$

Therefore, substituting Eqs. (7.1-92), (7.1-94), and (7.1-96) into Eq. (7.1-72) yields

$$\begin{aligned} S(f, f_X) = {}& 64a_3(f)x^6 + 16[a_2(f) - 6a_3(f)]x^4 \\ & + 4[a_1(f) - 4a_2(f) + 9a_3(f)]x^2 \\ & + a_0(f) - 2[a_1(f) - a_2(f) + a_3(f)], \quad |x| \le 1. \end{aligned}$$

$$(7.1\text{-}97)$$

We are now in a position to calculate the amplitude weights.

By comparing Eqs. (7.1-90) and (7.1-97), and equating coefficients of like powers of x, we obtain the following set of four equations in four unknowns:

$$64a_3(f) = 32x_0^6, \quad (7.1\text{-}98)$$

$$16[a_2(f) - 6a_3(f)] = -48x_0^4, \quad (7.1\text{-}99)$$

$$4[a_1(f) - 4a_2(f) + 9a_3(f)] = 18x_0^2, \quad (7.1\text{-}100)$$

$$a_0(f) - 2[a_1(f) - a_2(f) + a_3(f)] = -1. \quad (7.1\text{-}101)$$

By equating Eqs. (7.1-90) and (7.1-97), we are forcing the array polynomial given by Eq. (7.1-97) to have the same characteristics as the sixth-degree Chebyshev polynomial given by Eq. (7.1-90). Solving Eqs. (7.1-98) through (7.1-101) for the amplitude weights yields:

$$a_0(f) = 2x_0^2(5x_0^4 - 9x_0^2 + 4.5) - 1, \quad (7.1\text{-}102)$$

$$a_1(f) = x_0^2(7.5x_0^4 - 12x_0^2 + 4.5), \quad (7.1\text{-}103)$$

$$a_2(f) = 3x_0^4(x_0^2 - 1), \quad (7.1\text{-}104)$$

$$a_3(f) = 0.5x_0^6. \quad (7.1\text{-}105)$$

Substituting Eq. (7.1-88) into Eqs. (7.1-102) through (7.1-105) yields $a_0(f) = 7.135$, $a_1(f) = 6.236$, $a_2(f) = 4.057$, and $a_3(f) = 1.889$; and by dividing each weight by $a_0(f) = 7.135$, we obtain the following *normalized* values:

$$a_0(f) = 1, \tag{7.1-106}$$

$$a_{-1}(f) = a_1(f) = 0.874, \tag{7.1-107}$$

$$a_{-2}(f) = a_2(f) = 0.569, \tag{7.1-108}$$

and

$$a_{-3}(f) = a_3(f) = 0.265, \tag{7.1-109}$$

since the amplitude weights are even functions of the index n. Substituting the amplitude weights given by Eqs. (7.1-106) through (7.1-109) into Eq. (7.1-70) will produce a far-field beam pattern $S(f, f_X)$ that will have sidelobes 30 dB below the main lobe.

Now that the normalized amplitude weights are known, let us derive an expression for the 3-dB beamwidth of the vertical far-field beam pattern of $S(f, f_X)$. Since Eqs. (7.1-71) and (7.1-74) must be equal,

$$\pi f_X d = \pi u d / \lambda = \cos^{-1} x, \tag{7.1-110}$$

or

$$u = \sin \theta \cos \psi = \frac{\lambda}{\pi d} \cos^{-1} x. \tag{7.1-111}$$

Also, since the linear array lies along the X axis, the vertical beam pattern lies in the XZ plane (see Example 6.3-6). In the XZ plane, $u = \sin \theta$ when $\psi = 0$. If we let

$$u_+ = \sin \theta_+ > 0, \qquad \psi_+ = 0, \tag{7.1-112}$$

then from Eq. (7.1-111),

$$u_+ = \sin \theta_+ = \frac{\lambda}{\pi d} \cos^{-1} x_+, \tag{7.1-113}$$

where x_+ is the value of x that corresponds to the 3-dB-down point θ_+ and is given by the following expression:

$$\boxed{x_+ = \frac{1}{x_0} \cosh\left[\frac{1}{n} \cosh^{-1}\left(\frac{\sqrt{2}}{2}r\right)\right],} \tag{7.1-114}$$

where x_0 is given by Eq. (7.1-84). Equation (7.1-114) was obtained by solving $T_n(x_0 x_+) = \cosh[n \cosh^{-1}(x_0 x_+)] = \sqrt{2}r/2$ for x_+ [see Eq.

(7.1-73) for $|x| > 1$]. Therefore, since $\Delta\theta = 2\theta_+$ (see Example 6.3-6),

$$\boxed{\Delta\theta = 2\sin^{-1}\left(\frac{\lambda}{\pi d}\cos^{-1}x_+\right)} \qquad (7.1\text{-}115)$$

is the 3-dB beamwidth in degrees of the vertical far-field beam pattern of $S(f, f_X)$ in the XZ plane.

Continuing with our example, if we substitute Eqs. (7.1-85), (7.1-87), and (7.1-88) into Eq. (7.1-114), then

$$x_+ = 0.967, \qquad (7.1\text{-}116)$$

and upon substituting Eq. (7.1-116) into Eq. (7.1-115), we obtain

$$\Delta\theta = 2\sin^{-1}\left(0.258\frac{\lambda}{\pi d}\right). \qquad (7.1\text{-}117)$$

In order to evaluate Eq. (7.1-117), let the interelement spacing $d = \lambda/2$. Later, in Section 7.1.4, we shall show that $d < \lambda/2$ is a necessary condition for avoiding grating lobes under all conditions of beam steering. Therefore, substituting $d = \lambda/2$ into Eq. (7.1-117) yields

$$\Delta\theta = 18.9° \qquad (7.1\text{-}118)$$

as the 3-dB beamwidth of the vertical far-field beam pattern of $S(f, f_X)$, whose main-lobe level is 30 dB above the sidelobe level.

Let us finish our discussion of amplitude weighting by noting that if the amplitude weights are even functions of the index n, then the far-field beam pattern $S(f, f_X) = S(f, u/\lambda) \to S(f, u)$ will be a real and even function of direction cosine u, and if the amplitude weights are odd functions of the index n, then $S(f, u)$ will be an imaginary and odd function of u.

7.1.3 The Phased Array (Beam Steering)

Consider a linear array of N (odd) identical, equally spaced, complex-weighted, omnidirectional point elements lying along the X axis. From the product theorem, the far-field directivity function of this array is given by

$$D(f, f_X) = S(f, f_X) = \sum_{n=-N'}^{N'} c_n(f)\exp(+j2\pi f_X nd), \qquad (7.1\text{-}119)$$

since $E(f, f_X) = 1$ for an omnidirectional point element (electroacoustic transducer), N' is given by Eq. (7.1-20), and

$$c_n(f) = a_n(f) \exp[+j\theta_n(f)]. \tag{7.1-120}$$

Next, let $D(f, f_X)$ be the far-field beam pattern of the array when it is only amplitude weighted, and let $D'(f, f_X)$ be the far-field beam pattern of the array when it is complex weighted, that is,

$$D(f, f_X) = \sum_{n=-N'}^{N'} a_n(f) \exp(+j2\pi f_X nd) \tag{7.1-121}$$

and

$$D'(f, f_X) = \sum_{n=-N'}^{N'} a_n(f) \exp\{+j[2\pi f_X nd + \theta_n(f)]\}. \tag{7.1-122}$$

If the phase weights are given by [compare with Eqs. (6.3-114) through (6.3-117)]

$$\boxed{\theta_n(f) = -2\pi f_X' nd, \qquad n = -N', \ldots, 0, \ldots, N',} \tag{7.1-123}$$

where

$$\boxed{f_X' = \frac{u'}{\lambda} = \frac{\sin\theta'\cos\psi'}{\lambda},} \tag{7.1-124}$$

then substituting Eq. (7.1-123) into Eq. (7.1-122) yields

$$D'(f, f_X) = \sum_{n=-N'}^{N'} a_n(f) \exp[+j2\pi(f_X - f_X')nd], \tag{7.1-125}$$

or

$$D'(f, f_X) = D(f, f_X - f_X'). \tag{7.1-126}$$

Equation (7.1-126) can also be expressed as

$$D'(f, u) = D(f, u - u'), \tag{7.1-127}$$

since $D(f, f_X) = D(f, u/\lambda) \rightarrow D(f, u)$. Note that $\theta_n(f)$ given by Eq. (7.1-123) is a function of frequency f (wavelength λ) and is a *linear* phase variation (linear in the index n). Also note that $\theta_n(f)$ is an *odd* function of n [see Eqs. (7.1-40) and (7.1-41)]. Therefore, a linear phase variation applied across the length of the array will cause the beam pattern $D(f, u)$ to be steered in the direction $u = u'$ in direction-cosine space, which is equivalent to steering (tilting) the beam pattern to

$\theta = \theta'$ and $\psi = \psi'$ [see Eq. (7.1-124)]. Equation (7.1-127) indicates that the far-field directivity function $D'(f, u)$ is simply a translated, or shifted, version of $D(f, u)$. Recall from linear aperture theory that when a beam pattern $D(f, u)$ is steered, its shape and beamwidth remain unchanged when plotted as a function of direction cosine u. However, they do change when the beam pattern is plotted as a function of the spherical angles θ and ψ (see Section 6.3.3).

Finally, note that a phase shift $\theta_n(f)$ in radians is equivalent to a *time delay* τ_n in seconds, that is,

$$\theta_n(f) = 2\pi f \tau_n, \tag{7.1-128}$$

or

$$\tau_n = \frac{\theta_n(f)}{2\pi f}. \tag{7.1-129}$$

Substituting Eqs. (7.1-123) and (7.1-124) into Eq. (7.1-129) yields

$$\boxed{\tau_n = -u'nd/c, \qquad n = -N', \ldots, 0, \ldots, N',} \tag{7.1-130}$$

since $c = f\lambda$. According to Eq. (7.1-130), half of the time delays will be positive and half will be negative. In order to ensure that all of the time delays will be positive for practical applications, the constant $|u'|N'd/c$ should be added to all values of τ_n computed from Eq. (7.1-130) so that $\tau_{N'}$ equals zero when u' is positive, and $\tau_{-N'}$ equals zero when u' is negative. However, the use of time-delay circuits to do beam steering is really not necessary, since beam steering can be accomplished via phase weighting, which can be implemented by using digital signal processing (digital beamforming) (see Section 8.1). Figure 7.1-4 illustrates the equivalent operations of time delay and phase weighting. Since the phase weight $\theta_n(f)$ is a function of frequency f, an appropriate phase weight must be applied to *each* frequency component contained in the complex frequency spectrum $X_n(f)$ shown in Fig. 7.1-4b in order to produce the correct time delay τ_n. This must be done at each element in the array.

7.1.4 Far-Field Beam Patterns and the Discrete Fourier Transform

The far-field directivity function (beam pattern) of a linear array of N (odd) identical, equally spaced, complex-weighted, omnidirectional point elements lying along the X axis is given by [see Eq.(7.1-119)]

$$D(f, f_X) = \sum_{n=-N'}^{N'} c_n(f) \exp(+j2\pi f_X nd), \tag{7.1-131}$$

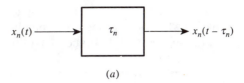

$$(a)$$

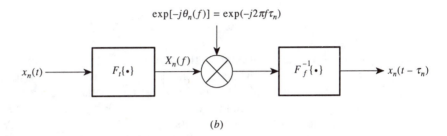

$$(b)$$

Figure 7.1-4. Illustration of the equivalent operations of (a) time delay and (b) phase weighting, where $F_t\{\cdot\}$ represents the forward Fourier transform with respect to time t and $F_f^{-1}\{\cdot\}$ represents the inverse Fourier transform with respect to frequency f.

where [see Eq. (7.1-20)]

$$N' = \frac{N-1}{2} \tag{7.1-132}$$

and

$$c_n(f) = a_n(f) \exp\left[+j\theta_n(f) \right]. \tag{7.1-133}$$

According to Eq. (7.1-131), $D(f, f_X)$ is evaluated at continuous values of the spatial frequency f_X. A more efficient way to compute $D(f, f_X)$ is to evaluate it at the following set of *discrete* values of f_X:

$$f_X = m\,\Delta f_X, \qquad m = -N',\ldots,0,\ldots,N', \tag{7.1-134}$$

where

$$\Delta f_X = \frac{1}{Nd} \tag{7.1-135}$$

is the *maximum* allowable spatial-frequency spacing. The reason for the particular range of values shown for the index m in Eq. (7.1-134) shall be explained shortly.

The index m is usually referred to as the *bin number*. Equation (7.1-135) is the spatial version of the maximum frequency spacing (or bin spacing) allowed in order to avoid aliasing when performing time-domain discrete Fourier transforms (DFTs). Upon substituting Eqs. (7.1-134) and (7.1-135) into Eq. (7.1-131), we obtain

$$D(f,m) = \sum_{n=-N'}^{N'} c_n(f)W_N^{mn}, \qquad m = -N', \ldots, 0, \ldots, N', \qquad (7.1\text{-}136)$$

where

$$W_N = \exp(+j2\pi/N). \qquad (7.1\text{-}137)$$

Note that Δf_X is suppressed in the argument of $D(f,m)$, that is, $D(f, m \, \Delta f_X) \rightarrow D(f,m)$. Therefore, evaluating the beam pattern $D(f,m)$ at bin number m corresponds to evaluating the beam pattern at spatial frequency $f_X = m \, \Delta f_X$. The summation on the right-hand side of Eq. (7.1-136) is the *discrete Fourier transform* (DFT) of the set of complex weights $c_n(f)$, $n = -N', \ldots, 0, \ldots, N'$. One can then use a *fast-Fourier-transform* (FFT) computer algorithm to compute the DFT, and hence the far-field beam pattern $D(f,m)$. Also note that the far-field beam pattern given by Eq. (7.1-136) is *periodic* with *period N*, that is,

$$D(f, m + N) = D(f, m). \qquad (7.1\text{-}138)$$

The periodicity of the DFT is the reason why the range of values for the index m need only cover one period, as shown in Eqs. (7.1-134) and (7.1-136). The reason that negative as well as positive values for m are used is to cover the entire visible region, as will be discussed later.

Most commonly available FFT algorithms assume that the total number of data points to be processed is equal to an integer power of two, that is,

$$N = 2^b, \qquad (7.1\text{-}139)$$

where b is a positive integer. For example, if $b = 4$, then $N = 16$. If N does not satisfy Eq. (7.1-139), then simply add enough zeros Z to the original data sequence until the total number of data points $N + Z$ is equal to an integer power of two. This is known as "zero padding" or "padding with zeros." Therefore, Eq. (7.1-136) can be rewritten as

$$D(f,m) = \sum_{n=-N'}^{N'} c_n(f)W_{N+Z}^{mn}, \qquad m = -N'', \ldots, 0, \ldots, N'', \qquad (7.1\text{-}140)$$

where

$$\Delta f_X = \frac{1}{(N + Z)d} \le \frac{1}{Nd}, \tag{7.1-141}$$

$$c_n(f) = 0, \qquad |n| > N', \tag{7.1-142}$$

$$W_{N+Z} = \exp\left(+j\frac{2\pi}{N + Z}\right), \tag{7.1-143}$$

and

$$N'' = \begin{cases} (N + Z)/2, & N + Z \text{ even}, & (7.1\text{-}144) \\ (N + Z - 1)/2, & N + Z \text{ odd}. & (7.1\text{-}145) \end{cases}$$

Since N is odd, Z must also be odd in order for $N + Z$ to be equal to an integer power of two, which is an even number. However, even if $N + Z$ does not equal an integer power of two, Eq. (7.1-140) can still be used to compute the far-field beam pattern $D(f, m)$. For example, if we want to evaluate $D(f, m)$ at twice as many points by reducing the bin spacing by a factor of two, then we set $Z = N$ so that $N + Z = 2N$, which is an even number, but is *not* an integer power of two, since N is odd in Eq. (7.1-140). And if $N + Z$ is odd, then N'' is given by Eq. (7.1-145).

Note that for the same value of the spatial frequency $f_X = m \Delta f_X$, the value of $D(f, m \Delta f_X) \to D(f, m)$ obtained from Eq. (7.1-140), where Δf_X is given by Eq. (7.1-141), is equal to the value of $D(f, m \Delta f_X) \to D(f, m)$ obtained from Eq. (7.1-136), where Δf_X is given by Eq. (7.1-135). Also note that the far-field beam pattern given by Eq. (7.1-140) is *periodic* with *period* $N + Z$, that is,

$$D(f, m + N + Z) = D(f, m). \tag{7.1-146}$$

In order to relate a bin number to values of the spherical angles θ and ψ, we first need to derive a relationship between the bin number m and the corresponding value of the direction cosine u_m. By referring to Eq. (7.1-134), we can write that

$$f_X = \frac{u}{\lambda} = m \Delta f_X, \tag{7.1-147}$$

or

$$u = u_m = m\lambda \Delta f_X. \tag{7.1-148}$$

Upon substituting Eq. (7.1-141) into Eq. (7.1-148), we obtain

$$
\boxed{u_m = \frac{m\lambda}{(N+Z)d}, \qquad m = -N'', \ldots, 0, \ldots, N'',}
\qquad (7.1\text{-}149)
$$

where, in general, $u_m = \sin\theta_m \cos\psi_m$. However, if we restrict ourselves to the vertical beam pattern, then $u_m = \pm\sin\theta_m$, since $\psi_m = 0$ or $\psi_m = \pi$. Similarly, if we restrict ourselves to the horizontal beam pattern, then $u_m = \cos\psi_m$, since $\theta_m = \pi/2$. Also, since the visible region corresponds to $-1 \le u_m \le 1$, and since u_m will take on negative values only for negative values of m, it is important to evaluate the far-field beam patterns given by Eqs. (7.1-136) and (7.1-140) at negative and positive bin numbers m.

The *normalized* far-field beam pattern of a linear array lying along the X axis is defined as follows [see Eq. (6.3-19)]:

$$
D_N(f, f_X) \triangleq \frac{D(f, f_X)}{D_{\max}},
\qquad (7.1\text{-}150)
$$

where $D_{\max}$ is the *normalization factor*, that is, it is the *maximum* value of the unnormalized far-field directivity function. The normalization factor for the unnormalized far-field beam patterns given by Eqs. (7.1-136) and (7.1-140) is

$$
D_{\max} = \sum_{n=-N'}^{N'} a_n(f),
\qquad (7.1\text{-}151)
$$

and as a result,

$$
D_N(f, m) = \frac{D(f, m)}{D_{\max}}.
\qquad (7.1\text{-}152)
$$

Figure 7.1-5 is a plot of the magnitude of the normalized far-field beam patterns of the rectangular, triangular, and cosine amplitude weights given by Eqs. (7.1-52) through (7.1-54). Figure 7.1-6 is a plot of the magnitude of the normalized far-field beam patterns of the Hanning, Hamming, and Blackman amplitude weights given by Eqs. (7.1-55) through (7.1-57). Figures 7.1-5 and 7.1-6 were obtained by using Eqs. (7.1-140), (7.1-151), and (7.1-152) with $N = 63$ and $Z = 193$, so that $N + Z = 256$. The plots are only shown for positive values of m, since the magnitudes of the far-field beam patterns are symmetric about $m = 0$.

Grating Lobes

Since the far-field beam pattern of a linear array of identical, equally spaced, complex-weighted, omnidirectional point elements is periodic, there is a potential

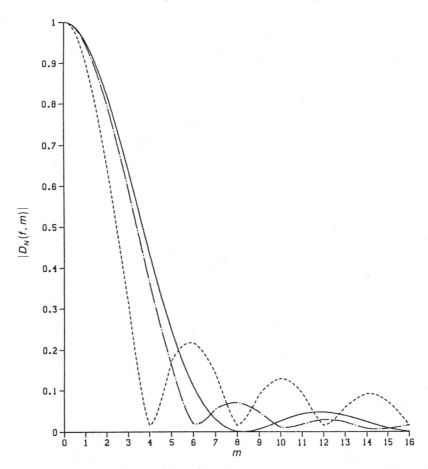

Figure 7.1-5. Magnitude of the normalized far-field beam patterns of the rectangular amplitude weights (dashed curve), triangular amplitude weights (solid curve), and cosine amplitude weights (dot-dash curve) plotted as a function of bin number m.

problem with *grating lobes*. Grating lobes are extraneous, unwanted main lobes that exist within the visible region. In order to guarantee only one main lobe within the visible region, steered in the desired direction, the interelement spacing d must be chosen properly. The criterion that d must satisfy can be derived by considering the far-field beam pattern of a linear array of N (odd) identical, equally spaced, complex-weighted (with linear phase weighting), omnidirectional point elements lying along the X axis given by [see Eq. (7.1-125)]

$$D(f, f_X) = \sum_{n=-N'}^{N'} a_n(f) \exp\left[+j2\pi(f_X - f_X')nd\right]. \qquad (7.1\text{-}153)$$

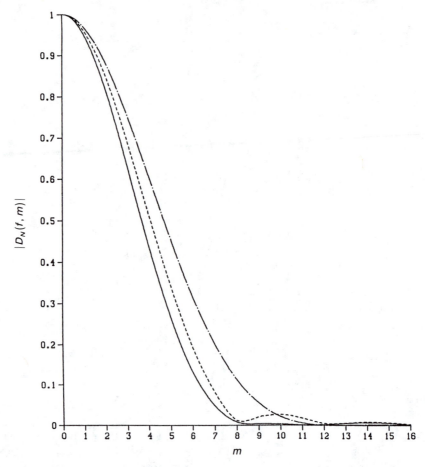

Figure 7.1-6. Magnitude of the normalized far-field beam patterns of the Hanning amplitude weights (dashed curve), Hamming amplitude weights (solid curve), and Blackman amplitude weights (dot-dash curve) plotted as a function of bin number m.

The maximum value of Eq. (7.1-153) occurs at $f_X = f_X'$ and is given by [see Eq. (7.1-151)]

$$D(f, f_X') = D_{max} = \sum_{n=-N'}^{N'} a_n(f). \qquad (7.1\text{-}154)$$

However, the maximum value of D_{max} also occurs at

$$f_X = f_X' + \frac{m}{d}, \qquad m = \pm 1, \pm 2, \pm 3, \ldots, \qquad (7.1\text{-}155)$$

since substituting Eq. (7.1-155) into Eq. (7.1-153) yields

$$D\left(f, f'_X + \frac{m}{d}\right) = \sum_{n=-N'}^{N'} a_n(f) \exp(+j2\pi mn) = \sum_{n=-N'}^{N'} a_n(f),$$

$$m = \pm 1, \pm 2, \pm 3, \ldots, \quad (7.1\text{-}156)$$

or

$$D\left(f, f'_X + \frac{m}{d}\right) = D_{\max}, \quad m = \pm 1, \pm 2, \pm 3, \ldots . \quad (7.1\text{-}157)$$

Therefore, if we express Eq. (7.1-155) in terms of direction cosines u and u', then we can state that grating lobes will exist at $u = u_g$, where

$$\boxed{u_g = u' + m\frac{\lambda}{d}, \quad m = \pm 1, \pm 2, \pm 3, \ldots,} \quad (7.1\text{-}158)$$

if $|u_g| \leq 1$, that is, if u_g is in the *visible region*.

If grating lobes can be avoided when a beam pattern is steered to end fire ($u' = \pm 1$), then they can be avoided for all possible directions of beam steering. If a beam pattern is steered to $u' = 1$, then from Eq. (7.1-158), the location of the first possible grating lobe within the visible region occurs when $m = -1$ and is given by

$$u_g = 1 - \frac{\lambda}{d}, \quad m = -1. \quad (7.1\text{-}159)$$

Now, if $d < \lambda/2$, then $|u_g| > 1$, which means that no grating lobes will exist in the visible region, even if we steer a beam pattern to end fire. However, since an array must be able to operate over a range of frequencies in general, if

$$\boxed{d < \lambda_{\min}/2,} \quad (7.1\text{-}160)$$

where $\lambda_{\min}$ is the minimum wavelength associated with the maximum frequency component $f_{\max} = c/\lambda_{\min}$, then *grating lobes will be avoided for all possible directions of beam steering*.

Figures 7.1-7 and 7.1-8 are polar plots of the magnitude of the normalized horizontal far-field beam patterns of a linear array of $N = 7$ identical, equally spaced, rectangular-amplitude-weighted, omnidirectional point elements lying along the X axis for beam tilt angles of $\psi' = 45°$ and $\psi' = 0°$ (end fire), respectively, for $d/\lambda = 1$, 0.5, and 0.45. Figure 7.1-7b shows that for a beam tilt angle of $\psi' = 45°$, $d = \lambda/2$ is sufficient to avoid grating lobes. However, for $\psi' = 0°$ (end fire), Fig. 7.1-8b shows that a grating lobe does exist for $d = \lambda/2$,

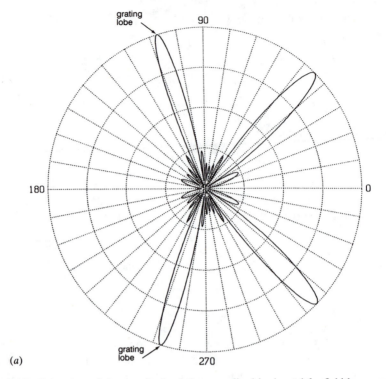

grating
lobe

90

180

0

grating
lobe

270

(a)

Figure 7.1-7. Polar plots of the magnitude of the normalized horizontal far-field beam patterns of a linear array of $N = 7$ identical, equally spaced, rectangular-amplitude-weighted, omnidirectional point elements lying along the X axis for a beam tilt angle of $\psi' = 45°$ and (a) $d/\lambda = 1$, (b) $d/\lambda = 0.5$, and (c) $d/\lambda = 0.45$.

and only when $d < \lambda/2$ does the grating lobe disappear [see Fig. 7.1-8c]. Figures 7.1-7 and 7.1-8 also show that as the ratio d/λ *decreases*, the main-lobe beamwidth *increases*. Recall from linear aperture theory that the main-lobe beamwidth will increase when the frequency (wavelength) is held constant and the length of the aperture is decreased, or when the length of the aperture is held constant and the frequency is decreased (wavelength is increased).

Example 7.1-6 (Spatial sampling theorem)
In this example we shall use the *spatial version* of the sampling theorem from communication theory to derive the criterion that the interelement spacing d must satisfy in order to avoid grating lobes for all possible directions of beam steering. The sampling theorem states that in order to avoid aliasing, a time-domain signal must be sampled at a

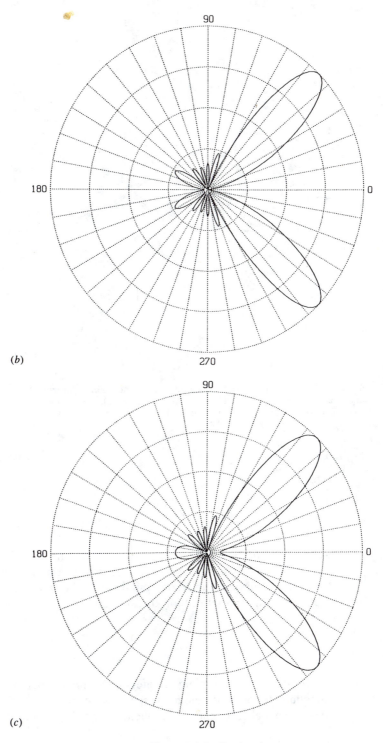

(b)

(c)

Figure 7.1-7. *Continued*

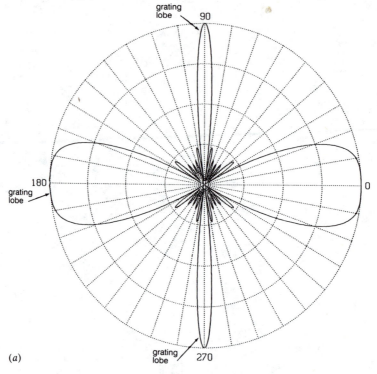

(a)

Figure 7.1-8. Polar plots of the magnitude of the normalized horizontal far-field beam patterns of a linear array of $N = 7$ identical, equally spaced, rectangular-amplitude-weighted, omnidirectional point elements lying along the X axis for a beam tilt angle of $\psi' = 0°$ (end fire) and (a) $d/\lambda = 1$, (b) $d/\lambda = 0.5$, and (c) $d/\lambda = 0.45$.

rate

$$f_S = 1/T_S \geq 2f_{\max}, \qquad (7.1\text{-}161)$$

where f_S is the sampling frequency, or sampling rate, in samples per second, T_S is the sampling period in seconds, and $f_{\max}$ is the highest frequency component in hertz contained in the frequency spectrum of the signal. The spatial version of Eq. (7.1-161) is

$$f_{X_S} = 1/d \geq 2f_{X_{\max}}, \qquad (7.1\text{-}162)$$

where f_{X_S} is the sampling frequency in samples per meter, d is the sampling period (interelement spacing) in meters, and $f_{X_{\max}}$ is the highest spatial frequency component in cycles per meter contained in

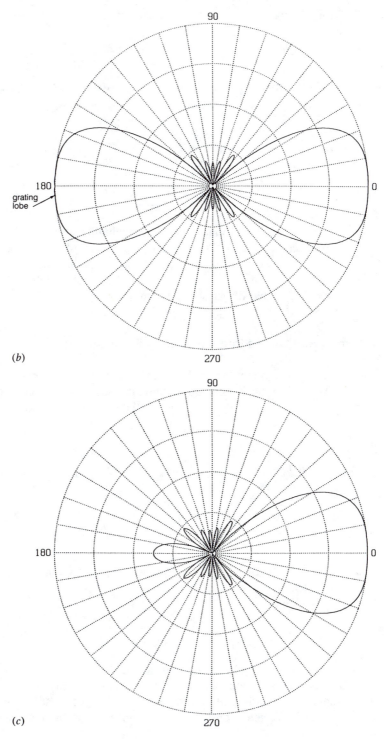

(b)

grating
lobe

(c)

Figure 7.1-8. *Continued*

the angular spectrum of the signal. Since

$$f_{X_{max}} = u_{max}/\lambda_{min}, \qquad (7.1\text{-}163)$$

substituting Eq. (7.1-163) into Eq. (7.1-162) and setting $u_{max} = 1$ yields

$$d \leq \lambda_{min}/2, \qquad (7.1\text{-}164)$$

or

$$\boxed{d < \lambda_{min}/2,} \qquad (7.1\text{-}165)$$

where λ_{min} corresponds to f_{max}, the highest frequency component in hertz contained in the frequency spectrum of the signal. Recall that $c = f\lambda$. Therefore, if the interelement spacing satisfies Eq. (7.1-165) [which is identical to Eq. (7.1-160)], then grating lobes will be avoided for all possible directions of beam steering.

7.2 Linear Arrays and Near-Field Directivity Functions

The near-field directivity function (beam pattern) of a *linear array* lying along the X axis, based on the Fresnel approximation [see Eqs. (6.2-35) and (6.2-40)], is given by the following one-dimensional spatial Fresnel transform [see Eqs. (6.4-12) through (6.4-14)]:

$$D(f, r, f_X) = F_{x_a}\left\{ A(f, x_a) \exp\left(-jk\frac{x_a^2}{2r} \right) \right\}$$

$$= \int_{-\infty}^{\infty} A(f, x_a) \exp\left(-jk\frac{x_a^2}{2r} \right) \exp(+j2\pi f_X x_a)\, dx_a, \quad (7.2\text{-}1)$$

where $A(f, x_a)$ is the complex frequency response (complex aperture function) of the array and

$$f_X = \frac{u}{\lambda} = \frac{\sin\theta\cos\psi}{\lambda}. \qquad (7.2\text{-}2)$$

Since the spatial frequency f_X can be expressed in terms of θ and ψ, the near-field directivity function can ultimately be expressed as a function of frequency f, the range r to a field point, and the spherical angles θ and ψ, that is, $D(f, r, f_X) \rightarrow D(f, r, \theta, \psi)$. We shall consider a field point to be in the near-field (Fresnel) region of an array if it is at a range

$$1.356R < r < \pi R^2/\lambda \qquad (7.2\text{-}3)$$

from the origin of the array, where R is the maximum radial extent of the array [see Eq. (6.2-40)]. The spherical coordinates (r, θ, ψ) are shown in Fig. 6.3-1. Therefore, in order to compute the near-field directivity function of an array, we must first specify a functional form for its complex aperture function.

Consider a linear array of N (odd) identical, equally spaced, complex-weighted, omnidirectional point elements lying along the X axis. The complex aperture function of this array is given by [see Eqs. (7.1-25) and (7.1-26)]

$$A(f, x_a) = \sum_{n=-N'}^{N'} c_n(f)\delta(x_a - nd), \qquad (7.2\text{-}4)$$

where

$$N' = \frac{N-1}{2}, \qquad (7.2\text{-}5)$$

$$c_n(f) = a_n(f)\exp[+j\theta_n(f)] \qquad (7.2\text{-}6)$$

is the frequency-dependent *complex weight* associated with element n, $a_n(f)$ and $\theta_n(f)$ are *real*, frequency-dependent amplitude and phase weights, respectively, and d is the *interelement spacing* in meters. Substituting Eq. (7.2-4) into Eq. (7.2-1) and making use of the sifting property of impulse functions yields the following expression for the near-field directivity function (based on the Fresnel approximation) of a linear array of N (odd) identical, equally spaced, complex-weighted, omnidirectional point elements lying along the X axis:

$$D(f, r, f_X) = \sum_{n=-N'}^{N'} c_n(f)\exp\left[-jk\frac{(nd)^2}{2r}\right]\exp(+j2\pi f_X nd). \qquad (7.2\text{-}7)$$

If a linear array lies along either the Y or the Z axis instead of the X axis, then simply replace x_a and f_X either with y_a and f_Y or with z_a and f_Z, respectively, in the appropriate equations, where the spatial frequencies f_Y and f_Z are given by Eqs. (6.2-44) and (6.2-45), respectively.

7.2.1 Beam Steering and Array Focusing

Let us investigate what happens to the near-field beam pattern given by Eq. (7.2-7) if the phase weights are given by [compare with Eqs. (6.4-22) through (6.4-26)]

$$\theta_n(f) = -2\pi f'_X nd + \frac{k}{2r'}(nd)^2, \qquad n = -N', \ldots, 0, \ldots, N', \qquad (7.2\text{-}8)$$

where

$$f'_X = \frac{u'}{\lambda} = \frac{\sin\theta' \cos\psi'}{\lambda} \qquad (7.2\text{-}9)$$

and

$$k = 2\pi f/c = 2\pi/\lambda. \qquad (7.2\text{-}10)$$

The first term on the right-hand side of Eq. (7.2-8) represents a linear phase variation (linear in n) across the length of the array and is responsible for beam steering, whereas the second term represents a quadratic phase variation (quadratic in n) and is responsible for focusing. If Eqs. (7.2-6) and (7.2-8) are substituted into Eq. (7.2-7), then

$$D(f,r,f_X) = \sum_{n=-N'}^{N'} a_n(f) \exp\left\{-jk\frac{(nd)^2}{2}\left[\frac{1}{r} - \frac{1}{r'}\right]\right\} \exp\left[+j2\pi(f_X - f'_X)nd\right],$$

$$(7.2\text{-}11)$$

and if Eq. (7.2-11) is evaluated at the near-field range $r = r'$, then Eq. (7.2-11) reduces to

$$D(f,r',f_X) = D(f,f_X - f'_X), \qquad (7.2\text{-}12)$$

where

$$D(f,f_X) = \sum_{n=-N'}^{N'} a_n(f) \exp(+j2\pi f_X nd) \qquad (7.2\text{-}13)$$

is the far-field beam pattern of the array when it is only amplitude weighted. Equation (7.2-12) indicates that the far-field beam pattern of the amplitude weights $a_n(f)$ has been *focused* to the near-field range $r = r'$ from the array, and *steered* in the direction $u = u'$ in direction-cosine space, which is equivalent to steering (tilting) the beam pattern to $\theta = \theta'$ and $\psi = \psi'$ [see Eq. (7.2-9)].

Example 7.2-1 [Beam steering and focusing in the near field (Fresnel region)]

As an example of beam steering and focusing in the near field (Fresnel region), consider a linear array of $N = 11$ identical, equally spaced, complex-weighted, omnidirectional point elements lying along the X axis. The array is operated at a frequency of $f = 1$ kHz, rectangular amplitude weights are used, and $c = 1500$ m/sec. Using half-wavelength interelement spacing, the maximum radial extent of the array is $R = 3.75$ m. Therefore, the Fresnel region begins at a range of $1.356R = 5.085$ m, and the range to the near-field–far-field boundary is $\pi R^2/\lambda = 29.452$ m. And finally, suppose we want to focus the beam pattern of the array to the near-field range of $r' = 8$ m.

Figures 7.2-1a, 7.2-2a, and 7.2-3a are polar plots of the magnitude of the *unfocused*, normalized horizontal near-field beam pattern of the array obtained by substituting only the beam steering term from Eq. (7.2-8) into Eq. (7.2-7) and evaluating the resulting expression at $r = 8$ m for beam tilt angles of $\psi' = 90°$ (broadside), 81°, and 72° (Fresnel angular limit of 18° off broadside), respectively. The results are normalized and then plotted. Similarly, Figs. 7.2-1b, 7.2-2b, and 7.2-3b are polar plots of the *focused*, normalized horizontal near-field beam pattern obtained by evaluating Eq. (7.2-11) at $r = r' = 8$ m for beam tilt angles of $\psi' = 90°$, 81°, and 72°, respectively. Note how the focused near-field beam patterns look like far-field beam patterns.

7.3 Array Gain

In this section we shall demonstrate the potential advantage of using an array of elements, as opposed to using a single element, with regard to the signal-to-noise ratio (SNR). Consider a linear array of N (odd) identical, equally spaced, complex-weighted, omnidirectional point elements lying along the X axis. The array is being used in the passive mode, that is, as a receiver. Assume that a sound source (target) is present in the fluid medium, in the far-field region of the array, radiating a random, wide-sense stationary (WSS) signal $s(t)$. In addition to the sound source, assume that there is random, WSS ambient noise present in the fluid medium as well. Let $s_i(t)$ and $n_i(t)$ represent the random, WSS *output signal* and *noise components* from element i, respectively, *after* amplitude and phase (time delay) weighting. Therefore,

$$s_i(t) = a_i s(t - \tau_A - \tau_i - \tau_i') \tag{7.3-1}$$

and

$$n_i(t) = a_i N_i(t - \tau_i'), \tag{7.3-2}$$

where a_i is the *constant* (frequency-independent) amplitude weight applied to element i; τ_A is the actual (true) time delay in seconds associated with the path length between the sound source and the center of the array;

$$\tau_i = u_s i d/c \tag{7.3-3}$$

is the relative time delay in seconds at element i (relative to the center element at $i = 0$) due to the direction of arrival of the plane wave radiated by the sound

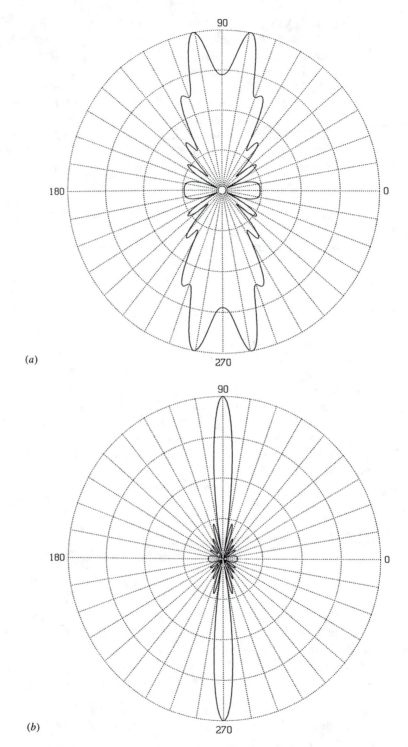

(a)

(b)

Figure 7.2-1. Polar plots of the magnitude of the (*a*) *unfocused* and (*b*) *focused* normalized horizontal near-field beam pattern of the array discussed in Example 7.2-1 for a beam tilt angle of $\psi' = 90°$ (broadside).

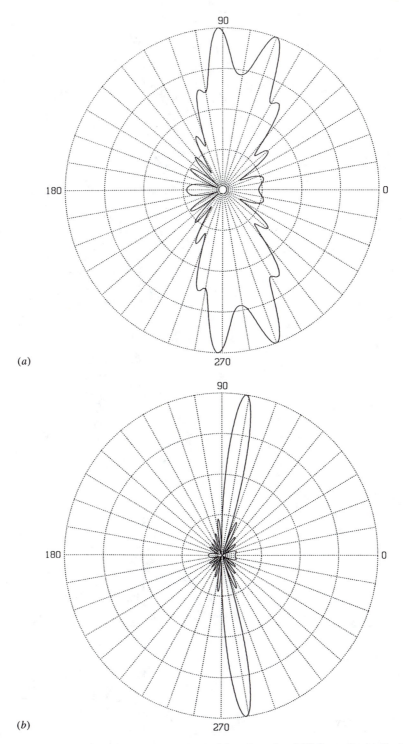

Figure 7.2-2. Polar plots of the magnitude of the (*a*) *unfocused* and (*b*) *focused* normalized horizontal near-field beam pattern of the array discussed in Example 7.2-1 for a beam tilt angle of $\psi' = 81°$.

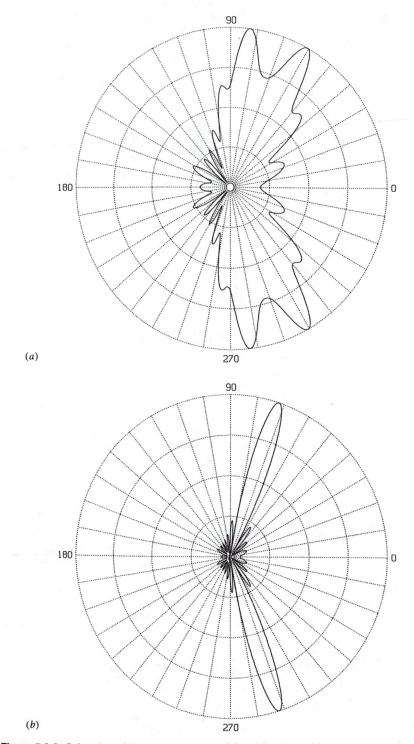

(a)

(b)

Figure 7.2-3. Polar plots of the magnitude of the (a) *unfocused* and (b) *focused* normalized horizontal near-field beam pattern of the array discussed in Example 7.2-1 for a beam tilt angle of $\psi' = 72°$.

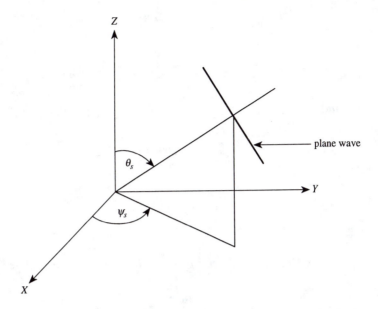

Figure 7.3-1. Plane wave incident upon a linear array lying along the X axis.

source, where (see Fig. 7.3-1)

$$u_s = \sin \theta_s \cos \psi_s; \qquad (7.3\text{-}4)$$

$$\tau_i' = -u'id/c \qquad (7.3\text{-}5)$$

is the time delay in seconds applied to element i in order to steer the beam in the direction (θ', ψ') [see Eq. (7.1-130)], where

$$u' = \sin \theta' \cos \psi'; \qquad (7.3\text{-}6)$$

d is the interelement spacing in meters; and c is the speed of sound in meters per second. Substituting Eqs. (7.3-3) and (7.3-5) into Eq. (7.3-1) yields

$$s_i(t) = a_i s\left(t - \tau_A + \frac{(u' - u_s)id}{c} \right), \qquad (7.3\text{-}7)$$

and substituting Eq. (7.3-5) into Eq. (7.3-2) yields

$$n_i(t) = a_i N_i\left(t + \frac{u'id}{c} \right). \qquad (7.3\text{-}8)$$

Next, let $s_T(t)$ and $n_T(t)$ represent the *total output signal* and *noise* from the array, that is,

$$s_T(t) = \sum_{i=-N'}^{N'} s_i(t) \tag{7.3-9}$$

and

$$n_T(t) = \sum_{i=-N'}^{N'} n_i(t), \tag{7.3-10}$$

where

$$N' = \frac{N-1}{2}. \tag{7.3-11}$$

The *output signal-to-noise power ratio of the array* is defined as follows:

$$\boxed{\text{SNR}_A \triangleq \frac{E\{|s_T(t)|^2\}}{E\{|n_T(t)|^2\}},} \tag{7.3-12}$$

where

$$E\{|s_T(t)|^2\} = \text{USP} + \sum_{i=-N'}^{N'} \sum_{\substack{j=-N' \\ i \neq j}}^{N'} E\{s_i(t)s_j^*(t)\} \tag{7.3-13}$$

it the *total output signal power* and

$$\text{USP} = \sum_{i=-N'}^{N'} E\{|s_i(t)|^2\} \tag{7.3-14}$$

is the *total output uncorrelated signal power*, and similarly,

$$E\{|n_T(t)|^2\} = \text{UNP} + \sum_{i=-N'}^{N'} \sum_{\substack{j=-N' \\ i \neq j}}^{N'} E\{n_i(t)n_j^*(t)\} \tag{7.3-15}$$

is the *total output noise power* and

$$\text{UNP} = \sum_{i=-N'}^{N'} E\{|n_i(t)|^2\} \tag{7.3-16}$$

is the *total output uncorrelated noise power*. The expressions $E\{s_i(t)s_j^*(t)\}$ and $E\{n_i(t)n_j^*(t)\}$ are the *output signal* and *noise correlation functions*, respectively. In Eqs. (7.3-12) through (7.3-16), $E\{\cdot\}$ is the *expectation* (or *ensemble average*) *operator*, and the asterisk denotes complex conjugate.

The *array gain* (AG) is defined as follows:

$$\boxed{AG \triangleq 10\log_{10}(SNR_A/SNR) \text{ dB,}} \qquad (7.3\text{-}17)$$

where

$$SNR = \frac{E\{|s_0(t)|^2\}}{E\{|n_0(t)|^2\}} \qquad (7.3\text{-}18)$$

is the *output signal-to-noise power ratio at the center element* $(i = 0)$ *in the array*. Therefore, the AG is a decibel measure of the increase in signal-to-noise ratio that is obtained by using an array of elements rather than a single element. It is important to note that under certain conditions, *no* increase in signal-to-noise ratio is obtained, that is, AG = 0 dB (see Examples 7.3-1 and 7.3-3). The total output uncorrelated signal and noise power terms USP and UNP, respectively, and the output signal and noise correlation functions $E\{s_i(t)s_j^*(t)\}$ and $E\{n_i(t)n_j^*(t)\}$, respectively, depend on the statistical properties of the signal and noise sound fields in which the array is placed. Therefore, in general, the same array will have different AG values when placed in different signal and noise sound fields.

Example 7.3-1 (No beam steering)

In this example we shall assume that no beam steering has been done, that is, assume that $\tau_i' = 0$ for all i. Therefore, Eqs. (7.3-1) and (7.3-2) reduce to

$$s_i(t) = a_i s(t - \tau_A - \tau_i) \qquad (7.3\text{-}19)$$

and

$$n_i(t) = a_i N_i(t), \qquad (7.3\text{-}20)$$

respectively.

Next, let us compute the output signal and noise correlation functions. With the use of Eq. (7.3-19), the output signal correlation function is given by

$$E\{s_i(t)s_j^*(t)\} = a_i a_j E\{s(t - \tau_A - \tau_i)s^*(t - \tau_A - \tau_j)\}, \qquad (7.3\text{-}21)$$

and since $s(t)$ was assumed to be WSS,

$$E\{s_i(t)s_j^*(t)\} = a_i a_j R_s(-\tau_i + \tau_j) = a_i a_j R_s(\tau_i - \tau_j), \quad (7.3\text{-}22)$$

where

$$R_s(\tau_i - \tau_j) = E\{s(\tau_i)s^*(\tau_j)\} \qquad (7.3\text{-}23)$$

is the autocorrelation function of the signal $s(t)$ radiated by the sound source, and is an even function of its argument. Recall that the autocorrelation function of a WSS random process has its maximum value at the origin, that is,

$$|R_s(\tau)| \le R_s(0), \qquad (7.3\text{-}24)$$

where τ is a time difference and

$$R_s(0) = E\{|s(t)|^2\} \qquad (7.3\text{-}25)$$

is the constant mean squared value of $s(t)$. Similarly, with the use of Eq. (7.3-20), the output noise correlation function is given by

$$E\{n_i(t)n_j^*(t)\} = a_i a_j E\{N_i(t)N_j^*(t)\}. \qquad (7.3\text{-}26)$$

If we assume that the time difference $\tau_i - \tau_j$ is large enough so that

$$R_s(\tau_i - \tau_j) = 0, \qquad i \ne j, \qquad (7.3\text{-}27)$$

then substituting Eq. (7.3-27) into Eq. (7.3-22) yields

$$E\{s_i(t)s_j^*(t)\} = 0, \qquad i \ne j, \qquad (7.3\text{-}28)$$

which indicates that *the output signal components from the different elements in the array have zero correlation*. If we also assume that

$$E\{N_i(t)N_j^*(t)\} = 0, \qquad i \ne j, \qquad (7.3\text{-}29)$$

then Eq. (7.3-26) reduces to

$$E\{n_i(t)n_j^*(t)\} = 0, \qquad i \ne j, \qquad (7.3\text{-}30)$$

which indicates that *the output noise components from the different elements in the array have zero correlation*. Therefore, substituting

Eqs. (7.3-14) and (7.3-28) into Eq. (7.3-13) yields

$$E\{|s_T(t)|^2\} = \text{USP} = \sum_{i=-N'}^{N'} E\{|s_i(t)|^2\}, \qquad (7.3\text{-}31)$$

and substituting Eqs. (7.3-16) and (7.3-30) into Eq. (7.3-15) yields

$$E\{|n_T(t)|^2\} = \text{UNP} = \sum_{i=-N'}^{N'} E\{|n_i(t)|^2\}. \qquad (7.3\text{-}32)$$

With the use of Eq. (7.3-19),

$$E\{|s_i(t)|^2\} = a_i^2 E\{|s(t - \tau_A - \tau_i)|^2\} = a_i^2 E\{|s(t)|^2\}, \quad (7.3\text{-}33)$$

since $s(t)$ was assumed to be WSS, and as a result, the mean squared value of $s(t)$ is constant. Therefore, upon substituting Eq. (7.3-33) into Eq. (7.3-31), we obtain the following expression for the total output signal power:

$$E\{|s_T(t)|^2\} = E\{|s(t)|^2\} \sum_{i=-N'}^{N'} a_i^2. \qquad (7.3\text{-}34)$$

With the use of Eq. (7.3-20),

$$E\{|n_i(t)|^2\} = a_i^2 E\{|N_i(t)|^2\}. \qquad (7.3\text{-}35)$$

If we further assume that the mean squared values of $N_i(t)$ for all the elements in the array are equal, then

$$E\{|N_i(t)|^2\} = E\{|N_0(t)|^2\} \qquad (7.3\text{-}36)$$

for all i. Therefore, substituting Eq. (7.3-36) into Eq. (7.3-35) yields

$$E\{|n_i(t)|^2\} = a_i^2 E\{|N_0(t)|^2\}, \qquad (7.3\text{-}37)$$

and upon substituting Eq. (7.3-37) into Eq. (7.3-32), we obtain the following expression for the total output noise power:

$$E\{|n_T(t)|^2\} = E\{|N_0(t)|^2\} \sum_{i=-N'}^{N'} a_i^2. \qquad (7.3\text{-}38)$$

Substituting Eqs. (7.3-34) and (7.3-38) into Eq. (7.3-12) yields the following expression for the output SNR of the array:

$$\text{SNR}_A = \frac{E\{|s(t)|^2\}}{E\{|N_0(t)|^2\}}. \qquad (7.3\text{-}39)$$

Substituting Eqs. (7.3-33) and (7.3-37) into Eq. (7.3-18) yields the following expression for the output SNR at the center element ($i = 0$) in the array:

$$\text{SNR} = \frac{E\{|s(t)|^2\}}{E\{|N_0(t)|^2\}}. \qquad (7.3\text{-}40)$$

And upon substituting Eqs. (7.3-39) and (7.3-40) into Eq. (7.3-17), we finally obtain

$$\text{AG} = 10 \log_{10} 1 = 0 \text{ dB}. \qquad (7.3\text{-}41)$$

An AG value of 0 dB means that there is no advantage, with regard to the signal-to-noise ratio, in using an array of elements rather than a single element [compare Eqs. (7.3-39) and (7.3-40)].

Example 7.3-2 (Beam steering)
In this example we shall assume that the far-field beam pattern of the array has been steered in the direction of the incident plane wave radiated by the sound source (target), that is, $u' = u_s$. With $u' = u_s$, Eq.(7.3-7) reduces to

$$s_i(t) = a_i s(t - \tau_A) \qquad (7.3\text{-}42)$$

for all i, which indicates that *all* of the output signal components are *in phase*. Note that if the sound source was in the near-field (Fresnel) region of the array, then beam steering and *focusing* would be required in order to cophase all of the output signal components. Therefore, with the use of Eq. (7.3-42), the output signal correlation function is given by

$$E\{s_i(t)s_j^*(t)\} = a_i a_j E\{|s(t - \tau_A)|^2\} = a_i a_j E\{|s(t)|^2\}, \quad (7.3\text{-}43)$$

since $s(t)$ was assumed to be a WSS random process. Upon substituting Eqs. (7.3-14), (7.3-42), and (7.3-43) into Eq. (7.3-13), we obtain the following expression for the total output signal power:

$$E\{|s_T(t)|^2\} = E\{|s(t)|^2\} \sum_{i=-N'}^{N'} a_i^2 + E\{|s(t)|^2\} \sum_{i=-N'}^{N'} \sum_{\substack{j=-N' \\ i \neq j}}^{N'} a_i a_j,$$

$$(7.3\text{-}44)$$

or

$$E\{|s_T(t)|^2\} = E\{|s(t)|^2\}\left[\sum_{i=-N'}^{N'} a_i\right]^2. \qquad (7.3\text{-}45)$$

Similarly, with the use of Eq. (7.3-2), the output noise correlation function is given by

$$E\{n_i(t)n_j^*(t)\} = a_i a_j E\{N_i(t-\tau_i')N_j^*(t-\tau_j')\}, \qquad (7.3\text{-}46)$$

and if $N_i(t)$ and $N_j(t)$ are jointly WSS, then

$$E\{n_i(t)n_j^*(t)\} = a_i a_j R_{N_i N_j}(-\tau_i' + \tau_j'). \qquad (7.3\text{-}47)$$

If we assume that

$$R_{N_i N_j}(-\tau_i' + \tau_j') = 0, \qquad i \neq j, \qquad (7.3\text{-}48)$$

then substituting Eq. (7.3-48) into Eq. (7.3-47) yields

$$E\{n_i(t)n_j^*(t)\} = 0, \qquad i \neq j, \qquad (7.3\text{-}49)$$

which indicates that *the output noise components from the different elements in the array have zero correlation*. In addition, if we further assume that the mean squared values of $N_i(t)$ are equal for all i so that Eq. (7.3-38) is applicable, then substituting Eqs. (7.3-45) and (7.3-38) into Eq. (7.3-12) yields the following expression for the output SNR of the array:

$$\text{SNR}_A = \frac{E\{|s(t)|^2\}\left(\sum_{i=-N'}^{N'} a_i\right)^2}{E\{|N_0(t)|^2\}\sum_{i=-N'}^{N'} a_i^2}. \qquad (7.3\text{-}50)$$

And upon substituting Eqs. (7.3-50) and (7.3-40) into Eq. (7.3-17), we finally obtain

$$\boxed{\text{AG} = 10\log_{10}\left(\frac{\left(\sum_{i=-N'}^{N'} a_i\right)^2}{\sum_{i=-N'}^{N'} a_i^2}\right) \text{ dB.}} \qquad (7.3\text{-}51)$$

If *rectangular amplitude weights* are used, that is, if $a_i = 1$ for all i, then Eq. (7.3-51) reduces to

$$\boxed{\text{AG} = 10\log_{10} N \text{ dB,}} \qquad (7.3\text{-}52)$$

where N is the total number of elements in the array. For example, if the number of elements is doubled and if Eq. (7.3-52) is applicable, then the AG is increased by 3 dB. Later, in Section 8.3.1, we shall show how a nonzero value for the AG can help to increase the probability of detecting a very weak signal with a low SNR at the output of a single element in an array.

Example 7.3-3 (Jamming)

In this example we shall assume that the sound source (target) is trying to *mask*, or *jam*, its own signal $s(t)$ by simultaneously placing in the fluid medium, at the same location, a *jammer* that produces a directional noise field $n(t)$ incident upon the array, traveling in the same direction $u_n = u_s$ as the signal $s(t)$. Therefore, the output noise component from element i, *after* amplitude and phase (time delay) weighting, is now given by

$$n_i(t) = a_i n(t - \tau_A - \tau_i - \tau_i')$$ (7.3-53)

or

$$n_i(t) = a_i n\left(t - \tau_A + \frac{(u' - u_s)id}{c}\right).$$ (7.3-54)

If the far-field beam pattern of the array is steered in the direction of the incident plane wave radiated by the sound source, that is, if $u' = u_s$, then not only will all the output signal components be cophased, but all the output noise components will be cophased as well. As a result, with $u' = u_s$, Eq. (7.3-54) reduces to

$$n_i(t) = a_i n(t - \tau_A)$$ (7.3-55)

for all i. Therefore, with the use of Eq. (7.3-55), if $n(t)$ is assumed to be a WSS random process, then the output noise correlation function is given by

$$E\{n_i(t)n_j^*(t)\} = a_i a_j E\{|n(t - \tau_A)|^2\} = a_i a_j E\{|n(t)|^2\}.$$ (7.3-56)

Upon substituting Eqs. (7.3-16), (7.3-55), and (7.3-56) into Eq. (7.3-15), we obtain

$$E\{|n_T(t)|^2\} = E\{|n(t)|^2\} \sum_{i=-N'}^{N'} a_i^2 + E\{|n(t)|^2\} \sum_{\substack{i=-N' \\ i \neq j}}^{N'} \sum_{j=-N'}^{N'} a_i a_j,$$

(7.3-57)

or

$$E\left\{|n_T(t)|^2\right\} = E\left\{|n(t)|^2\right\}\left[\sum_{i=-N'}^{N'} a_i\right]^2. \qquad (7.3\text{-}58)$$

Therefore, substituting Eqs. (7.3-45) and (7.3-58) into Eq. (7.3-12) yields

$$\text{SNR}_A = \frac{E\left\{|s(t)|^2\right\}}{E\left\{|n(t)|^2\right\}}, \qquad (7.3\text{-}59)$$

and upon substituting Eqs. (7.3-42) and (7.3-55) into Eq. (7.3-18), we obtain

$$\text{SNR} = \frac{a_0^2 E\left\{|s(t-\tau_A)|^2\right\}}{a_0^2 E\left\{|n(t-\tau_A)|^2\right\}} = \frac{E\left\{|s(t)|^2\right\}}{E\left\{|n(t)|^2\right\}}. \qquad (7.3\text{-}60)$$

Finally, substituting Eqs. (7.3-59) and (7.3-60) into Eq. (7.3-17) yields

$$\text{AG} = 10\log_{10} 1 = 0\text{ dB}. \qquad (7.3\text{-}61)$$

Let us end our discussion in this section by noting that if a random process has an expected value (average value) of zero, then a mean squared value is equal to a variance and a correlation function is equal to a covariance function.

7.4 Planar Arrays and Far-Field Directivity Functions

7.4.1 Product Theorem

The far-field directivity function (beam pattern) of a *planar array* lying in the XY plane is given by the following two-dimensional spatial Fourier transform [see Eq. (6.5-5)]:

$$D(f, f_X, f_Y) = F_{x_a} F_{y_a}\{A(f, x_a, y_a)\}$$

$$= \int_{-\infty}^{\infty}\int_{-\infty}^{\infty} A(f, x_a, y_a)\exp\left[+j2\pi(f_X x_a + f_Y y_a)\right] dx_a\, dy_a, \qquad (7.4\text{-}1)$$

where $A(f, x_a, y_a)$ is the complex frequency response (complex aperture function) of the array,

$$f_X = \frac{u}{\lambda} = \frac{\sin\theta\cos\psi}{\lambda}, \qquad (7.4\text{-}2)$$

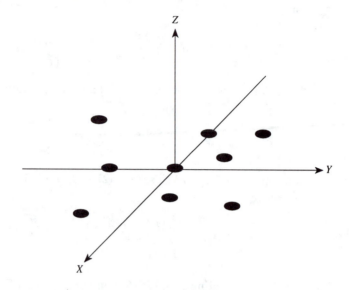

Figure 7.4-1. Planar array composed of an odd number of unevenly spaced elements lying in the XY plane.

and

$$f_Y = \frac{v}{\lambda} = \frac{\sin\theta\sin\psi}{\lambda}. \qquad (7.4\text{-}3)$$

Since the spatial frequencies f_X and f_Y can be expressed in terms of θ and ψ, the far-field directivity function can ultimately be expressed as a function of frequency f and the spherical angles θ and ψ, that is, $D(f, f_X, f_Y) \to D(f, \theta, \psi)$. We shall consider a field point to be in the far-field region of an array if it is at a range

$$r > \pi R^2/\lambda > 1.356R \qquad (7.4\text{-}4)$$

from the origin of the array, where R is the maximum radial extent of the array [see Eq. (6.2-39)]. The spherical coordinates (r, θ, ψ) are shown in Fig. 6.5-1. Therefore, in order to compute the far-field directivity function of an array, we must first specify a functional form for its complex aperture function.

Consider a planar array composed of an odd number $M \times N$ of elements lying in the XY plane. If the elements are not identical, that is, if each element has a different complex frequency response, and if the elements are not equally spaced, as shown in Fig. 7.4-1, then the complex frequency response (complex aperture function) of this array can be expressed as

$$A(f, x_a, y_a) = \sum_{m=-M'}^{M'} \sum_{n=-N'}^{N'} c_{mn}(f)e_{mn}(f, x_a - x_m, y_a - y_n), \quad (7.4\text{-}5)$$

where

$$M' = \frac{M-1}{2}, \tag{7.4-6}$$

$$N' = \frac{N-1}{2}, \tag{7.4-7}$$

M and N are the total odd numbers of elements in the X and Y directions, respectively;

$$c_{mn}(f) = a_{mn}(f)\exp[+j\theta_{mn}(f)] \tag{7.4-8}$$

is the frequency-dependent *complex weight* associated with element mn, where $a_{mn}(f)$ and $\theta_{mn}(f)$ are *real*, frequency-dependent amplitude and phase weights, respectively; $e_{mn}(f, x_a, y_a)$ is the complex frequency response (complex aperture function) of element mn, also known as the *element function*; and x_m and y_n are the x and y coordinates of the center of element mn. Note that, in general, $M \neq N$, $x_{-m} \neq x_m$, and $y_{-n} \neq y_n$. Equation (7.4-5) indicates that the complex frequency response of the array is equal to the *linear superposition* of the complex-weighted frequency responses of all the individual elements in the array. The complex weights are used to control the complex frequency response of the array and, thus, the array's far-field directivity function via amplitude and phase weighting. Substituting Eq. (7.4-5) into Eq. (7.4-1) yields the following expression for the far-field beam pattern:

$$D(f, f_X, f_Y) = \sum_{m=-M'}^{M'} \sum_{n=-N'}^{N'} c_{mn}(f) E_{mn}(f, f_X, f_Y)\exp[+j2\pi(f_X x_m + f_Y y_n)],$$
$$\tag{7.4-9}$$

where

$$E_{mn}(f, f_X, f_Y) = F_{x_a}F_{y_a}\{e_{mn}(f, x_a, y_a)\} \tag{7.4-10}$$

is the far-field beam pattern of element (electroacoustic transducer) mn.

If all the elements in the array are *identical* (i.e., all the elements have the same complex frequency response), but still not equally spaced, then Eq. (7.4-9) reduces to the *product theorem* for planar arrays:

$$\boxed{D(f, f_X, f_Y) = E(f, f_X, f_Y)S(f, f_X, f_Y),} \tag{7.4-11}$$

where

$$E(f, f_X, f_Y) = F_{x_a}F_{y_a}\{e(f, x_a, y_a)\} \tag{7.4-12}$$

is the far-field beam pattern of one of the identical elements in the array, and for M and N odd,

$$S(f, f_X, f_Y) = \sum_{m=-M'}^{M'} \sum_{n=-N'}^{N'} c_{mn}(f) \exp\left[+j2\pi(f_X x_m + f_Y y_n)\right], \quad (7.4\text{-}13)$$

where M' and N' are given by Eqs. (7.4-6) and (7.4-7), respectively. The product theorem states that the far-field directivity function $D(f, f_X, f_Y)$ of a planar array of *identical*, complex-weighted elements is equal to the product of $E(f, f_X, f_Y)$, which is the far-field directivity function of *one* of the identical elements in the array, and $S(f, f_X, f_Y)$, which is the far-field directivity function of an equivalent planar array of identical, complex-weighted, omnidirectional point elements. Although the product theorem requires identical elements, equal spacing of the elements is not required.

Next, let us take the inverse spatial Fourier transform of the product theorem given by Eq. (7.4-11). Doing so yields

$$A(f, x_a, y_a) = e(f, x_a, y_a) \underset{x_a, y_a}{**} s(f, x_a, y_a), \quad (7.4\text{-}14)$$

where $\underset{x_a, y_a}{**}$ denotes a two-dimensional convolution with respect to x_a and y_a, and

$$s(f, x_a, y_a) = F_{f_X}^{-1} F_{f_Y}^{-1}\{S(f, f_X, f_Y)\}$$

$$= \sum_{m=-M'}^{M'} \sum_{n=-N'}^{N'} c_{mn}(f)\delta(x_a - x_m)\delta(y_a - y_n) \quad (7.4\text{-}15)$$

is the complex frequency response (complex aperture function) of a planar array of $M \times N$ (odd) identical, complex-weighted, omnidirectional point elements. And since Eq. (7.4-15) is in the form of the impulse response of a two-dimensional FIR (finite impulse response) filter, a planar array of $M \times N$ (odd) identical, complex-weighted elements can be thought of as a two-dimensional spatial FIR filter, and $s(f, x_a, y_a)$ can be thought of as the spatial impulse response of the array. For example, if the element function (input signal) is the two-dimensional impulse function given by

$$e(f, x_a, y_a) = \delta(x_a)\delta(y_a), \quad (7.4\text{-}16)$$

then the aperture function (output signal) given by Eq. (7.4-14) reduces to

$$A(f, x_a, y_a) = \delta(x_a)\delta(y_a) \underset{x_a, y_a}{**} s(f, x_a, y_a) = s(f, x_a, y_a), \quad (7.4\text{-}17)$$

which is the spatial impulse response of the array.

Finally, if all the elements in the array are *equally spaced* in the X direction, and if all the elements in the array are *equally spaced* in the Y direction, then

$$x_m = md_X, \qquad m = -M', \ldots, 0, \ldots, M',$$

(7.4-18)

and

$$y_n = nd_Y, \qquad n = -N', \ldots, 0, \ldots, N',$$

(7.4-19)

where d_X and d_Y are the *interelement spacings* in meters in the X and Y directions, respectively. Note that, in general, $d_X \neq d_Y$.

Example 7.4-1 (Omnidirectional point elements)

Consider a planar array composed of an odd number $M \times N$ of identical, complex-weighted, omnidirectional point elements lying in the XY plane. The mathematical model that we shall use for the complex frequency response (complex aperture function) of an omnidirectional point element lying in the XY plane is given by Eq. (7.4-16). Since the element function given by Eq. (7.4-16) is separable in the rectangular coordinates x_a and y_a, substituting Eq. (7.4-16) into Eq. (7.4-12) yields

$$E(f, f_X, f_Y) = F_{x_a}\{\delta(x_a)\}F_{y_a}\{\delta(y_a)\} = 1.$$

(7.4-20)

Equation (7.4-20) indicates that the far-field beam pattern of an omnidirectional point element is equal to a constant, independent of the spherical angles θ and ψ, as expected. Therefore, upon substituting Eq. (7.4-20) into Eq. (7.4-11), we obtain the following expression for the far-field directivity function of a planar array composed of an odd number $M \times N$ of identical, complex-weighted, omnidirectional point elements lying in the XY plane:

$$D(f, f_X, f_Y) = S(f, f_X, f_Y)$$

(7.4-21)

where $S(f, f_X, f_Y)$ is given by Eq. (7.4-13).

Example 7.4-2 (Rectangular pistons)

Consider a planar array composed of an odd number $M \times N$ of identical, complex-weighted, rectangular pistons lying in the XY plane. Therefore, from Example 6.5-1, the element function is given by

$$e(f, x_a, y_a) = \text{rect}(x_a/L_X)\,\text{rect}(y_a/L_Y),$$

(7.4-22)

with corresponding unnormalized far-field beam pattern

$$E(f, f_X, f_Y) = L_X L_Y \, \text{sinc}(f_X L_X) \, \text{sinc}(f_Y L_Y), \qquad (7.4\text{-}23)$$

where $L_X L_Y$ is the area (in square meters) of an individual rectangular piston. Note that, in general, $L_X \neq L_Y$. Upon substituting Eq. (7.4-23) into Eq. (7.4-11), we obtain the following expression for the far-field directivity function of a planar array composed of an odd number $M \times N$ of identical, complex-weighted, rectangular pistons lying in the XY plane:

$$D(f, f_X, f_Y) = L_X L_Y \, \text{sinc}(f_X L_X) \, \text{sinc}(f_Y L_Y) \, S(f, f_X, f_Y),$$

$$(7.4\text{-}24)$$

where $S(f, f_X, f_Y)$ is given by Eq. (7.4-13).

If L_X and L_Y are small compared to a wavelength, then by referring to Example 7.1-1, Eq. (7.4-23) reduces to

$$E(f, f_X, f_Y) \approx L_X L_Y, \qquad L_X/\lambda \leq 0.0278, \quad L_Y/\lambda \leq 0.0278.$$

$$(7.4\text{-}25)$$

Since the far-field beam pattern given by Eq. (7.4-25) is equal to a constant, the element is omnidirectional, that is, its far-field beam pattern is independent of the spherical angles θ and ψ. In other words, the element radiates and/or receives equally well in all directions. As a result, the far-field beam pattern of the entire planar array reduces to

$$D(f, f_X, f_Y) \approx L_X L_Y S(f, f_X, f_Y), \qquad L_X/\lambda \leq 0.0278,$$

$$L_Y/\lambda \leq 0.0278. \quad (7.4\text{-}26)$$

Example 7.4-3 (Circular pistons)
Consider a planar array composed of an odd number $M \times N$ of identical, complex-weighted, circular pistons lying in the XY plane. Therefore, from Example 6.5-3, the element function is given by

$$e(f, x_a, y_a) = \text{circ}(r_a/a) = \begin{cases} 1, & r_a \leq a, \\ 0, & r_a > a, \end{cases} \qquad (7.4\text{-}27)$$

with corresponding unnormalized far-field beam pattern

$$E(f, \theta, \psi) = a \, \frac{J_1\!\left(\dfrac{2\pi a}{\lambda} \sin\theta \right)}{\dfrac{\sin\theta}{\lambda}}, \qquad (7.4\text{-}28)$$

where

$$r_a = \sqrt{x_a^2 + y_a^2} \qquad (7.4\text{-}29)$$

and a is the radius (in meters) of an individual piston. Upon substituting Eq. (7.4-28) into Eq. (7.4-11), we obtain the following expression for the far-field directivity function of a planar array composed of an odd number $M \times N$ of identical, complex-weighted, circular pistons lying in the XY plane:

$$D(f,\theta,\psi) = a\frac{J_1\left(\dfrac{2\pi a}{\lambda}\sin\theta\right)}{\dfrac{\sin\theta}{\lambda}}S(f,\theta,\psi), \qquad (7.4\text{-}30)$$

where $S(f,\theta,\psi)$ is given by Eq. (7.4-13).

Note that if the argument of the Bessel function is less than 1, that is, if

$$\frac{2\pi a}{\lambda}\sin\theta < 1, \qquad (7.4\text{-}31)$$

or, setting $\sin\theta = 1$, if

$$\frac{2\pi a}{\lambda} < 1, \qquad (7.4\text{-}32)$$

then

$$J_1\left(\frac{2\pi a}{\lambda}\sin\theta\right) \approx \frac{\pi a}{\lambda}\sin\theta, \qquad (7.4\text{-}33)$$

since

$$J_1(x) = \frac{x}{2} - \frac{2x^3}{2\times 4^2} + \frac{3x^5}{2\times 4^2\times 6^2} - \cdots. \qquad (7.4\text{-}34)$$

Substituting Eq. (7.4-33) into Eq. (7.4-28) yields

$$E(f,\theta,\psi) \approx \pi a^2, \qquad \frac{2\pi a}{\lambda} < 1, \qquad (7.4\text{-}35)$$

where πa^2 is the area (in square meters) of an individual piston. Equation (7.4-35) indicates that if the circumference $2\pi a$ of a piston is small compared to a wavelength, then its far-field beam pattern is equal to a constant. As a result, the element is omnidirectional, that is, its far-field beam pattern is independent of the spherical angles θ and ψ.

In other words, the element radiates and/or receives equally well in all directions. Therefore, the far-field beam pattern of the entire planar array reduces to

$$D(f, \theta, \psi) \approx \pi a^2 S(f, \theta, \psi), \qquad \frac{2\pi a}{\lambda} < 1, \qquad (7.4\text{-}36)$$

or

$$D(f, f_X, f_Y) \approx \pi a^2 S(f, f_X, f_Y), \qquad \frac{2\pi a}{\lambda} < 1, \qquad (7.4\text{-}37)$$

where $S(f, f_X, f_Y)$ is given by Eq. (7.4-13).

Separable Complex Weights

The product theorem for planar arrays given by Eq. (7.4-11) is a general expression. A somewhat simplified version of Eq. (7.4-11) can be obtained if the complex weights $c_{mn}(f)$ are *separable*, analogous to separable functions in complex aperture theory as discussed in Sections 6.5.2 and 6.5.3. If the complex weights are separable, then

$$c_{mn}(f) = c_m(f) w_n(f), \qquad (7.4\text{-}38)$$

where

$$c_m(f) = a_m(f) \exp\left[+j\theta_m(f)\right] \qquad (7.4\text{-}39)$$

is the frequency-dependent complex weight in the X direction, and

$$w_n(f) = b_n(f) \exp\left[+j\phi_n(f)\right] \qquad (7.4\text{-}40)$$

is the frequency-dependent complex weight in the Y direction, where $a_m(f)$ and $b_n(f)$ are real, frequency-dependent amplitude weights, and $\theta_m(f)$ and $\phi_n(f)$ are real, frequency-dependent phase weights.

Substituting Eq. (7.4-38) into Eq. (7.4-13) yields

$$S(f, f_X, f_Y) = S_X(f, f_X) S_Y(f, f_Y), \qquad (7.4\text{-}41)$$

where

$$S_X(f, f_X) = \sum_{m=-M'}^{M'} c_m(f) \exp(+j2\pi f_X x_m) \qquad (7.4\text{-}42)$$

is the far-field beam pattern of a linear array of M (odd) identical, complex-weighted, omnidirectional point elements lying along the X axis, and

$$S_Y(f, f_Y) = \sum_{n=-N'}^{N'} w_n(f) \exp(+j2\pi f_Y y_n) \qquad (7.4\text{-}43)$$

is the far-field beam pattern of a linear array of N (odd) identical, complex-weighted, omnidirectional point elements lying along the Y axis.

Similarly, substituting Eq. (7.4-38) into Eq. (7.4-15) yields

$$s(f, x_a, y_a) = s_X(f, x_a)s_Y(f, y_a), \qquad (7.4\text{-}44)$$

where

$$s_X(f, x_a) = F_{f_X}^{-1}\{S_X(f, f_X)\} = \sum_{m=-M'}^{M'} c_m(f)\delta(x_a - x_m) \qquad (7.4\text{-}45)$$

is the complex frequency response (complex aperture function) of a linear array of M (odd) identical, complex-weighted, omnidirectional point elements lying along the X axis, and

$$s_Y(f, y_a) = F_{f_Y}^{-1}\{S_Y(f, f_Y)\} = \sum_{n=-N'}^{N'} w_n(f)\delta(y_a - y_n) \qquad (7.4\text{-}46)$$

is the complex frequency response (complex aperture function) of a linear array of N (odd) identical, complex-weighted, omnidirectional point elements lying along the Y axis.

Therefore, upon substituting Eq. (7.4-41) into Eq. (7.4-11), we obtain the following form for the *product theorem* for planar arrays when the complex weights are *separable*:

$$\boxed{D(f, f_X, f_Y) = E(f, f_X, f_Y)S_X(f, f_X)S_Y(f, f_Y).} \qquad (7.4\text{-}47)$$

Example 7.4-4

Consider a linear array of M (odd) identical, complex-weighted, planar elements (such as rectangular or circular transducers) lying along the X axis as shown in Fig. 7.4-2. Since the elements lie along the X axis, the complex weights in the Y direction are given by

$$w_n(f) = \begin{cases} 1, & n = 0, \\ 0, & n \neq 0. \end{cases} \qquad (7.4\text{-}48)$$

Since the y coordinate of the center of the element at the origin is $y_0 = 0$, substituting Eq. (7.4-48) into Eq. (7.4-43) yields

$$S_Y(f, f_Y) = 1, \qquad (7.4\text{-}49)$$

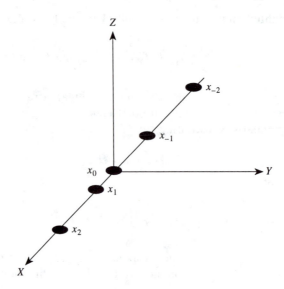

Figure 7.4-2. Linear array composed of an odd number of unevenly spaced planar elements lying along the X axis.

and as a result, the far-field directivity function of the array is given by

$$\boxed{D(f, f_X, f_Y) = E(f, f_X, f_Y)S_X(f, f_X).}$$ (7.4-50)

Example 7.4-5
Consider a linear array of N (odd) identical, complex-weighted, planar elements (such as rectangular or circular transducers) lying along the Y axis as shown in Fig. 7.4-3. Since the elements lie along the Y axis, the complex weights in the X direction are given by

$$c_m(f) = \begin{cases} 1, & m = 0, \\ 0, & m \neq 0. \end{cases}$$ (7.4-51)

Since the x coordinate of the center of the element at the origin is $x_0 = 0$, substituting Eq. (7.4-51) into Eq. (7.4-42) yields

$$S_X(f, f_X) = 1,$$ (7.4-52)

and as a result, the far-field directivity function of the array is given by

$$\boxed{D(f, f_X, f_Y) = E(f, f_X, f_Y)S_Y(f, f_Y).}$$ (7.4-53)

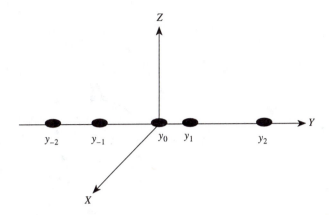

Figure 7.4-3. Linear array composed of an odd number of unevenly spaced planar elements lying along the Y axis.

7.4.2 Concentric Circular Arrays

Recall from planar aperture theory (see Section 6.5.3) that the far-field beam pattern of a planar, circular aperture lying in the XY plane with radius a meters and with complex frequency response (complex aperture function) expressed in terms of the polar coordinates r_a and ϕ_a is given by

$$D(f,\theta,\psi) = \int_0^{2\pi} \int_0^a A(f,r_a,\phi_a) \exp\left[+j\frac{2\pi r_a}{\lambda} \sin\theta \cos(\psi - \phi_a)\right] r_a \, dr_a \, d\phi_a.$$

$$(7.4\text{-}54)$$

Now consider a planar array of *concentric circular arrays* of identical, equally spaced, complex-weighted, omnidirectional point elements lying in the XY plane, as shown in Fig. 7.4-4. Recalling that the complex frequency response of an omnidirectional point element is modeled by an impulse function, the complex frequency response (complex aperture function) of this array can be expressed as

$$A(f,r_a,\phi_a) = \sum_{m=1}^{M} \sum_{n=1}^{N} c_{mn}(f) \frac{\delta(r_a - r_m)}{r_a} \delta(\phi_a - \phi_n), \qquad (7.4\text{-}55)$$

where M is the total number (even or odd) of concentric circular arrays; N is the total number (even or odd) of omnidirectional point elements per circular array;

$$c_{mn}(f) = a_{mn}(f) \exp\left[+j\theta_{mn}(f)\right] \qquad (7.4\text{-}56)$$

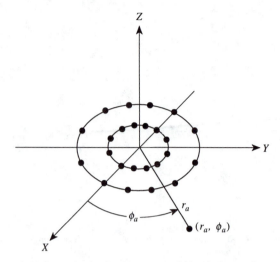

Figure 7.4-4. Planar array of concentric circular arrays of identical, equally spaced, omnidirectional point elements lying in the XY plane.

is the frequency-dependent *complex weight* associated with element mn, where $a_{mn}(f)$ and $\theta_{mn}(f)$ are *real*, frequency-dependent amplitude and phase weights, respectively; and (r_m, ϕ_n) are the polar coordinates of the center of element mn. Since the elements are equally spaced in the r and ϕ directions, we have

$$r_m = m\,\Delta r, \qquad (7.4\text{-}57)$$

where

$$\Delta r = a/M \qquad (7.4\text{-}58)$$

and a is the maximum radial extent of the array in meters, and

$$\phi_n = n\,\Delta\phi, \qquad (7.4\text{-}59)$$

where

$$\Delta\phi = 360° /N. \qquad (7.4\text{-}60)$$

Equation (7.4-55) indicates that the complex frequency response of the array is equal to the linear superposition of the complex-weighted frequency responses of all the individual elements in the array. Therefore, upon substituting Eq. (7.4-55) into Eq. (7.4-54), we obtain the following expression for the far-field beam pattern of a planar array of concentric circular arrays of identical, equally spaced, complex-weighted, omnidirectional point elements lying in the XY plane:

$$D(f,\theta,\psi) = \sum_{m=1}^{M} \sum_{n=1}^{N} c_{mn}(f) \exp\left[+j\frac{2\pi r_m}{\lambda} \sin\theta \cos(\psi - \phi_n)\right]. \quad (7.4\text{-}61)$$

7.4.3 The Phased Array (Beam Steering)

Consider a planar array of $M \times N$ (odd) identical, equally spaced, complex-weighted, omnidirectional point elements lying in the XY plane. From the product theorem, the far-field directivity function of this array is given by

$$D(f, f_X, f_Y) = S(f, f_X, f_Y)$$

$$= \sum_{m=-M'}^{M'} \sum_{n=-N'}^{N'} c_{mn}(f) \exp[+j2\pi(f_X md_X + f_Y nd_Y)], \quad (7.4\text{-}62)$$

since $E(f, f_X, f_Y) = 1$ for an omnidirectional point element (electroacoustic transducer), M' and N' are given by Eqs. (7.4-6) and (7.4-7), respectively, and

$$c_{mn}(f) = a_{mn}(f) \exp[+j\theta_{mn}(f)]. \quad (7.4\text{-}63)$$

Let $D(f, f_X, f_Y)$ be the far-field beam pattern of the array when it is only amplitude weighted, and let $D'(f, f_X, f_Y)$ be the far-field beam pattern of the array when it is complex weighted, that is,

$$D(f, f_X, f_Y) = \sum_{m=-M'}^{M'} \sum_{n=-N'}^{N'} a_{mn}(f) \exp[+j2\pi(f_X md_X + f_Y nd_Y)] \quad (7.4\text{-}64)$$

and

$$D'(f, f_X, f_Y)$$

$$= \sum_{m=-M'}^{M'} \sum_{n=-N'}^{N'} a_{mn}(f) \exp\{+j[2\pi f_X md_X + 2\pi f_Y nd_Y + \theta_{mn}(f)]\}.$$

$$(7.4\text{-}65)$$

If the phase weights are given by [compare with Eqs. (6.5-13) through (6.5-15)]

$$\theta_{mn}(f) = -2\pi f'_X md_X - 2\pi f'_Y nd_Y, \qquad m = -M',\ldots,0,\ldots,M',$$
$$n = -N',\ldots,0,\ldots,N', \quad (7.4\text{-}66)$$

where

$$f_X' = \frac{u'}{\lambda} = \frac{\sin\theta'\cos\psi'}{\lambda} \qquad (7.4\text{-}67)$$

and

$$f_Y' = \frac{v'}{\lambda} = \frac{\sin\theta'\sin\psi'}{\lambda}, \qquad (7.4\text{-}68)$$

then substituting Eq. (7.4-66) into Eq. (7.4-65) yields

$$D'(f, f_X, f_Y)$$

$$= \sum_{m=-M'}^{M'} \sum_{n=-N'}^{N'} a_{mn}(f) \exp\{+j2\pi[(f_X - f_X')md_X + (f_Y - f_Y')nd_Y]\},$$

$$(7.4\text{-}69)$$

or

$$D'(f, f_X, f_Y) = D(f, f_X - f_X', f_Y - f_Y'). \qquad (7.4\text{-}70)$$

Equation (7.4-70) can also be expressed as

$$D'(f, u, v) = D(f, u - u', v - v'). \qquad (7.4\text{-}71)$$

Note that $\theta_{mn}(f)$ given by Eq. (7.4-66) is a function of frequency f (wavelength λ) and is a *linear* phase variation (linear in m and n). Also note that $\theta_{mn}(f)$ is an *odd* function of m and n. Therefore, a linear phase variation applied across the array will cause the beam pattern $D(f, u, v)$ to be steered in the direction $u = u'$ and $v = v'$ in direction-cosine space, which is equivalent to steering (tilting) the beam pattern to $\theta = \theta'$ and $\psi = \psi'$. Equation (7.4-71) indicates that the far-field directivity function $D'(f, u, v)$ is simply a translated (shifted) version of $D(f, u, v)$. When the beam pattern $D(f, u, v)$ is steered, its shape and beamwidth remain unchanged when plotted as a function of direction cosines u and v. However, they do change when the beam pattern is plotted as a function of the spherical angles θ and ψ (see Section 6.3.3).

Finally, note that a phase shift $\theta_{mn}(f)$ in radians is equivalent to a *time delay* τ_{mn} in seconds, that is,

$$\theta_{mn}(f) = 2\pi f\tau_{mn}, \qquad (7.4\text{-}72)$$

or

$$\tau_{mn} = \frac{\theta_{mn}(f)}{2\pi f}. \tag{7.4-73}$$

Substituting Eqs. (7.4-66) through (7.4-68) into Eq. (7.4-73) yields

$$\tau_{mn} = -\frac{u'md_X}{c} - \frac{v'nd_Y}{c}, \qquad m = -M', \ldots, 0, \ldots, M',$$
$$n = -N', \ldots, 0, \ldots, N', \tag{7.4-74}$$

since $c = f\lambda$. According to Eq. (7.4-74), half of the time delays will be positive and half will be negative. In order to ensure that all of the time delays will be positive for practical applications, the constants $|u'|M'd_X/c$ and $|v'|N'd_Y/c$ should be added to all values of τ_{mn} computed from Eq. (7.4-74). However, as we mentioned earlier in our discussion of linear phased arrays, the use of time-delay circuits to do beam steering is really not necessary, since beam steering can be accomplished via phase weighting, which can be implemented by using digital signal processing (digital beamforming) (see Section 8.1). Figure 7.1-4 illustrates the equivalent operations of time delay and phase weighting.

7.5   Planar Arrays and Near-Field Directivity Functions

The near-field directivity function (beam pattern) of a *planar array* lying in the XY plane, based on the Fresnel approximation [see Eqs. (6.2-35) and (6.2-40)], is given by the following two-dimensional spatial Fresnel transform [see Eqs. (6.6-6) through (6.6-9)]:

$$D(f, r, f_X, f_Y) = F_{x_a} F_{y_a} \left\{ A(f, x_a, y_a) \exp\left[-jk\frac{(x_a^2 + y_a^2)}{2r}\right] \right\}$$

$$= \int_{-\infty}^{\infty} \int_{-\infty}^{\infty} A(f, x_a, y_a) \exp\left[-jk\frac{(x_a^2 + y_a^2)}{2r}\right]$$

$$\times \exp\left[+j2\pi(f_X x_a + f_Y y_a)\right] dx_a \, dy_a, \tag{7.5-1}$$

where $A(f, x_a, y_a)$ is the complex frequency response (complex aperture function) of the array,

$$f_X = \frac{u}{\lambda} = \frac{\sin\theta\cos\psi}{\lambda}, \tag{7.5-2}$$

and

$$f_Y = \frac{v}{\lambda} = \frac{\sin\theta\sin\psi}{\lambda}. \tag{7.5-3}$$

Since the spatial frequencies f_X and f_Y can be expressed in terms of θ and ψ, the near-field directivity function can ultimately be expressed as a function of frequency f, the range r to a field point, and the spherical angles θ and ψ, that is, $D(f, r, f_X, f_Y) \rightarrow D(f, r, \theta, \psi)$. We shall consider a field point to be in the near-field (Fresnel) region of an array if it is at a range

$$1.356R < r < \pi R^2/\lambda \tag{7.5-4}$$

from the origin of the array, where R is the maximum radial extent of the array [see Eq. (6.2-40)]. The spherical coordinates (r, θ, ψ) are shown in Fig. 6.5-1. Therefore, in order to compute the near-field directivity function of an array, we must first specify a functional form for its complex aperture function.

Consider a planar array of $M \times N$ (odd) identical, equally spaced, complex-weighted, omnidirectional point elements lying in the XY plane. The complex aperture function of this array is given by [see Eqs. (7.4-15), (7.4-18), and (7.4-19)]

$$A(f, x_a, y_a) = \sum_{m=-M'}^{M'} \sum_{n=-N'}^{N'} c_{mn}(f)\delta(x_a - md_X)\delta(y_a - nd_Y), \tag{7.5-5}$$

where

$$c_{mn}(f) = a_{mn}(f)\exp\left[+j\theta_{mn}(f)\right]. \tag{7.5-6}$$

Substituting Eq. (7.5-5) into Eq. (7.5-1) and making use of the sifting property of impulse functions yields the following expression for the near-field directivity function (based on the Fresnel approximation) of a planar array of $M \times N$ (odd) identical, equally spaced, complex-weighted, omnidirectional point elements lying in the XY plane:

$$\boxed{\begin{aligned} D(f, r, f_X, f_Y) &= \sum_{m=-M'}^{M'} \sum_{n=-N'}^{N'} c_{mn}(f)\exp\left\{-jk\frac{\left[(md_X)^2 + (nd_Y)^2\right]}{2r}\right\} \\ &\quad \times \exp\left[+j2\pi(f_X md_X + f_Y nd_Y)\right]. \end{aligned}}$$

$$\tag{7.5-7}$$

7.5.1 Beam Steering and Array Focusing

Let us investigate what happens to the near-field beam pattern given by Eq. (7.5-7) if the phase weights are given by [compare with Eqs. (6.6-11) through (6.6-14)]

$$
\theta_{mn}(f) = -2\pi f'_X md_X - 2\pi f'_Y nd_Y + k\frac{\left[(md_X)^2 + (nd_Y)^2\right]}{2r'},
$$

$$
m = -M',\ldots,0,\ldots,M',\quad n = -N',\ldots,0,\ldots,N',
$$

(7.5-8)

where

$$
f'_X = \frac{u'}{\lambda} = \frac{\sin\theta'\cos\psi'}{\lambda},
$$

(7.5-9)

$$
f'_Y = \frac{v'}{\lambda} = \frac{\sin\theta'\sin\psi'}{\lambda},
$$

(7.5-10)

and

$$
k = 2\pi f/c = 2\pi/\lambda.
$$

(7.5-11)

The first two terms on the right-hand side of Eq. (7.5-8) represent linear phase variations (linear in m and n) across the array and are responsible for beam steering, whereas the third term represents a quadratic phase variation (quadratic in m and n) and is responsible for focusing. If Eqs. (7.5-6) and (7.5-8) are substituted into Eq. (7.5-7), then

$$
D(f,r,f_X,f_Y) = \sum_{m=-M'}^{M'} \sum_{n=-N'}^{N'} a_{mn}(f)\exp\left\{-jk\frac{\left[(md_X)^2 + (nd_Y)^2\right]}{2}\left(\frac{1}{r} - \frac{1}{r'}\right)\right\}
$$

$$
\times \exp\{+j2\pi[(f_X - f'_X)md_X + (f_Y - f'_Y)nd_Y]\},
$$

(7.5-12)

and if Eq. (7.5-12) is evaluated at the near-field range $r = r'$, then it reduces to

$$
D(f,r',f_X,f_Y) = D(f,f_X - f'_X,f_Y - f'_Y),
$$

(7.5-13)

where

$$
D(f,f_X,f_Y) = \sum_{m=-M'}^{M'} \sum_{n=-N'}^{N'} a_{mn}(f)\exp[+j2\pi(f_X md_X + f_Y nd_Y)]
$$

(7.5-14)

is the far-field beam pattern of the array when it is only amplitude weighted.

Equation (7.5-13) indicates that the far-field beam pattern of the amplitude weights $a_{mn}(f)$ has been *focused* to the near-field range $r = r'$ from the array, and *steered* in the direction $u = u'$ and $v = v'$ in direction-cosine space, which is equivalent to steering (tilting) the beam pattern to $\theta = \theta'$ and $\psi = \psi'$.

7.6 Volume Arrays and Far-Field Directivity Functions

7.6.1 Cylindrical Arrays

Recall from Section 6.2.2 that the far-field directivity function (beam pattern) of an arbitrarily shaped volume aperture is given by the following three-dimensional spatial Fourier transform:

$$D(f, f_X, f_Y, f_Z) = \int_{-\infty}^{\infty} \int_{-\infty}^{\infty} \int_{-\infty}^{\infty} A(f, x_a, y_a, z_a)$$

$$\times \exp\left[+j2\pi(f_X x_a + f_Y y_a + f_Z z_a) \right] dx_a \, dy_a \, dz_a, \quad (7.6\text{-}1)$$

where $A(f, x_a, y_a, z_a)$ is the complex frequency response (complex aperture function) of the volume aperture, and

$$f_X = \frac{u}{\lambda} = \frac{\sin\theta \cos\psi}{\lambda}, \qquad (7.6\text{-}2)$$

$$f_Y = \frac{v}{\lambda} = \frac{\sin\theta \sin\psi}{\lambda}, \qquad (7.6\text{-}3)$$

and

$$f_Z = \frac{w}{\lambda} = \frac{\cos\theta}{\lambda}. \qquad (7.6\text{-}4)$$

Equation (7.6-1) can also be expressed in terms of the cylindrical coordinates (r_a, ϕ_a, z_a), shown in Fig. 7.6-1, by noting that

$$x_a = r_a \cos\phi_a, \qquad (7.6\text{-}5)$$

$$y_a = r_a \sin\phi_a, \qquad (7.6\text{-}6)$$

$$z_a = z_a, \qquad (7.6\text{-}7)$$

and

$$dx_a \, dy_a \, dz_a \rightarrow r_a \, dr_a \, d\phi_a \, dz_a. \qquad (7.6\text{-}8)$$

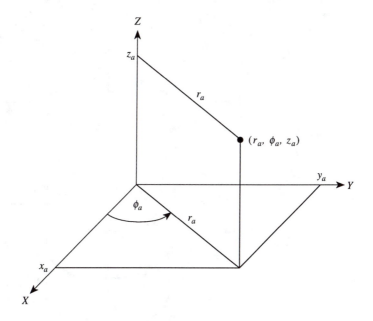

Figure 7.6-1. The cylindrical coordinates (r_a, ϕ_a, z_a).

Therefore, since

$$A(f, x_a, y_a, z_a) = A(f, r_a \cos \phi_a, r_a \sin \phi_a, z_a) \to A(f, r_a, \phi_a, z_a), \quad (7.6\text{-}9)$$

substituting Eqs. (7.6-2) through (7.6-9) into Eq. (7.6-1) yields

$$
\begin{aligned}
D(f, \theta, \psi) = &\int_{-\infty}^{\infty} \int_{0}^{2\pi} \int_{0}^{\infty} A(f, r_a, \phi_a, z_a) \\
&\times \exp\left\{ +j\frac{2\pi}{\lambda} \left[r_a \sin \theta \cos(\psi - \phi_a) + z_a \cos \theta \right] \right\} r_a \, dr_a \, d\phi_a \, dz_a.
\end{aligned}
$$

$$(7.6\text{-}10)$$

Now consider an array of identical, equally spaced, complex-weighted, omnidirectional point elements lying on the surface of a cylinder of radius a meters and length L meters, as shown in Fig. 7.6-2. Recalling that the complex frequency response of an omnidirectional point element is modeled by an impulse function,

566   7   Array Theory

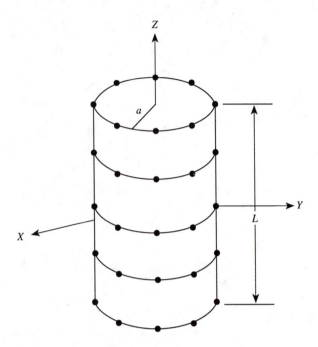

Figure 7.6-2. Cylindrical array of identical, equally spaced, omnidirectional point elements.

the complex frequency response (complex aperture function) of this array can be expressed as

$$A(f, r_a, \phi_a, z_a) = \frac{\delta(r_a - a)}{r_a} \sum_{m=1}^{M} \sum_{n=-N'}^{N'} c_{mn}(f)\delta(\phi_a - \phi_m)\delta(z_a - z_n),$$

$$(7.6\text{-}11)$$

where M is the total number (even or odd) of omnidirectional point elements per circular array;

$$N' = \frac{N-1}{2}, \qquad (7.6\text{-}12)$$

N is the total odd number of circular arrays;

$$c_{mn}(f) = a_{mn}(f) \exp\left[+j\theta_{mn}(f)\right] \qquad (7.6\text{-}13)$$

is the frequency-dependent *complex weight* associated with element mn, where $a_{mn}(f)$ and $\theta_{mn}(f)$ are *real*, frequency-dependent amplitude and phase weights, respectively; and (a, ϕ_m, z_n) are the cylindrical coordinates of the center of element mn. Since the elements are equally spaced in the ϕ and Z directions,

$$\phi_m = m \, \Delta\phi \qquad (7.6\text{-}14)$$

where

$$\Delta\phi = 360° / M, \qquad (7.6\text{-}15)$$

and

$$z_n = n \, \Delta z \qquad (7.6\text{-}16)$$

where

$$\Delta z = \frac{L}{N-1}, \qquad N \neq 1. \qquad (7.6\text{-}17)$$

Equation (7.6-11) indicates that the complex frequency response of the array is equal to the linear superposition of the complex-weighted frequency responses of all the individual elements in the array. Therefore, upon substituting Eq. (7.6-11) into Eq. (7.6-10), we obtain the following expression for the far-field beam pattern of a cylindrical array of identical, equally spaced, complex-weighted, omnidirectional point elements:

$$D(f, \theta, \psi) = \sum_{m=1}^{M} \sum_{n=-N'}^{N'} c_{mn}(f) \exp\left\{ +j \frac{2\pi}{\lambda} \left[a \sin\theta \cos(\psi - \phi_m) + z_n \cos\theta \right] \right\}.$$

$$(7.6\text{-}18)$$

Note that if $N = 1$, then $N' = 0$. And since $z_0 = 0$, Eq. (7.6-18) reduces to the far-field beam pattern of a single circular array of radius a meters.

If the complex weights are separable, then

$$c_{mn}(f) = c_m(f) w_n(f) \qquad (7.6\text{-}19)$$

where

$$c_m(f) = a_m(f) \exp[+j\theta_m(f)] \qquad (7.6\text{-}20)$$

is the frequency-dependent complex weight in the ϕ direction, and

$$w_n(f) = b_n(f) \exp[+j\phi_n(f)] \qquad (7.6\text{-}21)$$

is the frequency-dependent complex weight in the Z direction; where $a_m(f)$ and

$b_n(f)$ are real, frequency-dependent amplitude weights, and $\theta_m(f)$ and $\phi_n(f)$ are real, frequency-dependent phase weights. Substituting Eq. (7.6-19) into Eq. (7.6-18) yields

$$
\begin{aligned}
D(f, \theta, \psi) = &\sum_{m=1}^{M} c_m(f) \exp\left[+j\frac{2\pi a}{\lambda} \sin\theta \cos(\psi - \phi_m)\right] \\
&\times \sum_{n=-N'}^{N'} w_n(f) \exp(+j2\pi f_Z z_n),
\end{aligned}
$$

(7.6-22)

where

$$
f_Z = \frac{\cos\theta}{\lambda}.
$$

(7.6-23)

Once again note that if $N = 1$, then $N' = 0$. And since $z_0 = 0$, if $w_0(f) = 1$, then Eq. (7.6-22) also reduces to the far-field beam pattern of a single circular array of radius a meters.

If only certain elements in the cylindrical array are in operation at any given instant of time, the resulting far-field beam pattern can be obtained by setting $c_{mn}(f) = 0$ in Eq. (7.6-18), or $c_m(f) = 0$ and $w_n(f) = 0$ in Eq. (7.6-22), for those elements not in operation. Any single vertical column of elements or any set of several adjacent vertical columns of elements in the cylindrical array is known as a *stave*. Beam steering in the azimuthal, or ψ, direction can be accomplished by simply operating adjacent staves in a clockwise or counterclockwise fashion. This is known as *scanning*. Beam steering in the vertical, or θ, direction can be accomplished by using separable complex weights and by letting the phase weights in the ψ and Z directions be given by

$$
\theta_m(f) = -\frac{2\pi a}{\lambda} \sin\theta' \cos(\psi - \phi_m)
$$

(7.6-24)

and

$$
\phi_n(f) = -2\pi f_Z' z_n,
$$

(7.6-25)

respectively, where

$$
f_Z' = \frac{\cos\theta'}{\lambda}.
$$

(7.6-26)

7.6.2 Spherical Arrays

Equation (7.6-1) can also be expressed in terms of the spherical coordinates (r_a, γ_a, ϕ_a), shown in Fig. 7.6-3, by noting that

$$x_a = r_a \sin \gamma_a \cos \phi_a, \tag{7.6-27}$$

$$y_a = r_a \sin \gamma_a \sin \phi_a, \tag{7.6-28}$$

$$z_a = r_a \cos \gamma_a, \tag{7.6-29}$$

and

$$dx_a \, dy_a \, dz_a \rightarrow r_a^2 \sin \gamma_a \, dr_a \, d\gamma_a \, d\phi_a. \tag{7.6-30}$$

Therefore, since

$$A(f, x_a, y_a, z_a) = A(f, r_a \sin \gamma_a \cos \phi_a, r_a \sin \gamma_a \sin \phi_a, r_a \cos \gamma_a)$$

$$\rightarrow A(f, r_a, \gamma_a, \phi_a), \tag{7.6-31}$$

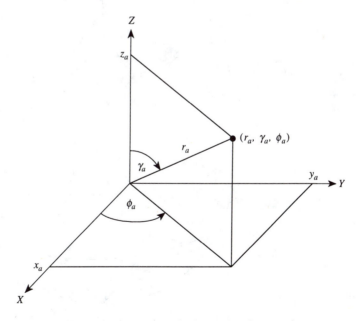

Figure 7.6-3. The spherical coordinates (r_a, γ_a, ϕ_a).

substituting Eqs. (7.6-2) through (7.6-4) and Eqs. (7.6-27) through (7.6-31) into Eq. (7.6-1) yields

$$
\begin{aligned}
D(f,\theta,\psi) = \int_0^{2\pi} \int_0^{\pi} \int_0^{\infty} & A(f,r_a,\gamma_a,\phi_a) \\
& \times \exp\left\{ +j\frac{2\pi r_a}{\lambda} \left[\sin\theta \sin\gamma_a \cos(\psi-\phi_a) + \cos\theta \cos\gamma_a\right] \right\} \\
& \times r_a^2 \sin\gamma_a \, dr_a \, d\gamma_a \, d\phi_a.
\end{aligned}
$$

(7.6-32)

Now consider an array of identical, equally spaced, complex-weighted, omnidirectional point elements lying on the surface of a sphere of radius a meters, as shown in Fig. 7.6-4. Recalling that the complex frequency response of an omnidirectional point element is modeled by an impulse function, the complex frequency

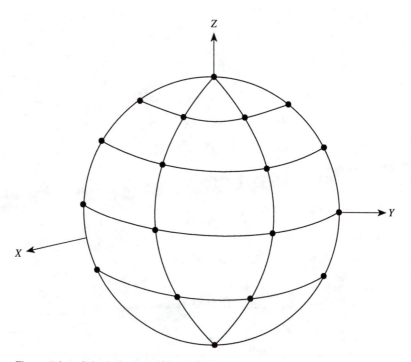

Figure 7.6-4. Spherical array of identical, equally spaced, omnidirectional point elements.

response (complex aperture function) of this array can be expressed as

$$A(f, r_a, \gamma_a, \phi_a) = \frac{\delta(r_a - a)}{r_a^2} \sum_{m=1}^{M} \sum_{n=1}^{N} c_{mn}(f) \frac{\delta(\gamma_a - \gamma_m)}{\sin \gamma_a} \delta(\phi_a - \phi_n), \quad (7.6\text{-}33)$$

where $c_{mn}(f)$, given by Eq. (7.6-13), is the frequency-dependent complex weight associated with element mn, and (a, γ_m, ϕ_n) are the spherical coordinates of the center of element mn. The parameters M and N can be even or odd. Since the elements are equally spaced in the γ and ϕ directions, we have

$$\gamma_m = (m - 1) \Delta\gamma, \quad (7.6\text{-}34)$$

where

$$\Delta\gamma = \frac{180°}{M - 1}, \quad M \neq 1, \quad (7.6\text{-}35)$$

and

$$\phi_n = n \Delta\phi, \quad (7.6\text{-}36)$$

where

$$\Delta\phi = 360° / N. \quad (7.6\text{-}37)$$

Equation (7.6-33) indicates that the complex frequency response of the array is equal to the linear superposition of the complex-weighted frequency responses of all the individual elements in the array. Therefore, upon substituting Eq. (7.6-33) into Eq. (7.6-32), we obtain the following expression for the far-field beam pattern of a spherical array of identical, equally spaced, complex-weighted, omnidirectional point elements:

$$D(f, \theta, \psi)$$
$$= \sum_{m=1}^{M} \sum_{n=1}^{N} c_{mn}(f) \exp\left\{ +j \frac{2\pi a}{\lambda} [\sin \theta \sin \gamma_m \cos(\psi - \phi_n) + \cos \theta \cos \gamma_m] \right\}.$$

$$(7.6\text{-}38)$$

If only certain elements in the spherical array are in operation at any given instant of time, the resulting far-field beam pattern can be obtained by setting $c_{mn}(f) = 0$ in Eq. (7.6-38) for those elements not in operation. Finally, note that if $M = 1$ and $\gamma_1 = 90°$, then Eq. (7.6-38) reduces to the far-field beam pattern of a single circular array of radius a meters.

Problems

7-1   Compute the normalized far-field directivity function $S_N(f, f_X)$ of
(a) Eq. (7.1-63a),
(b) Eq. (7.1-64).

Letting $N = 7$ and $d = \lambda/2$, and using the procedure discussed in Example 6.3-8, compute the 3-dB beamwidth Δu in direction-cosine space of the normalized far-field beam pattern
(c) from part (a),
(d) from part (b).

Compute the 3-dB beamwidth $\Delta\theta$ of the vertical beam pattern
(e) from part (c),
(f) from part (d).

7-2   Letting $N = 11$ and $d = \lambda/2$, repeat Problem 4-1(c), (d), (e), and (f).

7-3   Use the equations developed in Example 7.1-5 to compute the normalized Dolph-Chebyshev amplitude weights so that the ratio of the main-lobe level to the sidelobe level of $S(f, f_X)$ is
(a) 20 dB,
(b) 40 dB.
(c) Using $d = \lambda/2$, what are the beamwidths $\Delta\theta$ of the vertical beam patterns in parts (a) and (b)?

7-4   Suppose we are given a linear array of $N = 6$ identical, equally spaced elements lying along the X axis and it is desired to amplitude weight this array.
(a) Use the Dolph-Chebyshev method to compute the normalized amplitude weights so that the ratio of the main-lobe level to the sidelobe level of $S(f, f_X)$ is 30 dB.
(b) Using $d = \lambda/2$, what is the beamwidth $\Delta\theta$ of the vertical beam pattern of $S(f, f_X)$?

7-5   Consider a linear array of N (even) identical, equally spaced, complex-weighted, omnidirectional point elements lying along the X axis. Show that in order to do beam steering, the phase weights must be given by

$$\theta_n(f) = -2\pi f_X'(n - 0.5)d, \qquad n = 1, 2, \ldots, N/2,$$

and

$$\theta_{-n}(f) = -\theta_n(f), \qquad n = 1, 2, \ldots, N/2.$$

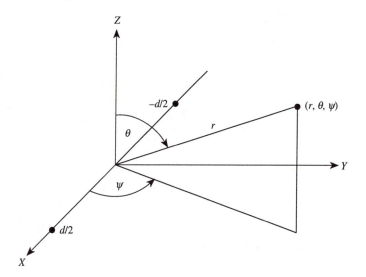

Figure P7-6

7-6   **The two-element interferometer.** Consider a linear array of $N = 2$ identical, complex-weighted, omnidirectional point elements as shown in Fig. P7-6.
   (a) Assume that only amplitude weighting is done, that is, $c_1(f) = a_1(f) = 1$ and $c_{-1}(f) = a_{-1}(f) = 1$. With this even-symmetry amplitude weighting, the linear array shown in Fig. P7-6 is known as a *two-element interferometer*. Find the unnormalized far-field beam pattern of this array as a function of θ and ψ.
   (b) Now we wish to steer the far-field beam pattern to $\theta = \theta'$ and $\psi = \psi'$. What are the equations for the phase weights that will accomplish this task?
   (c) What are the equations for the equivalent time delays?

7-7   **The dipole.** Consider a linear array of $N = 2$ identical, complex-weighted, omnidirectional point elements as shown in Fig. P7-6.
   (a) Assume that only amplitude weighting is done, that is, $c_1(f) = a_1(f) = 1$ and $c_{-1}(f) = a_{-1}(f) = -1$. With this odd-symmetry amplitude weighting, the linear array shown in Fig. P7-6 is known as a *dipole*. Find the unnormalized far-field beam pattern of this array as a function of θ and ψ.
   (b) Now we wish to steer the far-field beam pattern to $\theta = \theta'$ and $\psi = \psi'$. What are the equations for the phase weights that will accomplish this task?
   (c) What are the equations for the equivalent time delays?

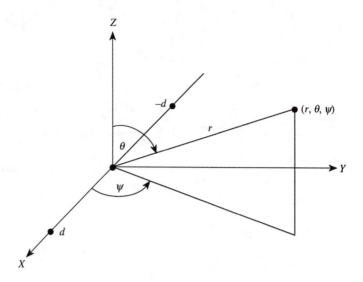

Figure P7-8

7-8   **The axial quadrupole.** Consider a linear array of $N = 3$ identical, equally spaced, complex-weighted, omnidirectional point elements as shown in Fig. P7-8.

(a) Assume that only amplitude weighting is done, that is, $c_1(f) = a_1(f) = 1$, $c_0(f) = a_0(f) = -2$, and $c_{-1}(f) = a_{-1}(f) = 1$. With this even-symmetry amplitude weighting, the linear array shown in Fig. P7-8 is known as an *axial quadrupole*. Find the unnormalized far-field beam pattern of this array as a function of θ and ψ.

(b) Now we wish to steer the far-field beam pattern to $\theta = \theta'$ and $\psi = \psi'$. What is the equation for the phase weights that will accomplish this task?

(c) What is the equation for the equivalent time delays?

7-9   Verify Eq. (7.1-146).

7-10   In order to steer the far-field beam pattern of a linear array of N (odd) identical, equally spaced, complex-weighted, omnidirectional point elements lying along the X axis, the equation for the phase weight to be applied at element n is given by

$$\theta_n(f) = -2\pi f_X' nd, \qquad n = -N', \ldots, 0, \ldots, N'. \qquad (7.1\text{-}123)$$

If we let [see Eq. (7.1-141)]

$$f'_X = m' \Delta f_X = \frac{m'}{(N + Z)d},$$

then

$$\theta_n(f) = -\frac{2\pi m' n}{N + Z}, \qquad n = -N', \ldots, 0, \ldots, N',$$

where m' is the bin where the main lobe is to be steered. As a result,

$$c_n(f) = a_n(f) \exp[+j\theta_n(f)] = a_n(f) W_{N+Z}^{-m'n},$$

where

$$W_{N+Z} = \exp\left[+j\frac{2\pi}{N + Z}\right]. \qquad (7.1\text{-}143)$$

Substituting this expression for the complex weight $c_n(f)$ into Eq. (7.1-140) yields

$$D(f, m) = \sum_{n=-N'}^{N'} a_n(f) W_{N+Z}^{(m-m')n}, \qquad m = -N'', \ldots, 0, \ldots, N'',$$

where N'' is given by Eqs. (7.1-144) and (7.1-145).

For *both* the rectangular and triangular amplitude weights and for the combinations of parameter values given by
(a) $N = 7$, $Z = 0$, $m' = 0$,
(b) $N = 7$, $Z = 25$, $m' = 0$,
(c) $N = 7$, $Z = 25$, $m' = -8$,
create a table for each part with values for the following column headings: bin number m; direction cosine u_m [use $d = \lambda/2$ in Eq. (7.1-149)]; bearing angle $\psi_m = \cos^{-1} u_m$ in degrees; and the magnitude of the normalized far-field beam pattern $D_N(f, m)$ obtained by substituting $D(f, m)$, as given in this problem, and Eq. (7.1-151) into Eq. (7.1-152). Also, for each part, plot $|D_N(f, m)|$ versus m.
(d) The closed-form expressions for the normalized far-field directivity functions $S_N(f, f_X)$ for the rectangular and triangular amplitude weights are given by [see Problem 7-1(a) and (b)]

$$S_N(f, f_X) = \frac{1}{N} \frac{\sin(\pi f_X N d)}{\sin(\pi f_X d)}$$

and

$$S_N(f, f_X) = \frac{4}{(N-1)^2} \frac{\sin^2[\pi f_X(N-1)d/2]}{\sin^2(\pi f_X d)},$$

respectively. Replacing f_X with

$$f_X - f_X' = (m - m')\,\Delta f_X = \frac{m - m'}{(N+Z)d},$$

let

$$D_{N_{REC}}(f, m) = \frac{1}{N} \frac{\sin(\pi[m - m']N/[N+Z])}{\sin(\pi[m - m']/[N+Z])}$$

and

$$D_{N_{TRI}}(f, m) = \frac{4}{(N-1)^2} \frac{\sin^2\left(\dfrac{\pi[m - m'][N-1]}{2[N+Z]}\right)}{\sin^2\left(\dfrac{\pi[m - m']}{N+Z}\right)}.$$

Check your results in parts (a) through (c) for *both* the rectangular and triangular amplitude weights by evaluating the magnitude of the closed-form expressions $D_{N_{REC}}(f, m)$ and $D_{N_{TRI}}(f, m)$.

7-11 Repeat Problem 7-10(a), (b), and (c) for the cosine, Hanning, Hamming, and Blackman amplitude windows.

7-12 Repeat Problem 7-10(a), (b), and (c) for the normalized Dolph-Chebyshev amplitude weights derived in Example 7.1-5 and Problem 7-3(a) and (b). However, compute and plot the magnitude of $D_N(f, m)$ in *decibels*. Compare the actual sidelobe levels with the theoretical design levels.

7-13 The spherical coordinates of a sound source with respect to a linear array of N (odd) identical, equally spaced, complex-weighted, omnidirectional point elements lying along the X axis have been determined to be $r_S = 63$ m, $\theta_S = 17°$, and $\psi_S = 51°$. It is also given that the length of the array is $L = 3$ m, $c = 1500$ m/sec, and $f = 15$ kHz.
(a) Is the sound source in the array's near field or far field?
(b) What is the equation for the phase weights that must be applied across the array if the array's far-field beam pattern is to be focused (if required) and steered to coordinates (r_S, θ_S, ψ_S)? Assume that the interelement spacing $d = \lambda/2$.

(c) The sound source changes its coordinates to $r_S = 120$ m, $\theta_S = 77°$, and $\psi_S = 118°$. Repeat parts (a) and (b).

(d) With $d = \lambda/2$, how many elements are in the array?

(e) Repeat parts (a) through (c) for a linear array lying along the Y axis.

7-14 Consider a linear array of $N = 5$ identical, equally spaced, complex-weighted, omnidirectional point elements lying along the Y axis.

(a) Assume that rectangular amplitude weights are used. What is the normalized far-field beam pattern of this array as a function of θ and ψ?

(b) Assume that triangular amplitude weights are used. What is the normalized far-field beam pattern of this array as a function of θ and ψ?

(c) Now we wish to steer the far-field beam pattern to $\theta = \theta'$ and $\psi = \psi'$. What is the equation for the phase weights that will accomplish this task?

(d) What is the equation for the equivalent time delays?

(e) Assume that this array must be designed to operate at any one of the following three frequencies: $f = 10$, 15, and 25 kHz. What criterion must the interelement spacing satisfy in order to avoid grating lobes under all conditions of beam steering regardless of which frequency is used? Use $c = 1500$ m/sec.

7-15 For $N = 7$, use Eq. (7.3-51) to compute the array gain for

(a) all six sets of amplitude weights given by Eqs. (7.1-52) through (7.1-57), and

(b) the normalized Dolph-Chebyshev amplitude weights computed in Example 7.1-5.

Hint: Use the fact that these amplitude weights are even functions of the index n.

7-16 If the complex weight $c_{mn}(f) = a_{mn}(f)\exp[+j\theta_{mn}(f)]$ obeys the relationship $c_{-m-n}(f) = c_{mn}^*(f)$, then what does Eq. (7.4-13) reduce to? Assume that the complex weight is *not* separable.

7-17 Consider a 3×5 planar array of identical, amplitude-weighted circular pistons lying in the XY plane with interelement spacings d_X and d_Y in the X and Y directions, respectively. The amplitude weights are separable with rectangular weighting in the X direction and triangular weighting in the Y direction. What is the normalized far-field beam pattern of this array as a function of θ and ψ?

7-18 Consider a linear array of $N = 5$ identical, equally spaced, amplitude-weighted rectangular pistons lying along the X axis. If triangular amplitude weights are used, then what is the normalized far-field beam pattern of this array as a function of θ and ψ?

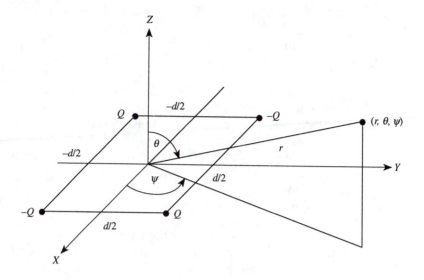

Figure P7-21

7-19 Consider a rectangular piston where $L_Y = L_X$. For the following set of frequencies, determine the criterion that the length L_X must satisfy in order for the rectangular piston to be considered as an omnidirectional element: $f = 1$, 5, and 10 kHz. Use $c = 1500$ m/sec.

7-20 Consider a circular piston with radius a meters. For the following set of frequencies, determine the criterion that the radius a must satisfy in order for the circular piston to be considered as an omnidirectional element: $f = 1$, 5, and 10 kHz. Use $c = 1500$ m/sec.

7-21 **The tesseral quadrupole.** Consider a planar array of four identical, equally spaced, omnidirectional point elements lying in the XY plane as shown in Fig. P7-21. With the amplitude weighting as shown, this array is known as a *tesseral quadrupole*. Find the unnormalized far-field beam pattern of this array as a function of θ and ψ.

7-22 The spherical coordinates of a sound source with respect to a planar array of $M \times N$ (odd) identical, equally spaced, complex-weighted, omnidirectional point elements lying in the XY plane have been determined to be $r_s = 50$ m, $\theta_s = 11°$, and $\psi_s = 49°$. It is also given that the lengths of the array in the X

and Y directions are $L_X = 1.95$ m and $L_Y = 3$ m, respectively, $c = 1500$ m/sec, and $f = 10$ kHz.

(a) Is the sound source in the array's near field or far field?

(b) What is the equation for the phase weights that must be applied across the array if the array's far-field beam pattern is to be focused (if required) and steered to coordinates (r_S, θ_S, ψ_S)? Assume that the interelement spacings are $d_X = \lambda/2$ and $d_Y = \lambda/2$.

(c) The sound source changes its coordinates to $r_S = 80$ m, $\theta_S = 79°$, and $\psi_S = 116°$. Repeat parts (a) and (b).

(d) With $d_X = \lambda/2$ and $d_Y = \lambda/2$, how many elements are in the array in the X and Y directions?

Bibliography

V. M. Albers, *Underwater Acoustics Handbook—II*, Pennsylvania State University Press, University Park, Pennsylvania, 1965.

F. J. Harris, "On the use of windows for harmonic analysis with the discrete Fourier transform," *Proc. IEEE*, **66**, 51–83 (1978).

W. C. Knight, R. G. Pridham, and S. M. Kay, "Digital signal processing for sonar," *Proc. IEEE*, **69**, 1451–1506 (1981).

B. D. Steinberg, *Principles of Aperture and Array System Design*, Wiley, New York, 1976.

Chapter 8

Signal Processing

8.1 FFT Beamforming for Planar Arrays

In both air and underwater acoustics applications, there are times when it is desired to be able to identify (classify) a sound source (target) and to estimate its direction relative to the origin of a coordinate system. For example, consider an array of electroacoustic transducers (elements) being used in the *passive mode*, "listening" for sound fields radiated by potential targets of interest. Based on the information contained in the output electrical signals from the individual elements in the array, it is desired to estimate both the frequency content (spectral lines) of the radiated sound field and the direction to the sound source relative to the center of the array. Estimating the spectral lines of the radiated sound field is referred to as doing a "target signature analysis," since, in general, different sound sources radiate energy at different sets of frequencies. One way to solve this problem is to use the method of FFT (fast-Fourier-transform) beamforming. This method provides the background necessary for the understanding of more sophisticated methods such as frequency-domain adaptive beamforming.

In order to provide the theoretical background for the method of FFT beamforming, we shall derive an expression for the complex frequency and angular spectrum of the output electrical signal $y(t, \mathbf{r})$ from a receive *planar array* when the acoustic field incident upon the array is a *plane wave with arbitrary time dependence*. Recall from Section 6.2.2 that $Y(\eta, \boldsymbol{\gamma})$, which is the complex frequency and angular spectrum of the output electrical signal $y(t, \mathbf{r})$ from a receive aperture, is given by the following three-dimensional convolution integral:

$$Y(\eta, \boldsymbol{\gamma}) = \int_{-\infty}^{\infty} Y_M(\eta, \boldsymbol{\beta}) D_R'(\eta, \boldsymbol{\gamma} - \boldsymbol{\beta}) \, d\boldsymbol{\beta}, \qquad (8.1\text{-}1)$$

where η represents output or received frequencies in hertz; $\boldsymbol{\gamma} = (\gamma_X, \gamma_Y, \gamma_Z)$ is a three-dimensional vector whose components are spatial frequencies in the X, Y, and Z directions, respectively; $Y_M(\eta, \boldsymbol{\beta})$ is the complex frequency and angular spectrum of the output acoustic signal from the fluid medium, which is also the input acoustic signal to the receive aperture; $\boldsymbol{\beta} = (\beta_X, \beta_Y, \beta_Z)$ is a three-dimensional vector whose components are spatial frequencies in the X, Y, and Z directions, respectively; $D_R'(\eta, \boldsymbol{\beta})$ is the far-field directivity function, or beam pattern, of the receive aperture; and $d\boldsymbol{\beta} = d\beta_X \, d\beta_Y \, d\beta_Z$.

If the receive aperture is a planar array of $M \times N$ (odd) identical, equally spaced, complex-weighted, omnidirectional point elements lying in the XY plane, then from the product theorem, the far-field directivity function of this array is given by (see Section 7.4.3)

$$D_R'(\eta, \boldsymbol{\beta}) = D_R'(\eta, \beta_X, \beta_Y, \beta_Z) = D_R'(\eta, \beta_X, \beta_Y) \qquad (8.1\text{-}2)$$

where

$$D_R'(\eta, \beta_X, \beta_Y) = \sum_{m=-M'}^{M'} \sum_{n=-N'}^{N'} c_{mn}(\eta) \exp\left[+j2\pi(\beta_X m d_X + \beta_Y n d_Y)\right] \qquad (8.1\text{-}3a)$$

is the far-field directivity function of the array when it is complex weighted,

$$D_R(\eta, \beta_X, \beta_Y) = \sum_{m=-M'}^{M'} \sum_{n=-N'}^{N'} a_{mn}(\eta) \exp\left[+j2\pi(\beta_X m d_X + \beta_Y n d_Y)\right] \qquad (8.1\text{-}3b)$$

is the far-field directivity function of the array when it is only amplitude weighted,

$$M' = \frac{M-1}{2}, \qquad (8.1\text{-}4)$$

$$N' = \frac{N-1}{2}, \qquad (8.1\text{-}5)$$

M and N are the total odd number of elements in the X and Y directions, respectively, and

$$c_{mn}(\eta) = a_{mn}(\eta) \exp\left[+j\theta_{mn}(\eta)\right] \qquad (8.1\text{-}6)$$

is the frequency-dependent *complex weight* associated with element *mn*, where $a_{mn}(\eta)$ and $\theta_{mn}(\eta)$ are *real*, frequency-dependent amplitude and phase weights, respectively. Note that the far-field directivity function of this receive array is only a function of the two spatial frequencies β_X and β_Y, and not β_Z, since the array is lying in the *XY* plane, where $z = 0$. Therefore, from Eq. (8.1-2),

$$D_R'(\eta, \boldsymbol{\gamma} - \boldsymbol{\beta}) = D_R'(\eta, \gamma_X - \beta_X, \gamma_Y - \beta_Y, \gamma_Z - \beta_Z)$$

$$= D_R'(\eta, \gamma_X - \beta_X, \gamma_Y - \beta_Y), \qquad (8.1\text{-}7)$$

where $D_R'(\eta, \beta_X, \beta_Y)$ is given by Eq. (8.1-3a). Equation (8.1-7) is one of the integrand factors appearing in Eq. (8.1-1).

Next, consider the remaining integrand factor appearing in Eq. (8.1-1), which is

$$Y_M(\eta, \boldsymbol{\beta}) = F_t F_{\mathbf{r}}\{y_M(t, \mathbf{r})\}, \qquad (8.1\text{-}8)$$

where $F_{\mathbf{r}}\{\cdot\}$ is shorthand notation for a three-dimensional spatial Fourier transform with respect to x, y, and z, that is, $F_{\mathbf{r}}\{\cdot\} = F_x F_y F_z\{\cdot\}$. If the sound source (target) is in a homogeneous fluid medium in the far-field region of the array, then the radiated acoustic field $y_M(t, \mathbf{r})$ incident upon the array can be modeled as a general plane-wave field with arbitrary time dependence propagating in the $-\hat{n}_0$ direction (see Fig. 8.1-1), that is,

$$y_M(t, \mathbf{r}) = g\left(t + \frac{\hat{n}_0 \cdot (\mathbf{r} - \mathbf{r}_0)}{c}\right) = g\left(t - \tau_A + \frac{\hat{n}_0 \cdot \mathbf{r}}{c}\right), \qquad (8.1\text{-}9)$$

where $g(t)$ is an *arbitrary* function of time;

$$\mathbf{r} = x\hat{x} + y\hat{y} + z\hat{z} \qquad (8.1\text{-}10)$$

is the position vector to a field point with rectangular coordinates (x, y, z);

$$\mathbf{r}_0 = r_0\hat{n}_0 \qquad (8.1\text{-}11)$$

is the position vector to the sound source, where r_0 is the range to the sound source;

$$\tau_A = \frac{\hat{n}_0 \cdot \mathbf{r}_0}{c} = \frac{r_0}{c} \qquad (8.1\text{-}12)$$

is the *actual*, or *true*, *time delay* in seconds associated with the path length between the sound source and the center of the array;

$$\hat{n}_0 = u_0\hat{x} + v_0\hat{y} + w_0\hat{z} \qquad (8.1\text{-}13)$$

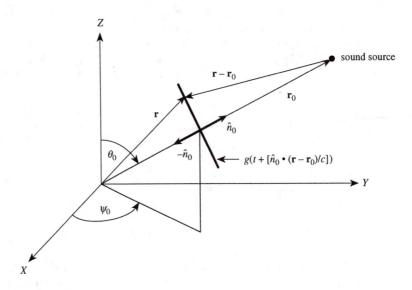

Figure 8.1-1. General plane-wave field $g(t + [\hat{n}_0 \cdot (\mathbf{r} - \mathbf{r}_0)/c])$ with arbitrary time dependence propagating in the $-\hat{n}_0$ direction.

is a unit vector where

$$u_0 = \sin \theta_0 \cos \psi_0, \qquad (8.1\text{-}14)$$

$$v_0 = \sin \theta_0 \sin \psi_0, \qquad (8.1\text{-}15)$$

and

$$w_0 = \cos \theta_0 \qquad (8.1\text{-}16)$$

are dimensionless direction cosines with respect to the X, Y, and Z axes, respectively; and c is the constant speed of sound (see Section 2.2.3 and Problem 2-4). As is illustrated in Fig. 8.1-1, the plane wave is traveling in the $-\hat{n}_0$ direction away from the target, and the target is located in the $+\hat{n}_0$ direction. Note that directions are measured with respect to the center of the array. With the use of Eq. (8.1-9), we shall evaluate Eq. (8.1-8) next.

The evaluation of Eq. (8.1-8) is a two-step process. The first step is to compute the time-domain Fourier transform. Since

$$Y_M(\eta, \mathbf{r}) = F_t\{y_M(t, \mathbf{r})\}, \qquad (8.1\text{-}17)$$

substituting Eq. (8.1-9) into Eq. (8.1-17) yields

$$Y_M(\eta, \mathbf{r}) = F_t\left\{ g\left(t - \tau_A + \frac{\hat{n}_0 \cdot \mathbf{r}}{c} \right) \right\}$$

$$= G(\eta) \exp(-j2\pi\eta\tau_A) \exp\left[+j2\pi\eta\left(\frac{\hat{n}_0 \cdot \mathbf{r}}{c} \right) \right], \qquad (8.1\text{-}18)$$

and upon substituting Eqs. (8.1-10) and (8.1-13) into Eq. (8.1-18), we obtain the following expression for the complex frequency spectrum of a general plane-wave field with arbitrary time dependence:

$$Y_M(\eta, x, y, z) = G(\eta) \exp(-j2\pi\eta\tau_A)$$

$$\times \exp(+j2\pi f_{X_0} x) \exp(+j2\pi f_{Y_0} y) \exp(+j2\pi f_{Z_0} z),$$

(8.1-19)

where $G(\eta) = F_t\{g(t)\}$ is the complex frequency spectrum of the signal $g(t)$ radiated by the sound source, $c = \eta\lambda$, and

$$f_{X_0} = u_0/\lambda,$$

(8.1-20)

$$f_{Y_0} = v_0/\lambda,$$

(8.1-21)

and

$$f_{Z_0} = w_0/\lambda,$$

(8.1-22)

are the spatial frequencies in the X, Y, and Z directions, respectively, with units of cycles per meter (see Section 2.2.1), associated with the direction of propagation of the plane wave.

The second step in the evaluation of Eq. (8.1-8) is to compute the spatial-domain Fourier transform. Since

$$Y_M(\eta, \boldsymbol{\beta}) = F_r\{Y_M(\eta, \mathbf{r})\},$$

(8.1-23)

substituting Eq. (8.1-19) into Eq. (8.1-23) yields

$$Y_M(\eta, \boldsymbol{\beta}) = Y_M(\eta, \beta_X, \beta_Y, \beta_Z) = G(\eta) \exp(-j2\pi\eta\tau_A) \, F_x\{\exp(+j2\pi f_{X_0} x)\}$$

$$\times F_y\{\exp(+j2\pi f_{Y_0} y)\} F_z\{\exp(+j2\pi f_{Z_0} z)\},$$

(8.1-24)

or

$$Y_M(\eta, \boldsymbol{\beta}) = Y_M(\eta, \beta_X, \beta_Y, \beta_Z) = G(\eta) \exp(-j2\pi\eta\tau_A)$$

$$\times \delta(\beta_X + f_{X_0}) \delta(\beta_Y + f_{Y_0}) \delta(\beta_Z + f_{Z_0}),$$

(8.1-25)

which is the complex frequency and angular spectrum of a general plane-wave field with arbitrary time dependence. We are now in a position to evaluate the three-dimensional convolution integral given by Eq. (8.1-1).

Substituting Eqs. (8.1-7) and (8.1-25) into Eq. (8.1-1) yields

$$Y(\eta, \gamma_X, \gamma_Y) = \int_{-\infty}^{\infty} \int_{-\infty}^{\infty} \int_{-\infty}^{\infty} G(\eta) \exp(-j2\pi\eta\tau_A) D_R'(\eta, \gamma_X - \beta_X, \gamma_Y - \beta_Y)$$

$$\times \delta(\beta_X + f_{X_0})\delta(\beta_Y + f_{Y_0})\delta(\beta_Z + f_{Z_0}) \, d\beta_X \, d\beta_Y \, d\beta_Z, \quad (8.1\text{-}26)$$

or

$$\boxed{Y(\eta, \gamma_X, \gamma_Y) = G(\eta) \exp(-j2\pi\eta\tau_A) D_R'(\eta, \gamma_X + f_{X_0}, \gamma_Y + f_{Y_0}),} \quad (8.1\text{-}27)$$

since

$$\int_{-\infty}^{\infty} \delta(\beta_Z + f_{Z_0}) \, d\beta_Z = 1, \quad (8.1\text{-}28)$$

where

$$\gamma_X = \frac{u}{\lambda} = \frac{\sin\theta\cos\psi}{\lambda} \quad (8.1\text{-}29)$$

and

$$\gamma_Y = \frac{v}{\lambda} = \frac{\sin\theta\sin\psi}{\lambda} \quad (8.1\text{-}30)$$

are spatial frequencies in the X and Y directions, respectively, with units of cycles per meter, and D_R' is given by Eq. (8.1-3a). Equation (8.1-27) is the complex frequency and angular spectrum of the output electrical signal from a planar array of $M \times N$ (odd) identical, equally spaced, complex-weighted, omnidirectional point elements lying in the XY plane when the acoustic field incident upon the array is a general plane-wave field with arbitrary time dependence. Note that if we were to calculate the complex frequency and angular spectrum of a general plane-wave field with arbitrary time dependence in the XY plane where $z = 0$, then Eq. (8.1-25) would become

$$Y_M(\eta, \beta_X, \beta_Y, \beta_Z) = G(\eta) \exp(-j2\pi\eta\tau_A) \, \delta(\beta_X + f_{X_0})\delta(\beta_Y + f_{Y_0})\delta(\beta_Z),$$

$$(8.1\text{-}31)$$

since setting $z = 0$ in Eq. (8.1-24) yields

$$F_z\{1\} = \delta(\beta_Z). \quad (8.1\text{-}32)$$

Equation (8.1-27) would still be valid, since

$$\int_{-\infty}^{\infty} \delta(\beta_Z) \, d\beta_Z = 1. \tag{8.1-33}$$

If linear phase weighting is done for purposes of beam steering in order to maximize the array gain, then the real, frequency-dependent phase weight $\theta_{mn}(\eta)$ appearing in Eq. (8.1-6) is given by

$$\theta_{mn}(\eta) = -2\pi f_X' m d_X - 2\pi f_Y' n d_Y, \qquad m = -M', \ldots, 0, \ldots, M',$$

$$n = -N', \ldots, 0, \ldots, N', \quad (8.1\text{-}34)$$

where

$$f_X' = \frac{u'}{\lambda} = \frac{\sin \theta' \cos \psi'}{\lambda} \tag{8.1-35}$$

and

$$f_Y' = \frac{v'}{\lambda} = \frac{\sin \theta' \sin \psi'}{\lambda} \tag{8.1-36}$$

are the spatial frequencies in the X and Y directions, respectively, with units of cycles per meter, associated with beam steering. In this case, Eq. (8.1-27) becomes

$$\boxed{\begin{aligned} Y(\eta, \gamma_X, \gamma_Y) &= G(\eta) \exp(-j2\pi\eta\tau_A) \\ &\times D_R\Big(\eta, \gamma_X - \big[f_X' - f_{X_0}\big], \gamma_Y - \big[f_Y' - f_{Y_0}\big]\Big), \end{aligned}} \tag{8.1-37}$$

or, in direction-cosine space,

$$\boxed{Y(\eta, u, v) = G(\eta) \exp(-j2\pi\eta\tau_A) \, D_R(\eta, u - [u' - u_0], v - [v' - v_0]),} \tag{8.1-38}$$

where D_R is given by Eq. (8.1-3b). If beam steering is done correctly, that is, if

$$u' = u_0 \tag{8.1-39}$$

and

$$v' = v_0, \tag{8.1-40}$$

so that the far-field beam pattern of the array is steered in the direction of the incident plane wave radiated by the sound source, then all of the output electrical signals from all of the elements in the planar array will be in phase, and as a result, the array gain will be maximized (see Examples 7.3-2 and 8.1-2). With the use of Eqs. (8.1-39) and (8.1-40), and for a given frequency η, the maximum value of the magnitude of Eq. (8.1-38) will be located at $u = 0$ and $v = 0$. If the array gain is not maximized, then the signal radiated by the sound source may not be detected (see Example 8.3-1). Note that if maximizing the array gain is not an issue (e.g., in the absence of noise), and if beam steering is not done, that is, if $u' = 0$ and $v' = 0$, then we can still estimate u_0 and v_0, since in this case, for a given frequency η, the maximum value of the magnitude of Eq. (8.1-38) will be located at $u = -u_0$ and $v = -v_0$.

Equations (8.1-39) and (8.1-40) imply that the values of the direction cosines u_0 and v_0 associated with the direction of propagation of the plane wave are now known. We are thus in a position to estimate both the frequency content (spectral lines) of the radiated sound field and the direction to the sound source (target) relative to the center of the array.

The spectral lines (frequency content) of the sound field radiated by the sound source are represented by the complex frequency spectrum $G(\eta)$. For example, if $g(t)$ is equal to the sum of three different time-harmonic components, that is, if

$$g(t) = \alpha_1 \exp(+j2\pi\eta_1 t) + \alpha_2 \exp(+j2\pi\eta_2 t) + \alpha_3 \exp(+j2\pi\eta_3 t), \quad (8.1\text{-}41)$$

then

$$G(\eta) = \alpha_1 \delta(\eta - \eta_1) + \alpha_2 \delta(\eta - \eta_2) + \alpha_3 \delta(\eta - \eta_3), \quad (8.1\text{-}42)$$

where α_1, α_2, and α_3 are the amplitudes associated with the frequency components η_1, η_2, and η_3, respectively.

After the values of the direction cosines u_0 and v_0 have been determined, the *angles of arrival*—or, in other words, the direction to the sound source relative to the center of the array—can be computed. Since

$$\sin\theta_0 = \sqrt{u_0^2 + v_0^2}, \quad (8.1\text{-}43)$$

the angle θ_0 to the sound source, measured with respect to the positive Z axis of the array (see Fig. 8.1-1), is given by

$$\boxed{\theta_0 = \sin^{-1}\left(\sqrt{u_0^2 + v_0^2}\right).} \quad (8.1\text{-}44)$$

The azimuthal or bearing angle ψ_0 to the sound source, measured in the counter-clockwise direction with respect to the positive X axis of the array (see Fig. 8.1-1), can be obtained by forming the ratio

$$v_0/u_0 = \tan \psi_0, \tag{8.1-45}$$

so that

$$\boxed{\psi_0 = \tan^{-1}(v_0/u_0).} \tag{8.1-46}$$

FFT Beamforming

We have just demonstrated that if we compute the frequency and angular spectrum of the output electrical signal from a receive planar array, then we can estimate both the frequency content (spectral lines) of the radiated sound field and the direction to the sound source (target) relative to the center of the array. The next question is how can we compute the frequency and angular spectrum. As was mentioned previously, one way to solve this problem is to use the method of FFT beamforming.

We begin the discussion of this method by once again considering the receive aperture to be a planar array of $M \times N$ (odd) identical, equally spaced, complex-weighted, omnidirectional point elements lying in the XY plane. The output electrical signal $y(t, \mathbf{r})$ from element mn in the array can be expressed as $y(t, md_X, nd_Y)$, since element mn is located at $x = md_X$, $y = nd_Y$, and $z = 0$. Next, sample the output signal from element mn in time, so that $y(t, md_X, nd_Y)$ $\rightarrow y(lT_S, md_X, nd_Y)$, where T_S is the sampling period in seconds. The expression $y(lT_S, md_X, nd_Y)$ is a mathematical representation of the time-sampled output electrical signal from one element in the planar array. In order to proceed further, we must mathematically represent the *sampled* version of $y(t, \mathbf{r})$ from *all* of the elements in the planar array. Let this sampled version be denoted by $y_S(t, \mathbf{r}) = y_S(t, x, y)$, where

$$y_S(t, x, y) = \sum_{m=-M'}^{M'} \sum_{n=-N'}^{N'} \sum_{l=-\infty}^{\infty} y(lT_S, md_X, nd_Y) \delta(t - lT_S)$$

$$\times \delta(x - md_X) \delta(y - nd_Y) \tag{8.1-47}$$

and M' and N' are given by Eqs. (8.1-4) and (8.1-5), respectively. Equation (8.1-47) is simply a three-dimensional generalization of one-dimensional time-domain impulse sampling.

The frequency and angular spectrum of $y_S(t, x, y)$ is given by

$$Y_S(\eta, \gamma_X, \gamma_Y) = F_t F_x F_y \{y_S(t, x, y)\}, \tag{8.1-48}$$

and upon substituting Eq. (8.1-47) into Eq. (8.1-48), we obtain

$$Y_S(\eta, \gamma_X, \gamma_Y) = \sum_{m=-M'}^{M'} \sum_{n=-N'}^{N'} c_{mn}(\eta) \sum_{l=-\infty}^{\infty} y(lT_S, md_X, nd_Y) \exp(-j2\pi\eta lT_S)$$

$$\times \exp(+j2\pi\gamma_X md_X) \exp(+j2\pi\gamma_Y nd_Y), \tag{8.1-49}$$

where the frequency-dependent complex weight $c_{mn}(\eta)$ is introduced in order to allow for beamforming and is given by Eq. (8.1-6). Note that the complex weight is just a complex number in a computer program, that is, electronic amplifiers and time-delay circuits do not have to be used in order to implement amplitude and phase weighting. Also note that the innermost summation over the time index l is the discrete-time Fourier transform of $y(lT_S, md_X, nd_Y)$. Later, we shall express Eq. (8.1-49) as the three-dimensional discrete Fourier transform (DFT) of the sequence of numbers or data points $\{y(lT_S, md_X, nd_Y)\}$.

The next step in the analysis is to assume, as before, that the sound source is in a homogeneous fluid medium in the far-field region of the array, so that the radiated acoustic field incident upon the array can be modeled as a general plane-wave field with arbitrary time dependence as given by Eq. (8.1-9). As a result, the acoustic field incident upon element mn, which is located at $x = md_X$, $y = nd_Y$, and $z = 0$, can be expressed as

$$y_M(t, md_X, nd_Y) = g\left(t - \tau_A + \frac{u_0 md_X + v_0 nd_Y}{c}\right). \tag{8.1-50}$$

If it is further assumed that the output electrical signal from element mn is *directly proportional* to the input acoustical signal (this is reasonable if the electroacoustic transducers are operating as linear filters), then by suppressing the constant of proportionality, we can write that

$$y(t, md_X, nd_Y) = y_M(t, md_X, nd_Y), \tag{8.1-51}$$

and with the use of Eq. (8.1-50),

$$y(lT_S, md_X, nd_Y) = g\left(lT_S - \tau_A + \frac{u_0 md_X + v_0 nd_Y}{c}\right). \tag{8.1-52}$$

Therefore, the sampled output electrical signals $y(lT_S, md_X, nd_Y)$ are directly proportional to time and spatial samples of the incident acoustic field.

If we substitute Eq. (8.1-52) into Eq. (8.1-49) and let

$$l'T_S = lT_S - \tau_A + \frac{u_0 md_X + v_0 nd_Y}{c}, \tag{8.1-53}$$

then

$$Y_S(\eta, \gamma_X, \gamma_Y) = \sum_{l'=-\infty}^{\infty} g(l'T_S) \exp(-j2\pi\eta l'T_S)$$

$$\times D_R'(\eta, \gamma_X + f_{X_0}, \gamma_Y + f_{Y_0}) \exp(-j2\pi\eta\tau_A), \quad (8.1\text{-}54)$$

and if we replace l' with l and assume that only L (odd) time samples are available, then

$$Y_S(\eta, \gamma_X, \gamma_Y) = \sum_{l=-L'}^{L'} g(lT_S) \exp(-j2\pi\eta lT_S)$$

$$\times D_R'(\eta, \gamma_X + f_{X_0}, \gamma_Y + f_{Y_0}) \exp(-j2\pi\eta\tau_A), \quad (8.1\text{-}55)$$

where

$$L' = \frac{L-1}{2}, \qquad\qquad (8.1\text{-}56)$$

and D_R' is the far-field beam pattern of the planar array given by Eq. (8.1-3a). If linear phase weighting is done for purposes of beam steering, then Eq. (8.1-55) can be expressed in direction-cosine space as follows:

$$
\boxed{
\begin{aligned}
Y_S(\eta, u, v) &= \sum_{l=-L'}^{L'} g(lT_S) \exp(-j2\pi\eta lT_S) \\
&\times D_R(\eta, u - [u' - u_0], v - [v' - v_0]) \exp(-j2\pi\eta\tau_A),
\end{aligned}
}
$$

$$(8.1\text{-}57)$$

where D_R is given by Eq. (8.1-3b). The only difference between the *sampled* output frequency and angular spectrum given by Eq. (8.1-57), and the *theoretical* output frequency and angular spectrum given by Eq. (8.1-38), is that the theoretical complex frequency spectrum $G(\eta)$ in Eq. (8.1-38) is replaced by the discrete-time Fourier transform of the time samples $g(lT_S)$ in Eq. (8.1-57).

Therefore, based on our previous discussion concerning the physical significance of the theoretical output frequency and angular spectrum given by Eq. (8.1-38), if a sound source (target) is in a homogeneous fluid medium in the far-field region of a planar array of point elements so that the radiated acoustic field incident upon the array can be modeled as a general plane-wave field with arbitrary time dependence, then both the frequency content (spectral lines) of the radiated sound field and the direction to the sound source (target) relative to the center of the array can be obtained by processing the time-domain samples of the output electrical signals from the elements in the array according to Eq. (8.1-49).

A more efficient way to evaluate Eq. (8.1-49) is to transform it into a *three-dimensional* DFT by discretizing η, γ_X, and γ_Y as follows (see Section 7.1.4):

$$\eta = q\,\Delta\eta, \tag{8.1-58}$$

$$\gamma_X = r\,\Delta\gamma_X, \tag{8.1-59}$$

and

$$\gamma_Y = s\,\Delta\gamma_Y, \tag{8.1-60}$$

where

$$\Delta\eta = \frac{1}{LT_S}, \tag{8.1-61}$$

$$\Delta\gamma_X = \frac{1}{(M + ZM)d_X}, \tag{8.1-62}$$

$$\Delta\gamma_Y = \frac{1}{(N + ZN)d_Y}, \tag{8.1-63}$$

and q, r, and s are integers. The parameters ZM and ZN are also integers and are used for "zero padding" (see Section 7.1.4). Substituting Eqs. (8.1-58) through (8.1-63) into Eq. (8.1-49) and assuming that only L (odd) time samples are available yields

$$Y_S(q,r,s) = \sum_{n=-N'}^{N'} \sum_{m=-M'}^{M'} c_{mn}(q) \sum_{l=-L'}^{L'} y(l,m,n) W_L^{-ql} W_{M+ZM}^{rm} W_{N+ZN}^{sn},$$

$$q = -L',\dots,0,\dots,L', \quad r = -M'',\dots,0,\dots,M'', \quad s = -N'',\dots,0,\dots,N'',$$

$$\tag{8.1-64}$$

where $\Delta\eta$, $\Delta\gamma_X$, and $\Delta\gamma_Y$ have been suppressed in the argument of Y_S; $\Delta\eta$ has been suppressed in the argument of c_{mn}; T_S, d_X, and d_Y have been suppressed in the argument of y; and where

$$L' = \frac{L-1}{2}, \tag{8.1-65}$$

$$M' = \frac{M-1}{2}, \tag{8.1-66}$$

$$N' = \frac{N-1}{2}, \tag{8.1-67}$$

$$W_L = \exp(+j2\pi/L), \tag{8.1-68}$$

$$W_{M+ZM} = \exp\left[+j\frac{2\pi}{M+ZM}\right], \tag{8.1-69}$$

$$W_{N+ZN} = \exp\left[+j\frac{2\pi}{N+ZN}\right], \tag{8.1-70}$$

$$M'' = \begin{cases} (M+ZM)/2, & M+ZM \quad \text{even,} \\ (M+ZM-1)/2, & M+ZM \quad \text{odd,} \end{cases} \tag{8.1-71}$$

$$N'' = \begin{cases} (N+ZN)/2, & N+ZN \quad \text{even,} \\ (N+ZN-1)/2, & N+ZN \quad \text{odd.} \end{cases} \tag{8.1-72}$$

Processing the output data $y(l,m,n)$ from the planar array according to Eq. (8.1-64) is known as *FFT beamforming*, since one can use a FFT computer algorithm to compute a DFT. The term "beamforming" implies that amplitude and phase weighting are done via the complex weights $c_{mn}(q)$. In order to avoid grating lobes under all conditions of beam steering, the interelement spacings d_X and d_Y must satisfy $d_X < \lambda_{min}/2$ and $d_Y < \lambda_{min}/2$, where λ_{min} is the minimum wavelength associated with the maximum expected frequency.

In order to evaluate Eq. (8.1-64) numerically, rewrite it in terms of three one-dimensional DFTs as follows:

$$Y_S(q,r,s) = \sum_{n=-N'}^{N'} Y_S(q,r,n)W_{N+ZN}^{sn}, \qquad s = -N'',\dots,0,\dots,N'', \tag{8.1-73}$$

where

$$Y_S(q,r,n) = \sum_{m=-M'}^{M'} Y_S(q,m,n)W_{M+ZM}^{rm}, \qquad r = -M'',\dots,0,\dots,M'', \tag{8.1-74}$$

and

$$Y_S(q, m, n) = c_{mn}(q) \sum_{l=-L'}^{L'} y(l, m, n) W_L^{-ql}, \qquad q = -L', \ldots, 0, \ldots, L'.$$

(8.1-75)

Equation (8.1-75) is the complex-weighted time-domain DFT (or complex frequency spectrum) of the output electrical signal at element mn in the planar array. Recall that the output electrical signal from each element in the array is directly proportional to the signal $g(t)$ radiated by a sound source (target) [see Eq. (8.1-52)]. Example 8.1-1 addresses the problem of how fast the output electrical signals, and hence $g(t)$, should be sampled in order to avoid aliasing.

Example 8.1-1

In this example we shall derive equations that will allow us to calculate the sampling rate to be used and the number of data points to be collected per element in an array in order to avoid aliasing. We begin the analysis by noting that in any real-world problem, the signal $g(t)$ radiated by a sound source (target) will have finite energy in a finite time interval. A signal $g(t)$ is said to have finite energy E_g in the time interval $|t| \leq T_0/2$ if

$$E_g = \int_{-T_0/2}^{T_0/2} |g(t)|^2 \, dt < \infty.$$

(8.1-76)

Any signal $g(t)$ with finite energy in the time interval $|t| \leq T_0/2$ can be approximated in that interval—in the *minimum-mean-squared-error* sense—by a *finite Fourier series* representation. The signal $g(t)$ may be real or complex, and periodic or nonperiodic. If $g(t)$ is real, then it can be represented by the following *real trigonometric form* of the finite Fourier series:

$$g(t) = a_0 + 2 \sum_{q=1}^{K} a_q \cos(2\pi q f_0 t + \theta_q), \qquad |t| \leq \frac{T_0}{2}, \quad (8.1\text{-}77)$$

where T_0 is the *length of the data record*, or the *fundamental period*, in seconds;

$$f_0 = 1/T_0$$

(8.1-78)

is the *fundamental frequency*, or *first harmonic*, in hertz; a_0 is the *average value*, or *dc component*, of $g(t)$;

$$c_q = a_q \exp(+j\theta_q) = \frac{1}{T_0} \int_{-T_0/2}^{T_0/2} g(t) \exp(-j2\pi q f_0 t) \, dt \qquad (8.1\text{-}79)$$

is the *complex Fourier series coefficient* of the *qth harmonic* (correspond-ing to the spectral line at $q f_0$ hertz); a_q and θ_q are *real* amplitude and phase components, respectively; and K is the *total number of harmonics* used in the representation. In general, as K increases, the mean squared error of the approximation decreases. If $g(t)$ is periodic with period T_0, then a finite Fourier series representation is valid over all time. Equation (8.1-77) is also known as a *trigonometric polynomial*.

The real signal $g(t)$ can also be represented by the following *complex form* of the finite Fourier series:

$$g(t) = \sum_{q=-K}^{K} c_q \exp(+j2\pi q f_0 t), \qquad |t| \le \frac{T_0}{2}. \qquad (8.1\text{-}80)$$

However, if $g(t)$ is a complex signal, for example, if $g(t)$ is a complex envelope (see Section 8.2), then only the complex form of the finite Fourier series given by Eq. (8.1-80) can be used.

From the sampling theorem we know that in order to avoid aliasing, a signal must be sampled at a rate

$$\boxed{f_S = \frac{1}{T_S} \ge 2f_{\max},} \qquad (8.1\text{-}81)$$

where f_S is the *sampling frequency (sampling rate)* in samples per second, T_S is the *sampling period* in seconds, and $f_{\max}$ is the *highest* frequency component (in hertz) contained in the frequency spectrum of the signal. The *minimum* sampling rate is called the *Nyquist rate* and is equal to $2f_{\max}$ samples per second. However, in practical applications, it is always a good idea to sample at a rate greater than the Nyquist rate.

One way to guarantee the value of $f_{\max}$ in order to avoid aliasing is to pass the output electrical signals from the elements through *low-pass pre-filters* before sampling. The pre-filters will filter out all frequency components above $f_{\max}$. Since $g(t)$ is given by Eqs. (8.1-77) or (8.1-80),

$$\boxed{f_{\max} = K f_0 = K/T_0,} \qquad (8.1\text{-}82)$$

and as a result,

$$f_S \geq 2Kf_0 = 2K/T_0, \tag{8.1-83}$$

$$\frac{1}{T_S} \geq \frac{2K}{T_0}, \tag{8.1-84}$$

$$\frac{T_0}{T_S} \geq 2K, \tag{8.1-85}$$

and

$$\frac{T_S}{T_0} \leq \frac{1}{2K}. \tag{8.1-86}$$

Since the length of the data record can be expressed as

$$T_0 = (L - 1)T_S, \tag{8.1-87}$$

where L is the total number of time samples, substituting Eq. (8.1-85) into Eq. (8.1-87) yields

$$L - 1 = \frac{T_0}{T_S} \geq 2K, \tag{8.1-88}$$

or

$$\boxed{L \geq 2K + 1.} \tag{8.1-89}$$

Equation (8.1-89) stipulates the total number of time samples that must be taken per element—in terms of the total number of harmonics K—in order to avoid aliasing.

Next, taking the reciprocal of Eq. (8.1-88) yields

$$\frac{T_S}{T_0} = \frac{1}{L - 1} \leq \frac{1}{2K}. \tag{8.1-90}$$

Since

$$\frac{1}{L} < \frac{1}{L - 1} \leq \frac{1}{2K}, \tag{8.1-91}$$

then

$$\frac{1}{L} < \frac{1}{2K}. \tag{8.1-92}$$

Therefore, *if we let*

$$\boxed{\frac{T_S}{T_0} = \frac{1}{L}},\qquad (8.1\text{-}93)$$

then substituting Eq. (8.1-92) into Eq. (8.1-93) yields

$$\frac{T_S}{T_0} < \frac{1}{2K}, \qquad (8.1\text{-}94)$$

which satisfies Eq. (8.1-86), and hence the sampling theorem. Equation (8.1-93) can also be expressed as follows:

$$\boxed{f_0 = \frac{1}{T_0} = \frac{1}{LT_S} = \frac{f_S}{L}}. \qquad (8.1\text{-}95)$$

The sampling rate f_S can be computed according to Eq. (8.1-95). Note that the frequency domain DFT bins are separated by f_0 hertz as given by Eq. (8.1-95) [see Eqs. (8.1-58) and (8.1-61)]. As can be seen from Eq. (8.1-95), in order to increase the frequency resolution (i.e., decrease the DFT bin separation f_0), one must increase the length of the data record (i.e., the fundamental period), T_0. Also note that if the minimum number of time samples required per element is computed according to Eq. (8.1-89) so that $L = 2K + 1$, then the corresponding minimum sampling rate $f_S = Lf_0$ computed according to Eq. (8.1-95) will always be greater than the Nyquist rate, which is desirable for practical applications.

Example 8.1-2 (Cophasing via beam steering using phase weights)
In this example we shall demonstrate how to cophase all the output electrical signals from all the elements in a planar array by beam steering using phase weights. Since $y(l, m, n)$ is shorthand notation for $y(lT_S, md_X, nd_Y)$, substituting Eq. (8.1-52) into Eq. (8.1-75) yields

$$Y_S(q, m, n) = c_{mn}(q) \sum_{l=-L'}^{L'} g(lT_S - \tau_A + \tau_{mn})W_L^{-ql},$$

$$q = -L',\ldots,0,\ldots, L', \quad (8.1\text{-}96)$$

where

$$\tau_{mn} = \frac{u_0 m d_X + v_0 n d_Y}{c} \tag{8.1-97}$$

is the *relative time delay* in seconds at element mn [relative to the center element at $(m = 0, n = 0)$] due to the direction of arrival of the plane wave radiated by the sound source. In order to proceed further, we note that the complex Fourier series coefficients c_q given by Eq. (8.1-79) can also be computed as follows:

$$c_q = \frac{1}{L} DFT\{g(lT_S)\} = \frac{1}{L} \sum_{l=-L'}^{L'} g(lT_S) W_L^{-ql},$$

$$q = -L', \ldots, 0, \ldots, L'. \tag{8.1-98}$$

Since $g(t)$ has Fourier series coefficients c_q, we know from the properties of Fourier series that the time-shifted signal $g(t - \tau_A + \tau_{mn})$ will have Fourier series coefficients $c_q \exp(-j2\pi q f_0 \tau_A) \exp(+j2\pi q f_0 \tau_{mn})$. Therefore,

$$c_q \exp(-j2\pi q f_0 \tau_A) \exp(+j2\pi q f_0 \tau_{mn})$$

$$= \frac{1}{L} DFT\{g(lT_S - \tau_A + \tau_{mn})\}$$

$$= \frac{1}{L} \sum_{l=-L'}^{L'} g(lT_S - \tau_A + \tau_{mn}) W_L^{-ql},$$

$$q = -L', \ldots, 0, \ldots, L'. \tag{8.1-99}$$

Substituting Eqs. (8.1-99), (8.1-97), and (8.1-6) into Eq. (8.1-96) yields

$$\boxed{\begin{aligned} Y_S(q, m, n) &= L c_q a_{mn}(q) \exp[+j\theta_{mn}(q)] \\ &\times \exp\left(+j2\pi q f_0 \frac{u_0 m d_X + v_0 n d_Y}{c}\right) \exp(-j2\pi q f_0 \tau_A), \\ & \qquad\qquad q = -L', \ldots, 0, \ldots, L'. \end{aligned}}$$

$$\tag{8.1-100}$$

Note that besides c_q and τ_A, both u_0 and v_0 are also unknown *a priori*. They are, in fact, those two quantities that need to be determined in order to estimate the direction to the sound source.

Equation (8.1-100) is directly proportional to the complex Fourier series coefficients c_q, and thus, to the complex frequency spectrum of $g(t)$ at $\eta = qf_0$ hertz as measured at element mn in the array. Also note that the phase of Eq. (8.1-100) is different at each element. As a result, the output electrical signals at each element are out of phase.

If beam steering is done correctly, that is, if $u' = u_0$ and $v' = v_0$, so that the phase weights are given by

$$\theta_{mn}(q) = -2\pi q f_0 \frac{u_0 m d_X + v_0 n d_Y}{c}, \qquad (8.1\text{-}101)$$

then substituting Eq. (8.1-101) into Eq. (8.1-100) yields

$$Y_S(q, m, n) = L c_q a_{mn}(q) \exp(-j2\pi q f_0 \tau_A), \qquad q = -L', \ldots, 0, \ldots, L'. \qquad (8.1\text{-}102)$$

The phase of Eq. (8.1-102) is the same at each element. Therefore, the output electrical signals at each element are in phase, and as a result, the array gain is maximized. Later, in Section 8.3, we shall discuss how to estimate not only τ_A, but also ϕ_A, which are the actual round-trip time delay and Doppler shift of a sound source, respectively.

8.2   Complex Envelopes

In this section we shall discuss the representation of a real bandpass signal in terms of its *complex envelope*. As shall be shown later, the complex envelope is, in general, a *complex signal* with a *low-pass (baseband)* frequency spectrum. One of the main advantages of expressing real bandpass signals in terms of their complex envelopes is that it provides for a simple representation of amplitude- and angle-modulated carriers, which is very useful for doing analysis. For example, in Section 8.3 we shall discuss the *auto-ambiguity function*, which is used as a measure of the range and Doppler resolving capabilities of different transmitted electrical signals (amplitude- and angle-modulated carriers) used in active sonar systems. The complex envelope of a signal is required in order to compute its ambiguity function.

Consider an arbitrary real bandpass signal $x(t)$ whose amplitude spectrum $|X(f)|$ is centered about $f = \pm f_c$ hertz, where

$$X(f) = F_i\{x(t)\} = \int_{-\infty}^{\infty} x(t) \exp(-j2\pi ft)\, dt, \qquad (8.2\text{-}1)$$

and f_c is referred to as the *center* or *carrier frequency* (see Fig. 8.2-1). The *complex*

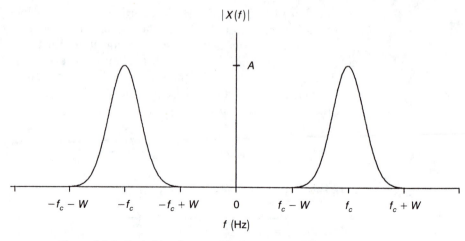

Figure 8.2-1. Typical bandpass amplitude spectrum with bandwidth $2W$ hertz.

envelope of $x(t)$, denoted by $\tilde{x}(t)$, is defined as follows:

$$\boxed{\tilde{x}(t) \triangleq x_p(t) \exp(-j2\pi f_c t),} \tag{8.2-2}$$

where

$$\boxed{x_p(t) = x(t) + j\hat{x}(t)} \tag{8.2-3}$$

is the *pre-envelope*, or *analytic signal*, of $x(t)$, and $\hat{x}(t)$ is the *Hilbert transform* of $x(t)$. From Eq. (8.2-2) we can write that

$$x_p(t) = \tilde{x}(t) \exp(+j2\pi f_c t), \tag{8.2-4}$$

and from Eq. (8.2-3),

$$x(t) = \mathrm{Re}\{x_p(t)\}, \tag{8.2-5}$$

where Re means "take the real part." Substituting Eq. (8.2-4) into Eq. (8.2-5) yields

$$\boxed{x(t) = \mathrm{Re}\{\tilde{x}(t) \exp(+j2\pi f_c t)\}.} \tag{8.2-6}$$

Therefore, given a complex envelope $\tilde{x}(t)$, the corresponding real bandpass signal $x(t)$ can be obtained from Eq. (8.2-6).

The *envelope* of $x(t)$, denoted by $E(t)$, is defined as follows:

$$E(t) \triangleq \text{Abs}\{|\tilde{x}(t)|\} = \text{Abs}\{|x_p(t)|\} \geq 0, \qquad (8.2\text{-}7)$$

where Abs means "take the absolute value." At first glance it would seem that taking the absolute value of the magnitude of a complex signal is redundant. However, since the magnitude of either the complex envelope or the pre-envelope is, in general, a function of time that can take on both positive and negative values, in order to ensure that the envelope $E(t)$ is nonnegative (analogous to the output from an envelope detector), we must also take the absolute value.

As can be seen from Eqs. (8.2-2) and (8.2-3), in order to compute the complex envelope of $x(t)$, we first need to compute the Hilbert transform of $x(t)$. One way to compute the Hilbert transform is to perform the following inverse Fourier transform (see Problem 8-1 for an alternative approach):

$$\hat{x}(t) = F_f^{-1}\{-j\,\text{sgn}(f)\,X(f)\}, \qquad (8.2\text{-}8)$$

where

$$\text{sgn}(f) = \begin{cases} 1, & f > 0, \\ 0, & f = 0, \\ -1, & f < 0, \end{cases} \qquad (8.2\text{-}9)$$

is the *signum* or *sign function*, and $X(f)$ is given by Eq. (8.2-1). Similarly, from Eq. (8.2-8),

$$\hat{X}(f) = F_t\{\hat{x}(t)\} = -j\,\text{sgn}(f)\,X(f). \qquad (8.2\text{-}10)$$

The Hilbert transform is basically a 90° phase shifter. It is theoretically possible to compute the Hilbert transform of a signal by passing it through a linear filter that provides a 90° phase lag at all positive frequencies and a 90° phase lead at all negative frequencies. For example, from Eqs. (8.2-8) and (8.2-9), for $f > 0$,

$$\hat{x}(t) = F_f^{-1}\{-jX(f)\} = F_f^{-1}\{\exp(-j\pi/2)\,X(f)\}, \qquad f > 0, \quad (8.2\text{-}11)$$

which corresponds to a 90° phase lag, and for $f < 0$,

$$\hat{x}(t) = F_f^{-1}\{+jX(f)\} = F_f^{-1}\{\exp(+j\pi/2)\,X(f)\}, \qquad f < 0, \quad (8.2\text{-}12)$$

which corresponds to a 90° phase lead. If we designate the Hilbert transform by

$H\{\cdot\}$, and since $f_c > 0$, then it can be shown that

$$H\{\cos(2\pi f_c t + \theta_0)\} = \sin(2\pi f_c t + \theta_0) = \cos\left(2\pi f_c t + \theta_0 - \frac{\pi}{2}\right) \quad (8.2\text{-}13)$$

and

$$H\{\sin(2\pi f_c t + \theta_0)\} = -\cos(2\pi f_c t + \theta_0) = \sin\left(2\pi f_c t + \theta_0 - \frac{\pi}{2}\right), \quad (8.2\text{-}14)$$

where θ_0 is an arbitrary, constant phase angle in radians.

Next we shall demonstrate that the complex envelope of a real bandpass signal has a low-pass (baseband) frequency spectrum. Taking the Fourier transform with respect to t of Eq. (8.2-2) yields

$$\boxed{\tilde{X}(f) = X_p(f + f_c),} \quad (8.2\text{-}15)$$

where

$$\tilde{X}(f) = F_t\{\tilde{x}(t)\} \quad (8.2\text{-}16)$$

and

$$X_p(f) = F_t\{x_p(t)\}. \quad (8.2\text{-}17)$$

And taking the Fourier transform with respect to t of Eq. (8.2-3) yields

$$X_p(f) = X(f) + j\hat{X}(f). \quad (8.2\text{-}18)$$

Substituting Eq. (8.2-10) into Eq. (8.2-18) yields

$$\boxed{X_p(f) = [1 + \operatorname{sgn}(f)] X(f),} \quad (8.2\text{-}19)$$

and upon substituting Eq. (8.2-9) into Eq. (8.2-19),

$$\boxed{X_p(f) = \begin{cases} 2X(f), & f > 0, \\ X(0) = 0, & f = 0, \\ 0, & f < 0, \end{cases}} \quad (8.2\text{-}20)$$

where $X(0) = 0$ because $x(t)$ was assumed to be a bandpass signal (see Fig. 8.2-1).

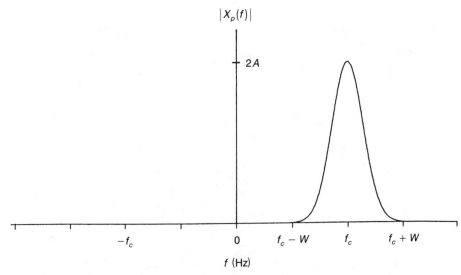

Figure 8.2-2. Typical one-sided pre-envelope amplitude spectrum with bandwidth $2W$ hertz.

The frequency spectrum of the pre-envelope given by Eq. (8.2-20) is known as a *one-sided spectrum*, since it only contains positive frequency components (see Fig. 8.2-2). Note that the maximum value of $|X_p(f)|$ is $2A$, compared with A for $|X(f)|$. Finally, by referring to Eq. (8.2-15), it can be seen that the frequency spectrum of the complex envelope can be obtained by shifting the frequency spectrum of the pre-envelope to the left by an amount equal to the carrier frequency f_c, as shown in Fig. 8.2-3. Figure 8.2-3 illustrates the fact that the frequency spectrum of the complex envelope of a real bandpass signal is low-pass (baseband), which is another reason for representing a real bandpass signal in terms of its complex envelope. For example, suppose we want to compute the frequency spectrum of the real bandpass signal $x(t)$ by performing a DFT on the time samples $x(lT_S)$, where T_S is the sampling period. If we first transform $x(t)$ into $\tilde{x}(t)$, then we can sample the low-pass complex envelope $\tilde{x}(t)$ at a much lower rate than the bandpass signal $x(t)$ (see Example 8.2-2). The resulting amplitude spectrum of the DFT of the complex envelope $\tilde{x}(t)$ can be related to the amplitude spectrum of the DFT of the bandpass signal $x(t)$ by comparing Figs. 8.2-1 and 8.2-3 (also see Example 8.2-4).

Before we demonstrate the calculation of the complex envelope of an amplitude- and angle-modulated carrier in Example 8.2-1, we need the following property of the Hilbert transform: if

$$x(t) = x_{\mathrm{LP}}(t)\, x_{\mathrm{BP}}(t), \tag{8.2-21}$$

where $x_{\mathrm{LP}}(t)$ is a low-pass signal and $x_{\mathrm{BP}}(t)$ is a bandpass signal, and if the

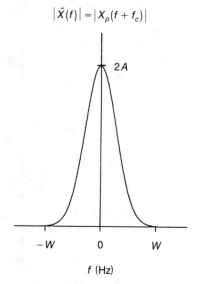

$$\left|\tilde{X}(f)\right| = \left|X_p(f + f_c)\right|$$

2A

−W   0   W

f (Hz)

Figure 8.2-3. Typical low-pass (baseband) complex envelope amplitude spectrum with bandwidth W hertz.

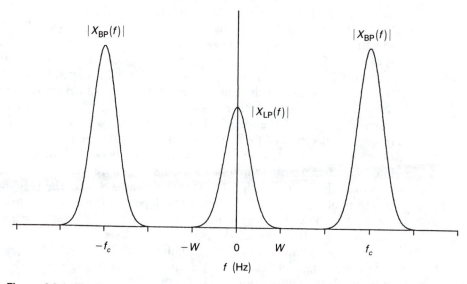

$\left|X_{BP}(f)\right|$ $\left|X_{BP}(f)\right|$

$\left|X_{LP}(f)\right|$

$-f_c$ $-W$ 0 W f_c

f (Hz)

Figure 8.2-4. The low-pass amplitude spectrum $|X_{LP}(f)|$ and the bandpass amplitude spectrum $|X_{BP}(f)|$ do not overlap, as shown.

amplitude spectra $|X_{\mathrm{LP}}(f)|$ and $|X_{\mathrm{BP}}(f)|$ do not overlap (see Fig. 8.2-4), then

$$\boxed{\hat{x}(t) = x_{\mathrm{LP}}(t)\hat{x}_{\mathrm{BP}}(t),}$$

(8.2-22)

that is, only the bandpass signal is Hilbert transformed.

**Example 8.2-1 (Complex envelope of an amplitude-
and angle-modulated carrier)**
In this example we shall compute the complex envelope of the following amplitude- and angle-modulated carrier:

$$x(t) = a(t)\cos[2\pi f_c t + \theta(t)],$$

(8.2-23)

where $a(t)$ and $\theta(t)$ are real amplitude- and angle-modulating signals, respectively, $\cos(2\pi f_c t)$ is the carrier waveform, and f_c is the carrier frequency in hertz. The signal $\theta(t)$ is also known as the *phase deviation*. Before proceeding further, let us briefly review some of the basic terminology and equations associated with angle modulation. This review will be helpful for our discussion of the linear frequency-modulated (LFM) pulse in Section 8.3.2.

The *instantaneous phase* (in radians) of $x(t)$ is defined as the argument of the cosine function, that is,

$$\boxed{\theta_i(t) \triangleq 2\pi f_c t + \theta(t).}$$

(8.2-24)

The *instantaneous radian frequency* (in radians per second) of $x(t)$ is defined as

$$\boxed{\omega_i(t) \triangleq \frac{d}{dt}\theta_i(t) = 2\pi f_c + \frac{d}{dt}\theta(t),}$$

(8.2-25)

and the *instantaneous frequency* (in hertz) is defined as

$$\boxed{f_i(t) \triangleq \frac{1}{2\pi}\frac{d}{dt}\theta_i(t) = f_c + \frac{1}{2\pi}\frac{d}{dt}\theta(t).}$$

(8.2-26)

The time derivative $d\theta(t)/dt$ is known as the *frequency deviation* in radians per second.

The two basic types of angle modulation are *phase modulation* (PM) and *frequency modulation* (FM). Phase modulation implies that the phase deviation of the carrier is directly proportional to some *message* or *modulating signal* $m(t)$, that is,

$$\theta(t) = D_p m(t), \qquad (8.2\text{-}27)$$

where D_p is the *phase-deviation constant* in radians per unit of $m(t)$. Frequency modulation implies that the frequency deviation of the carrier is directly proportional to $m(t)$, that is,

$$\frac{d}{dt}\theta(t) = D_f m(t), \qquad (8.2\text{-}28)$$

where D_f is the *frequency-deviation constant* in radians per second, per unit of $m(t)$. As a result, the phase deviation of a frequency-modulated carrier is given by

$$\theta(t) = \theta(t_0) + D_f \int_{t_0}^{t} m(\alpha)\, d\alpha, \qquad (8.2\text{-}29)$$

where $\theta(t_0)$ is the phase deviation at $t = t_0$.

Let us return to the problem of computing the complex envelope of $x(t)$ as given by Eq. (8.2-23). By making use of the trigonometric identity

$$\cos(\alpha + \beta) = \cos \alpha \cos \beta - \sin \alpha \sin \beta, \qquad (8.2\text{-}30)$$

Eq. (8.2-23) can be rewritten as

$$\boxed{x(t) = x_c(t) \cos(2\pi f_c t) - x_s(t) \sin(2\pi f_c t),} \qquad (8.2\text{-}31)$$

where

$$\boxed{x_c(t) = a(t) \cos \theta(t)} \qquad (8.2\text{-}32)$$

and

$$\boxed{x_s(t) = a(t) \sin \theta(t)} \qquad (8.2\text{-}33)$$

are known as the *cosine* and *sine components* of $x(t)$, respectively. The functions $x_c(t)$ and $x_s(t)$ are also known as the *in-phase* and *quadra-ture-phase components*, respectively. If both $x_c(t)$ and $x_s(t)$ are low-pass (baseband) signals with bandwidth W hertz, that is, if

$$|X_c(f)| = 0, \quad |f| > W, \qquad (8.2\text{-}34)$$

and

$$|X_s(f)| = 0, \quad |f| > W, \qquad (8.2\text{-}35)$$

then, with the use of Eqs. (8.2-22), (8.2-13), and (8.2-14), the Hilbert transform of Eq. (8.2-31) is given by

$$\boxed{\hat{x}(t) = x_c(t) \sin(2\pi f_c t) + x_s(t) \cos(2\pi f_c t)} \qquad (8.2\text{-}36)$$

provided that $f_c > W$, since

$$F_t\{\cos(2\pi f_c t)\} = 0.5\delta(f - f_c) + 0.5\delta(f + f_c) \qquad (8.2\text{-}37)$$

and

$$F_t\{\sin(2\pi f_c t)\} = -j0.5\delta(f - f_c) + j0.5\delta(f + f_c). \qquad (8.2\text{-}38)$$

Substituting Eqs. (8.2-31) and (8.2-36) into Eq. (8.2-3) yields the pre-envelope

$$x_p(t) = [x_c(t) + jx_s(t)] \exp(+j2\pi f_c t), \qquad (8.2\text{-}39)$$

and upon substituting Eq. (8.2-39) into Eq. (8.2-2), we finally obtain the complex envelope

$$\boxed{\tilde{x}(t) = x_c(t) + jx_s(t).} \qquad (8.2\text{-}40)$$

Or, in terms of amplitude and phase,

$$\boxed{\tilde{x}(t) = a(t) \exp[+j\theta(t)],} \qquad (8.2\text{-}41)$$

where it can be shown that

$$a(t) = \sqrt{x_c^2(t) + x_s^2(t)} \qquad (8.2\text{-}42)$$

and

$$\theta(t) = \tan^{-1}\left(\frac{x_s(t)}{x_c(t)}\right) \qquad (8.2\text{-}43)$$

by direct substitution of Eqs. (8.2-32) and (8.2-33) into Eqs. (8.2-42) and (8.2-43). In addition, if we substitute Eq. (8.2-41) into Eq. (8.2-7), then we obtain

$$E(t) = |a(t)| \geq 0, \qquad (8.2\text{-}44)$$

which is the envelope of $x(t)$.

The form of the complex envelope given by Eq. (8.2-41) provides for a convenient way to represent the real amplitude- and angle-modulated carrier given by Eq. (8.2-23). Note that if Eq. (8.2-41) is substituted into Eq. (8.2-6), then we obtain Eq. (8.2-23). The function $x(t)$ given by Eq. (8.2-23) is known as a *narrowband waveform* when $a(t)$ and $\theta(t)$ are slowly varying functions of time compared to $\cos(2\pi f_c t)$, that is, when $a(t)$ and $\theta(t)$ are low-pass signals. Narrowband implies that the bandwidth of $x(t)$, which is $2W$ (see Fig. 8.2-1), divided by the carrier frequency f_c, is much less than unity, that is, $2W/f_c \ll 1$.

The complex envelope given by Eq. (8.2-40) was obtained by evaluating the mathematical definition given by Eq. (8.2-2). However, in practical signal-processing applications, the complex envelope given by Eq. (8.2-40) can be obtained by passing the real amplitude- and angle-modulated carrier given by Eq. (8.2-23) through the *quadrature demodulator system* shown in Fig. 8.2-5. As indicated in Fig. 8.2-5, the output signals from this system are the cosine and sine components given by Eqs. (8.2-32) and (8.2-33), respectively. The cosine component is the output of the I (in-phase) channel and is the real part of the complex envelope. The sine component is the output of the Q (quadrature-phase) channel and is the imaginary part of the complex envelope.

In order for the quadrature demodulator system shown in Fig. 8.2-5 to compute the complex envelope of the input amplitude- and angle-modulated carrier, the carrier frequency of the input signal must be known *a priori*. However, if the amplitude and angle-modulating signals of the input signal are unknown *a priori*, they can be determined from the cosine and sine components by using Eqs. (8.2-42) and (8.2-43), respectively.

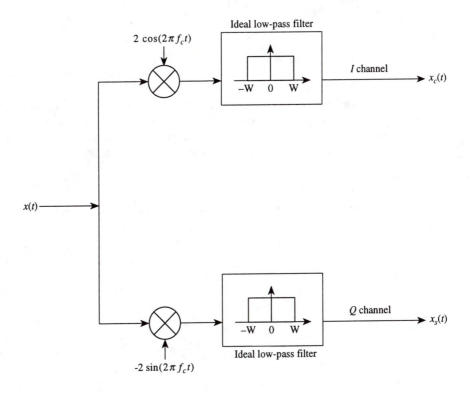

Figure 8.2-5. Quadrature decomposition of the real amplitude- and angle-modulated carrier given by Eq. (8.2-23).

Example 8.2-2   (Bandpass sampling theorem)
Since the complex envelope $\tilde{x}(t)$ is a low-pass (baseband) signal and is bandlimited to W hertz (see Fig. 8.2-3), the *low-pass sampling theorem* states that we can reconstruct $\tilde{x}(t)$ from its sampled values $\tilde{x}(nT_S)$ as follows:

$$\tilde{x}(t) = \sum_{n=-\infty}^{\infty} \tilde{x}(nT_S) \operatorname{sinc}\left(\frac{t - nT_S}{T_S}\right), \qquad (8.2\text{-}45)$$

where

$$f_S = \frac{1}{T_S} \geq 2W \qquad (8.2\text{-}46)$$

is the *sampling frequency* (*sampling rate*) in samples per second, T_S is the *sampling period* in seconds, and

$$\text{sinc}(x) \triangleq \frac{\sin(\pi x)}{\pi x}. \tag{8.2-47}$$

The *minimum* sampling rate is called the *Nyquist rate* and is equal to $2W$ samples per second. Recall that the real bandpass signal $x(t)$ is related to its complex envelope $\tilde{x}(t)$ by Eq. (8.2-6). Substituting Eq. (8.2-45) into Eq. (8.2-6) and using Eq. (8.2-40) yields

$$x(t) = \sum_{n=-\infty}^{\infty} \text{Re}\{[x_c(nT_S) + jx_s(nT_S)]\exp(+j2\pi f_c t)\}\text{sinc}\left(\frac{t - nT_S}{T_S}\right), \tag{8.2-48}$$

or, upon expanding and taking the real part,

$$x(t) = \sum_{n=-\infty}^{\infty} [x_c(nT_S)\cos(2\pi f_c t) - x_s(nT_S)\sin(2\pi f_c t)]$$
$$\times \text{sinc}\left(\frac{t - nT_S}{T_S}\right). \tag{8.2-49}$$

Equation (8.2-49) is the *bandpass sampling theorem*. It states that we can reconstruct a real bandpass signal [i.e., the amplitude- and angle-modulated carrier given by Eq. (8.2-23)] from sampled values of its cosine and sine components. Therefore, instead of sampling $x(t)$ directly at the high rate of $f_S \geq 2(f_c + W)$ samples per second, we need only sample $x_c(t)$ and $x_s(t)$ at the lower rate of $f_S \geq 2W$ samples per second. However, *two* waveforms must be sampled, namely, the cosine and sine components. Therefore, during any T-second interval, $L \geq 2T(2W) = 4TW$ samples must be taken.

Example 8.2-3 (Signal energy and time-average power)
In this example we shall relate the energy of the real amplitude- and angle-modulated carrier $x(t)$ given by Eq. (8.2-23) to the energy of its

complex envelope $\tilde{x}(t)$ given by Eq. (8.2-41). We shall then use the energy relationship to compute the time-average power of $x(t)$.

The *energy* E_x in joules of the real signal $x(t)$ is, by definition,

$$E_x \triangleq \int_{-\infty}^{\infty} x^2(t)\, dt = \int_{-\infty}^{\infty} a^2(t) \cos^2\left[2\pi f_c t + \theta(t)\right] dt. \qquad (8.2\text{-}50)$$

To be more precise, Eq. (8.2-50) is really the *energy per ohm* of $x(t)$. For example, if $x(t)$ is the current flowing through a one-ohm resistor, then $x^2(t)$ is the *instantaneous power* (energy per unit time), and as a result, Eq. (8.2-50) corresponds to the total energy dissipated in the one-ohm resistor.

Similarly, the *energy* $E_{\tilde{x}}$ in joules of the complex envelope $\tilde{x}(t)$ is, by definition,

$$E_{\tilde{x}} \triangleq \int_{-\infty}^{\infty} |\tilde{x}(t)|^2\, dt = \int_{-\infty}^{\infty} a^2(t)\, dt, \qquad (8.2\text{-}51)$$

where use has been made of Eq. (8.2-41). From Eqs. (8.2-2) and (8.2-3),

$$|\tilde{x}(t)|^2 = x^2(t) + \hat{x}^2(t). \qquad (8.2\text{-}52)$$

Substituting Eq. (8.2-52) into Eq. (8.2-51) yields

$$E_{\tilde{x}} = E_x + E_{\hat{x}}, \qquad (8.2\text{-}53)$$

where

$$E_{\hat{x}} = \int_{-\infty}^{\infty} \hat{x}^2(t)\, dt \qquad (8.2\text{-}54)$$

is the energy in joules of the Hilbert transform $\hat{x}(t)$. With the use of *Parseval's theorem*, Eq. (8.2-54) can be expressed as

$$E_{\hat{x}} = \int_{-\infty}^{\infty} |\hat{X}(f)|^2\, df, \qquad (8.2\text{-}55)$$

and upon substituting Eq. (8.2-10) into Eq. (8.2-55), we obtain

$$E_{\hat{x}} = \int_{-\infty}^{\infty} |X(f)|^2\, df = E_x, \qquad (8.2\text{-}56)$$

that is, $x(t)$ and its Hilbert transform $\hat{x}(t)$ have equal energy. Substituting Eq. (8.2-56) into Eq. (8.2-53) finally yields the desired result

$$E_x = 0.5E_{\hat{x}}. \tag{8.2-57}$$

It is much easier, in general, to compute the energy of the real amplitude- and angle-modulated carrier $x(t)$ given by Eq. (8.2-23) by using Eqs. (8.2-57) and (8.2-51) instead of Eq. (8.2-50).

Since in any practical application $x(t)$ has finite duration, so that $x(t) = 0$ for $|t| > T/2$, the *time-average power* P_{avg} in watts of $x(t)$ during the time interval T seconds is given by

$$P_{avg} = \frac{E_x}{T} = \frac{E_{\hat{x}}}{2T}, \tag{8.2-58}$$

where now

$$E_x = \int_{-T/2}^{T/2} x^2(t)\, dt = \int_{-T/2}^{T/2} a^2(t)\cos^2[2\pi f_c t + \theta(t)]\, dt \tag{8.2-59}$$

and

$$E_{\hat{x}} = \int_{-T/2}^{T/2} |\tilde{x}(t)|^2\, dt = \int_{-T/2}^{T/2} a^2(t)\, dt. \tag{8.2-60}$$

Note that because of Eqs. (8.2-57), (8.2-58), and (8.2-60), the angle-modulating signal $\theta(t)$ does *not* contribute to the energy and, hence, to the time-average power of $x(t)$. Finally, if the energies used in Eq. (8.2-58) are energies per ohm, then Eq. (8.2-58) corresponds to a time-average power per ohm.

Knowing the time-average power of an electrical signal is important, since electroacoustic transducers have limits on the maximum time-average power that they can handle before being damaged.

Example 8.2-4 (Frequency-spectrum relationships)
In this example we shall relate the complex frequency spectrum $X(f)$ of the real bandpass signal $x(t)$ to the complex frequency spectrum $\tilde{X}(f)$ of its complex envelope $\tilde{x}(t)$. We begin by rewriting Eq. (8.2-6) as follows:

$$x(t) = 0.5[\tilde{x}(t)\exp(+j2\pi f_c t) + \tilde{x}^*(t)\exp(-j2\pi f_c t)], \tag{8.2-61}$$

since for any arbitrary complex quantity Z,

$$\text{Re}\{Z\} = 0.5(Z + Z^*). \tag{8.2-62}$$

Taking the Fourier transform of both sides of Eq. (8.2-61) with respect to t yields

$$\boxed{X(f) = 0.5\left[\tilde{X}(f - f_c) + \tilde{X}^*(-[f + f_c])\right],} \tag{8.2-63}$$

since

$$F_t\{\tilde{x}(t)\exp(\pm j2\pi f_c t)\} = \tilde{X}(f \mp f_c) \tag{8.2-64}$$

and

$$F_t\{\tilde{x}^*(t)\} = \tilde{X}^*(-f). \tag{8.2-65}$$

Equation (8.2-63) is the desired general result. Note that for an amplitude-modulated carrier where $\theta(t) = 0$, $x(t)$ given by Eq. (8.2-23) reduces to

$$x(t) = a(t)\cos(2\pi f_c t). \tag{8.2-66}$$

Equation (8.2-66) is known as a *double-sideband suppressed-carrier* (DSBSC) waveform. When $\theta(t) = 0$, the complex envelope given by Eq. (8.2-41) reduces to the real amplitude-modulating signal, that is,

$$\tilde{x}(t) = a(t). \tag{8.2-67}$$

When $\tilde{x}(t)$ is real and is given by Eq. (8.2-67),

$$\tilde{X}^*(-f) = \tilde{X}(f) = A(f), \tag{8.2-68}$$

and so the general result given by Eq. (8.2-63) reduces to

$$X(f) = 0.5[A(f - f_c) + A(f + f_c)], \tag{8.2-69}$$

which is the complex frequency spectrum of Eq. (8.2-66).

Example 8.2-5 (Orthogonality relationships)
In this example we shall show that a signal $x(t)$ and its Hilbert transform $\hat{x}(t)$ are *orthogonal*, that is, their *inner product* is equal to zero. In addition, we shall also show that the cosine and sine components $x_c(t)$ and $x_s(t)$ of the real amplitude- and angle-modulated carrier $x(t)$ given by Eq. (8.2-23) are orthogonal.

Consider the following inner product of $x(t)$ and $\hat{x}(t)$:

$$\langle x(t), \hat{x}(t) \rangle \triangleq \int_{-\infty}^{\infty} x(t) \hat{x}^*(t) \, dt = \int_{-\infty}^{\infty} X(f) \hat{X}^*(f) \, df \quad (8.2\text{-}70)$$

where use has been made of Parseval's theorem. Substituting Eq. (8.2-10) into Eq. (8.2-70) yields

$$\langle x(t), \hat{x}(t) \rangle = j \int_{-\infty}^{\infty} \text{sgn}(f) |X(f)|^2 \, df = 0, \quad (8.2\text{-}71)$$

since the integrand in Eq. (8.2-71) is an odd function of frequency f, being equal to the product of the odd function $\text{sgn}(f)$ and even function $|X(f)|^2$. Therefore, from Eqs. (8.2-70) and (8.2-71),

$$\boxed{\langle x(t), \hat{x}(t) \rangle = \langle X(f), \hat{X}(f) \rangle = 0,} \quad (8.2\text{-}72)$$

which indicates that $x(t)$ and its Hilbert transform $\hat{x}(t)$ are orthogonal, and that the Fourier transforms of two orthogonal functions are also orthogonal.

Next we shall show that the cosine and sine components of the real amplitude- and angle-modulated carrier given by Eq. (8.2-23) are also orthogonal. The energy of the complex envelope $\tilde{x}(t)$, originally defined by Eq. (8.2-51), can also be expressed as

$$E_{\tilde{x}} \triangleq \int_{-\infty}^{\infty} |\tilde{x}(t)|^2 \, dt = \int_{-\infty}^{\infty} |\tilde{X}(f)|^2 \, df \quad (8.2\text{-}73)$$

or, with the use of Eq. (8.2-40),

$$E_{\tilde{x}} = \int_{-\infty}^{\infty} \left[x_c^2(t) + x_s^2(t) \right] dt = \int_{-\infty}^{\infty} \left[|X_c(f)|^2 + |X_s(f)|^2 \right] df,$$

$$(8.2\text{-}74)$$

where use has been made of Parseval's theorem in Eqs. (8.2-73) and (8.2-74). Equating the right-hand sides of Eqs. (8.2-73) and (8.2-74) yields

$$\int_{-\infty}^{\infty} |\tilde{X}(f)|^2 \, df = \int_{-\infty}^{\infty} \left[|X_c(f)|^2 + |X_s(f)|^2 \right] df. \quad (8.2\text{-}75)$$

Since

$$\tilde{X}(f) = X_c(f) + jX_s(f) \quad (8.2\text{-}76)$$

and

$$|\tilde{X}(f)|^2 = \tilde{X}(f)\tilde{X}^*(f), \qquad (8.2\text{-}77)$$

substituting Eq. (8.2-76) into Eq. (8.2-77) yields

$$|\tilde{X}(f)|^2 = |X_c(f)|^2 + 2\,\mathrm{Re}\{-jX_c(f)X_s^*(f)\} + |X_s(f)|^2, \quad (8.2\text{-}78)$$

and upon substituting Eq. (8.2-78) into Eq. (8.2-75), it can be seen that we must have

$$\int_{-\infty}^{\infty} X_c(f)X_s^*(f)\,df = \int_{-\infty}^{\infty} x_c(t)x_s^*(t)\,dt = 0 \qquad (8.2\text{-}79)$$

in order for the equality in Eq. (8.2-75) to hold, where use has been made of Parseval's theorem and the fact that the integral of the real part of a complex quantity is equal to the real part of the integral of the complex quantity. Therefore, from Eq. (8.2-79),

$$\boxed{\langle x_c(t), x_s(t)\rangle = \langle X_c(f), X_s(f)\rangle = 0,} \qquad (8.2\text{-}80)$$

which indicates that the cosine and sine components and their Fourier transforms are orthogonal.

8.3 The Auto-Ambiguity Function

8.3.1 Derivation and Interpretation

The *auto-ambiguity function* is used as a measure of the range and Doppler resolving capabilities of different transmitted electrical signals (amplitude- and angle-modulated carriers) used in active sonar (and radar) systems. We shall derive and provide an interpretation of the auto-ambiguity function by considering the following *binary hypothesis-testing problem* of trying to detect a slowly fluctuating point-target return in the presence of noise:

$$H_0: \quad \tilde{r}(t) = \tilde{n}(t), \qquad\qquad -\infty < t < \infty, \qquad (8.3\text{-}1)$$

$$H_1: \quad \tilde{r}(t) = \tilde{y}_T(t) + \tilde{n}(t), \qquad -\infty < t < \infty, \qquad (8.3\text{-}2)$$

where

$$\tilde{y}_T(t) = c\tilde{x}(t - \tau_A)\exp(+j2\pi\phi_A t) \qquad (8.3\text{-}3)$$

and

$$c = a \exp(+j\theta). \hspace{3cm} (8.3\text{-}4)$$

Hypothesis H_0, the *null* hypothesis, states that the complex envelope of the *received signal* $\tilde{r}(t)$ is equal to the complex envelope of the *noise* $\tilde{n}(t)$. Hypothesis H_1 states that the complex envelope of the received signal $\tilde{r}(t)$ is equal to the sum of the complex envelopes of the *target return* $\tilde{y}_T(t)$ and noise $\tilde{n}(t)$. It is assumed that $\tilde{y}_T(t)$ and $\tilde{n}(t)$ are zero-mean, statistically independent, *Gaussian* random processes and that $\tilde{n}(t)$ is wide-sense stationary (WSS), low-pass (baseband) white noise. Therefore, the real noise signal $n(t)$ must be a zero-mean, Gaussian, WSS, bandpass, white-noise random process. The target return given by Eq. (8.3-3) is modeled as a complex-weighted, time- and frequency-shifted replica of the complex envelope of the *transmitted electrical signal* $\tilde{x}(t)$. The complex weight c given by Eq. (8.3-4) is assumed to be a zero-mean, complex, Gaussian random variable that accounts for random amplitude attenuation and random phase shift. The magnitude a is *Rayleigh distributed*, and the phase θ is *uniformly distributed*. Therefore, the magnitude and phase are statistically independent random variables. It is also assumed that the parameters τ_A and ϕ_A, which are the *actual round-trip time delay* and *Doppler shift* of the target, respectively, are unknown, nonrandom parameters to be estimated. Note that the complex envelope of an amplitude- and angle-modulated received carrier $r(t)$ can be obtained by passing $r(t)$ through the quadrature demodulator system shown in Fig. 8.2-5.

In order to decide whether or not a target return is present, we shall process $\tilde{r}(t)$ with the receiver shown in Fig. 8.3-1. The function $\tilde{g}(t)$ is referred to as the *processing waveform* and is yet unspecified. The receiver performs the following test: choose hypothesis H_1 if

$$|\tilde{l}|^2 > \gamma, \hspace{3cm} (8.3\text{-}5)$$

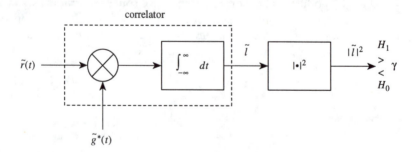

Figure 8.3-1. Correlator followed by a magnitude-squared operation.

and choose hypothesis H_0 otherwise, where

$$\tilde{l} = \langle \tilde{r}(t), \tilde{g}(t) \rangle \triangleq \int_{-\infty}^{\infty} \tilde{r}(t)\tilde{g}^*(t)\, dt \qquad (8.3\text{-}6)$$

is the output from the correlator and is, in fact, the *inner product* of $\tilde{r}(t)$ and $\tilde{g}(t)$. The symbol $\langle \cdot , \cdot \rangle$ indicates inner product. The real scalar $|\tilde{l}|^2$ is referred to as the *sufficient statistic* or as the *test statistic*, since it contains all the information necessary to make a decision. The *decision threshold* γ is commonly chosen to satisfy the *Neyman-Pearson criterion*, which is to maximize the probability of detection for a given probability of false alarm.

Let us begin the analysis of the receiver shown in Fig. 8.3-1 by computing its *output* signal-to-noise power ratio (SNR), which is defined as follows:

$$\text{SNR} \triangleq \frac{E\{|\tilde{l}_T|^2\}}{E\{|\tilde{l}_N|^2\}}, \qquad (8.3\text{-}7)$$

where

$$\tilde{l}_T = \langle \tilde{y}_T(t), \tilde{g}(t) \rangle = \int_{-\infty}^{\infty} \tilde{y}_T(t)\tilde{g}^*(t)\, dt \qquad (8.3\text{-}8)$$

is the output from the correlator due to processing the target return, and

$$\tilde{l}_N = \langle \tilde{n}(t), \tilde{g}(t) \rangle = \int_{-\infty}^{\infty} \tilde{n}(t)\tilde{g}^*(t)\, dt \qquad (8.3\text{-}9)$$

is the output from the correlator due to processing the noise. Equations (8.3-8) and (8.3-9) are also the inner products of the target return and processing waveform and of the noise and processing waveform, respectively.

As can be seen from Eq. (8.3-7), in order to compute the output SNR, we must compute the mean squared values of $\tilde{l}_T$ and $\tilde{l}_N$. To do that, we must specify a functional form for $\tilde{g}(t)$. Therefore, let the processing waveform be given by

$$\tilde{g}(t) = \tilde{x}(t - \hat{\tau})\exp(+j2\pi\hat{\phi}t), \qquad (8.3\text{-}10)$$

which is identical in form to the target return given by Eq. (8.3-3). The parameters $\hat{\tau}$ and $\hat{\phi}$ are *estimates* of the actual round-trip time delay τ_A and Doppler shift ϕ_A of the target, respectively.

Substituting Eqs. (8.3-3) and (8.3-10) into Eq. (8.3-8) yields

$$\tilde{l}_T = c\int_{-\infty}^{\infty} \tilde{x}(t - \tau_A)\tilde{x}^*(t - \hat{\tau})\exp\left[-j2\pi(\hat{\phi} - \phi_A)t\right] dt. \qquad (8.3\text{-}11)$$

If we let $t' = t - \tau_A$ in Eq. (8.3-11), then

$$\tilde{l}_T = cX(\tau, \phi) \exp(-j2\pi\phi\tau_A), \tag{8.3-12}$$

where

$$
\begin{aligned}
X(\tau, \phi) &= \langle \tilde{x}(t), \tilde{x}(t - \tau) \exp(+j2\pi\phi t) \rangle \\
&= \int_{-\infty}^{\infty} \tilde{x}(t)\tilde{x}^*(t - \tau) \exp(-j2\pi\phi t)\, dt \\
&= F_t\{\tilde{x}(t)\tilde{x}^*(t - \tau)\}
\end{aligned}
\tag{8.3-13}
$$

is the *unnormalized auto-ambiguity function* of $\tilde{x}(t)$,

$$\tau = \hat{\tau} - \tau_A \tag{8.3-14}$$

is the *error* in seconds in estimating τ_A, and

$$\phi = \hat{\phi} - \phi_A \tag{8.3-15}$$

is the *error* in hertz in estimating ϕ_A. Substituting Eq. (8.3-4) into Eq. (8.3-12) and taking the magnitude squared of both sides of the resulting expression yields

$$|\tilde{l}_T|^2 = a^2 |X(\tau, \phi)|^2. \tag{8.3-16}$$

As a result, the mean squared value of $\tilde{l}_T$, which is given by

$$E\{|\tilde{l}_T|^2\} = E\{a^2\} |X(\tau, \phi)|^2, \tag{8.3-17}$$

is directly proportional to the magnitude squared of the auto-ambiguity function. The auto-ambiguity function has its maximum value at $\tau = 0$ and $\phi = 0$, that is,

$$X(0,0) = \langle \tilde{x}(t), \tilde{x}(t) \rangle = \int_{-\infty}^{\infty} |\tilde{x}(t)|^2\, dt = E_{\tilde{x}}, \tag{8.3-18}$$

which is the energy of $\tilde{x}(t)$. From Eqs. (8.3-14) and (8.3-15), it can be seen that $\tau = 0$ and $\phi = 0$ implies that the estimates $\hat{\tau}$ and $\hat{\phi}$ are equal to the actual or true values τ_A and ϕ_A. We can easily verify that $E_{\tilde{x}}$ is the maximum value of the auto-ambiguity function by making use of the *Schwarz inequality* given by

$$\left| \int_{-\infty}^{\infty} u(\alpha) v(\alpha) \, d\alpha \right|^2 \leq \int_{-\infty}^{\infty} |u(\alpha)|^2 \, d\alpha \int_{-\infty}^{\infty} |v(\alpha)|^2 \, d\alpha, \qquad (8.3\text{-}19)$$

where $u(\alpha)$ and $v(\alpha)$ are arbitrary finite-energy complex functions. Taking the magnitude squared of both sides of Eq. (8.3-13) and making use of the Schwarz inequality given by Eq. (8.3-19) yields

$$|X(\tau, \phi)|^2 \leq \int_{-\infty}^{\infty} |\tilde{x}(t)|^2 \, dt \int_{-\infty}^{\infty} |\tilde{x}^*(t-\tau)|^2 \, dt \qquad (8.3\text{-}20)$$

or, from Eq. (8.3-18),

$$|X(\tau, \phi)|^2 \leq |X(0,0)|^2 = E_{\tilde{x}}^2. \qquad (8.3\text{-}21)$$

Let us compute the mean squared value of $\tilde{l}_N$ next. Taking the magnitude squared of both sides of Eq. (8.3-9) yields

$$|\tilde{l}_N|^2 = \tilde{l}_N \tilde{l}_N^* = \int_{-\infty}^{\infty} \int_{-\infty}^{\infty} \tilde{g}^*(t) \tilde{n}(t) \tilde{n}^*(t') \tilde{g}(t') \, dt \, dt', \qquad (8.3\text{-}22)$$

so that

$$E\{|\tilde{l}_N|^2\} = \int_{-\infty}^{\infty} \int_{-\infty}^{\infty} \tilde{g}^*(t) R_{\tilde{n}}(t, t') \tilde{g}(t') \, dt \, dt', \qquad (8.3\text{-}23)$$

where

$$R_{\tilde{n}}(t, t') = E\{\tilde{n}(t) \tilde{n}^*(t')\} \qquad (8.3\text{-}24)$$

is the autocorrelation function of $\tilde{n}(t)$. Since it was assumed that $\tilde{n}(t)$ is a WSS, low-pass (baseband), white-noise random process, its power spectral density function is given by

$$S_{\tilde{n}}(f) = \begin{cases} N_0, & |f| \leq B, \\ 0, & |f| > B, \end{cases} \qquad (8.3\text{-}25)$$

where the constant N_0 has units of watts per hertz, or joules, and B is the bandwidth in hertz of the low-pass filters in the quadrature demodulator system. As we shall discuss later, B is, in general, *not* equal to the bandwidth W of $\tilde{x}(t)$. In addition, because of the WSS assumption,

$$R_{\tilde{n}}(t, t') = R_{\tilde{n}}(\Delta t), \qquad (8.3\text{-}26)$$

where

$$R_{\tilde{n}}(\Delta t) = F_f^{-1}\{S_{\tilde{n}}(f)\} = \int_{-\infty}^{\infty} S_{\tilde{n}}(f) \exp(+j2\pi f \Delta t)\, df \qquad (8.3\text{-}27)$$

and

$$\Delta t = t - t'. \qquad (8.3\text{-}28)$$

Therefore, substituting Eq. (8.3-25) into Eq. (8.3-27) yields

$$R_{\tilde{n}}(\Delta t) = 2BN_0 \operatorname{sinc}(2B\,\Delta t) \qquad (8.3\text{-}29)$$

where

$$\operatorname{sinc}(x) \triangleq \frac{\sin(\pi x)}{\pi x}. \qquad (8.3\text{-}30)$$

The variance $\sigma_{\tilde{n}}^2$ of the zero-mean noise $\tilde{n}(t)$ can be obtained by evaluating Eq. (8.3-29) at $\Delta t = 0$. Doing so yields

$$\sigma_{\tilde{n}}^2 = R_{\tilde{n}}(0) = E\{|\tilde{n}(t)|^2\} = 2BN_0, \qquad (8.3\text{-}31)$$

where $\sigma_{\tilde{n}}^2$ corresponds to an average power with units of watts.
   Substituting Eq. (8.3-29) into Eq. (8.3-23) yields

$$E\{|\tilde{l}_N|^2\} = 2BN_0 \int_{-\infty}^{\infty} \tilde{f}(t)\tilde{g}^*(t)\, dt \qquad (8.3\text{-}32)$$

where

$$\tilde{f}(t) = \tilde{g}(t) \underset{t}{*} \operatorname{sinc}(2Bt) = \int_{-\infty}^{\infty} \tilde{g}(t')\operatorname{sinc}[2B(t-t')]\, dt'. \qquad (8.3\text{-}33)$$

Using Parseval's theorem, we can write that

$$\int_{-\infty}^{\infty} \tilde{f}(t)\tilde{g}^*(t)\, dt = \int_{-\infty}^{\infty} \tilde{F}(f)\tilde{G}^*(f)\, df, \qquad (8.3\text{-}34)$$

where, from Eq. (8.3-33),

$$\tilde{F}(f) = \tilde{G}(f)F_t\{\operatorname{sinc}(2Bt)\}. \qquad (8.3\text{-}35)$$

Since

$$F_t\{\operatorname{sinc}(2Bt)\} = \frac{1}{2B}\operatorname{rect}\left(\frac{f}{2B}\right), \qquad (8.3\text{-}36)$$

where

$$\text{rect}\left(\frac{f}{2B}\right) = \begin{cases} 1, & |f| \le B, \\ 0, & |f| > B, \end{cases} \tag{8.3-37}$$

substituting Eqs. (8.3-35) through (8.3-37) into Eq. (8.3-34) yields

$$\int_{-\infty}^{\infty} \tilde{f}(t)\tilde{g}^*(t)\, dt = \frac{1}{2B} \int_{-B}^{B} |\tilde{G}(f)|^2\, df. \tag{8.3-38}$$

However, the processing waveform $\tilde{g}(t)$ is given by Eq. (8.3-10). Therefore, taking the Fourier transform of both sides of Eq. (8.3-10) with respect to t yields

$$\tilde{G}(f) = \tilde{X}(f - \hat{\phi})\exp\left[-j2\pi(f - \hat{\phi})\hat{\tau}\right], \tag{8.3-39}$$

and upon substituting Eq. (8.3-39) into Eq. (8.3-38), we obtain

$$\int_{-\infty}^{\infty} \tilde{f}(t)\tilde{g}^*(t)\, dt = \frac{1}{2B} \int_{-B}^{B} |\tilde{X}(f - \hat{\phi})|^2\, df. \tag{8.3-40}$$

Recall that the frequency spectrum $\tilde{X}(f)$ is baseband with bandwidth W hertz (see Fig. 8.2-3). If the bandwidth B of the noise $\tilde{n}(t)$, which is equal to the bandwidth of the low-pass filters in the quadrature demodulator system, satisfies the relationship

$$\boxed{B \ge W + |\hat{\phi}_{\max}|,} \tag{8.3-41}$$

where $\hat{\phi}_{\max}$ represents the estimate (positive or negative in value) of the *maximum* expected target Doppler shift, then Eq. (8.3-40) simplifies to

$$\int_{-\infty}^{\infty} \tilde{f}(t)\tilde{g}^*(t)\, dt = \frac{1}{2B} \int_{-\infty}^{\infty} |\tilde{X}(f)|^2\, df = \frac{1}{2B} \int_{-\infty}^{\infty} |\tilde{x}(t)|^2\, dt, \tag{8.3-42}$$

or

$$\int_{-\infty}^{\infty} \tilde{f}(t)\tilde{g}^*(t)\, dt = \frac{E_{\tilde{x}}}{2B}, \tag{8.3-43}$$

where $E_{\tilde{x}}$ is the energy of $\tilde{x}(t)$. Therefore, substituting Eq. (8.3-43) into Eq. (8.3-32) yields

$$\boxed{E\{|\tilde{I}_N|^2\} = N_0 E_{\tilde{x}}.} \tag{8.3-44}$$

By substituting Eqs. (8.3-17) and (8.3-44) into Eq. (8.3-7), we finally obtain the following expression for the *output* signal-to-noise power ratio (SNR) for the receiver shown in Fig. 8.3-1:

$$\text{SNR} = |X_N(\tau,\phi)|^2 E\{a^2\} E_{\tilde{x}}/N_0,$$
(8.3-45)

where

$$X_N(\tau,\phi) \triangleq \frac{X(\tau,\phi)}{E_{\tilde{x}}}$$
(8.3-46)

is the *normalized* auto-ambiguity function, so that [see Eq. (8.3-18)]

$$X_N(0,0) = 1.$$
(8.3-47)

Because of the assumption of Gaussian statistics, the decision threshold γ and the error performance of this receiver for a Neyman-Pearson test are given by (see Appendix 8A)

$$\gamma = E_{\tilde{x}} N_0 \ln(1/P_{FA})$$
(8.3-48)

and

$$P_D = P_{FA}^{1/(1+\text{SNR})},$$
(8.3-49)

respectively, where P_D and P_{FA} are the *probabilities of detection and false alarm*, respectively. From Eq. (8.3-48) it can be seen that γ is inversely proportional to P_{FA}. Therefore, as $P_{FA} \to 0$, $\gamma \to \infty$. And from Eq. (8.3-49), it can be seen that as SNR increases, P_D increases for a given P_{FA}. Therefore, as SNR $\to \infty$, $P_D \to P_{FA}^0 = 1$. Similarly, as SNR $\to 0$, $P_D \to P_{FA}$.

Taking the logarithm (base 10) of both sides of Eq. (8.3-49) yields

$$\log_{10} P_D = \frac{1}{1+\text{SNR}} \log_{10} P_{FA},$$
(8.3-50)

and upon solving for the SNR, we obtain

$$\text{SNR} = \frac{\log_{10} P_{FA}}{\log_{10} P_D} - 1.$$
(8.3-51)

Equation (8.3-51) can be used to compute the SNR that is required in order to obtain a desired P_D for a given P_{FA}. Note that the SNR computed according to Eq. (8.3-51) is *dimensionless*.

Let us now provide an interpretation of the normalized auto-ambiguity function. From Eq. (8.3-45) it can be seen that the SNR is directly proportional to $|X_N(\tau, \phi)|^2$. When $\tau = 0$ and $\phi = 0$, that is, when $\hat{\tau} = \tau_A$ and $\hat{\phi} = \phi_A$, the SNR is a maximum, as would be expected, with value [see Eqs. (8.3-45) and (8.3-47)]

$$\text{SNR}_{\max} = E\{a^2\}E_{\tilde{x}}/N_0. \qquad (8.3\text{-}52)$$

Substituting Eq. (8.3-52) into Eq. (8.3-49) yields the corresponding maximum value for the probability of detection, which is desirable, since we want the probability of declaring that we detected a target with the *correct* round-trip time delay (range) and Doppler values to be maximized. Similarly, whenever $\hat{\tau} \neq \tau_A$ and $\hat{\phi} \neq \phi_A$, we would like the SNR to decrease as the errors in the estimates of τ_A and ϕ_A increase. A small SNR results in a small P_D [see Eq. (8.3-49)], which is desirable, since whenever we make large estimation errors, we want the probability of declaring that we detected a target with the *wrong* round-trip time delay (range) and Doppler values to be small.

To ensure this kind of performance, and hence to avoid any *ambiguity* concerning the estimates of τ_A and ϕ_A, the normalized auto-ambiguity function should have as narrow a main lobe about the origin ($\tau = 0$, $\phi = 0$) as possible, so that as the errors in the estimates of τ_A and ϕ_A increase, the value of $|X_N(\tau, \phi)|$ decreases rapidly, and as a result, both SNR and P_D decrease rapidly. In addition, it is also desirable that $|X_N(\tau, \phi)|$ have sidelobe levels as low as possible.

If it is assumed that τ_A and ϕ_A are known constants, and if we let

$$\tilde{g}(t) = \tilde{x}(t - \tau_A) \exp(+j2\pi\phi_A t), \qquad (8.3\text{-}53)$$

then the receiver shown in Fig. 8.3-1, with processing waveform $\tilde{g}(t)$ given by Eq. (8.3-53), is the *optimal log-likelihood ratio test* for detecting signals with random amplitude and phase, where $|\tilde{l}|^2$ is the test statistic. The decision threshold and the error performance of this receiver for a Neyman-Pearson test are given by Eqs. (8.3-48) and (8.3-49), respectively, with output SNR given by Eq. (8.3-52).

Example 8.3-1 (Array gain and the probability of detection)
In this example we shall discuss the relationship between array gain and the probability of detection. In order to take array gain into account, consider processing the output electrical signals from all the elements in a linear array of N (odd) identical, equally spaced, complex-weighted, omnidirectional point elements lying along the X axis. If rectangular amplitude weights are used and correct beam steering is done so that the output electrical signals from all the elements in the array are in

phase, then the complex envelopes of the target return and noise can be expressed as follows (see Section 7.3):

$$\tilde{y}_T(t) = \sum_{i=-N'}^{N'} \tilde{s}_i(t) \tag{8.3-54}$$

and

$$\tilde{n}(t) = \sum_{i=-N'}^{N'} \tilde{n}_i(t), \tag{8.3-55}$$

where

$$\tilde{s}_i(t) = c\tilde{x}(t - \tau_A)\exp(+j2\pi\phi_A t) \tag{8.3-56}$$

is the complex envelope of the output signal component from element i,

$$\tilde{n}_i(t) = \tilde{N}_i(t - \tau_i') \tag{8.3-57}$$

is the complex envelope of the output noise component from element i, N' is given by Eq. (7.3-11), and τ_i' is the time delay (in seconds) applied to element i in order to beam-steer.

If we make the same assumptions and follow the same approach used in this section, and if we further assume that the output noise components from the different elements in the array have zero correlation and that the noise is identically distributed at all elements (see Example 7.3-2), then the output signal-to-noise power ratio for the receiver shown in Fig. 8.3-1 due to processing the complex envelopes of the received signals from a linear array of elements is given by

$$\boxed{\text{SNR}_A = N \times \text{SNR},} \tag{8.3-58}$$

where N is the total odd number of elements in the array and SNR is given by Eq. (8.3-45). Note that if $N = 1$, then $\text{SNR}_A = \text{SNR}$, which is the output signal-to-noise power ratio due to processing the complex envelope of the received signal from a single element in the array. Since the array gain (AG) is defined as (see Section 7.3)

$$\text{AG} \triangleq 10\log_{10}(\text{SNR}_A/\text{SNR})\ \text{dB}, \tag{8.3-59}$$

substituting Eq. (8.3-58) into Eq. (8.3-59) yields

$$\text{AG} = 10\log_{10} N\ \text{dB} \tag{8.3-60}$$

as expected (see Example 7.3-2).

It can also be shown that the decision threshold for a Neyman-Pearson test due to processing the complex envelopes of the received signals from a linear array of elements is given by

$$\boxed{\gamma_A = N\gamma,}$$

(8.3-61)

where γ is given by Eq. (8.3-48). In addition,

$$\boxed{P_D = P_{FA}^{1/(1+SNR_A)},}$$

(8.3-62)

and upon solving for SNR_A, we obtain

$$\boxed{SNR_A = \frac{\log_{10} P_{FA}}{\log_{10} P_D} - 1.}$$

(8.3-63)

Equation (8.3-63) can be used to compute the SNR_A that is required in order to obtain a desired P_D for a given P_{FA}. Note that the SNR_A computed according to Eq. (8.3-63) is *dimensionless*.

Once SNR_A has been determined from Eq. (8.3-63), and if the dimensionless SNR is also known, then solving for N in Eq. (8.3-58) yields

$$\boxed{N = SNR_A/SNR,}$$

(8.3-64)

which is the total number of elements required in order to obtain a desired P_D for a given P_{FA}. Note that the value of N obtained from Eq. (8.3-64) may have to be *rounded up* to the nearest *odd* number for our linear array. Once N has been determined, and if the interelement spacing d is known (e.g., $d \leq \lambda_{min}/2$), then the overall length of the array that is required in order to satisfy the P_D and P_{FA} constraints is given by $L = (N - 1)d$. Equation (8.3-64) is a *dimensionless version of array gain*.

8.3.2 The Normalized Auto-Ambiguity Functions of Rectangular-Envelope CW and LFM Pulses

In this section we shall derive and discuss the normalized auto-ambiguity functions of two very common transmitted signals used in active sonar systems, namely, the rectangular-envelope continuous-wave (CW) pulse and the rectangular-envelope linear frequency-modulated (LFM) pulse. Since the auto-ambiguity function involves the complex envelope of the transmitted signal, different signals have different ambiguity functions, and hence different range and Doppler resolving capabilities.

Rectangular-Envelope CW Pulse

A *CW pulse* is nothing more than a *finite-duration* DSBSC waveform given by [see Eq. (8.2-66)]

$$x(t) = a(t) \cos(2\pi f_c t) \operatorname{rect}(t/T), \qquad (8.3\text{-}65)$$

where

$$\operatorname{rect}(t/T) = \begin{cases} 1, & |t| \le T/2, \\ 0, & |t| > T/2, \end{cases} \qquad (8.3\text{-}66)$$

is the rectangular amplitude window, and T is the *pulse length* in seconds. By referring to Eqs. (8.2-41) and (8.2-44), it can be seen that the complex envelope and the envelope of Eq. (8.3-65) are given by

$$\tilde{x}(t) = a(t) \operatorname{rect}(t/T) \qquad (8.3\text{-}67)$$

and

$$E(t) = |a(t)| \operatorname{rect}(t/T), \qquad (8.3\text{-}68)$$

respectively (see Fig. 8.3-2*a*).

If the amplitude-modulating signal $a(t)$ is equal to a positive constant A, then

$$x(t) = A \cos(2\pi f_c t) \operatorname{rect}(t/T) \qquad (8.3\text{-}69)$$

is known as a *rectangular-envelope CW pulse* with complex envelope and envelope given by

$$\tilde{x}(t) = A \operatorname{rect}(t/T) \qquad (8.3\text{-}70)$$

and

$$E(t) = A \operatorname{rect}(t/T), \qquad (8.3\text{-}71)$$

respectively (see Fig. 8.3-2*b*). The energy of the complex envelope given by

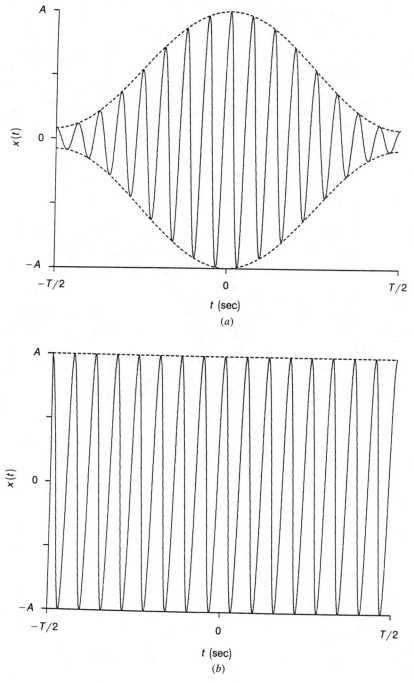

Figure 8.3-2. A CW pulse with (*a*) an arbitrary envelope $E(t) = |a(t)| \geq 0$, and (*b*) a rectangular envelope $E(t) = A > 0$.

Eq. (8.3-70) is

$$E_{\tilde{x}} \triangleq \int_{-\infty}^{\infty} |\tilde{x}(t)|^2 \, dt = \int_{-T/2}^{T/2} A^2 \, dt = A^2 T,$$

(8.3-72)

and as a result, the energy and time-average power of a rectangular-envelope CW pulse are given by

$$E_x = 0.5 E_{\tilde{x}} = 0.5 A^2 T$$

(8.3-73)

and

$$P_{\text{avg}} = \frac{E_x}{T} = \frac{E_{\tilde{x}}}{2T} = \frac{A^2}{2},$$

(8.3-74)

respectively (see Example 8.2-3).

The normalized auto-ambiguity function of a rectangular-envelope CW pulse will be derived next. Substituting Eq. (8.3-70) into Eq. (8.3-13) yields

$$X(\tau, \phi) = A^2 \int_{-\infty}^{\infty} \text{rect}\left(\frac{t}{T}\right) \text{rect}\left(\frac{t-\tau}{T}\right) \exp(-j2\pi\phi t) \, dt,$$

(8.3-75)

or

$$X(\tau, \phi) = A^2 F_t \left\{ \text{rect}\left(\frac{t}{T}\right) \text{rect}\left(\frac{t-\tau}{T}\right) \right\}.$$

(8.3-76)

By referring to Fig. 8.3-3, one can see that for $0 \le \tau \le T$, Eq. (8.3-75) reduces to

$$X(\tau, \phi) = A^2 \int_{(-T/2)+\tau}^{T/2} \exp(-j2\pi\phi t) \, dt, \qquad 0 \le \tau \le T,$$

(8.3-77)

or

$$X(\tau, \phi) = \begin{cases} A^2(T-\tau) \, \text{sinc}[\phi(T-\tau)] \exp(-j\pi\phi\tau), & 0 \le \tau \le T, \\ 0, & \tau > T. \end{cases}$$

(8.3-78)

Similarly, for $-T \le \tau \le 0$, Eq. (8.3-75) reduces to

$$X(\tau, \phi) = A^2 \int_{-T/2}^{(T/2)+\tau} \exp(-j2\pi\phi t) \, dt, \qquad -T \le \tau \le 0,$$

(8.3-79)

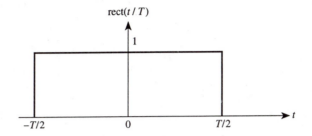

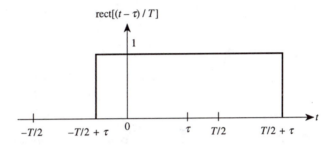

Figure 8.3-3. Illustration of the two rectangle functions appearing in the integrand of Eq. (8.3-75).

or

$$X(\tau,\phi) = \begin{cases} A^2(T + \tau)\,\text{sinc}\big[\phi(T + \tau)\big]\exp(-j\pi\phi\tau), & -T \leq \tau \leq 0, \\ 0, & \tau < -T. \end{cases}$$

$$(8.3\text{-}80)$$

Both Eq. (8.3-78) and Eq. (8.3-80) can be represented by the following single expression:

$$X(\tau,\phi) = \begin{cases} A^2T\left(1 - \dfrac{|\tau|}{T}\right)\text{sinc}\left[\phi T\left(1 - \dfrac{|\tau|}{T}\right)\right]\exp(-j\pi\phi\tau), & |\tau| \leq T, \\ 0, & |\tau| > T. \end{cases}$$

$$(8.3\text{-}81)$$

Equation (8.3-81) is the *unnormalized* auto-ambiguity function of a rectangular-envelope CW pulse. Substituting Eqs. (8.3-81) and (8.3-72) into Eq. (8.3-46) yields the following expression for the *normalized auto-ambiguity function of a rectangu-lar-envelope CW pulse*:

$$
X_N(\tau,\phi) = \begin{cases} \left(1 - \dfrac{|\tau|}{T}\right) \text{sinc}\left[\phi T\left(1 - \dfrac{|\tau|}{T}\right)\right] \exp(-j\pi\phi\tau), & |\tau| \le T, \\ 0, & |\tau| > T. \end{cases}
$$

$$(8.3\text{-}82)$$

By setting $\phi = 0$ in Eq. (8.3-82) and taking the magnitude of the resulting expression, we obtain the following expression for the *round-trip time-delay* (*range*) *profile*:

$$
|X_N(\tau,0)| = \begin{cases} 1 - \dfrac{|\tau|}{T}, & |\tau| \le T, \\ 0, & |\tau| > T, \end{cases}
$$

$$(8.3\text{-}83)$$

which is the equation for a triangle. Similarly, by setting $\tau = 0$ in Eq. (8.3-82) and taking the magnitude of the resulting expression, we obtain the following expression for the *Doppler profile*:

$$
|X_N(0,\phi)| = |\text{sinc}(\phi T)|.
$$

$$(8.3\text{-}84)$$

By inspecting Eq. (8.3-83), it can be seen that the width of the main lobe of the ambiguity function along the τ axis is *directly proportional to the pulse length T*, since the range profile decreases to zero at $\tau = \pm T$ seconds. However, by inspecting Eq. (8.3-84), it can be seen that the width of the main lobe of the ambiguity function along the ϕ axis is *inversely proportional to T*, since the locations of the first zero crossings of the Doppler profile are at $\phi = \pm 1/T$ hertz. A single parameter, the pulse length T, controls both the round-trip time delay (range) and Doppler resolving capabilities of a rectangular-envelope CW pulse. One can make the width of the main lobe of the ambiguity function arbitrarily narrow in either direction by varying T, but not in both directions simultaneously. As a result, *a long-duration CW pulse has poor range resolution but good Doppler resolution, and a short-duration CW pulse has good range resolution but poor Doppler resolution.* If we are going to improve our range and Doppler estimates *simultaneously*, we must try a more complicated signal that contains several parameters that we can vary. A rectangular-envelope, linear frequency-modulated (LFM) pulse is such a signal and will be discussed next. However, before doing so, recall that besides wanting the ambiguity function to have as narrow a main lobe about the origin as possible, we also want it to have sidelobe levels as low as possible.

The sidelobe levels of an ambiguity function can be reduced by using an amplitude-modulating function $a(t)$ other than rectangular (i.e., a constant), analogous to using different amplitude weights to reduce the sidelobe levels of the far-field beam pattern of an array. In fact, one can use the same continuous amplitude windows discussed in Section 6.3.1 for $a(t)$. However, as with far-field beam patterns, there is a tradeoff, since as the sidelobe levels decrease, the width of the main lobe of the ambiguity function generally increases.

Rectangular-Envelope LFM Pulse

If we begin with a finite-duration amplitude- and angle-modulated carrier

$$x(t) = a(t)\cos\left[2\pi f_c t + \theta(t)\right] \text{rect}(t/T), \tag{8.3-85}$$

and if we let the angle-modulating signal, or phase deviation, be given by

$$\theta(t) = D_p t^2, \tag{8.3-86}$$

so that the frequency deviation

$$\frac{d}{dt}\theta(t) = 2D_p t \tag{8.3-87}$$

is equal to a *linear* function of time, then the signal

$$x(t) = a(t)\cos\left(2\pi f_c t + D_p t^2\right)\text{rect}(t/T) \tag{8.3-88}$$

is known as a *linear frequency-modulated (LFM) pulse*, where D_p is the phase-deviation constant. If we let $D_p = 0$, then Eq. (8.3-88) reduces to the CW pulse given by Eq. (8.3-65). By referring to Eqs. (8.2-41) and (8.2-44), it can be seen that the complex envelope and the envelope of Eq. (8.3-88) are given by

$$\tilde{x}(t) = a(t)\exp\left(+jD_p t^2\right)\text{rect}(t/T) \tag{8.3-89}$$

and

$$E(t) = |a(t)|\,\text{rect}(t/T), \tag{8.3-90}$$

respectively (see Fig. 8.3-4a).

If the amplitude-modulating signal $a(t)$ is equal to a positive constant A, then

$$x(t) = A\cos\left(2\pi f_c t + D_p t^2\right)\text{rect}(t/T) \tag{8.3-91}$$

is known as a *rectangular-envelope LFM pulse*, and its complex envelope and

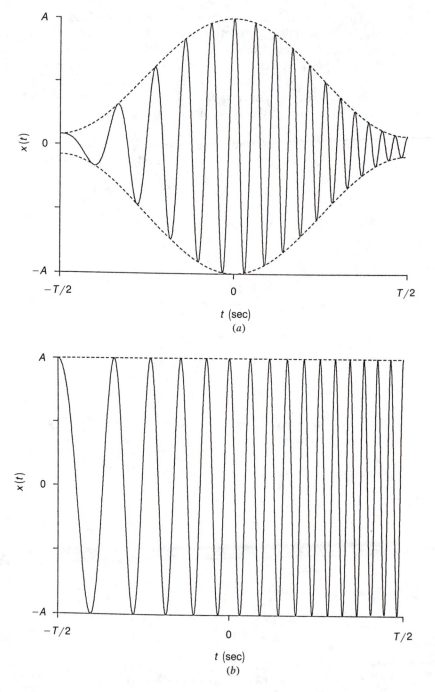

Figure 8.3-4. A LFM pulse with (*a*) an arbitrary envelope $E(t) = |a(t)| \geq 0$, and (*b*) a rectangular envelope $E(t) = A > 0$.

envelope are given by

$$\tilde{x}(t) = A \exp\left(+jD_p t^2\right) \text{rect}(t/T)$$

(8.3-92)

and

$$E(t) = A \, \text{rect}(t/T),$$

(8.3-93)

respectively (see Fig. 8.3-4b). If we let $D_p = 0$, then Eq. (8.3-91) reduces to the rectangular-envelope CW pulse given by Eq. (8.3-69). The energy of the complex envelope given by Eq. (8.3-92) is

$$E_{\tilde{x}} \triangleq \int_{-\infty}^{\infty} |\tilde{x}(t)|^2 \, dt = \int_{-T/2}^{T/2} A^2 \, dt = A^2 T,$$

(8.3-94)

and as a result, the energy and time-average power of a rectangular-envelope LFM pulse are given by

$$E_x = 0.5 E_{\tilde{x}} = 0.5 A^2 T$$

(8.3-95)

and

$$P_{\text{avg}} = \frac{E_x}{T} = \frac{E_{\tilde{x}}}{2T} = \frac{A^2}{2},$$

(8.3-96)

respectively (see Example 8.2-3), which are equal to the energy and time-average power of a rectangular-envelope CW pulse [see Eqs. (8.3-73) and (8.3-74)].

The instantaneous phase in radians and the instantaneous frequency in hertz of Eqs. (8.3-88) and (8.3-91) are given by

$$\theta_i(t) = \begin{cases} 2\pi f_c t + D_p t^2, & |t| \le T/2, \\ 0, & |t| > T/2, \end{cases}$$

(8.3-97)

and

$$f_i(t) = \begin{cases} f_c + (D_p/\pi)t, & |t| \le T/2, \\ 0, & |t| > T/2, \end{cases}$$

(8.3-98)

respectively (see Fig. 8.3-5). From Fig. 8.3-5 it can be seen that with the phase-

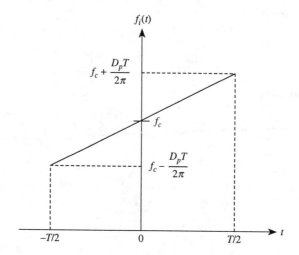

Figure 8.3-5. Instantaneous frequency of a LFM pulse with phase-deviation constant $D_p > 0$.

deviation constant $D_p > 0$, the amount of frequency sweep is equal to

$$f_c + \frac{D_p T}{2\pi} - \left(f_c - \frac{D_p T}{2\pi} \right) = \frac{D_p T}{\pi}. \qquad (8.3\text{-}99)$$

The expression $|D_p|T/\pi$ is referred to as the *swept bandwidth* in hertz. A LFM pulse is also known as a *chirp pulse*. When the phase-deviation constant $D_p > 0$, a LFM pulse is called an "up chirp" because the instantaneous frequency $f_i(t)$ increases with time; and when $D_p < 0$, it is called a "down chirp" because $f_i(t)$ decreases with time.

The normalized auto-ambiguity function of the rectangular-envelope LFM pulse will be derived next. Substituting Eq. (8.3-92) into Eq. (8.3-13) yields

$$X(\tau, \phi) = A^2 \exp\left(-jD_p\tau^2\right) \int_{-\infty}^{\infty} \text{rect}\left(\frac{t}{T}\right) \text{rect}\left(\frac{t-\tau}{T}\right) \exp\left[-j2\pi\left(\phi - \frac{D_p\tau}{\pi}\right)t\right] dt,$$

$$(8.3\text{-}100)$$

or

$$X(\tau, \phi) = A^2 \exp\left(-jD_p\tau^2\right) F_t\left\{ \text{rect}\left(\frac{t}{T}\right) \text{rect}\left(\frac{t-\tau}{T}\right) \exp\left(+j2D_p\tau t\right) \right\}. \quad (8.3\text{-}101)$$

If we designate Eq. (8.3-75) for the rectangular-envelope CW pulse as $X_{\text{CW}}(\tau, \phi)$, then Eq. (8.3-100) can be expressed as

$$X(\tau, \phi) = X_{\mathrm{CW}}\left(\tau, \phi - \frac{D_p\tau}{\pi}\right)\exp\left(-jD_p\tau^2\right). \qquad (8.3\text{-}102)$$

Therefore, if we replace ϕ with $\phi - (D_p\tau/\pi)$ in Eq. (8.3-81) for the unnormalized auto-ambiguity function of a rectangular-envelope CW pulse, and then substitute the resulting expression into Eq. (8.3-102), we obtain

$$X(\tau, \phi) = \begin{cases} A^2 T\left(1 - \dfrac{|\tau|}{T}\right)\operatorname{sinc}\left[\left(\phi - \dfrac{D_p\tau}{\pi}\right)(T - |\tau|)\right] \\ \times \exp(-j\pi\phi\tau)\exp\left(-jD_p\tau^2\right), & |\tau| \le T, \\ 0, & |\tau| > T, \end{cases} \qquad (8.3\text{-}103)$$

which is the *unnormalized* auto-ambiguity function of a rectangular-envelope LFM pulse. Substituting Eqs. (8.3-103) and (8.3-94) into Eq. (8.3-46) yields the following expression for the *normalized auto-ambiguity function of a rectangular-envelope LFM pulse*:

$$X_N(\tau, \phi) = \begin{cases} \left(1 - \dfrac{|\tau|}{T}\right)\operatorname{sinc}\left[\left(\phi - \dfrac{D_p\tau}{\pi}\right)(T - |\tau|)\right] \\ \times \exp(-j\pi\phi\tau)\exp\left(-jD_p\tau^2\right), & |\tau| \le T, \\ 0, & |\tau| > T. \end{cases} \qquad (8.3\text{-}104)$$

Note that if we let $D_p = 0$, then Eq. (8.3-104) reduces to Eq. (8.3-82) for a rectangular-envelope CW pulse.

By setting $\phi = 0$ in Eq. (8.3-104) and taking the magnitude of the resulting expression, we obtain the following expression for the *round-trip time-delay (range) profile*:

$$|X_N(\tau, 0)| = \begin{cases} \left(1 - \dfrac{|\tau|}{T}\right)\left|\operatorname{sinc}\left(\dfrac{D_p\tau(T - |\tau|)}{\pi}\right)\right|, & |\tau| \le T, \\ 0, & |\tau| > T, \end{cases} \qquad (8.3\text{-}105)$$

since $\operatorname{sinc}(-x) = \operatorname{sinc}(x)$. Similarly, by setting $\tau = 0$ in Eq. (8.3-104) and taking the magnitude of the resulting expression, we obtain the following expression for the *Doppler profile*:

$$|X_N(0, \phi)| = |\operatorname{sinc}(\phi T)|, \qquad (8.3\text{-}106)$$

which is identical to the Doppler profile of a rectangular-envelope CW pulse given by Eq. (8.3-84).

Since sinc(± 1) = 0, the range profile given by Eq. (8.3-105) has its first zero crossings along the τ axis when the magnitude of the argument of the sinc function is equal to 1, that is,

$$\left| \frac{D_p \tau (T - |\tau|)}{\pi} \right| = 1, \qquad (8.3\text{-}107)$$

or

$$|\tau|^2 - T|\tau| + \frac{\pi}{|D_p|} = 0. \qquad (8.3\text{-}108)$$

Using the quadratic formula to solve Eq. (8.3-108) yields

$$|\tau| = \frac{T}{2} \pm \frac{1}{2} \sqrt{T^2 - \frac{4\pi}{|D_p|}}, \qquad (8.3\text{-}109)$$

and upon choosing the minus sign (since we want the locations of the *first* zero crossings), Eq. (8.3-109) can be rewritten as

$$|\tau| = \frac{T}{2} \left(1 - \sqrt{1 - \frac{4\pi}{|D_p|T^2}} \right), \qquad (8.3\text{-}110)$$

where

$$\boxed{\frac{4\pi}{|D_p|T^2} \leq 1} \qquad (8.3\text{-}111)$$

must be satisfied *at all times* in order to ensure that $|\tau|$ is real and not complex. However, if we impose the more stringent condition that

$$\boxed{\frac{4\pi}{|D_p|T^2} \leq 0.1,} \qquad (8.3\text{-}112)$$

then we can use the first two terms in a binomial expansion of the square root in Eq. (8.3-110). Doing so yields

$$|\tau| \approx \frac{T}{2} \left(1 - 1 + \frac{2\pi}{|D_p|T^2} \right), \qquad (8.3\text{-}113)$$

$$|\tau| \approx \frac{\pi}{|D_p|T}, \qquad (8.3\text{-}114)$$

or

$$\tau \approx \pm \frac{\pi}{|D_p|T} \text{ sec.} \qquad (8.3\text{-}115)$$

Equation (8.3-115) indicates that the width of the main lobe of the ambiguity function along the τ axis is *inversely proportional to the swept bandwidth*, since the locations of the first zero crossings of the range profile are equal to plus and minus the reciprocal of the swept bandwidth when Eq. (8.3-112) is satisfied.

By inspecting Eq. (8.3-106), it can be seen that the width of the main lobe of the ambiguity function along the ϕ axis is *inversely proportional to the pulse length T*, since the locations of the first zero crossings of the Doppler profile are at

$$\phi = \pm \frac{1}{T} \text{ Hz.} \qquad (8.3\text{-}116)$$

By inspecting Eqs. (8.3-115) and (8.3-116), it can also be seen that the width of the main lobe of the ambiguity function along both the τ and ϕ axes is *inversely proportional to the pulse length T*.

Now that we have two parameters that we can vary, namely, the phase-deviation constant D_p and the pulse length T, we can simultaneously control both the round-trip time delay (range) and Doppler resolving capabilities of a rectangular-envelope LFM pulse. For example, first choose T to provide the desired Doppler resolution [see Eq. (8.3-116)], and then choose D_p to provide the desired range resolution [see Eq. (8.3-115)]. However, keep in mind that Eq. (8.3-112) must also be satisfied in order for Eq. (8.3-115) to be valid.

Let us conclude the discussion in this section by noting that the magnitude of the normalized auto-ambiguity function of a rectangular-envelope LFM pulse given by Eq. (8.3-104) has sinc(x)-type profiles along every line with a constant τ value, that is, along every line parallel to and including the ϕ axis. For a given value of τ, the maximum value of the corresponding profile is $1 - (|\tau|/T)$, and this maximum value is located at

$$\phi = D_p \tau / \pi, \qquad (8.3\text{-}117)$$

which is the equation of a straight line in the $\tau\phi$ plane that passes through the origin. Therefore, according to Eq. (8.3-117), when $D_p > 0$ ("up chirp"), $|X_N(\tau, \phi)|$ will be concentrated mainly in the first and third quadrants of the $\tau\phi$ plane; and when $D_p < 0$ ("down chirp"), $|X_N(\tau, \phi)|$ will be concentrated mainly in the second and fourth quadrants.

8.4 Time Compression/Stretch Factor, Time Delay, and Doppler Shift Expressions

Consider the bistatic scattering geometry shown in Fig. 8.4-1, where T and R represent the transmit and receive apertures (arrays), respectively; S represents the point scatterer or target; $\mathbf{V}_T$, $\mathbf{V}_R$, and $\mathbf{V}_S$ are the *constant* velocity vectors (i.e., there are no accelerations) of the transmit and receive apertures and scatterer, respectively; $\hat{n}_T$ and $\hat{n}_R$ are unit vectors in the directions of propagation of the transmitted and received waves, respectively; and R_{0_T} and R_{0_R} are the initial ranges of the point scatterer or target from the transmit and receive apertures, respectively, when transmission begins at time $t = 0$. Let $\tilde{x}(t)$ represent the complex envelope of the transmitted signal at the transmit aperture (array), and let $\tilde{y}(t)$ represent the complex envelope of the received signal or target return at the receive aperture (array). Therefore, with respect to the bistatic geometry shown in Fig. 8.4-1, $\tilde{y}(t)$ can be modeled as a time- and frequency-shifted replica of $\tilde{x}(t)$ as follows [compare with Eq. (8.3-3)]:

$$\tilde{y}(t) = \tilde{x}\big(s[t - \tau_A]\big)\exp(+j2\pi\phi_A t), \tag{8.4-1}$$

where s is the dimensionless *time compression/stretch factor* given by

$$s = \frac{(1 - \beta_{ST})(1 - \beta_{RR})}{(1 - \beta_{SR})(1 - \beta_{TT})}, \tag{8.4-2}$$

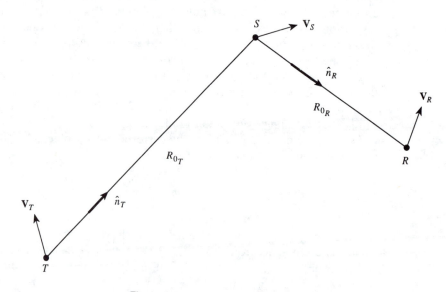

Figure 8.4-1. Bistatic scattering geometry.

where

$$\beta_{ST} = \mathbf{V}_S \cdot \hat{n}_T/c, \tag{8.4-3}$$

$$\beta_{RR} = \mathbf{V}_R \cdot \hat{n}_R/c, \tag{8.4-4}$$

$$\beta_{SR} = \mathbf{V}_S \cdot \hat{n}_R/c, \tag{8.4-5}$$

and

$$\beta_{TT} = \mathbf{V}_T \cdot \hat{n}_T/c; \tag{8.4-6}$$

τ_A is the *actual* (*true*) *time delay* in seconds given by

$$\tau_A = \frac{\dfrac{1 - \beta_{SR}}{1 - \beta_{ST}} R_{0_T} + R_{0_R}}{(1 - \beta_{RR})c}; \tag{8.4-7}$$

ϕ_A is the *actual* (*true*) *Doppler shift* in hertz given by

$$\phi_A = -(1 - s)f_c, \tag{8.4-8}$$

where f_c is the carrier frequency in hertz of the real amplitude- and angle-modulated carrier given by

$$x(t) = \mathrm{Re}\{\tilde{x}(t) \exp(+j2\pi f_c t)\}; \tag{8.4-9}$$

and c is the constant speed of sound in meters per second. Note that Eqs. (8.4-3) through (8.4-6) are *normalized radial components of velocity*—normalized by the speed of sound—and as a result, they are dimensionless quantities.

Equations (8.4-1) through (8.4-8) are general expressions that pertain to both bistatic and monostatic (backscatter) geometries. For a monostatic geometry, set

$$\mathbf{V}_R = \mathbf{V}_T, \tag{8.4-10}$$

$$\hat{n}_R = -\hat{n}_T, \tag{8.4-11}$$

and

$$R_{0_R} = R_{0_T} = R_0 \tag{8.4-12}$$

in Eqs. (8.4-3) through (8.4-7). Keep in mind that all velocity vectors and unit vectors are three-dimensional in general, that is, they generally have three nonzero components.

There are three ranges of values for the time compression/stretch factor s. When $s = 1$, this is an indication that there is *no* time compression or time stretch, and as a result, there is *no* change in signal bandwidth. Also, when $s = 1$, $\phi_A = 0$, that is, there is *no* Doppler shift [see Eq. (8.4-8)]. When $s > 1$, this is an indication of *time compression*, that is, the duration of the received signal given by

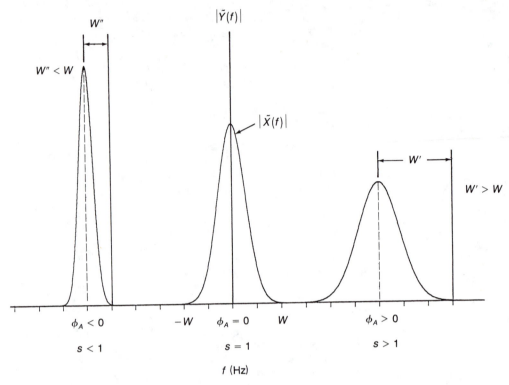

Figure 8.4-2. Illustration of Eq. (8.4-13) for the three cases $s = 1$, $s > 1$, and $s < 1$.

Eq. (8.4-1) is *decreased*, resulting in an *increase* in signal bandwidth. Also, when $s > 1$, $\phi_A > 0$, which indicates a *positive* Doppler shift. Finally, when $s < 1$, this is an indication of *time stretch*, that is, the duration of the received signal is *increased*, resulting in a *decrease* in signal bandwidth. Also, when $s < 1$, $\phi_A < 0$, which indicates a *negative* Doppler shift. The three ranges of values for the time compression/stretch factor and its effect on the amplitude spectrum of the complex envelope of the received signal given by Eq. (8.4-1) is illustrated in Fig. 8.4-2 where it can be shown that

$$|\tilde{Y}(f)| = \frac{1}{|s|}\left|\tilde{X}\left(\frac{f - \phi_A}{s}\right)\right|. \tag{8.4-13}$$

Recall that the complex envelope $\tilde{x}(t)$ is a low-pass (baseband) signal, so that $|\tilde{X}(f)|$ is centered about $f = 0$ Hz.

Example 8.4-1 (Bistatic geometry)
Referring to the bistatic scattering geometry shown in Fig. 8.4-1, assume that neither the transmit nor the receive aperture (array) is in

motion. Therefore, with $V_T = 0$ and $V_R = 0$, the time compression/stretch factor given by Eq. (8.4-2) reduces to

$$s = \frac{1 - \beta_{ST}}{1 - \beta_{SR}} = \frac{1 - (\mathbf{V}_S \cdot \hat{n}_T/c)}{1 - (\mathbf{V}_S \cdot \hat{n}_R/c)}, \qquad (8.4\text{-}14)$$

since $\beta_{TT} = 0$ and $\beta_{RR} = 0$. Similarly, with the use of Eq. (8.4-14), and since $\beta_{RR} = 0$, the time delay given by Eq. (8.4-7) reduces to

$$\tau_A = \frac{1}{s}\left[\tau_0 - (1 - s)\frac{R_{0_R}}{c}\right], \qquad (8.4\text{-}15)$$

where s is given by Eq. (8.4-14) and

$$\tau_0 = \frac{R_{0_T} + R_{0_R}}{c} \qquad (8.4\text{-}16)$$

is the bistatic time delay in seconds when neither the point scatterer (target) nor the transmit and receive apertures are in motion. Finally, substituting Eq. (8.4-14) into Eq. (8.4-8) yields

$$\phi_A = -\frac{\left[(\hat{n}_T - \hat{n}_R) \cdot \mathbf{V}_S\right] f_c}{c\left[1 - (\mathbf{V}_S \cdot \hat{n}_R/c)\right]}, \qquad (8.4\text{-}17)$$

and if the speed of the point scatterer is much less than the speed of sound, that is, if

$$\frac{|\mathbf{V}_S|}{c} \ll 1, \qquad (8.4\text{-}18)$$

then the Doppler shift expression given by Eq. (8.4-17) reduces to

$$\phi_A \approx -\left[(\hat{n}_T - \hat{n}_R) \cdot \mathbf{V}_S\right]\frac{f_c}{c}. \qquad (8.4\text{-}19)$$

Example 8.4-2 [Monostatic (backscatter) geometry]
Referring to the monostatic (backscatter) geometry shown in Fig. 8.4-3 assume that the transmit/receive aperture (array) is not in motion. Therefore, with $V_R = V_T = 0$ and with the use of Eqs. (8.4-10) through

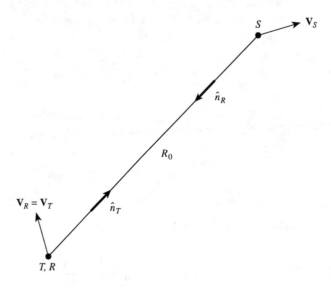

Figure 8.4-3. Monostatic (backscatter) geometry.

(8.4-12), the results obtained in Example 8.4-1 further reduce as follows:

$$s = \frac{1 - \beta_{ST}}{1 - \beta_{SR}} = \frac{1 - (\mathbf{V}_S \cdot \hat{n}_T/c)}{1 + (\mathbf{V}_S \cdot \hat{n}_T/c)}, \tag{8.4-20}$$

$$\tau_A = \frac{\tau_0}{2}\left(1 + \frac{1}{s}\right), \tag{8.4-21}$$

where s is given by Eq. (8.4-20) and

$$\tau_0 = 2R_0/c \tag{8.4-22}$$

is the monostatic time delay in seconds when neither the point scatterer nor the transmit and receive apertures are in motion, and

$$\phi_A = -\frac{2(\mathbf{V}_S \cdot \hat{n}_T)f_c}{c[1 + (\mathbf{V}_S \cdot \hat{n}_T/c)]}. \tag{8.4-23}$$

If Eq. (8.4-18) is satisfied, then Eq. (8.4-23) reduces to

$$\phi_A \approx -2(\mathbf{V}_S \cdot \hat{n}_T)\frac{f_c}{c}. \tag{8.4-24}$$

These results are particularly easy to interpret. For example, when the point scatterer (target) is moving *away* from the transmit/receive aperture (array), the radial component of velocity $\mathbf{V}_S \cdot \hat{n}_T > 0$, which means that $s < 1$ (*time stretch*—the duration of the received signal is *increased*), $\tau_A > \tau_0$, and $\phi_A < 0$, which is physically correct. Similarly, when the point scatterer (target) is moving *toward* the transmit/receive aperture (array), the radial component of velocity $\mathbf{V}_S \cdot \hat{n}_T < 0$, which means that $s > 1$ (*time compression*—the duration of the received signal is *decreased*), $\tau_A < \tau_0$, and $\phi_A > 0$, which is also physically correct.

Problems

8-1   Equation (8.2-10) can be rewritten as follows:

$$\hat{X}(f) = X(f)H(f),$$

where

$$H(f) = -j\operatorname{sgn}(f)$$

is the transfer function of a Hilbert-transform filter. Show that the impulse response of a Hilbert-transform filter is given by

$$h(t) = \frac{1}{\pi t}.$$

Hint:

$$H(f) = \lim_{a \to 0} \begin{cases} -j\exp(-af), & f > 0, \\ +j\exp(+af), & f < 0, \end{cases}$$

where $a > 0$.

   Therefore, the Hilbert transform of $x(t)$ can be expressed as the following time-domain convolution integral:

$$\hat{x}(t) = x(t) \underset{t}{*} h(t) = x(t) \underset{t}{*} \frac{1}{\pi t},$$

or

$$\hat{x}(t) = \frac{1}{\pi} \int_{-\infty}^{\infty} \frac{x(\tau)}{t - \tau}\, d\tau.$$

8-2   Use Eq. (8.2-8) to compute the Hilbert transform of the following signals:
(a) $\cos(2\pi f_c t + \theta_0)$,
(b) $\sin(2\pi f_c t + \theta_0)$.

Compare your answers with Eqs. (8.2-13) and (8.2-14), respectively.

8-3   If the input signal to the quadrature demodulator system shown in Fig. 8.2-5 is the amplitude- and angle-modulated carrier given by Eq. (8.2-23), then show that the output signals from this system are the cosine and sine components given by Eqs. (8.2-32) and (8.2-33), respectively.

8-4   Let

$$\Delta = \frac{E\{|\tilde{l}|^2|H_1\} - E\{|\tilde{l}|^2|H_0\}}{E\{|\tilde{l}|^2|H_0\}},$$

where

$$H_0: \quad \tilde{l} = \tilde{l}_N,$$

$$H_1: \quad \tilde{l} = \tilde{l}_T + \tilde{l}_N,$$

where $\tilde{l}_T$ and $\tilde{l}_N$ are given by Eqs. (8.3-8) and (8.3-9), respectively. If we assume that $\tilde{y}_T(t)$ and $\tilde{n}(t)$ are uncorrelated and that $\tilde{n}(t)$ has zero mean, then show that Δ is equal to the output SNR definition given by Eq. (8.3-7).

8-5   Consider the binary hypothesis-testing problem discussed in Section 8.3.1 and the receiver shown in Fig. 8.3-1. For $P_{FA} = 0.1$, find the required output SNR in decibels for this receiver so that
(a) $P_D = 0.9$,
(b) $P_D = 0.95$,
(c) and $P_D = 0.99$.

Hint: In order to express the SNR in decibels, use $10\log_{10}$ SNR dB.
(d) Repeat parts (a) through (c) for $P_{FA} = 0.01$.

8-6   With respect to Example 8.3-1:
(a) How many array elements are required in order to achieve $P_D = 0.99$ for $P_{FA} = 0.01$ if SNR $= -10$ dB?
(b) Find the corresponding value of AG in decibels.

8-7   Design a rectangular-envelope, "down chirp", LFM pulse with a range resolving capability of ± 1 m and a Doppler resolving capability of ± 1 Hz as measured by the locations of the first zero crossings of the round-trip time delay (range) and Doppler profiles, respectively, of the corresponding nor-malized auto-ambiguity function. In addition, the pulse must have unit energy. Use $\tau = 2r/c$ and $c = 1500$ m/sec.

(a) What should the pulse length, phase deviation constant, and amplitude of the pulse be in order to satisfy these requirements? Also, check to make sure that Eq. (8.3-112) is satisfied.

(b) What is the resulting swept bandwidth in hertz?

Using your answers for parts (a) and (b), and for a carrier frequency $f_c = 10$ kHz,

(c) compute the instantaneous phase in degrees at $t = -T/2$, 0, and $T/2$ seconds.

(d) compute the instantaneous frequency in hertz at $t = -T/2$, 0, and $T/2$ seconds.

8-8   Verify Eq. (8.4-13).

8-9   Consider a monostatic (backscatter) geometry where $\mathbf{V}_R = \mathbf{V}_T = \mathbf{0}$, $R_0 = 1$ km, $\hat{n}_T = \sqrt{\frac{1}{2}}\hat{y} + \sqrt{\frac{1}{2}}\hat{z}$, $\mathbf{V}_S = -\hat{x} + \hat{y} + 2\hat{z}$, $c = 1500$ m/sec, and $f_c = 10$ kHz.

(a) Compute the time compression/stretch factor, round-trip time delay, and Doppler shift.

(b) Repeat part (a) using $\mathbf{V}_S = 2\hat{x} - 3\hat{y}$.

8-10 Consider a monostatic (backscatter) geometry where $\mathbf{V}_R = \mathbf{V}_T = -\hat{y}$, $R_0 = 2$ km, $\hat{n}_T = -\sqrt{\frac{1}{2}}\hat{y} + \sqrt{\frac{1}{2}}\hat{z}$, $\mathbf{V}_S = 4\hat{y}$, $c = 1500$ m/sec, and $f_c = 15$ kHz.

(a) Compute the time compression/stretch factor, round-trip time delay, and Doppler shift.

(b) Repeat part (a) using $\hat{n}_T = \sqrt{\frac{1}{2}}\hat{y} - \sqrt{\frac{1}{2}}\hat{z}$.

Appendix 8A

In this appendix we shall derive both the decision threshold and the error performance given by Eqs. (8.3-48) and (8.3-49), respectively. Since the receiver shown in Fig. 8.3-1 uses the quantity $|\tilde{l}|^2$ to test against the threshold γ, we need to find the *conditional probability density functions* $p_0(|\tilde{l}|^2 | H_0)$ and $p_1(|\tilde{l}|^2 | H_1)$, as explained next.

Since the Neyman-Pearson criterion, which is to maximize the probability of detection P_D for a given probability of false alarm P_{FA}, is a special case of the *Bayes criterion*, which is to *minimize the average cost or risk*, a likelihood-ratio test (LRT) will maximize the P_D for a given P_{FA}. In our problem, the LRT is as follows: choose hypothesis H_1 if

$$\text{LR}(|\tilde{l}|^2) \geq \gamma_0, \tag{8A-1}$$

and choose hypothesis H_0 otherwise, where

$$\mathrm{LR}(|\tilde{l}|^2) \triangleq \frac{p_1(|\tilde{l}|^2 | H_1)}{p_0(|\tilde{l}|^2 | H_0)} \tag{8A-2}$$

is known as the *likelihood ratio* (LR), and the threshold γ_0 is chosen to satisfy the P_{FA} constraint. Both of the conditional probability density functions appearing in Eq. (8A-2) are also known as *likelihood functions*.

Under hypotheses H_0 and H_1, the output of the correlator $\tilde{l}$ can be expressed as

$$H_0: \quad \tilde{l} = \tilde{l}_N, \tag{8A-3}$$

$$H_1: \quad \tilde{l} = \tilde{l}_T + \tilde{l}_N, \tag{8A-4}$$

where $\tilde{l}_T$ and $\tilde{l}_N$ are given by Eqs. (8.3-8) and (8.3-9), respectively. Since $\tilde{l}_T$ and $\tilde{l}_N$ are equal to inner products (an inner product is a *linear transformation*), and since $\tilde{y}_T(t)$ and $\tilde{n}(t)$ were assumed to be complex, zero-mean, statistically independent, Gaussian random processes, and the processing waveform $\tilde{g}(t)$ is a complex, *deterministic* function, $\tilde{l}_T$ and $\tilde{l}_N$ are complex, zero-mean, statistically independent, Gaussian random variables. Therefore, under hypotheses H_0 and H_1, the output of the correlator $\tilde{l}$ is a complex, zero-mean, Gaussian random variable.

Since it can be shown that under hypotheses H_0 and H_1, $|\tilde{l}|^2$ is an *exponential* random variable, we have

$$p_0(|\tilde{l}|^2 | H_0) = \frac{1}{E\{|\tilde{l}|^2 | H_0\}} \exp\left[-\frac{|\tilde{l}|^2}{E\{|\tilde{l}|^2 | H_0\}}\right], \quad |\tilde{l}|^2 \geq 0, \tag{8A-5}$$

where [see Eqs. (8A-3) and (8.3-44)]

$$E\{|\tilde{l}|^2 | H_0\} = N_0 E_{\tilde{x}}, \tag{8A-6}$$

and

$$p_1(|\tilde{l}|^2 | H_1) = \frac{1}{E\{|\tilde{l}|^2 | H_1\}} \exp\left[-\frac{|\tilde{l}|^2}{E\{|\tilde{l}|^2 | H_1\}}\right], \quad |\tilde{l}|^2 \geq 0, \tag{8A-7}$$

where [see Eq. (8A-4)]

$$E\{|\tilde{l}|^2 | H_1\} = E\{|\tilde{l}_T|^2\} + E\{|\tilde{l}_N|^2\}, \tag{8A-8}$$

since

$$E\{\tilde{l}_T \tilde{l}_N^*\} = 0, \qquad (8A-9)$$

because $\bar{y}_T(t)$ and $\bar{n}(t)$ were assumed to be zero-mean, statistically independent, random processes. Substituting Eqs. (8.3-17) and (8.3-44) into Eq. (8A-8) yields

$$E\{|\tilde{l}|^2 | H_1\} = E\{a^2\} |X(\tau, \phi)|^2 + N_0 E_{\tilde{x}}, \qquad (8A-10)$$

or

$$E\{|\tilde{l}|^2 | H_1\} = E_{\tilde{x}} N_0 (1 + \text{SNR}), \qquad (8A-11)$$

where [see Eqs. (8.3-45) and (8.3-46)]

$$\text{SNR} = |X_N(\tau, \phi)|^2 E\{a^2\} E_{\tilde{x}} / N_0. \qquad (8A-12)$$

Now that we have expressions for the two conditional probability density functions, we can form the likelihood ratio. Substituting Eqs. (8A-5) and (8A-7) into Eq. (8A-2) yields

$$\text{LR}(|\tilde{l}|^2) = \frac{E\{|\tilde{l}|^2 | H_0\}}{E\{|\tilde{l}|^2 | H_1\}} \exp\left[|\tilde{l}|^2 \left(\frac{E\{|\tilde{l}|^2 | H_1\} - E\{|\tilde{l}|^2 | H_0\}}{E\{|\tilde{l}|^2 | H_0\} E\{|\tilde{l}|^2 | H_1\}}\right)\right]. \qquad (8A-13)$$

Substituting Eq. (8A-13) into Eq. (8A-1) and taking the natural logarithm of both sides of the resulting equation yields the following *log-likelihood ratio test* (see Fig. 8.3-1): choose hypothesis H_1 if

$$|\tilde{l}|^2 \geq \gamma, \qquad (8A-14)$$

and choose hypothesis H_0 otherwise, where the threshold γ is chosen to satisfy the P_{FA} constraint and is related to the threshold γ_0 by

$$\gamma = \frac{E\{|\tilde{l}|^2 | H_0\} E\{|\tilde{l}|^2 | H_1\}}{E\{|\tilde{l}|^2 | H_1\} - E\{|\tilde{l}|^2 | H_0\}} \ln\left(\frac{E\{|\tilde{l}|^2 | H_1\}}{E\{|\tilde{l}|^2 | H_0\}} \gamma_0\right), \qquad E\{|\tilde{l}|^2 | H_1\} \neq E\{|\tilde{l}|^2 | H_0\}.$$

$$(8A-15)$$

Note that γ_0 can be obtained by evaluating the likelihood ratio given by Eq. (8A-13) at $|\tilde{l}|^2 = \gamma$, that is,

$$\text{LR}(\gamma) = \gamma_0, \qquad (8A-16)$$

where γ is given by Eq. (8A-15).

Since

$$P_{FA} = Pr\left(|\tilde{l}|^2 \geq \gamma \big| H_0\right) = \int_{\gamma}^{\infty} p_0\left(|\tilde{l}|^2 \big| H_0\right) d|\tilde{l}|^2, \qquad (8A\text{-}17)$$

substituting Eq. (8A-5) into Eq. (8A-17) yields

$$P_{FA} = \exp\left[-\frac{\gamma}{E\{|\tilde{l}|^2 | H_0\}}\right]. \qquad (8A\text{-}18)$$

Taking the natural logarithm of both sides of Eq. (8A-18) yields

$$\gamma = E\{|\tilde{l}|^2 | H_0\} \ln(1/P_{FA}), \qquad (8A\text{-}19)$$

and upon substituting Eq. (8A-6) into Eq. (8A-19), we obtain the decision threshold

$$\boxed{\gamma = E_{\tilde{x}} N_0 \ln(1/P_{FA}).} \qquad (8A\text{-}20)$$

We shall conclude this appendix by deriving the error performance. Since

$$P_D = Pr\left(|\tilde{l}|^2 \geq \gamma \big| H_1\right) = \int_{\gamma}^{\infty} p_1\left(|\tilde{l}|^2 \big| H_1\right) d|\tilde{l}|^2, \qquad (8A\text{-}21)$$

substituting Eq. (8A-7) into Eq. (8A-21) yields

$$P_D = \exp\left[-\frac{\gamma}{E\{|\tilde{l}|^2 | H_1\}}\right], \qquad (8A\text{-}22)$$

and upon substituting Eq. (8A-11) into Eq. (8A-22), we obtain

$$P_D = \exp\left\{-\frac{\gamma}{E_{\tilde{x}} N_0(1 + SNR)}\right\}. \qquad (8A\text{-}23)$$

Substituting Eq. (8A-6) into Eq. (8A-18) yields

$$P_{FA} = \exp\left[-\frac{\gamma}{E_{\tilde{x}} N_0}\right], \qquad (8A\text{-}24)$$

and by comparing Eqs. (8A-23) and (8A-24), it can be seen that Eq. (8A-23) can be rewritten as

$$P_{\mathrm{D}} = P_{\mathrm{FA}}^{1/(1+\mathrm{SNR})}, \qquad (8\text{A-}25)$$

which is the error performance of the receiver shown in Fig. 8.3-1.

Bibliography

A. B. Baggeroer, "Sonar signal processing," in *Applications of Digital Signal Processing* (edited by A. V. Oppenheim), pp. 331–437, Prentice-Hall, Englewood Cliffs, New Jersey, 1978.

C. E. Cook and M. Bernfeld, *Radar Signals*, Academic Press, New York, 1967.

L. W. Couch II, *Digital and Analog Communication Systems*, 3rd ed., Macmillan, New York, 1990.

W. B. Davenport, Jr., *Probability and Random Processes*, McGraw-Hill, New York, 1970.

S. Haykin, Communication Systems, 2nd ed., Wiley, New York, 1983.

W. C. Knight, R. G. Pridham, and S. M. Kay, "Digital signal processing for sonar," *Proc. IEEE*, **69**, 1451–1506 (1981).

N. L. Owsley, "Sonar array processing," in *Array Signal Processing* (edited by S. Haykin), pp. 115–193, Prentice-Hall, Englewood Cliffs, New Jersey, 1985.

A. Papoulis, *Signal Analysis*, McGraw-Hill, New York, 1977.

H. Stark, F. B. Tuteur, and J. B. Anderson, *Modern Electrical Communications*, 2nd ed., Prentice-Hall, Englewood Cliffs, New Jersey, 1988.

H. L. Van Trees, *Detection, Estimation, and Modulation Theory, Part III*, Wiley, New York, 1971.

A. D. Whalen, *Detection of Signals in Noise*, Academic Press, New York, 1971.

L. J. Ziomek, "Comments on the generalized ambiguity function," *IEEE Trans. Acoust., Speech, and Signal Processing*, **ASSP-30**, 117–119 (1982).

Chapter 9

Fundamentals of Linear, Time-Variant, Space-Variant Filters and the Propagation of Small-Amplitude Acoustic Signals

9.1 Impulse Response and Transfer Function

Since the propagation of small-amplitude acoustic signals in an unbounded or bounded fluid medium can be described by a linear wave equation, we can treat an unbounded or bounded fluid medium as a linear filter, as was done in Section 2.7.1. In this chapter we shall show that the principles of linear, time-variant, space-variant filter theory and time-domain and spatial-domain Fourier transforms provide a consistent, logical, and straightforward mathematical framework, known as the *coupling equations*, for the solution of small-amplitude acoustic pulse-propagation problems. The coupling equations and pulse propagation will be discussed in Section 9.4. A linear, time-variant, space-variant filter is one that satisfies the principle of superposition (i.e., homogeneity and additivity) but whose own properties change with time and space. Before we can discuss the coupling equations and pulse propagation in Section 9.4, we need to discuss some of the fundamentals of linear, time-variant, space-variant filter theory in Sections 9.1 through 9.3.

A *linear, time-variant, space-variant filter*, as depicted in Fig. 9.1-1, can be characterized by its *time-variant, space-variant impulse response* $h(t, \mathbf{r}; t_0, \mathbf{r}_0)$ which describes the response of the filter at time t and spatial location $\mathbf{r} = (x, y, z)$ due to the application of a unit-amplitude impulse at time t_0 and spatial location $\mathbf{r}_0 = (x_0, y_0, z_0)$. The impulse response is also called the *Green's function*. If the filter is *causal*, then $h(t, \mathbf{r}; t_0, \mathbf{r}_0) = 0$ for $t < t_0$, that is, the filter cannot respond before the application of an input. The relationship between the input signal $x(t, \mathbf{r})$ and the output signal $y(t, \mathbf{r})$ for a linear, time-variant, space-variant filter is

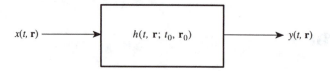

Figure 9.1-1. Illustration of a linear, time-variant, space-variant filter.

given by

$$y(t,\mathbf{r}) = \int_{-\infty}^{\infty} \int_{-\infty}^{\infty} x(t_0,\mathbf{r}_0)h(t,\mathbf{r};t_0,\mathbf{r}_0)\, dt_0\, d\mathbf{r}_0, \qquad (9.1\text{-}1)$$

where $d\mathbf{r}_0 = dx_0\, dy_0\, dz_0$. Therefore, the right-hand side of Eq. (9.1-1) is shorthand notation for a fourfold integral. The input-output relationship given by Eq. (9.1-1) is applicable to *any* linear, time-variant, space-variant filter, whether the filter is meant to model small-amplitude acoustic wave propagation in fluid media or any other physical system whose input and output are related by a linear partial differential equation. Note that if Eq. (9.1-1) is rewritten as

$$y(t,\mathbf{r}) = \int_{-\infty}^{\infty} \int_{-\infty}^{\infty} x(\alpha,\zeta)h(t,\mathbf{r};\alpha,\zeta)\, d\alpha\, d\zeta, \qquad (9.1\text{-}2)$$

and if the input is a unit-amplitude impulse applied at time $\alpha = t_0$ and spatial location $\zeta = \mathbf{r}_0$, that is, if

$$x(\alpha,\zeta) = \delta(\alpha - t_0,\zeta - \mathbf{r}_0), \qquad (9.1\text{-}3)$$

then substituting Eq. (9.1-3) into Eq. (9.1-2) and making use of the sifting property of impulse functions yields

$$y(t,\mathbf{r}) = \int_{-\infty}^{\infty} \int_{-\infty}^{\infty} \delta(\alpha - t_0,\zeta - \mathbf{r}_0)h(t,\mathbf{r};\alpha,\zeta)\, d\alpha\, d\zeta = h(t,\mathbf{r};t_0,\mathbf{r}_0). \quad (9.1\text{-}4)$$

Example 9.1-1
Different forms of the input-output relationship given by Eq. (9.1-1) can be obtained by making the following simplifying assumptions:

(1) If a linear filter is *time invariant* and *space invariant*, then

$$h(t,\mathbf{r};t_0,\mathbf{r}_0) = h(t - t_0,\mathbf{r} - \mathbf{r}_0), \qquad (9.1\text{-}5)$$

that is, the impulse response depends only on the *time difference* $\tau = t - t_0$ and the *vector spatial difference* $\mathbf{R} = \mathbf{r} - \mathbf{r}_0$. The time difference is a measure of how long ago the impulse was applied, or the travel time between source and field points. The vector spatial difference corresponds to the distance and direction between source and field points. The magnitude of the vector spatial difference, $|\mathbf{R}| = |\mathbf{r} - \mathbf{r}_0|$, is a measure of how far away the impulse was applied, or the line-of-sight distance traveled between source and field points. Substituting Eq. (9.1-5) into Eq. (9.1-1) yields

$$y(t, \mathbf{r}) = \int_{-\infty}^{\infty} \int_{-\infty}^{\infty} x(t_0, \mathbf{r}_0) h(t - t_0, \mathbf{r} - \mathbf{r}_0) \, dt_0 \, d\mathbf{r}_0, \quad (9.1\text{-}6)$$

which is a multidimensional convolution integral, as expected.

For example, it was shown in Section 2.7.1 that the impulse response of an ideal (nonviscous), unbounded, homogeneous, fluid medium was given by [see Eq. (2.7-28)]

$$
\begin{aligned}
h_M(t, \mathbf{r}; t_0, \mathbf{r}_0) &= h_M(t - t_0, \mathbf{r} - \mathbf{r}_0) \\
&= -\frac{1}{4\pi |\mathbf{r} - \mathbf{r}_0|} \delta\left[t - \left(t_0 + \frac{|\mathbf{r} - \mathbf{r}_0|}{c} \right) \right].
\end{aligned}
\quad (9.1\text{-}7)
$$

Equation (9.1-7) indicates that the medium filter is time invariant (no motion) and space invariant (unbounded, homogeneous medium). It was also shown in Section 2.7.1 that if we identify the source distribution $x_M(t, \mathbf{r})$ as the *input signal* $x(t, \mathbf{r})$ to the filter and the velocity potential $\varphi(t, \mathbf{r})$ as the *output signal* $y(t, \mathbf{r})$ from the filter, then substituting Eq. (9.1-7) into Eq. (9.1-6) yields a solution to the *linear*, three-dimensional, lossless, inhomogeneous wave equation [see Eqs. (2.7-1) and (2.7-30)].

(2) If a linear filter is *time invariant* and *space variant*, then

$$h(t, \mathbf{r}; t_0, \mathbf{r}_0) = h(t - t_0, \mathbf{r}; \mathbf{r}_0), \quad\quad (9.1\text{-}8)$$

and as a result, Eq. (9.1-1) reduces to

$$y(t, \mathbf{r}) = \int_{-\infty}^{\infty} \int_{-\infty}^{\infty} x(t_0, \mathbf{r}_0) h(t - t_0, \mathbf{r}; \mathbf{r}_0) \, dt_0 \, d\mathbf{r}_0. \quad (9.1\text{-}9)$$

(3) If a linear filter is *time variant* and *space invariant*, then

$$h(t, \mathbf{r}; t_0, \mathbf{r}_0) = h(t, \mathbf{r} - \mathbf{r}_0; t_0), \quad\quad (9.1\text{-}10)$$

and as a result, Eq. (9.1-1) reduces to

$$y(t,\mathbf{r}) = \int_{-\infty}^{\infty}\int_{-\infty}^{\infty} x(t_0,\mathbf{r}_0)h(t,\mathbf{r}-\mathbf{r}_0;t_0)\,dt_0\,d\mathbf{r}_0. \quad (9.1\text{-}11)$$

Finally, note that a linear filter may be space variant in one or two directions, but space invariant in the other (see Example 9.4-2).

Analogous to the complex frequency response or transfer function $H(f)$ of a linear, time-invariant filter is the *time-variant, space-variant complex frequency response* or *transfer function* $H(t,\mathbf{r};f,\mathbf{v})$ of a linear, time-variant, space-variant filter, which is defined as follows:

$$\boxed{\begin{aligned}H(t,\mathbf{r};f,\mathbf{v}) &\triangleq \int_{-\infty}^{\infty}\int_{-\infty}^{\infty} h(t,\mathbf{r};t_0,\mathbf{r}_0)\exp\left[-j2\pi f(t-t_0)\right]\\ &\times\exp\left[+j2\pi\mathbf{v}\cdot(\mathbf{r}-\mathbf{r}_0)\right]dt_0\,d\mathbf{r}_0,\end{aligned}}$$

$$(9.1\text{-}12)$$

where f corresponds to *input (transmitted) frequencies* in hertz, and $\mathbf{v} = (f_X, f_Y, f_Z)$ is a three-dimensional vector whose components are the *input (transmitted) spatial frequencies* in cycles per meter in the X, Y, and Z directions, respectively. Recall from Section 2.2.1 that the spatial frequencies in the X, Y, and Z directions are related to the direction cosines in those directions, respectively, which can be expressed in terms of the spherical angles θ and ψ. Also recall from Section 2.2.1 that propagation-vector components and spatial frequencies are related by a factor of 2π, and as a result, spatial frequencies describe the direction of wave propagation. Therefore, the input spatial frequencies represent the *initial* directions of wave propagation of the acoustic field at the source. It is very important to note that the transfer function defined by Eq. (9.1-12) is *not* equal to the time-domain and spatial-domain Fourier transform with respect to t_0 and $\mathbf{r}_0 = (x_0, y_0, z_0)$ of the impulse response.

Similarly,

$$\boxed{\begin{aligned}h(t,\mathbf{r};t_0,\mathbf{r}_0) &= \int_{-\infty}^{\infty}\int_{-\infty}^{\infty} H(t,\mathbf{r};f,\mathbf{v})\exp\left[+j2\pi f(t-t_0)\right]\\ &\times\exp\left[-j2\pi\mathbf{v}\cdot(\mathbf{r}-\mathbf{r}_0)\right]df\,d\mathbf{v},\end{aligned}}$$

$$(9.1\text{-}13)$$

where $d\mathbf{v} = df_X\,df_Y\,df_Z$. Equation (9.1-13) is *not* the inverse time-domain and spatial-domain Fourier transform of the transfer function.

Example 9.1-2 (**Transfer function of a linear, time-invariant, space-variant filter**)

If a linear filter is time invariant and space variant, then its corresponding transfer function can be obtained by substituting Eq. (9.1-8) into Eq. (9.1-12), yielding

$$H(t,\mathbf{r};f,\boldsymbol{\nu}) = \int_{-\infty}^{\infty}\int_{-\infty}^{\infty} h(t-t_0,\mathbf{r};\mathbf{r}_0)\exp[-j2\pi f(t-t_0)]$$

$$\times \exp[+j2\pi\boldsymbol{\nu}\cdot(\mathbf{r}-\mathbf{r}_0)]\,dt_0\,d\mathbf{r}_0. \qquad (9.1\text{-}14)$$

If we let $\tau = t - t_0$, then $dt_0 = -d\tau$, and as a result, Eq. (9.1-14) reduces to

$$H(t,\mathbf{r};f,\boldsymbol{\nu}) = H(f,\mathbf{r};\boldsymbol{\nu}) = \int_{-\infty}^{\infty}\int_{-\infty}^{\infty} h(\tau,\mathbf{r};\mathbf{r}_0)\exp(-j2\pi f\tau)$$

$$\times\exp[+j2\pi\boldsymbol{\nu}\cdot(\mathbf{r}-\mathbf{r}_0)]\,d\tau\,d\mathbf{r}_0.$$

$$(9.1\text{-}15)$$

Equation (9.1-15) will be used later in Example 9.4-1.

Example 9.1-3 (**Transfer function of a linear, time-invariant, space-invariant filter**)

If a linear filter is time invariant and space invariant, then its corresponding transfer function can be obtained by substituting Eq. (9.1-5) into Eq. (9.1-12), yielding

$$H(t,\mathbf{r};f,\boldsymbol{\nu}) = \int_{-\infty}^{\infty}\int_{-\infty}^{\infty} h(t-t_0,\mathbf{r}-\mathbf{r}_0)\exp[-j2\pi f(t-t_0)]$$

$$\times \exp[+j2\pi\boldsymbol{\nu}\cdot(\mathbf{r}-\mathbf{r}_0)]\,dt_0\,d\mathbf{r}_0. \qquad (9.1\text{-}16)$$

If we let $\tau = t - t_0$ and $\mathbf{R} = \mathbf{r} - \mathbf{r}_0$, then $dt_0 = -d\tau$ and $d\mathbf{r}_0 = -d\mathbf{R}$, and as a result, Eq. (9.1-16) reduces to

$$H(t,\mathbf{r};f,\boldsymbol{\nu}) = H(f,\boldsymbol{\nu})$$

$$= \int_{-\infty}^{\infty}\int_{-\infty}^{\infty} h(\tau,\mathbf{R})\exp(-j2\pi f\tau)\exp(+j2\pi\boldsymbol{\nu}\cdot\mathbf{R})\,d\tau\,d\mathbf{R},$$

$$(9.1\text{-}17)$$

or

$$H(f, \mathbf{v}) = F_t F_{\mathbf{r}}\{h(t, \mathbf{r})\}$$

$$= \int_{-\infty}^{\infty} \int_{-\infty}^{\infty} h(t, \mathbf{r}) \exp(-j2\pi ft) \exp(+j2\pi \mathbf{v} \cdot \mathbf{r}) \, dt \, d\mathbf{r}$$

(9.1-18)

which is the time-domain and spatial-domain Fourier transform with respect to t and $\mathbf{r} = (x, y, z)$ of the impulse response $h(t, \mathbf{r})$, where $F_{\mathbf{r}}\{\cdot\} = F_x F_y F_z\{\cdot\}$ and $d\mathbf{r} = dx\, dy\, dz$. The impulse response $h(t, \mathbf{r})$ describes the response of the filter at time t and spatial location $\mathbf{r} = (x, y, z)$ due to the application of a unit-amplitude impulse at time $t_0 = 0$ and spatial location $\mathbf{r}_0 = \mathbf{0} = (0, 0, 0)$. If the filter is *causal*, then $h(t, \mathbf{r}) = 0$ for $t < 0$, that is, the filter cannot respond before the application of an input.

Similarly, the impulse response is equal to the inverse time-domain and spatial-domain Fourier transform of the transfer function $H(f, \mathbf{v})$, that is,

$$h(t, \mathbf{r}) = F_f^{-1} F_{\mathbf{v}}^{-1}\{H(f, \mathbf{v})\}$$

$$= \int_{-\infty}^{\infty} \int_{-\infty}^{\infty} H(f, \mathbf{v}) \exp(+j2\pi ft) \exp(-j2\pi \mathbf{v} \cdot \mathbf{r}) \, df \, d\mathbf{v},$$

(9.1-19)

where $F_{\mathbf{v}}^{-1}\{\cdot\} = F_{f_x}^{-1} F_{f_y}^{-1} F_{f_z}^{-1}\{\cdot\}$ and $d\mathbf{v} = df_x\, df_y\, df_z$.

For example, it was shown in Section 2.7.1 that the transfer function of an ideal (nonviscous), unbounded, homogeneous, fluid medium was given by [see Eq. (2.7-13)]

$$H_M(t, \mathbf{r}; f, \mathbf{v}) = H_M(f, \mathbf{v})$$

$$= \frac{1}{(2\pi f/c)^2 - \left[(2\pi f_x)^2 + (2\pi f_Y)^2 + (2\pi f_Z)^2\right]}.$$

(9.1-20)

Equation (9.1-20) indicates that the medium filter is time invariant (no motion) and space invariant (unbounded, homogeneous medium). Equation (9.1-20) is the transfer function that corresponds to the impulse response given by Eq. (9.1-7).

In order to use Eq. (9.1-12) to calculate the time-variant, space-variant transfer function, the time-variant, space-variant impulse response must be known. However, even if the impulse response is not known *a priori*, the transfer function can still be obtained by using a time-harmonic plane wave as the input signal to a linear, time-variant, space-variant filter. Therefore, if we let

$$x(t,\mathbf{r}) = \exp(+j2\pi ft)\exp(-j\mathbf{k}\cdot\mathbf{r}) = \exp(+j2\pi ft)\exp(-j2\pi\mathbf{v}\cdot\mathbf{r}), \quad (9.1\text{-}21)$$

where $\mathbf{k} = 2\pi\mathbf{v}$ is the propagation vector, then substituting Eq. (9.1-21) into Eq. (9.1-1) yields

$$y(t,\mathbf{r}) = \int_{-\infty}^{\infty}\int_{-\infty}^{\infty} h(t,\mathbf{r};t_0,\mathbf{r}_0)\exp(+j2\pi ft_0)\exp(-j2\pi\mathbf{v}\cdot\mathbf{r}_0)\,dt_0\,d\mathbf{r}_0. \quad (9.1\text{-}22)$$

Since Eq. (9.1-12) can be rewritten as

$$H(t,\mathbf{r};f,\mathbf{v}) = \exp(-j2\pi ft)\exp(+j2\pi\mathbf{v}\cdot\mathbf{r})$$

$$\times \int_{-\infty}^{\infty}\int_{-\infty}^{\infty} h(t,\mathbf{r};t_0,\mathbf{r}_0)\exp(+j2\pi ft_0)\exp(-j2\pi\mathbf{v}\cdot\mathbf{r}_0)\,dt_0\,d\mathbf{r}_0,$$

$$(9.1\text{-}23)$$

substituting Eq. (9.1-23) into Eq. (9.1-22) yields

$$\boxed{y(t,\mathbf{r}) = H(t,\mathbf{r};f,\mathbf{v})\exp(+j2\pi ft)\exp(-j2\pi\mathbf{v}\cdot\mathbf{r}).} \quad (9.1\text{-}24)$$

Therefore, a time-harmonic plane wave is an *eigenfunction* of a linear, time-variant, space-variant filter, and the time-variant, space-variant transfer function is an *eigenvalue*.

Recall that the relationship between the input signal $x(t,\mathbf{r})$ and the output signal $y(t,\mathbf{r})$ for a linear, time-variant, space-variant filter is given by Eq. (9.1-1). An alternative expression for the output signal can be obtained as follows. If $X(f,\mathbf{v})$ is the frequency and angular spectrum of the input signal, then

$$x(t,\mathbf{r}) = F_f^{-1}F_{\mathbf{v}}^{-1}\{X(f,\mathbf{v})\}$$

$$= \int_{-\infty}^{\infty}\int_{-\infty}^{\infty} X(f,\mathbf{v})\exp(+j2\pi ft)\exp(-j2\pi\mathbf{v}\cdot\mathbf{r})\,df\,d\mathbf{v}. \quad (9.1\text{-}25)$$

Next, since a linear, time-variant, space-variant filter can be represented by a *linear operator* $L\{\cdot\}$ which operates on input signals that are functions of time and space, the output signal from a filter can be expressed as

$$y(t,\mathbf{r}) = L\{x(t,\mathbf{r})\}. \quad (9.1\text{-}26)$$

Therefore, if the input signal is given by Eq. (9.1-25), then the output signal is given by

$$y(t,\mathbf{r}) = \int_{-\infty}^{\infty} \int_{-\infty}^{\infty} X(f,\boldsymbol{v}) L\{\exp(+j2\pi ft)\exp(-j2\pi\boldsymbol{v}\cdot\mathbf{r})\}\, df\, d\boldsymbol{v}. \quad (9.1\text{-}27)$$

Since

$$H(t,\mathbf{r};f,\boldsymbol{v})\exp(+j2\pi ft)\exp(-j2\pi\boldsymbol{v}\cdot\mathbf{r}) = L\{\exp(+j2\pi ft)\exp(-j2\pi\boldsymbol{v}\cdot\mathbf{r})\},$$

$$(9.1\text{-}28)$$

substituting Eq. (9.1-28) into Eq. (9.1-27) yields

$$\boxed{y(t,\mathbf{r}) = \int_{-\infty}^{\infty} \int_{-\infty}^{\infty} X(f,\boldsymbol{v}) H(t,\mathbf{r};f,\boldsymbol{v})\exp(+j2\pi ft)\exp(-j2\pi\boldsymbol{v}\cdot\mathbf{r})\, df\, d\boldsymbol{v}.}$$

$$(9.1\text{-}29)$$

9.2 Bifrequency Function

An additional system function, the *bifrequency function*, can also be used to describe a linear, time-variant, space-variant filter. The bifrequency function is defined as the time-domain and spatial-domain Fourier transform with respect to t and $\mathbf{r} = (x, y, z)$ of the time-variant, space-variant transfer function:

$$\boxed{\begin{aligned} B(\phi,\boldsymbol{\kappa};f,\boldsymbol{v}) &\triangleq F_t F_{\mathbf{r}}\{H(t,\mathbf{r};f,\boldsymbol{v})\} \\ &= \int_{-\infty}^{\infty}\int_{-\infty}^{\infty} H(t,\mathbf{r};f,\boldsymbol{v})\exp(-j2\pi\phi t)\exp(+j2\pi\boldsymbol{\kappa}\cdot\mathbf{r})\, dt\, d\mathbf{r} \end{aligned}}$$

$$(9.2\text{-}1)$$

where ϕ corresponds to the rate of change of the transfer function in hertz and $\boldsymbol{\kappa} = (\kappa_X, \kappa_Y, \kappa_Z)$ is a three-dimensional vector whose components are spatial frequencies that correspond to the rate of change of the transfer function in cycles per meter. The values of ϕ are a measure of how rapidly the filter changes with time, and the values of the components of $\boldsymbol{\kappa}$ are a measure of how rapidly the filter changes from point to point in space. Similarly,

$$H(t, \mathbf{r}; f, \mathbf{v}) = F_\phi^{-1} F_\kappa^{-1} \{ B(\phi, \kappa; f, \mathbf{v}) \}$$

$$= \int_{-\infty}^{\infty} \int_{-\infty}^{\infty} B(\phi, \kappa; f, \mathbf{v}) \exp(+j2\pi\phi t) \exp(-j2\pi\kappa \cdot \mathbf{r}) \, d\phi \, d\kappa,$$

$$(9.2\text{-}2)$$

where $F_\kappa^{-1} \{ \cdot \} = F_{\kappa_X}^{-1} F_{\kappa_Y}^{-1} F_{\kappa_Z}^{-1} \{ \cdot \}$ and $d\kappa = d\kappa_X \, d\kappa_Y \, d\kappa_Z$.

Example 9.2-1 (Bifrequency function of a linear, time-invariant, space-invariant filter)

If a linear filter is time invariant and space invariant, then from Eq. (9.1-17), the transfer function is given by

$$H(t, \mathbf{r}; f, \mathbf{v}) = H(f, \mathbf{v}). \qquad (9.2\text{-}3)$$

Substituting Eq. (9.2-3) into Eq. (9.2-1) yields

$$B(\phi, \kappa; f, \mathbf{v}) = H(f, \mathbf{v}) \int_{-\infty}^{\infty} \exp(-j2\pi\phi t) \, dt \int_{-\infty}^{\infty} \exp(+j2\pi\kappa \cdot \mathbf{r}) \, d\mathbf{r}$$

$$(9.2\text{-}4)$$

$$= H(f, \mathbf{v}) F_t\{1\} F_\mathbf{r}\{1\}, \qquad (9.2\text{-}5)$$

or, since

$$F_t\{1\} = \delta(\phi) \qquad (9.2\text{-}6)$$

and

$$F_\mathbf{r}\{1\} = \delta(\kappa), \qquad (9.2\text{-}7)$$

$$\boxed{B(\phi, \kappa; f, \mathbf{v}) = H(f, \mathbf{v}) \delta(\phi) \delta(\kappa).} \qquad (9.2\text{-}8)$$

Equation (9.2-8) indicates that the bifrequency function only exists at $\phi = 0$ and $\kappa = 0$, and is equal to the time-invariant, space-invariant transfer function $H(f, \mathbf{v})$. A value of $\phi = 0$ means that the filter does *not* change with time, and a value of $\kappa = 0$ means that the filter does *not* change from point to point in space.

9.3 Output Frequency and Angular Spectrum

The frequency and angular spectrum of the output signal from a linear, time-variant, space-variant filter is given by

$$Y(\eta, \boldsymbol{\beta}) = F_t F_{\mathbf{r}}\{y(t, \mathbf{r})\} = \int_{-\infty}^{\infty}\int_{-\infty}^{\infty} y(t, \mathbf{r}) \exp(-j2\pi\eta t) \exp(+j2\pi\boldsymbol{\beta}\cdot\mathbf{r})\, dt\, d\mathbf{r},$$

(9.3-1)

where η corresponds to *output* (*received*) *frequencies* in hertz, and $\boldsymbol{\beta} = (\beta_X, \beta_Y, \beta_Z)$ is a three-dimensional vector whose components are the *output* (*received*) *spatial frequencies* in cycles per meter. The output spatial frequencies represent the *final* directions of wave propagation of the acoustic field incident upon a receive aperture (array). Substituting Eq. (9.1-29) into Eq. (9.3-1) yields

$$Y(\eta, \boldsymbol{\beta}) = \int_{-\infty}^{\infty}\int_{-\infty}^{\infty} X(f, \boldsymbol{v}) \int_{-\infty}^{\infty}\int_{-\infty}^{\infty} H(t, \mathbf{r}; f, \boldsymbol{v}) \exp\left[-j2\pi(\eta - f)t\right]$$

$$\times \exp\left[+j2\pi(\boldsymbol{\beta} - \boldsymbol{v})\cdot\mathbf{r}\right] dt\, d\mathbf{r}\, df\, d\boldsymbol{v}, \qquad (9.3\text{-}2)$$

and by letting

$$\phi = \eta - f \qquad (9.3\text{-}3)$$

and

$$\boldsymbol{\kappa} = \boldsymbol{\beta} - \boldsymbol{v} \qquad (9.3\text{-}4)$$

in the bifrequency function given by Eq. (9.2-1), Eq. (9.3-2) can be rewritten as

$$\boxed{Y(\eta, \boldsymbol{\beta}) = \int_{-\infty}^{\infty}\int_{-\infty}^{\infty} X(f, \boldsymbol{v}) B(\eta - f, \boldsymbol{\beta} - \boldsymbol{v}; f, \boldsymbol{v})\, df\, d\boldsymbol{v},} \qquad (9.3\text{-}5)$$

which is in the form of a *multidimensional* "*frequency-domain*" *convolution integral*. The term "frequency-domain" is in quotes because the integration in Eq. (9.3-5) is over frequencies in hertz and spatial frequencies in cycles per meter. From Eq. (9.3-3) it can be seen that the output frequency η is equal to the sum of the input frequency f and the variation in frequency ϕ due to the time-variant property of the filter, that is,

$$\eta = f + \phi. \qquad (9.3\text{-}6)$$

For example, ϕ may represent a Doppler shift in hertz.

Similarly, from Eq. (9.3-4), it can be seen that the output spatial-frequency vector $\boldsymbol{\beta}$ is equal to the sum of the input spatial-frequency vector $\boldsymbol{\nu}$ and the variation in spatial-frequency vector $\boldsymbol{\kappa}$ due to the space-variant property of the filter, that is,

$$\boldsymbol{\beta} = \boldsymbol{\nu} + \boldsymbol{\kappa}. \tag{9.3-7}$$

For example, $\boldsymbol{\kappa}$ may represent a change in the direction of wave propagation due to refraction when the speed of sound is a function of position.

The multidimensional convolution integral given by Eq. (9.3-5) indicates that a linear, time-variant, space-variant filter will *spread* the input frequency and angular spectrum $X(f, \boldsymbol{\nu})$ in both frequencies in hertz and spatial frequencies in cycles per meter.

Example 9.3-1 (Output frequency spectrum from a linear, time-variant, space-variant filter)
The frequency spectrum of the output signal from a linear, time-variant, space-variant filter is given by

$$Y(\eta, \mathbf{r}) = F_t\{y(t, \mathbf{r})\} = \int_{-\infty}^{\infty} y(t, \mathbf{r}) \exp(-j2\pi\eta t)\, dt, \tag{9.3-8}$$

where η corresponds to output frequencies in hertz. Substituting Eq. (9.1-29) into Eq. (9.3-8) yields

$$\boxed{Y(\eta, \mathbf{r}) = \int_{-\infty}^{\infty}\int_{-\infty}^{\infty} X(f, \boldsymbol{\nu}) B(\eta - f, \mathbf{r}; f, \boldsymbol{\nu}) \exp(-j2\pi\boldsymbol{\nu}\cdot\mathbf{r})\, df\, d\boldsymbol{\nu},}$$

$$\tag{9.3-9}$$

where

$$B(\phi, \mathbf{r}; f, \boldsymbol{\nu}) = F_t\{H(t, \mathbf{r}; f, \boldsymbol{\nu})\} = \int_{-\infty}^{\infty} H(t, \mathbf{r}; f, \boldsymbol{\nu}) \exp(-j2\pi\phi t)\, dt.$$

$$\tag{9.3-10}$$

Equation (9.3-10) is *not* the fully transformed bifrequency function, since only the time-domain Fourier transform of the time-variant, space-variant transfer function was taken [see Eq. (9.2-1)].

Example 9.3-2 (Output frequency and angular spectrum from a linear, time-invariant, space-invariant filter)

If a linear filter is time invariant and space invariant, then the bifrequency function is given by Eq. (9.2-8). Substituting Eq. (9.2-8) into Eq. (9.3-5) yields

$$Y(\eta, \beta) = \int_{-\infty}^{\infty} \int_{-\infty}^{\infty} X(f, \nu) H(f, \nu) \delta(\eta - f) \delta(\beta - \nu) \, df \, d\nu,$$

$$(9.3\text{-}11)$$

and by using the sifting property of impulse functions,

$$Y(\eta, \beta) = X(\eta, \beta) H(\eta, \beta). \qquad (9.3\text{-}12)$$

Furthermore, since the bifrequency function of a linear, time-invariant, space-invariant filter given by Eq. (9.2-8) predicts that $\phi = 0$ and $\kappa = 0$, Eqs. (9.3-6) and (9.3-7) reduce to $\eta = f$ and $\beta = \nu$, respectively. Therefore, Eq. (9.3-12) can be rewritten as

$$\boxed{Y(f, \nu) = X(f, \nu) H(f, \nu),} \qquad (9.3\text{-}13)$$

which is the output frequency and angular spectrum from a linear, time-invariant, space-invariant filter.

9.4 Coupling Equations and Pulse Propagation

In Sections 9.1 through 9.3, we discussed some of the fundamentals of the theory of linear, time-variant, space-variant filters. Before we can discuss the pulse-propagation problem in this section, we also need to discuss the *coupling equations*. The coupling equations are based on the above theory as well as complex aperture theory (Chapter 6) and array theory (Chapter 7). They provide a consistent, logical, and straightforward mathematical framework for the solution of problems in small-amplitude acoustic pulse propagation.

The following set of equations correspond to Figs. 9.4-1 and 9.4-2 and are referred to as the *coupling equations*, since they couple the transmitted and received electrical signals to the transfer function of a fluid medium (e.g., air or

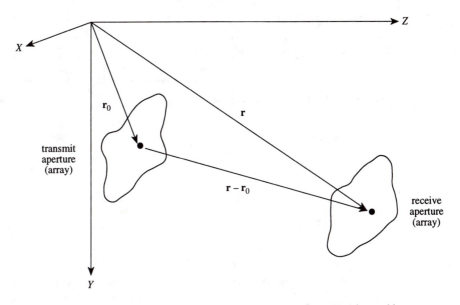

Figure 9.4-1. Illustration of a typical underwater acoustic propagation problem.

water) via the transmit and receive apertures:

$$y(t,\mathbf{r}) = \int_{-\infty}^{\infty} Y_M(\eta,\mathbf{r}) A_R(\eta,\mathbf{r}) \exp(+j2\pi\eta t)\, d\eta, \qquad (9.4\text{-}1)$$

$$Y_M(\eta,\mathbf{r}) = \int_{-\infty}^{\infty}\int_{-\infty}^{\infty} X_M(f,\mathbf{v}) B_M(\eta - f,\mathbf{r}; f,\mathbf{v}) \exp(-j2\pi\mathbf{v}\cdot\mathbf{r})\, df\, d\mathbf{v},$$

$$(9.4\text{-}2)$$

$$X_M(f,\mathbf{v}) = \int_{-\infty}^{\infty} X(f,\boldsymbol{\alpha}) D_T(f,\mathbf{v} - \boldsymbol{\alpha})\, d\boldsymbol{\alpha}, \qquad (9.4\text{-}3)$$

$$B_M(\phi,\mathbf{r}; f,\mathbf{v}) = F_t\{H_M(t,\mathbf{r}; f,\mathbf{v})\} = \int_{-\infty}^{\infty} H_M(t,\mathbf{r}; f,\mathbf{v}) \exp(-j2\pi\phi t)\, dt,$$

$$(9.4\text{-}4)$$

$$X(f,\boldsymbol{\alpha}) = F_t F_\mathbf{r}\{x(t,\mathbf{r})\} = \int_{-\infty}^{\infty}\int_{-\infty}^{\infty} x(t,\mathbf{r}) \exp(-j2\pi f t) \exp(+j2\pi\boldsymbol{\alpha}\cdot\mathbf{r})\, dt\, d\mathbf{r},$$

$$(9.4\text{-}5)$$

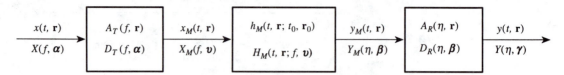

Figure 9.4-2. Block diagram representation of the underwater acoustic propagation problem illustrated in Fig. 9.4-1.

and

$$D_T(f, \boldsymbol{\alpha}) = F_{\mathbf{r}}\{A_T(f, \mathbf{r})\} = \int_{-\infty}^{\infty} A_T(f, \mathbf{r}) \exp(+j2\pi\boldsymbol{\alpha} \cdot \mathbf{r}) \, d\mathbf{r}, \quad (9.4\text{-}6)$$

where $y(t, \mathbf{r})$ is the time-domain output electrical signal (pulse) from the receive aperture (array) at time t and spatial location $\mathbf{r} = (x, y, z)$ [see Eq. (6.1-15)]; $Y_M(\eta, \mathbf{r})$ is the complex frequency spectrum of the output acoustic field $y_M(t, \mathbf{r})$ from the fluid medium filter [see Eq. (9.3-9)], which is also the acoustic field incident upon the receive aperture; $A_R(\eta, \mathbf{r})$ is the complex receive aperture function, which is the complex frequency response at spatial location $\mathbf{r}$ of the receive aperture; $X_M(f, \boldsymbol{v})$ is the complex frequency and angular spectrum of the input acoustic signal to the fluid medium $x_M(t, \mathbf{r})$ [see Eq. (6.2-59)]; $X(f, \boldsymbol{\alpha})$ is the complex frequency and angular spectrum of the transmitted electrical signal $x(t, \mathbf{r})$ [see Eq. (6.2-50)]; $D_T(f, \boldsymbol{\alpha})$ is the far-field directivity function, or beam pattern, of the complex transmit aperture function $A_T(f, \mathbf{r})$, which is the complex frequency response at spatial location $\mathbf{r}$ of the transmit aperture (array) [see Eq. (6.2-56)]; and $H_M(t, \mathbf{r}; f, \boldsymbol{v})$ is the time-variant, space-variant transfer function of the fluid-medium filter.

It is also important to emphasize that the input and output fluid-medium signals, $x_M(t, \mathbf{r})$ and $y_M(t, \mathbf{r})$, are related by the *linear wave equation*

$$\nabla^2 y_M(t, \mathbf{r}) - \frac{1}{c^2(\mathbf{r})} \frac{\partial^2}{\partial t^2} y_M(t, \mathbf{r}) = x_M(t, \mathbf{r}), \quad (9.4\text{-}7)$$

and that the solution of Eq. (9.4-7) is given by

$$y_M(t, \mathbf{r}) = \int_{-\infty}^{\infty} \int_{-\infty}^{\infty} x_M(t_0, \mathbf{r}_0) h_M(t, \mathbf{r}; t_0, \mathbf{r}_0) \, dt_0 \, d\mathbf{r}_0. \quad (9.4\text{-}8)$$

Example 9.4-1 (Acoustic pulse propagation in a time-invariant, space-variant ocean)

In this example we shall demonstrate how the coupling equations reduce if we treat the ocean medium as a linear, time-invariant, space-variant filter. A time-invariant ocean-medium filter implies that both the transmit and receive apertures (arrays) illustrated in Fig. 9.4-1 are motionless, and that no other motion is being modeled. A space-variant ocean-medium filter implies that the speed of sound is a function of position and/or that the ocean medium is bounded (e.g., by the ocean surface and bottom).

If the ocean-medium filter is in fact time-invariant, then its time-variant, space-variant transfer function reduces to [see Eq. (9.1-15)]

$$H_M(t,\mathbf{r};f,\boldsymbol{\nu}) = H_M(f,\mathbf{r};\boldsymbol{\nu}). \tag{9.4-9}$$

Substituting Eq. (9.4-9) into Eq. (9.4-4) yields [see Eq. (9.2-6)]

$$B_M(\phi,\mathbf{r};f,\boldsymbol{\nu}) = H_M(f,\mathbf{r};\boldsymbol{\nu})\delta(\phi), \tag{9.4-10}$$

and upon substituting Eq. (9.4-10) into Eq. (9.4-2),

$$Y_M(\eta,\mathbf{r}) = \int_{-\infty}^{\infty} X_M(\eta,\boldsymbol{\nu})H_M(\eta,\mathbf{r};\boldsymbol{\nu})\exp(-j2\pi\boldsymbol{\nu}\cdot\mathbf{r})\,d\boldsymbol{\nu}. \tag{9.4-11}$$

However, when a linear filter is time-invariant, the output frequencies η are equal to the input frequencies f, that is, $\eta = f$ (see Example 9.3-2). For example, if there is no motion, then there is no Doppler shift. Therefore, Eqs. (9.4-1) and (9.4-11) can be rewritten as

$$y(t,\mathbf{r}) = \int_{-\infty}^{\infty} Y_M(f,\mathbf{r})A_R(f,\mathbf{r})\exp(+j2\pi ft)\,df \tag{9.4-12}$$

and

$$Y_M(f,\mathbf{r}) = \int_{-\infty}^{\infty} X_M(f,\boldsymbol{\nu})H_M(f,\mathbf{r};\boldsymbol{\nu})\exp(-j2\pi\boldsymbol{\nu}\cdot\mathbf{r})\,d\boldsymbol{\nu}, \tag{9.4-13}$$

respectively. Equations (9.4-12), (9.4-13), (9.4-3), (9.4-5), and (9.4-6) represent the formal solution of the pulse-propagation problem for a time-invariant, space-variant ocean medium.

Since in most practical situations an identical input electrical signal (pulse) is applied to all elements in the transmit array *before* the application of complex weights for beam steering purposes, it can then be shown that Eq. (9.4-3) reduces to (see Example 6.2-1)

$$X_M(f,\mathbf{v}) = X(f)D_T(f,\mathbf{v}),\qquad(9.4\text{-}14)$$

where $X(f)$ is the complex frequency spectrum of the transmitted electrical signal. Therefore, substituting Eq. (9.4-14) into Eq. (9.4-13) yields the following expression for the complex frequency spectrum of the acoustic field incident upon the receive aperture (array) at frequency f and spatial location $\mathbf{r} = (x, y, z)$:

$$Y_M(f,\mathbf{r}) = X(f)H(f,\mathbf{r}),\qquad(9.4\text{-}15)$$

where the *overall system complex frequency response* is given by

$$H(f,\mathbf{r}) = \int_{-\infty}^{\infty} D_T(f,\mathbf{v})H_M(f,\mathbf{r};\mathbf{v})\exp(-j2\pi\mathbf{v}\cdot\mathbf{r})\,d\mathbf{v}\quad(9.4\text{-}16)$$

or, upon expanding Eq. (9.4-16),

$$
\begin{aligned}
H(f,\mathbf{r}) &= H(f, x, y, z)\\
&= \int_{-\infty}^{\infty}\int_{-\infty}^{\infty}\int_{-\infty}^{\infty} D_T(f, f_X, f_Y, f_Z)H_M(f, x, y, z; f_X, f_Y, f_Z)\\
&\quad \times \exp\big[-j2\pi(f_X x + f_Y y + f_Z z)\big]\,df_X\,df_Y\,df_Z.
\end{aligned}
\quad(9.4\text{-}17)
$$

And upon substituting Eq. (9.4-15) into Eq. (9.4-12), we obtain the following expression for the time-domain output electrical signal (pulse) from the receive aperture (array) at time t and spatial location $\mathbf{r} = (x, y, z)$:

$$y(t,\mathbf{r}) = \int_{-\infty}^{\infty} X(f)H(f,\mathbf{r})A_R(f,\mathbf{r})\exp(+j2\pi ft)\,df,\quad(9.4\text{-}18)$$

or

$$y(t,\mathbf{r}) = F_f^{-1}\{Y(f,\mathbf{r})\} = \int_{-\infty}^{\infty} Y(f,\mathbf{r})\exp(+j2\pi ft)\,df,\quad(9.4\text{-}19)$$

where

$$Y(f,\mathbf{r}) = Y_M(f,\mathbf{r})A_R(f,\mathbf{r}) = X(f)H(f,\mathbf{r})A_R(f,\mathbf{r})\quad(9.4\text{-}20)$$

is the complex frequency spectrum of the output electrical signal (pulse) from the receive aperture (array) at frequency f and spatial location $\mathbf{r} = (x, y, z)$.

Example 9.4-2 (**Acoustic pulse propagation in a time-invariant, depth-variant ocean**)

In this example we shall make several additional simplifying assumptions in order to demonstrate how the coupling equations for the time-invariant, space-variant ocean-medium model derived in Example 9.4-1 reduce even further. First, besides being time-invariant, let the ocean-medium filter be *space-invariant* in the cross-range (X) and down-range (Z) directions, but *space-variant* in the depth (Y) direction (see Fig. 9.4-3). This implies that the speed of sound is an arbitrary function of depth y only—that is, $c(\mathbf{r}) = c(y)$—and/or that the ocean is bounded by the ocean surface and bottom. Because of these additional assumptions, the impulse response (Green's function) of the ocean-medium filter is given by

$$h_M(t,\mathbf{r};t_0,\mathbf{r}_0) = h_M(t - t_0, x - x_0, y, z - z_0; y_0) \quad (9.4\text{-}21)$$

with corresponding transfer function [see Eq. (9.1-12)]

$$H_M(t,\mathbf{r};f,\boldsymbol{\nu}) = H_M(f, f_X, y, f_Z; f_Y). \quad (9.4\text{-}22)$$

Since the ocean-medium filter in this example is space variant only in the Y direction, with the use of Eq. (9.4-22), the time-invariant, space-

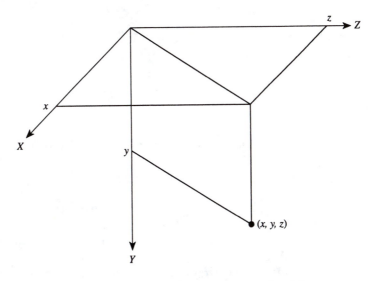

Figure 9.4-3. The rectangular coordinates (x, y, z).

variant ocean-medium transfer function in Example 9.4-1 reduces to

$$H_M(f, \mathbf{r}; \mathbf{v}) = H_M(f, x, y, z; f_X, f_Y, f_Z) = H_M(f, f_X, y, f_Z; f_Y).$$

(9.4-23)

Second, let the transmit aperture depicted in Fig. 9.4-1 be a planar array of $MT \times NT$ (odd) complex-weighted point elements, centered at $(x_0 = 0, \ y_0 = y_T, \ z_0 = 0)$ and lying in the XY plane. Therefore, the complex transmit aperture function is given by

$$A_T(f, x, y, z) = A_T(f, x, y)$$

$$= \sum_{m'=-MT'}^{MT'} \sum_{n'=-NT'}^{NT'} c_{m'n'}(f) \delta(x - x_0) \delta(y - y_0)$$

(9.4-24)

with corresponding far-field directivity function (beam pattern)

$$D_T(f, f_X, f_Y, f_Z) = D_T(f, f_X, f_Y)$$

$$= \sum_{m'=-MT'}^{MT'} \sum_{n'=-NT'}^{NT'} c_{m'n'}(f) \exp(+j2\pi f_X x_0)$$

$$\times \exp(+j2\pi f_Y y_0),$$

(9.4-25)

where

$$MT' = \frac{MT - 1}{2},$$

(9.4-26)

$$NT' = \frac{NT - 1}{2},$$

(9.4-27)

$c_{m'n'}(f)$ is the frequency-dependent complex weight at element $m'n'$,

$$x_0 = m'd_{XT},$$

(9.4-28)

$$y_0 = y_T + n'd_{YT},$$

(9.4-29)

and d_{XT} and d_{YT} are the interelement spacings (in meters) in the X and Y directions, respectively. Note that a planar array automatically includes linear arrays and a single omnidirectional point element as special cases. Substituting Eqs. (9.4-23) and (9.4-25) into Eq. (9.4-17) yields

$$H(f, x, y, z) = \int_{-\infty}^{\infty} \int_{-\infty}^{\infty} \sum_{m'=-MT'}^{MT'} \sum_{n'=-NT'}^{NT'} c_{m'n'}(f)$$

$$\times \int_{-\infty}^{\infty} H_M(f, f_X, y, f_Z; f_Y) \exp[-j2\pi f_Y(y - y_0)] \, df_Y$$

$$\times \exp[-j2\pi f_X(x - x_0)] \exp(-j2\pi f_Z z) \, df_X \, df_Z,$$

$$(9.4\text{-}30)$$

or [see Eq. (9.1-13)],

$$H(f, x, y, z) = \int_{-\infty}^{\infty} \int_{-\infty}^{\infty} \sum_{m'=-MT'}^{MT'} \sum_{n'=-NT'}^{NT'} c_{m'n'}(f) H_M(f, f_X, y, f_Z; y_0)$$

$$\times \exp[-j2\pi f_X(x - x_0)] \exp(-j2\pi f_Z z) \, df_X \, df_Z,$$

$$(9.4\text{-}31)$$

where $H_M(f, f_X, y, f_Z; y_0)$ *is the solution of the one-dimensional, inhomogeneous Helmholtz equation in the Y direction* (see Example 9.4-3). The overall system complex frequency response given by Eq. (9.4-31) is, in fact, *the full-wave solution of the three-dimensional, inhomogeneous Helmholtz equation when the speed of sound is an arbitrary function of depth y only.*

Third, let the receive aperture depicted in Fig. 9.4-1 also be a planar array of $MR \times NR$ (odd) complex-weighted point elements, centered at (x_R, y_R, z_R) and parallel to the XY plane. Therefore, the complex receive aperture function is given by

$$A_R(f, x, y, z) = \sum_{m=-MR'}^{MR'} \sum_{n=-NR'}^{NR'} w_{mn}(f) \delta[x - (x_R + md_{XR})]$$

$$\times \delta[y - (y_R + nd_{YR})] \delta(z - z_R), \qquad (9.4\text{-}32)$$

where

$$MR' = \frac{MR - 1}{2}, \qquad (9.4\text{-}33)$$

$$NR' = \frac{NR - 1}{2}, \qquad (9.4\text{-}34)$$

$w_{mn}(f)$ is the frequency-dependent complex weight at element mn, and d_{XR} and d_{YR} are the interelement spacings in meters in the X and Y directions, respectively. Substituting Eq. (9.4-32) into Eq. (9.4-18) yields the following expression for the time-domain output electrical signal (pulse) at element mn in the receive array *after beam steering*, that is, after applying the complex weight $w_{mn}(f)$:

$$y(t, m, n, z_R) = F_f^{-1}\{Y(f, m, n, z_R)\}$$

$$= \int_{-\infty}^{\infty} Y(f, m, n, z_R) \exp(+j2\pi ft)\, df,$$

$$m = -MR', \ldots, 0, \ldots, MR', \quad n = -NR', \ldots, 0, \ldots, NR',$$

$$(9.4\text{-}35)$$

where

$$Y(f, m, n, z_R) = w_{mn}(f)\, Y_M(f, m, n, z_R),$$

$$m = -MR', \ldots, 0, \ldots, MR', \quad n = -NR', \ldots, 0, \ldots, NR',$$

$$(9.4\text{-}36)$$

is the complex frequency spectrum of the output electrical signal (pulse) at element mn in the receive array *after beam steering*, and [see Eq. (9.4-15)]

$$Y_M(f, m, n, z_R) = X(f) H(f, x_R + md_{XR}, y_R + nd_{YR}, z_R),$$

$$m = -MR', \ldots, 0, \ldots, MR', \quad n = -NR', \ldots, 0, \ldots, NR',$$

$$(9.4\text{-}37)$$

is the complex frequency spectrum of the acoustic field incident upon element mn in the receive array.

Cylindrical Coordinates

The overall system complex frequency response given by Eq. (9.4-31) can be expressed in terms of the cylindrical coordinates (r, ϕ, y) by using the following transformation equations (see Figs. 9.4-4 and 9.4-5):

$$x - x_0 = x - m'd_{XT} = r \cos \phi, \tag{9.4-38}$$

$$y = y, \tag{9.4-39}$$

and

$$z = r \sin \phi, \tag{9.4-40}$$

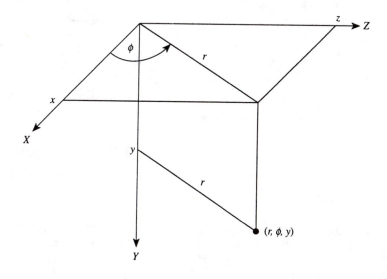

Figure 9.4-4. The cylindrical coordinates (r, ϕ, y).

where

$$r = \sqrt{(x - x_0)^2 + z^2} = \sqrt{(x - m'd_{XT})^2 + z^2} \qquad (9.4\text{-}41)$$

is the horizontal range between element $m'n'$ in the transmit array and the field point, and

$$\phi = \tan^{-1}\left(\frac{z}{x - x_0}\right) = \tan^{-1}\left(\frac{z}{x - m'd_{XT}}\right) \qquad (9.4\text{-}42)$$

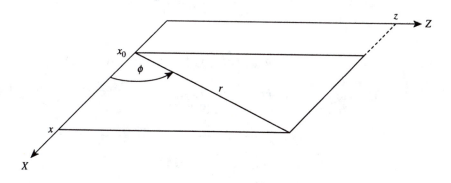

Figure 9.4-5. Relationship between the rectangular coordinates x, x_0, and z and the cylindrical coordinates r and ϕ in the XZ plane.

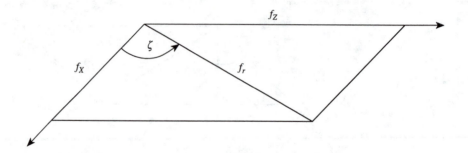

Figure 9.4-6. Relationship between the spatial frequencies f_X, f_Z, and f_r.

is the corresponding azimuthal angle. The spatial frequencies must also be transformed, using the following set of equations (see Fig. 9.4-6):

$$f_X = f_r \cos \zeta, \tag{9.4-43}$$

$$f_Y = f_Y, \tag{9.4-44}$$

and

$$f_Z = f_r \sin \zeta, \tag{9.4-45}$$

where

$$f_r = \sqrt{f_X^2 + f_Z^2} \tag{9.4-46}$$

is the spatial frequency in the radial direction, and

$$df_X \, df_Z \rightarrow f_r \, df_r \, d\zeta. \tag{9.4-47}$$

In addition,

$$H_M(f, f_X, y, f_Z; y_0) \rightarrow H_M(f, f_r, y; y_0). \tag{9.4-48}$$

Therefore, upon substituting Eqs. (9.4-38) through (9.4-40), (9.4-43), (9.4-45), (9.4-47), and (9.4-48) into Eq. (9.4-31), we obtain

$$H(f, r, \phi, y) = 2\pi \int_0^\infty \sum_{m' = -MT'}^{MT'} \sum_{n' = -NT'}^{NT'} c_{m'n'}(f) H_M(f, f_r, y; y_0) J_0(2\pi f_r r) f_r \, df_r,$$

$$\tag{9.4-49}$$

where

$$J_0(2\pi f_r r) = \frac{1}{2\pi} \int_0^{2\pi} \exp\left[-j2\pi f_r r \cos(\phi - \zeta)\right] d\zeta \qquad (9.4\text{-}50)$$

is the zeroth-order Bessel function of the first kind, r is given by Eq. (9.4-41), and y_0 is given by Eq. (9.4-29).

Example 9.4-3

In this example we shall show that the function $H_M(f, f_X, y, f_Z; y_0)$ or, in cylindrical coordinates, $H_M(f, f_r, y; y_0)$, is the solution of the one-dimensional, inhomogeneous Helmholtz equation in the Y direction for a linear ocean-medium filter that is time invariant and space invariant in the cross-range (X) and down-range (Z) directions, but space variant in the depth (Y) direction. The function $H_M(f, f_X, y, f_Z; y_0)$ or $H_M(f, f_r, y; y_0)$ is required in order to evaluate the overall system complex frequency response given by Eq. (9.4-31) or, in cylindrical coordinates, Eq. (9.4-49).

We begin the analysis by setting the source distribution or input acoustic signal to the fluid medium equal to a unit-amplitude impulse applied at time t_0 and position $\mathbf{r}_0 = (x_0, y_0, z_0)$, that is,

$$x_M(t, \mathbf{r}) = \delta(t - t_0, \mathbf{r} - \mathbf{r}_0) = \delta(t - t_0)\delta(\mathbf{r} - \mathbf{r}_0). \qquad (9.4\text{-}51)$$

Substituting Eq. (9.4-51) and $c(\mathbf{r}) = c(y)$ into the linear wave equation given by Eq. (9.4-7) and expanding yields

$$\frac{\partial^2}{\partial x^2} y_M(t, x, y, z) + \frac{\partial^2}{\partial y^2} y_M(t, x, y, z) + \frac{\partial^2}{\partial z^2} y_M(t, x, y, z)$$

$$-\frac{1}{c^2(y)} \frac{\partial^2}{\partial t^2} y_M(t, x, y, z) = \delta(t - t_0)\delta(x - x_0)\delta(y - y_0)\delta(z - z_0).$$

$$(9.4\text{-}52)$$

Since the linear ocean-medium filter is time invariant and space invariant in the X and Z directions, the output frequencies η are equal to the input frequencies f, and the output spatial frequencies β_X and β_Z are equal to the input spatial frequencies f_X and f_Z (see Example 9.3-2). Therefore, taking the time-domain Fourier transform with respect to t and the spatial-domain Fourier transforms with respect to x

and z of Eq. (9.4-52) yields (see Appendices 2B and 2C)

$$\frac{d^2}{dy^2} Y_M(f, f_X, y, f_Z) + B_Y^2(y) Y_M(f, f_X, y, f_Z)$$

$$= \exp(-j2\pi f t_0) \exp(+j2\pi f_X x_0) \, \delta(y - y_0) \exp(+j2\pi f_Z z_0),$$

$$(9.4\text{-}53)$$

where

$$Y_M(f, f_X, y, f_Z) = F_t F_x F_z \{ y_M(t, x, y, z) \}$$

$$= \int_{-\infty}^{\infty} \int_{-\infty}^{\infty} \int_{-\infty}^{\infty} y_M(t, x, y, z) \exp(-j2\pi ft) \, dt$$

$$\times \exp(+j2\pi f_X x) \exp(+j2\pi f_Z z) \, dx \, dz \quad (9.4\text{-}54)$$

is the frequency and angular spectrum (in the X and Z directions only) of the output acoustic signal from the ocean-medium filter,

$$B_Y(y) = \sqrt{k^2(y) - (k_X^2 + k_Z^2)} \qquad (9.4\text{-}55)$$

is the *output* depth-dependent propagation-vector component in the Y direction, and

$$k(y) = 2\pi f / c(y) \qquad (9.4\text{-}56)$$

is the depth-dependent wave number. Note that

$$k_X = 2\pi f_X, \qquad (9.4\text{-}57)$$

$$B_Y(y) = 2\pi \beta_Y(y), \qquad (9.4\text{-}58)$$

and

$$k_Z = 2\pi f_Z, \qquad (9.4\text{-}59)$$

where $\beta_Y(y)$ is the *output* depth-dependent spatial frequency in the Y direction.

However, since the ocean-medium filter is time invariant and space invariant in the X and Z directions, and the input acoustic signal is the unit-amplitude impulse given by Eq. (9.4-51), the solution of the wave equation given by Eq. (9.4-52) is equal to the following impulse response [see Eq. (9.4-21)]:

$$y_M(t, x, y, z) = h_M(t - t_0, x - x_0, y, z - z_0; y_0). \quad (9.4\text{-}60)$$

Substituting Eq. (9.4-60) into Eq. (9.4-54) yields

$$
\begin{aligned}
Y_M(f, f_X, y, f_Z) \\
&= F_t F_x F_z\{h_M(t - t_0, x - x_0, y, z - z_0; y_0)\} \\
&= \int_{-\infty}^{\infty} \int_{-\infty}^{\infty} \int_{-\infty}^{\infty} h_M(t - t_0, x - x_0, y, z - z_0; y_0) \exp(-j2\pi ft)\, dt \\
&\quad \times \exp(+j2\pi f_X x) \exp(+j2\pi f_Z z)\, dx\, dz \\
&= H_M(f, f_X, y, f_Z; y_0)\exp(-j2\pi ft_0)\exp(+j2\pi f_X x_0)\exp(+j2\pi f_Z z_0),
\end{aligned}
$$

(9.4-61)

and upon substituting Eq. (9.4-61) into Eq. (9.4-53), we obtain

$$
\frac{d^2}{dy^2} H_M(f, f_X, y, f_Z; y_0) + B_Y^2(y) H_M(f, f_X, y, f_Z; y_0) = \delta(y - y_0),
$$

(9.4-62)

where

$$
B_Y(y) = \begin{cases} \sqrt{k^2(y) - (k_X^2 + k_Z^2)}, & k_X^2 + k_Z^2 \leq k^2(y), \\ -j\sqrt{k_X^2 + k_Z^2 - k^2(y)}, & k_X^2 + k_Z^2 > k^2(y). \end{cases}
$$

(9.4-63)

In cylindrical coordinates,

$$
\frac{d^2}{dy^2} H_M(f, f_r, y; y_0) + B_Y^2(y) H_M(f, f_r, y; y_0) = \delta(y - y_0),
$$

(9.4-64)

where

$$
B_Y(y) = \begin{cases} \sqrt{k^2(y) - k_r^2}, & k_r^2 \leq k^2(y), \\ -j\sqrt{k_r^2 - k^2(y)}, & k_r^2 > k^2(y), \end{cases}
$$

(9.4-65)

$$
k_r^2 = k_X^2 + k_Z^2,
$$

(9.4-66)

$$
k_r = 2\pi f_r.
$$

(9.4-67)

is the propagation-vector component in the horizontal radial direction, and f_r is the spatial frequency in the horizontal radial direction.

Since the ocean-medium filter is space variant in the depth (Y) direction, the input and output propagation-vector components (spatial frequencies) in the Y direction are, in general, *not* equal. Therefore, let us end our discussion in this example by deriving an equation that relates the input and output propagation-vector components (spatial frequencies) in the Y direction.

Recall from Chapter 5 that the wave number $k(y)$ can be expressed as

$$k(y) = k_0 n(y) = 2\pi f/c(y),$$ (9.4-68)

where

$$k_0 = 2\pi f/c_0$$ (9.4-69)

is the reference wave number,

$$c_0 = c(y_0)$$ (9.4-70)

is the speed of sound at the source depth y_0, and

$$n(y) = c_0/c(y)$$ (9.4-71)

is the dimensionless index of refraction. Note that

$$k(y_0) = k_0.$$ (9.4-72)

Therefore, at the depth of the sound source,

$$k_X^2 + k_Y^2 + k_Z^2 = k^2(y_0) = k_0^2,$$ (9.4-73)

or

$$k_X^2 + k_Z^2 = k_0^2 - k_Y^2,$$ (9.4-74)

where

$$k_Y = 2\pi f_Y$$ (9.4-75)

is the *input* propagation-vector component in the Y direction, and f_Y is the *input* spatial frequency in the Y direction. Finally, substituting Eqs. (9.4-68) and (9.4-74) into Eq. (9.4-55) yields

$$B_Y(y) = \sqrt{k_Y^2 + k_0^2[n^2(y) - 1]}.$$ (9.4-76)

Note that at the source depth y_0,

$$B_Y(y_0) = k_Y, \qquad (9.4\text{-}77)$$

since $n(y_0) = 1$.

Bibliography

P. A. Bello, "Characterization of randomly time-variant linear channels," *IEEE Trans. Communication Systems*, **11**, 360–393 (1963).

A. W. Ellinthorpe and A. H. Nuttall, "Theoretical and empirical results on the characterization of undersea acoustic channels," IEEE First Annual Communication Convention, 1965.

T. Kailath, "Channel characterization: Time-variant dispersive channels," in *Lectures on Communication System Theory* (edited by E. J. Baghdady), McGraw-Hill, New York, 1961.

R. S. Kennedy, *Fading Dispersive Communication Channels*, Wiley-Interscience, New York, 1969.

R. Laval, "Sound propagation effects on signal processing," in *Signal Processing*, (edited by J. W. R. Griffiths, P. L. Stocklin, and C. Van Schooneveld), Academic Press, New York, 1973.

R. Laval, "Time-frequency-space generalized coherence and scattering functions," in *Aspects of Signal Processing*, Part I, (edited by G. Tacconi), D. Reidel, Dordrecht, Holland, 1977.

D. Middleton, "A Statistical theory of reverberation and similar first-order scattered fields. Part I: Waveforms and the general process," *IEEE Trans. Inf. Theory*, **IT-13**, 372–392 (1967).

D. Middleton, "Channel modeling and threshold signal processing in underwater acoustics: An analytical overview," *IEEE J. Oceanic Engineering*, **OE-12**, 4–28 (1987).

K. A. Sostrand, "Mathematics of the time-varying channel," in *Proceedings of the NATO Advanced Study Institute on Signal Processing with emphasis on underwater acoustics*, Vol. II, Enschede, The Netherlands, 1968.

L. A. Zadeh, "Frequency analysis of variable networks," *Proc. IRE*, **38**, 291–299 (1950).

L. A. Zadeh, "Correlation functions and power spectra in variable networks," *Proc. IRE*, **38**, 1342–1345 (1950).

L. A. Zadeh and C. A. Desoer, *Linear System Theory*, McGraw-Hill, New York, 1963.

L. J. Ziomek, *Underwater Acoustics—A Linear Systems Theory Approach*, Academic Press, Orlando, Florida, 1985.

Symbols and Abbreviations

$A(f, \mathbf{r}_a)$	complex frequency response (amplitude and phase) at spatial location $\mathbf{r}_a$ of an aperture, also known as a complex aperture function
$A_T(f, \mathbf{r}_T)$	complex frequency response (amplitude and phase) at spatial location $\mathbf{r}_T$ of a transmit aperture, also known as a complex transmit aperture function
$A_R(\eta, \mathbf{r}_R)$	complex frequency response (amplitude and phase) at spatial location $\mathbf{r}_R$ of a receive aperture, also known as a complex receive aperture function
$a(f, \mathbf{r}_a)$	amplitude of the complex frequency response at spatial location $\mathbf{r}_a$ of an aperture
$a_{mn}(f)$	real, frequency-dependent amplitude weight associated with element (electroacoustic transducer) mn in a planar array
$a_n(f)$	real, frequency-dependent amplitude weight associated with element (electroacoustic transducer) n in a linear array
AG	array gain in dB
$\alpha, \alpha(f)$	attenuation coefficient in Np/m
$\alpha_X, \alpha_Y, \alpha_Z$	attenuation coefficients in Np/m in the X, Y, and Z directions, respectively
$\alpha', \alpha'(f)$	attenuation coefficient in dB/m
b	ray parameter in sec/m
B_T, B_E	isothermal and adiabatic bulk modulus (elasticity) of a fluid, respectively, in Pa
$B(\phi, \boldsymbol{\kappa}; f, \boldsymbol{\nu})$	bifrequency function of a linear, time-variant, space-variant filter

$c, c(\mathbf{r})$	speed of sound in m/sec
c_0	reference speed of sound in m/sec at the source location
$c_{g_X}, c_{g_Y}, c_{g_Z}$	group speeds in m/sec in the X, Y, and Z directions, respectively
$c_{p_X}, c_{p_Y}, c_{p_Z}$	phase speeds in m/sec in the X, Y, and Z directions, respectively
$c_{mn}(f)$	frequency-dependent complex weight associated with element (electroacoustic transducer) mn in a planar array
$c_n(f)$	frequency-dependent complex weight associated with element (electroacoustic transducer) n in a linear array
D	directivity (dimensionless) of either a single electroacoustic transducer or an array of electroacoustic transducers
D_f	frequency-deviation constant in radians per second, per unit of $m(t)$, where $m(t)$ is some message or modulating signal
D_p	phase-deviation constant in radians per unit of $m(t)$, where $m(t)$ is some message or modulating signal
DFT	discrete Fourier transform
DI	directivity index in dB of either a single electroacoustic transducer or an array of electroacoustic transducers
$D(f, r, \boldsymbol{\alpha})$	near-field directivity function, or beam pattern, of a complex aperture function
$D(f, \boldsymbol{\alpha})$	far-field directivity function, or beam pattern, of a complex aperture function
$D_T(f, r, \boldsymbol{\alpha})$	near-field directivity function, or beam pattern, of a complex transmit aperture function
$D_T(f, \boldsymbol{\alpha})$	far-field directivity function, or beam pattern, of a complex transmit aperture function
$D_R(\eta, \boldsymbol{\beta})$	far-field directivity function, or beam pattern, of a complex receive aperture function
$E(t)$	envelope of the real bandpass signal $x(t)$
E_x	energy, in joules, of the real signal $x(t)$
$E_{\hat{x}}$	energy, in joules, of the Hilbert transform $\hat{x}(t)$
$E_{\tilde{x}}$	energy, in joules, of the complex envelope $\tilde{x}(t)$
$e_{mn}(f, x_a, y_a)$	complex frequency response or complex aperture function of element (electroacoustic transducer) mn in a planar array lying in the XY plane, also known as the element function
$e_n(f, x_a)$	complex frequency response or complex aperture function of element (electroacoustic transducer) n in a linear array lying along the X axis, also known as the element function
f	frequency in Hz
f_c	"center" or "carrier" frequency in Hz
$f_i(t)$	instantaneous frequency in Hz
f_n	cutoff frequency in Hz for the nth normal mode
f_S	sampling frequency, or sampling rate, in samples per second
f_0	fundamental frequency, or first harmonic, in Hz
FFT	fast Fourier transform

f_X, f_Y, f_Z	spatial frequencies in cycles/m in the X, Y, and Z directions, respectively
g	constant sound-speed gradient in $\sec^{-1}$
$g_f(\mathbf{r}\mid\mathbf{r}_0)$	time-independent free-space Green's function, or spatial impulse response, of an unbounded, homogeneous, lossless fluid medium
$h(t, \mathbf{r}; t_0, \mathbf{r}_0)$	impulse response of a linear, time-variant, space-variant filter at time t and spatial location $\mathbf{r} = (x, y, z)$ due to the application of a unit-amplitude impulse at time t_0 and spatial location $\mathbf{r}_0 = (x_0, y_0, z_0)$. The impulse response is also called the Green's function.
$H(t, \mathbf{r}; f, \boldsymbol{\nu})$	complex frequency response, or transfer function, of a linear, time-variant, space-variant filter
$h_m^{(1)}(\zeta), h_m^{(2)}(\zeta)$	mth-order spherical Hankel functions of the first and second kind, respectively, also known as spherical Bessel functions of the third kind
$H_n^{(1)}(\zeta), H_n^{(2)}(\zeta)$	nth-order Hankel functions of the first and second kind, respectively, also known as Bessel functions of the third kind
$\mathbf{I}(t, \mathbf{r})$	instantaneous intensity vector in W/m^2
$\mathbf{I}_{\text{avg}}(\mathbf{r})$	time-average intensity vector in W/m^2
$I_{\text{avg}}(\mathbf{r})$	time-average intensity in W/m^2
I_{ref}	reference intensity in W/m^2
$j_m(\zeta)$	mth-order spherical Bessel function of the first kind
$J_n(\zeta)$	nth-order Bessel function of the first kind
$k, k(\mathbf{r})$	wave number in rad/m
k_0	reference wave number in rad/m
$\mathbf{k}, \mathbf{k}(\mathbf{r})$	propagation vector in rad/m
$K(\mathbf{r})$	complex wave number in rad/m
$\mathbf{K}$	complex propagation vector in rad/m
$k_X, k_X(\mathbf{r}), k_Y,$ $k_Y(\mathbf{r}), k_Z, k_Z(\mathbf{r})$	propagation-vector components in rad/m in the X, Y, and Z directions, respectively
K_X, K_Y, K_Z	complex propagation-vector components in rad/m in the X, Y, and Z directions, respectively
κ_T, κ_E	isothermal and adiabatic compressibility of the fluid, respectively, in Pa^{-1}
λ	wavelength in m
$\lambda_X, \lambda_Y, \lambda_Z$	wavelengths, in m, in the X, Y, and Z directions, respectively
μ	coefficient of viscosity, or shear viscosity, of a fluid medium in Pa-sec
μ_v	coefficient of bulk (volume) viscosity of a fluid medium in Pa-sec
n_{12}, N_{12}	real and complex dimensionless indices of refraction, respectively, between media I and II, where the incident wave is in medium I, since the number 1 appears first

$p, p(t, \mathbf{r})$	acoustic pressure in pascals (Pa), where 1 Pa = 1 N/m^2, where 1 N = 1 kg-m/sec^2
$p_f(\mathbf{r})$	spatial-dependent part of the time-harmonic acoustic pressure in pascals (Pa)
$p_0, p_0(\mathbf{r})$	equilibrium or ambient pressure in pascals (Pa)
P_{avg}	time-average power in W
P_D	probability of detection
P_{FA}	probability of false alarm
$P_m(\cos\theta)$	Legendre polynomial of degree m
$P_m^n(\cos\theta)$	associated Legendre function of degree m and order n of the first kind, also known as an associated Legendre polynomial
P_{ref}	root-mean-square (rms) reference pressure in pascals (Pa)
$\varphi(t, \mathbf{r})$	scalar velocity potential in m^2/sec
$\varphi_f(\mathbf{r})$	spatial-dependent part of the scalar, time-harmonic velocity potential in m^2/sec
ϕ_A	actual (true) Doppler shift of a sound source (target) in Hz
$\rho_0, \rho_0(\mathbf{r})$	ambient or equilibrium density of a fluid medium in kg/m^3
$\rho_0 c$	characteristic impedance of a fluid medium in rayls, where 1 rayl = 1 Pa-sec/m
$\Delta\psi$	3-dB beamwidth, in degrees, of the normalized horizontal far-field beam pattern
$Q(t, \mathbf{r})$	source distribution in sec^{-1} (volume flow rate per unit volume of fluid) at time t and position $\mathbf{r} = (x, y, z)$
$Q_m(\cos\theta)$	Legendre function of the second kind of degree m
$Q_m^n(\cos\theta)$	associated Legendre function of degree m and order n of the second kind
R_c	radius of curvature in m
$R_{12}, R_{12}', R_{12}'', R_{12}'''$	plane-wave velocity potential, acoustic pressure, time-average intensity, and time-average power reflection coefficients, respectively, where the subscript 12 is meant to indicate that the boundary is between media I and II and that the incident wave is in medium I, since the number 1 appears first
s	condensation (dimensionless), or arc length (path length) in m, or time compression/stretch factor (dimensionless)
SIL	sound-intensity level in dB
SL	source level in dB
SNR	signal-to-noise power ratio (dimensionless)
SNR_A	signal-to-noise power ratio (dimensionless) due to processing signals from an array of elements
SPL	sound-pressure level in dB
σ_s	scattering cross section in m^2
T	pulse length in sec
TL	transmission loss in dB
T_0	data record length or fundamental period in sec
T_s	sampling period in sec

$T_{12}, T'_{12}, T''_{12}, T'''_{12}$	plane-wave velocity potential, acoustic pressure, time-average intensity, and time-average power transmission coefficients, respectively, where the subscript 12 is meant to indicate that the boundary is between media I and II and that the incident wave is in medium I, since the number 1 appears first
$\theta(f, \mathbf{r}_a)$	phase of the complex frequency response at spatial location $\mathbf{r}_a$ of an aperture
θ_c	critical angle of incidence
$\theta_i(t)$	instantaneous phase in radians
θ_I	angle of intromission
$\theta_{mn}(f)$	real, frequency-dependent phase weight associated with element (electroacoustic transducer) mn in a planar array
$\theta_n(f)$	real, frequency-dependent phase weight associated with element (electroacoustic transducer) n in a linear array
$\Delta\theta$	3-dB beamwidth, in degrees, of the normalized vertical far-field beam pattern
τ	travel time or time delay in sec
τ_A	actual (true) round-trip time delay of a sound source (target) in sec
u	acoustic fluid (particle) speed in m/sec
$u_f(\mathbf{r})$	spatial-dependent part of the time-harmonic acoustic fluid (particle) speed in m/sec
$\mathbf{u}, \mathbf{u}(t, \mathbf{r})$	acoustic fluid (particle) velocity vector in m/sec
$\mathbf{u}_f(\mathbf{r})$	spatial-dependent part of the time-harmonic acoustic fluid (particle) velocity vector in m/sec
$u, u(\mathbf{r}), v, v(\mathbf{r}), w, w(\mathbf{r})$	dimensionless direction cosines with respect to the X, Y, and Z axes, respectively
Δu	dimensionless 3-dB beamwidth in direction cosine u space
$W(\mathbf{r})$	the eikonal in m
$\tilde{x}(t)$	complex envelope of the real bandpass signal $x(t)$
$x_p(t)$	pre-envelope, or analytic signal, of the real bandpass signal $x(t)$
$\hat{x}(t)$	Hilbert transform of $x(t)$
$x_c(t), x_s(t)$	cosine and sine components of $x(t)$, respectively, also known as the in-phase and quadrature-phase components, respectively
$x_M(t, \mathbf{r})$	input acoustic signal to the fluid medium or the source distribution in $\sec^{-1}$ (volume flow rate per unit volume of fluid) at time t and position $\mathbf{r} = (x, y, z)$
$X(\tau, \phi), X_N(\tau, \phi)$	unnormalized and normalized auto-ambiguity function of $x(t)$, respectively
$y_m(\zeta)$	mth-order spherical Bessel function of the second kind
$Y_n(\zeta)$	nth-order Bessel function of the second kind
z_{SD}	skin depth in m

Index